HISTORY OF PHILOSOPHY OF SCIENCE
NEW TRENDS AND PERSPECTIVES

VIENNA CIRCLE INSTITUTE YEARBOOK [2001]

9

VIENNA CIRCLE INSTITUTE YEARBOOK [2001]
9

Institut 'Wiener Kreis'
Society for the Advancement of the Scientific World Conception

Series-Editor:
Friedrich Stadler
Director, Institut 'Wiener Kreis'
and University of Vienna, Austria

The titles published in this series are listed at the end of this volume.

HISTORY OF PHILOSOPHY OF SCIENCE

NEW TRENDS AND PERSPECTIVES

Edited by

MICHAEL HEIDELBERGER
University of Tübingen, Germany

and

FRIEDRICH STADLER
*University of Vienna and
Institute Vienna Circle, Austria*

KLUWER ACADEMIC PUBLISHERS
DORDRECHT / BOSTON / LONDON

A C.I.P. Catalogue record for this book is available from the Library of Congress.

ISBN 1-4020-0509-1
Series ISSN 0929-6328

Published by Kluwer Academic Publishers,
P.O. Box 17, 3300 AA Dordrecht, The Netherlands.

Sold and distributed in North, Central and South America
by Kluwer Academic Publishers,
101 Philip Drive, Norwell, MA 02061, U.S.A.

In all other countries, sold and distributed
by Kluwer Academic Publishers,
P.O. Box 322, 3300 AH Dordrecht, The Netherlands.

Printed on acid-free paper

Gedruckt mit Förderung des Österreichischen Bundesministeriums
für Bildung, Wissenschaft und Kultur
Printed with financial support of the Austrian Ministry for Education, Science and
Culture

In cooperation with the *University of Vienna, Center for Interdisciplinary Research/
Zentrum für überfakultäre Forschung*

Printed in the Netherlands

EDITORIAL

This volume contains the invited papers and a selection from the 150 papers that were presented at the *Third International History of Philosophy of Science Conference (HOPOS 2000),* – held at the Vienna University Campus in Vienna, July 6-9, 2000.

This Third Biennial Meeting of the *History of Philosophy of Science Working Group* was the first HOPOS conference to take place in Europe. It was jointly organized with the *Institute Vienna Circle* and the *Center for Interdisciplinary Research* and the *Department of Contemporary History* of the *University of Vienna.* The meeting brought together researchers in the history of philosophy of science from throughout Europe, America and elsewhere in Vienna, a most important site for the emergence of modern philosophy of science in the days of Mach, Boltzmann and the Vienna Circle. The conference was not easy to organize, and the political turn in Austria at the beginning of 2000 certainly did not make it any easier. After long and intense deliberations, the Program Committee, together with the Steering Committee of HOPOS, decided to issue a declaration in response to the political situation in Austria and to organize two additional special panels. This side program focused on the political context and cultural relevance of philosophy of science, especially with regard to Logical Empiricism from the 1920s to the present. These contributions will be published in a separate volume (*Wissenschaftsphilosophie und Politik. From Central Europe to America Before and After 1938)* co-edited by us and published by Springer Verlag (Wien – New York 2002).

HOPOS 2000 turned out to be an important event, promoting the convergence and cooperation of European and American scholars in the various fields of the history of philosophy of science and reflecting a diversity of scholarly approaches. Accordingly, the reader will find research-oriented contributions on the philosophy of science since Kant and Leibniz, on natural philosophy of the 19th century, on Hilbert's program, and on the Vienna Circle. The contributions of the four invited speakers including the keynote speech and 8th Vienna Circle lecture by Michael Friedman are of more general scope. They deal with ancient and early modern as well as contemporary philosophy of science and concern the development of the notion of law as well as other issues.

The general part of this Yearbook features two studies by renowned scholars: one on Skinner and Bridgman and the other on Wittgenstein's sociology of science. It also contains a section with review essays and reviews in the context of recent philosophy of science. The past and forthcoming activities of the

Vienna Circle Institute are documented as well. We also have the sad duty to commemorate three outstanding scholars, old friends and collaborators of the Institute who died recently: Marie Jahoda, Wesley C. Salmon, and Paul Neurath. Their life and work will remain cornerstones for the scientific community at large, but also for our own research.

As Chairs of HOPOS 2000 we would like to thank the members of the Program Committee (Gary Hatfield, Don Howard, Cassandra Pinnick, Joëlle Proust, G.A. John Rogers, Rose Mary Sargent, Thomas Uebel, and Catherine Wilson) as well as the History of Philosophy of Science Working Group, our local organizing committee (Angelika Rzihacek, Daria Mascha, Margit Kurka, Robert Kaller, and Marianne Ertl). We would also like to express our appreciation to all collaborators and sponsors of the *Third International History of Philosophy of Science Conference* (Austrian Ministry of Science and Research, City of Vienna, the University of Vienna, the Bank Austria), last not least the editorial team of this Yearbook (Hartwig Jobst, Camilla Nielsen, Erich Papp) and the review editor Michael Stöltzner.

Tübingen/Vienna,

Michael Heidelberger
(University of Tübingen)

Friedrich Stadler
(University of Vienna,
and Vienna Circle Institute)

TABLE OF CONTENTS

A. HISTORY OF PHILOSOPHY OF SCIENCE – NEW TRENDS AND PERSPECTIVES

I

II

B. GENERAL PART

REPORT – DOCUMENTATION

REVIEW ESSAYS

REVIEWS

ACTIVITIES OF THE IVC

OBITUARIES

MARIA LUISA DALLA CHIARA
ROBERTO GIUNTINI*

ON THE NOTION OF "LAW"

The term "law" appears in different contexts with different meanings. We are used to speaking of *natural laws, legal laws, moral laws, aesthetic laws, historical laws*. Such a linguistic convention has represented a constant phenomenon through the history of civilization. Is there any deep common root among all these different uses and meanings?

I. LAWS OF NATURE

The expression "νόμοι τῆς φύσεως" (*laws of nature*) can be found in Democritus, the Sophists, Plato, Aristotle. Its main meaning seems to be: "something that cannot be escaped", a kind of ἀνάγκη (*fate*). Sometimes "νόμοι τῆς φύσεως" were opposed to "human laws", which are possibly violated.

In the wake of the modern scientific revolution, the term "law of nature" was currently used by Kepler, Galileo, Newton, Leibniz. In a letter to Benedetto Castelli, Galileo writes:

La natura inesorabile e immutabile non trasgredisce mai i termini delle *leggi* impostele. (Nature, which is inexorable and immutable, never violates the *laws* that have been imposed to her).

We find here again a juxtaposition of *human laws* (which are possibly violated) and *natural laws* (which cannot be escaped).

What is a *natural law*? Galileo proposes what seems to our eyes a "quite modern" definition:

A natural law represents a constant mathematical relation among varying values of physical quantities, in the framework of a class of similar phenomena.

Sometimes Galileo speaks of *constant proportions among phenomena*, which have been associated to *numerical values* by means of *measurements*. The celebrated *law of falling bodies*

$$d = (1/2)gt^2$$

1

M. Heidelberger and F. Stadler (eds.), History of Philosophy and Science, 1–11.
© 2002 *Kluwer Academic Publishers. Printed in the Netherlands.*

represents an example of this kind. The law asserts that a constant mathematical relation holds between the values of two *physical quantities*: 1) the *distance d* covered by a falling body; 2) the *time t* used to cover such a distance (in a vacuum-situation).

In a sense, a natural law represents a *quid facti* against a *quid iuris* that may be imposed by an authority. However, natural laws are far from being mere descriptions of *pure facts* because of the essential role played by mathematics. According to Galileo, "le certe e necessarie dimostrazioni" (the certain and nec-essary proofs) are not less important than "le sensate esperienze" (the sense experiences):

L'esperienza serve a piegare quelli alla mente dei quali non arrivano le necessità delle dimostrazioni geometriche.
(Experience serves to band those whose mind is unable to grasp the necessity of geo-metrical proofs).

A quite similar position has been defended by Newton: *natural laws* represent *certainties* that are derived by means of *mathematical arguments* and that are verified by *observations*. A translation into a contemporary language might be:

a natural law is a proposition that can be proved in a theory verified by experiments.

What is still missing in Galileo and in Newton? Three basic ideas of contem-porary science:

1) Any measurement is accompanied by a characteristic *precision*.
2) Any *scientific law* is associated to a *domain of validity*.
3) Generally, a *law* is a *probabilistic assertion*.

In a contemporary perspective, *natural laws* can be formally characterized in the framework of what has been called *the semantic approach to empirical sciences* ([5], [8], [9], [4]).

According to van Fraassen, the basic *slogan* of the semantic approach can be sketched as follows:

presenting a theory means presenting a class of models.

But how can *models* be characterized [4]?

Take a physical theory *T* (say classical or relativistic or quantum mechanics). A *model M* of *T* can be described as a *structure* consisting of three parts:

1) a *mathematical part;*
2) an *experimental part;*
3) a *translation,* that gives a mathematical interpretation to all elements of the experimental part.

We can write:

$$M = <Math, Exp, Transl>,$$

where:

1) the mathematical part *Math* corresponds to a *mathematical model*, in the standard sense of *logical semantics;*

2) the experimental part *Exp* consists of a set of *physical systems*, accompanied by an appropriate set of *physical quantities*, which are supposed to be *operationally defined*;

3) *Transl* represents the mathematical interpretation.

In this framework one can define the concept of *empirical truth* with respect to a given model *M*. For instance, Galileo's law of falling bodies

$$d = (1/2)gt^2$$

turns out to be *empirically true* with respect to an appropriate class of *models* of classical mechanics.

Roughly, the *empirical-truth definition* can be sketched as follows. Consider a sentence *A* asserting that a given mathematical relation holds among the values of some physical quantities (say, the law of falling bodies):

A will be *empirically true* with respect to a model *M* if, and only if, whenever we perform a measurement of the physical quantities (which *A* refers to) in any physical system (of *M*), we obtain some results that satisfy the mathematical relation expressed by *A*, *up to a given precision* (where the precision depends on the operational definitions of the quantities in question).

Such a semantics permits us to formally analyze *precisions and approximations, domains of validity* and *probabilistic results*. On this basis one can understand how it is possible that two *rival theories* are at the same time *empirically true* with respect to a certain domain of phenomena, even if their logical conjunction turns out to be contradictory!

Let us think, for instance, of the case of classical mechanics and quantum theory. Such puzzling situations can be hardly described in the framework of classical semantics, where the *non-contradiction principle* (*a sentence and its negation cannot be at the same time true*) is supposed to be always valid.

According to van Fraassen ([8]), the semantic approach leads to a strong criticism of the very notion of *law of nature*, which is based on a metaphysical "pre-kantian" conception of *absolute truth*.

Van Fraassen claims:

there are no laws of Nature, there are only laws of models!

Is such an "extremistic conclusion" justified? It seems to us that the intuitive idea of *law of nature* (which is currently used by the working scientists) can be preserved and justified even in the framework of a semantic (non-metaphysical) approach. Laws of nature can be regarded as *sentences that are empirically true with respect to all the models of an accepted theory*. In fact, most physicists usually speak of "laws of nature" just in this sense, and not according to a meta-physical commitment, which seems to be somewhat far from a scientific attitude.

II. ALETHIC AND DEONTIC MODALITIES

What can be said about the relationship between *natural laws* and *legal norms?* A main difference can be explained in terms of *logical modalities.*

Let us first distinguish two kinds of *modal operators*:

- the operator *physically necessary* (or according to a physical law);
- the operator *legally obligatory* (or according to a legal norm).

We will simply write: *necessary* and *obligatory*, respectively. *necessary* represents an *alethic modality*, while *obligatory* corresponds to a *deontic modality.* The basic difference between the two cases is illustrated by the following logical relations:

- *Necessary A implies A.*

In other words: all that is necessary does happen. Physical laws cannot be escaped!

However:

- *Obligatory A does not imply A.*

Legal norms are possibly violated!

The distinction between *alethic* and *deontic modalities* is clearly reflected in some natural languages by means of appropriate verbs. In English, the verb *"ought"* mainly corresponds to the modality *obligatory,* while: *"must"* mainly corresponds to *necessary.* Similarly in German we have the distinction between: *"sollen"* and *"müssen".*

Also the *dual* modalities are linguistically distinguished both in English and in German. Consider the modal operators: *permitted* and *possible.* Clearly, *permitted* means: *not obligatory not;* while *possible* means: *not necessary not.* In English, the verb *"may"* mainly corresponds to the modality *permitted,* while *"can"* mainly corresponds to *possible.* Similarly in German we have the distinction between *"dürfen"* and *"können".*

The result of a confusion between *deontic* and *alethic modalities* has been ironically commented in Busch's *Max und Moritz*:

ES KANN NICHT SEIN WAS NICHT SEIN DARF !

In other words:

IF A IS NOT PERMITTED, THEN A IS NOT POSSIBLE !

Strangely enough such modal distinctions are not precisely reflected either in the Latin languages (Latin, Italian, French, Spanish, ...) or in most Slavish languages (Russian, Polish, ...).

Logicians have proposed a natural semantic analysis of the behaviour of different *modal operators*. The conceptual tool that has been successfully used is represented by the Leibnizian notion of *possible world*.

Possible worlds are not necessarily bound to a metaphysical interpretation. As shown by the development of contemporary semantics, the concept of *possible world* represents a quite flexible notion, which can be used for different applications. For instance, in the sense of:

- context or circumstance;
- time-instant or time-interval;
- physical situation or physical system in a given state;
- information system.

Suppose a class of possible worlds $i, j, k, ...$ Two worlds i and j may be *accessible*. We will say the the world j is *accessible* to the world i, whenever j represents a *possible alternative for* i.

On this basis one can give a truth definition for modal sentences:

- ***Necessary* A** *is true in the world* i if and only if
  A is true in all the worlds that are accessible to i.

Similarly:

- ***Obligatory* A** *is true in the world* i if and only if
  A is true in all the worlds that are accessible to i.

In this framework, it is useful to distinguish two semantic hypotheses:

I) Any world is accessible to itself
   (in other words, the accessibility relation is *reflexive*).

II) A world might be inaccessible to itself
   (the accessibility relation is *not* generally *reflexive*).

Our first hypothesis gives rise to an alethic modality:

Necessary A *implies* A.

If A is true in all the worlds that are accessible to i, then A must be true also in i (because the accessibility relation is reflexive).

Our second hypothesis, instead, gives rise to a deontic modality:

***Obligatory* A** *does not imply* A.

It may happen:

A is true in all the worlds that are accessible to i.

However:

A is not true in the actual world *i*.

"NO SMOKING" is obligatory. In spite of this, somebody does smoke!

From an intuitive point of view, the accessible worlds of a deontic semantics can be regarded as *idealized* worlds, where all that happens is governed by well extablished norms. By paraphrasing *Max und Moritz* we might say:

> *ES GESCHIEHT NICHTS WAS NICHT GESCHEHEN DARF!*

In other words:
> *IF A, THEN A IS PERMITTED!*

Of course, such a happy situation generally occurs only in our idealized accessible worlds, not in the *actual world.*

A question arises: do physical laws always correspond to *alethic-like* rather than to *deontic-like modalities*? According to our semantic analysis such a question means: how should an accessibility relation between *physical possible worlds* behave in a reasonable way? Is it reasonable to admit an application of possible world semantics to physics, where a possible world might be inaccessible to itself?

In the applications of possible world semantics to physics, possible worlds may represent:

- *physical situations;*
- *physical* systems *in* well *determined states;*
- *pieces of* information corresponding to the observer's knowledge about the system under investigation.

In fact, similarly to the deontic semantics, physicists often refer to *idealized physical situations*, where some *real parameters*, which do have an important role in the *actual physical world*, are neglected. Some typical parameters that are frequently neglected are for instance: friction, air resistance, the influence of the environment.

The actual physical world represents a "magmatic" situation, where too many irrelevant parameters give rise to *disturbing effects*. The physical situations which our experiments refer to are better represented by some *idealized physical situations.* This happens both in the case of concrete and of *Gedankenexperiments.* As is well known, Galileo making experiments with falling bodies from the Pisa Tower is only a legend!

On this basis one might conclude that the notions of *legal norm* and of *natural law* are less far from each other than it is usually supposed.

III. QUANTUM LOGIC AND HISTORIOGRAPHY

Let us now turn to a discussion of the concepts of *historical truth* and of *historical law*. We would like to propose a comparison that might appear *prima facie* somewhat puzzling: the use of *possible worlds* in the logic of quantum theory and in historiography.

The logic that has been termed *quantum logic*, was created in the thirties by Birkhoff and von Neumann [1], as a natural abstraction from the mathematical formalism of quantum theory. In the seventies, logicians proposed a characterization of quantum logic in the framework of a particular form of possible world semantics [3]. The main idea can be sketched as follows. In quantum theoretical applications, possible worlds are identified with *pieces of information*, corresponding to the observer's knowledge about the quantum system under investigation (for instance, an electron). As a limit case, a possible world *i* might correspond to *a non contradictory maximal knowledge*, which cannot be consistently extended to a more precise information. Pieces of information of this kind are usually called *pure states*. In the formalism of quantum theory, pure states are mathematically represented as particular vectors in an appropriate abstract space (which is called a *Hilbert space*).

What about the accessibility relation? *Accessibility* here means *logical compatibility*: two pieces of information *i* and *j* are *accessible* if, and only if, they do not have contradictory consequences. In the Hilbert space geometry, two accessible pure states will correspond to a pair of *non-orthogonal vectors*. On this basis, our accessibility relation turns out to be reflexive and symmetric.

Let us now illustrate the characteristic *quantum logical truth conditions*. The basic semantic relation is the following:

> *an information i forces a sentence A to be true.*

> (we also say that *i verifies* or *forces A*).

A sentence *A* may be either *atomic* (when it cannot be decomposed into simpler sentential parts) or *molecular* (compound by means of the logical connectives *not, and, or*). Generally, an atomic sentence *A* will assert that the value of a *physical quantity* belongs to a certain set of possible values.

Let us first consider the case of atomic sentences:

- an information *i* verifies *A* if, and only if,

for any information *j* accessible to *i* there exists an accessible information *k* that verifies *A*.

From an intuitive point of view, such a condition represents a kind of *stability* requirement.

Suppose now the case of a negated sentence *not A:*

- an information *i* verifies *not A* if, and only if,
  i has no accessible information *j* that verifies *A*.

In other words: *i* verifies *not A* when it is impossible to transform *i* into a compatible information *j* that verfies *A*. As a consequence, an information *i* will not generally *decide* a sentence *A*, which may remain strongly indeterminate for *i*:

i verifies neither *A* nor *not A*.

The semantic *tertium non datur* fails, in accordance with the probabilistic character of quantum theory.

What about conjunction and disjunction? Conjunction has a classical behaviour, governed by the usual truth-table. In other words:

- an information *i* verifies a conjunction *A and B* if, and only if,
  i verifies both members.

At the same time, like in classical logic, disjunction is supposed to be defined in terms of negation and conjunction by means of the so-called *de Morgan law:*

- *i* verifies *A or B* if, and only if, *i* verifies *not (not A and not B)*.

Since negation does not have the usual classical meaning here, the characteristic truth-table for the connective *or* turns out to be violated: the truth of a disjunction does not require the truth of at least one member. As a consequence, the following situation is possible:

i verifies *A or B*; however *i* verifies neither *A* nor *B* !

This peculiar behaviour of disjunction corresponds to a fairly typical quantum situation. In quantum theory, one can cite a number of examples where an alternative is determined and true, even if both members are strongly indeterminate (and hence not true). For instance, any electron has in any direction either *spin up* or *spin down*. However, owing to the *uncertainty relations*, an electron that has a well determined spin value in the x direction (say value *up*) cannot have a well-determined spin value in the y direction (neither *up*, nor *down*).

Birkhoff and von Neumann's quantum logic represents, in a sense, a "semi-aristotelian logic". As we have seen, the *semantic tertium non datur* fails: a sentence is not necessarily either true or false. However, the *non-contradiction principle* still holds. New forms of quantum logic that are even "more non-Aristotelian" arise in the framework of the so-called *unsharp approach to quantum theory* (first proposed by Ludwig and developed by Mittelstaedt, Busch, Lahti, Bugajski, Beltrametti, Cattaneo and many others).

Why move to the unsharp approach? From an intuitive point of view, one can say that such approach represents, in a sense, an important step towards a kind of second degree of "fuzziness" or "ambiguity". We can try and illustrate the difference between the standard and the unsharp approach by referring to a non-scientific example. Let us consider the two following sentences, which apparently have no definite truth-value: ·

I)   *Hamlet is 1.70 meters tall.*

II) *Brutus is an honourable man.*

The semantic uncertainty involved in our first example seems to depend on the *logical incompleteness* of the *individual concept* associated with the name "Hamlet". In other words: the property "being 1.70 meters tall" is certainly a sharp property. However, our concept of Hamlet is unable to decide whether such a property is satisfied or not. As opposed to real persons, literary characters have a number of indeterminate properties. The semantic uncertainty involved in our second example, is mainly caused by the ambiguity of the concept "honourable". What does "being honourable" mean? Just recall how the ambiguity of the concept "honourable" plays an important role in the famous Mark Antony's monologue in Shakespeare's "Julius Caesar". Orthodox quantum theory and quantum logic take into consideration only examples of the first kind: properties are sharp, while all semantic uncertainties are due to the incompleteness of the individual concepts, which correspond to pure states of quantum objects. Unsharp quantum theory, by contrast, also investigates examples of the second kind. In Hilbert space quantum theory, physical properties that may be ambiguous and unsharp are mathematically represented by particular operators, called *effects*. At the same time, sharp properties correspond to *projection operators*, which are limit-cases of effects. An extreme case of an unsharp property is represented by the *semi-transparent effect*, to which any physical state assigns probability 1/2. This represents a kind of paradigmatic example of a totally indeterminate physical property.

In the possible world semantics for unsharp quantum logic, an information *i* may correspond to ambiguous knowledge. As a consequence, pieces of information turn out to be not necessarily compatible with themselves: reflexivity breaks down. On this basis, it turns out that an information may verify at the same time a sentence *not A* and its negation *not A*. The non-contradiction principle is violated! The result is a *para-consistent* (or *fuzzy* or *unsharp*) form of quantum logic.

A similar semantic behaviour can be recognized in the case of historiography. Also historians naturally interact with possible worlds, at least implicitly. The role of possible worlds in historical research has been investigated in [6], [7]. We will use here a slightly different semantic approach. Similarly to *physical states*, even *historical sources* determine pieces of information about possible state of affairs. From an abstract point of view, in the simplest case, a source can be idealized as a set of sentences:

a source *i* will assert a sentence *A* when *A* is contained in *i*.

For instance, the source *The Lives of the Twelve Caesars* asserts the sentence:

> *Evident miracles had announced to Julius Caesar his violent death.*

More generally, a source may be also represented by some non-linguistic objects (monuments, tools, tombs, ...). As a consequence, the relation "*i* asserts *A*" might have a weaker meaning, like "*A* is *confirmed, testified* by *i*".

Similarly to a quantum logical situation, two sources may be either *compatible* or *incompatible*. We will say that two sources *i* and *j* are *compatible* when it is not the case that *i* asserts a sentence *A* whereas *j* asserts the negation *not A*. Since a source is not necessarily consistent, we will argue that some sources might be incompatible with themselves. Apparently, the presence of a local contradiction is not a sufficient reason for a given source to be completely unreliable. Consequently, like in the case of unsharp quantum logic, the accessibility relation between pieces of information turns out to be symmetric, but generally not reflexive.

Let *I* represent a set of sources that are available to a given historian. When will our historian accept the *historical truth* of a given statement? At first sight, one is tempted to take into consideration the two following possibilities:

I)   All sources assert *A*.

II)   At least one source asserts *A*.

However, our first choice seems too strong: many events that are sometimes considered "historical facts" are not asserted by all sources. At the same time, our second choice appears too weak; some sources might be not completely reliable. For instance, why should we trust Svetonius when he asserts that "Evident miracles had announced to Julius Caesar his violent death"?

An intermediate choice between I) and II), which seems quite reasonable from the intuitive point of view, naturally leads to a quantum logic like truth condition:

> a source *i* forces the truth of an atomic sentence *A* if and only if
> for any source *j* compatible with *i*, there exists a compatible source
> *k* that asserts *A*.

In other words, *A* represents a kind of *stable assertion*. As regards the case of compound sentences, again the quantum logical truth conditions appear to be quite natural. In particular, one might mention a number of examples in historical contexts, where an alternative is true, whereas both members are strongly indeterminate. Finally let us ask: when will a historian accept the historical truth of a given sentence? The natural answer in our abstract semantics is the following: a historian will accept the truth of *A*, when the truth of *A* is forced by a set of sources that are judged reliable by our historian. On this basis one can conclude that truths discovered by historians and by quantum physicists turn out to have some strong formal similarities!

What about the concept of "historical law"? A "natural temptation" might suggest the following simple proposal:

Historical laws are represented by sentences that turn out to be true in all the considered models.

Is such a characterization too naive? Not necessarily, if our sentences represent some general probabilistic assertions.

We are aware that *prima facie* such formal comparisons might appear somewhat artificial. However, it seems to us that a winning trend of contemporary epistemology is to try and look for some common *formal patterns* in different fields of knowledge. The examples we have discussed lead in this direction.

NOTES

*   We thank Giuliano Toraldo di Francia for some useful suggestions.

REFERENCES

[1]   G. Birkhoff, J. von Neumann, "The Logic of Quantum Mechanics", *Annals of Mathematics*, 37(1936), 823-843.
[2]   M.L. Dalla Chiara and R. Giuntini, "Quantum Logical Semantics, Historical Truths and Interpretations in Art", *Proceedings of the Conference "Einstein meets Magritte"*, Kluwer, 1999.
[3]   M.L. Dalla Chiara and R.Giuntini, *Quantum Logics*, in *Handbook of Philosophical Logic*, Kluwer (new edition, to appear).
[4]   M.L. Dalla Chiara and G. Toraldo di Francia, "A Formal Analysis of Physical Theories", in G. Toraldo di Francia (ed.), *Problems in the Foundations of Physics*, North-Holland, Amsterdam, 1979
[5]   P.Suppes, *Studies on the Methodology and Foundations of Science*, Reidel, Dordrecht, 1969.
[6]   G. Toraldo di Francia, *Un Universo Troppo Semplice*, Feltrinelli, Milano, 1990.
[7]   G. Toraldo di Francia, "Historical Truth", *Foundations of Science*, 1-3 (1995-1996), 417-425.
[8]   B. Van Fraassen, *Laws and Symmtery*, Clarendon Press, Oxford, 1989.
[9]   R. Wójcicki, "Set Theoretic Representations of Empirical Phenomena", *Journal of Philosophical Logic*, 3 (1974), 337-43.

Maria Luisa Dalla Chiara
Dipartimento di Filosofia
Università degli Studi
Via Bolognese 52
I-50139 Firenze
Italy
dachiara@risc.idg.fi.cnr.it

Roberto Giuntini
Dipartimento di Scienze Pedagogiche e Filosofiche
Università di Cagliari
Via Is Mirrionis 1
I-09123 Cagliari
Italy
giuntini@philos.unifi.it

MARINA FRASCA-SPADA

HUME ON SENSE IMPRESSIONS AND OBJECTS

This essay is on the nature and roles of sense impressions and objects in Hume's account of perception in the *Treatise of Human Nature*. I start by considering how Hume introduces sense impressions at the beginning of the *Treatise* and show that, although he explains the distinction between impressions and ideas on the basis of their different strength and liveliness, the crucial difference between them is in fact that ideas are copies of impressions, while impressions do not, in turn, copy anything. They are what ideas represent, the objects of our thought. But if impressions are non-representative, how can Hume talk about 'objects' at all – in fact, what are Humean 'objects'? This problem is the subject of the present discussion.

I.

Hume introduces what he calls "perceptions of the human mind" in real time, as the sensations, thoughts and emotions that his reader has while perusing page 1 of his *Treatise*. Impressions are both the perceptions arising from sight and touch – the sight of these black marks on this white page, the feeling of the crispness of the corner of the page between my fingers and of the bulk of the fat volume I am holding in my hand, etc. – and "the pleasure or uneasiness" occasioned by the reading. "All our sensations, passions and emotions, as they make their first appearance to the soul" are impressions.[1] These perceptions are immediate and compelling, commanding my instant unreserved assent: so it simply does not cross my mind that I may not really be seeing the colours and shapes, and feeling the mass and textures that constitute my old copy of Hume's *Treatise*. Impressions, as he puts it, are the perceptions "which enter with most force and violence".

Impressions are constantly accompanied by a second kind of perceptions: ideas, which are the mental copies of former impressions. "By *ideas*", Hume says, "I mean the faint images of [impressions] in thinking and reasoning". For example, at this moment they include everything which is going on in my mind, except all that I listed a moment ago under the heading "impressions": my reactions to Hume's "discourse", and presumably, as far as "images" *stricto sensu* go, that of the dark green cover of my copy of Hume's *Treatise* (I cannot see it while reading the book, but I remember it distinctly from former sense impressions).[2]

13

M. Heidelberger and F. Stadler (eds.), History of Philosophy and Science, 13–24.
© 2002 *Kluwer Academic Publishers. Printed in the Netherlands.*

This is Hume's so-called copy principle: all our ideas are faithful, if fainter, copies of former impressions. This principle presupposes that the umbrella term for all mental contents is not "ideas", as was standard within the Lockean frame of reference, but rather "perceptions"; Lockean ideas are, here, only one of the two kinds of perceptions. Hume is explicit in stating that this distinction between impressions and ideas introduces something new in contemporary philosophical idiom: "I here make use of these terms, *impressions* and *ideas*, in a sense different from what is usual, and I hope this liberty will be allowed me". But then he also suggests – in a way which is somewhat typical of contemporary criticisms of Locke[3] – that perhaps this appearance of originality is illusory, by adding: "Perhaps I rather restore the word, idea, to its original sense, from which Mr. *Locke* had perverted it, in making it stand for all perceptions" (T2n).

A notorious problem with this distinction has to do with the exact characterisation of the two kinds of perceptions relative to one another. Hume's first approach to this issue is apparently straightforward: "the difference betwixt these consists in the degrees of force and liveliness, with which they strike upon the mind, and make their way into our thought or consciousness" (T1).[4] But the possible objections to this way of establishing the distinction are obvious. There are counterexamples with a solid early modern tradition at least since Descartes: as Hume is quick to admit, ideas can, sometimes, be pretty lively and strong. Take the classic case, mentioned by Hume, of very vivid dreams or of hallucinations. These are, in his view, ideas, and yet it would be very hard to tell them from impressions on the basis of their liveliness and strength. On the other hand, many impressions may be particularly calm and faint (T2). Strength and liveliness provide a criterion which is both too vague to justify the distinction, and too ambiguous to make it actually work.[5] Moreover it is easy to construct cases of simple ideas which are not, in fact, copied from corresponding impressions by imagining a continuous gradation of a quality (Hume's choice is the colour blue) with just one gap – wouldn't our mind be able to fill the gap...?[6]

It is clear that Hume is well aware of all this. But, he insists, it does not really matter, for we know very well the difference, as he puts it, between feeling and thinking; and the so-called copy principle simply says that feeling always comes before, is more compelling than, and provides the material for the corresponding thinking; and viceversa. Hume is a famously artful writer. What he is saying here is not all that clear, and far from uncontroversial; but his rhetoric is reassuring, as if what he is saying were nothing challenging or less than obvious: after Locke there can be no serious doubt that all our ideas derive from experience.[7]

In fact, the implications of what he is saying are far-reaching. Impressions are prior in time to the corresponding ideas; they are constantly conjoined with ideas; we are unable to form ideas without the corresponding impressions: so we can conclude, in conformity with Hume's discussion of causation later in the book, that impressions *cause* ideas. In this sense, the so-called copy principle is Hume's version of Locke's theory of ideas: but while Lockean ideas represent or conform to objects, and are caused by a physical process through which objects affect our nerves and brain,[8] Humean ideas copy impressions, and are caused by processes that Hume declares he does not intend to discuss because "the examination of our sensations belongs more to anatomists and natural philosophers than to moral" (T8).[9] The pattern of systematic differences within the structural similarity of the two theories could be followed up in detail on a number of issues.[10] To put all this slightly differently, Hume's approach suggests a Malebranchean reading of Locke's account of knowledge in which the world is duplicated: roughly speaking, the real world of external objects is reproduced in the mind as a world of ideas which represent or conform to those objects. Hume's copy principle is a version of this duplication of the world. All perceptions, he says, come in pairs: "All perceptions of the mind are double, and appear both as impressions and as ideas" (T2-3). But, unlike Locke, Hume regards both worlds as composed of perceptions.

The single most revealing feature of Hume's impressions is that by this term he refers not only to our sensations, but also, as we have seen, to pains and pleasures, and to yet another variety of original perceptions, those "passions and emotions" (T1) which are the subject of Book 2 of the *Treatise*. This is how he explains what sense impressions, bodily pains and pleasures, and impressions of reflection have in common:

... every impressions, external and internal, passions, affections, sensations, pains and pleasures, are originally on the same footing; and [...] whatever other differences we may observe among them, they appear, all of them, in their true colours, as impressions or perceptions (T190).

So my visual impressions of the black marks on the white page of Hume's *Treatise* are on a par with, say, the pain I feel if the spine of the book strikes me on the head, and with my feeling of confidence in reading the familiar first lines of the book: all these kinds of perceptions are immediate, and immediately transparent to my mind. In Hume's own words, "Every thing that enters the mind, being in *reality* a perception, 'tis impossible any thing shou'd to *feeling* appear different" (T190). Of course, this complete transparency of impressions is more evident in the cases of the pains and pleasures and of the passions than in the case of sense impressions – pleasures and pains, like the desire or disgust and hope or fear to which our impressions and ideas give rise, are sparked off, or at least accompanied, by sense impressions or ideas, but there is no sense in which they represent these impressions or ideas, or indeed anything else; nor is there any sense in which they could be deceitful. But it is sense impressions Hume is primarily interested in here, and the conclusion he is angling at has to do specifically with sense impressions: this immediacy is the reason why, he writes, our senses simply cannot be prone to the "kind of fallacy and illusion" necessary for them to suggest that their own objects, that is, sense impressions, exist externally and independently of ourselves.

Some readers have found that Hume's grouping together of sense impressions, bodily pains and pleasures, and passions is odd. But in my reading this oddity is deliberate: the way Hume restored "ideas" to its original, pre-Lockean meaning is by grouping together the non-representational mental contents – all sensations, including bodily pains and pleasures, and passions – under the label "impressions", and all the representational ones under the label "ideas". What this discussion of the immediacy of impressions suggests is that Hume's sense impressions too are meant, like bodily pains and pleasures and like the impressions of reflexion, to be regarded as non-representational: rather than being representations of objects, they themselves *are* the objects of our common experience.

II.

I have suggested that Hume's emphasis on the originality, immediacy, and innateness of sense impressions may be regarded as showing that Humean sense impressions are non-representational. Of course Hume is not saying that there is nothing in the world but our minds. In fact, often he seems to assume just the opposite. Let us consider a concrete case. Take Hume's apparently innocent reference to a visual object, a spot of ink:

Put a spot of ink upon paper, fix your eye upon that spot, and retire to such a distance, that, at last you lose sight of it; 'tis plain, that the moment before it vanish'd the image or impression was perfectly indivisible. 'Tis not for want of rays of light striking on our eyes, that the minute parts of distant bodies convey not any sensible impression; but because they are remov'd beyond that distance, at which their impressions were reduc'd to a *minimum*, and were incapable of any farther diminution. A microscope or telescope, which renders them visible, produces not any new rays of light, but only spreads those, which always flow'd from them ... (T27-28).

Here Hume is discussing the conceivability of indivisibles of space, and arguing that, far from being counterintuitive and paradoxical, minima are familiar objects of our perception. In particular, he is showing that we are actually acquainted with them through vision: the spot of ink, just before disappearing, is a unit of sight, that is, an indivisible of space immediately present to our eyes. There are several difficulties in this passage. I have treated extensively elsewhere the various and thorny problems specifically connected with the idea of space and with the nature, conceivability etc. of visual and spatial indivisibles.[11] Here I would like to mention only three questions. First, what comes across most immediately is that Hume is using a (broadly defined) Lockean approach to descrive the process of sensation: the whole "minute parts" story, with the addition of the natural philosophical talk of rays of light and optical instruments is in apparent tension with Hume's open and repeated refusal I cited earlier to engage with either natural philosophy, or anatomical matters.[12] I shall come back to this tension later, at the end of this section. A second equally obvious feature of this passage is that this supposed phenomenology of a visual experience seems to be based on the commonsensical assumption of the existence out there of such things as spots of ink, rays of light and optical instruments.[13] But if impressions are all we have as the material of our knowledge of the world, how can we ever conceive and talk about anything like "a spot of ink"? Especially so, since the spot of ink cannot be easily identified with any one visual impression, being in fact constituted by *a series of* impressions. Now, as a reader noted, "What is *the* 'spot' if the "impressions" form a series?"[14] And this is not all, for as we are going to see, Hume entitles us to wonder, even more radically: how can impressions ever "form series" at all?

I intend to focus specifically on this last question; but it may be useful to introduce it by elaborating on the more general question of Hume's use of objective language. Hume's terminological ambiguity is notorious. In connection with our present problem, in a stimulating recent study Marjorie Greene presents a systematic examination of Hume's uses of the term "object" in the *Treatise*. These can be classified, she suggests, under three main headings. The first is "intentional objects" or "targets of attention" – that is, what ideas, passions and perhaps occasionally even sense impressions are about. The second use of "object" is as synonym of "sense impression". This is, according to Greene, the canonical sense of Hume's "object" and the very core of his empiricist project; and from what I have said so far it is clear that on the whole I agree with her on this point. The third use of "objects" is in the "objective sense", meaning external objects – and she notices that, rather surprisingly, this third category appears to be the best represented, with over 120 instances![15] Hume's use of objective terminology in connection with the question of external objects is indeed a *locus classicus* of Hume scholarship; and Hume's scanty, but unmistakable references to "unknown efficacies", "secret force and energy of causes", "conceal'd causes", "secret operations", "unknown and mysterious connexions", and so on, has been for the last 15 years at the centre of a lively scholarly debate relative to his analysis of the idea of cause and effect.[16] Let me briefly outline the main point of this debate, which is of some direct relevance to my present purposes because its focus is on a problem analogous to mine.

Hume famously pointed out that our idea of a necessary connexion between cause and effect is not copied from a sense impression of power in the cause, or of interaction, or of necessary connexion between the cause and the effect – it is a fact, Hume maintains, that we have no such sense impression. Rather, the idea of necessary connexion is copied from the feeling of easy and inescapable transition from the sense impression of the "cause" to a strong and lively idea of the "effect", once we have acquired the habit of experiencing this particular kind of "cause" as constantly conjoyned with this particular kind of "effect". The alleged unknown power, or secret spring, or necessary connexion between the two are, however, totally unavailable to us. And yet, there are various references to them in Hume's text. Don't they seem to indicate that he did believe in the existence, out there, of necessary connexions, however inaccessible to us, between causes and effects? And exactly how inaccessible are they – can we at least meaning-fully talk about them? Etc.[17]

As I said, the question of how impressions can ever "form series" poses a problem which is at the same time very similar to this, and more basic – so, not with our notions of causal links among objects, but rather with the links in virtue of which we can think of the objects themselves as such. All our particular per-ceptions, Hume says, "are different, and distinguishable, and separable from each other, and may be separately consider'd, and may exist separately, and have no need of any thing to support their existence ..." (T252). In other words, we take it for granted that impressions are grouped together to make up objects, while each of them is, in fact, completely self-contained. So it is not just the connection, necessary or otherwise, between distinct and causally linked objects which is on the line of fire here, but also the connection between sense impres-sions which we should regard as independent existences, and of which, on the contrary, we think as different appearances *of the same thing*.

If each of our sightings of the retiring spot of ink is a new and independent existence, how do we come to see them as forming the series which we further unify under the term "a spot of ink"? Hume's answer to this question is to be found in the section of the *Treatise* devoted to "the scepticism with regard to the senses". Here Hume considers the origin of our ideas of the continued, and of the distinct existence of objects, establishing that these are separate, if strictly linked questions. What is the faculty responsible for our idea of the continued existence of objects? Certainly not our senses, for "they cannot operate beyond the extent, in which they really operate" (T191); nor, as we have seen, can they mislead us in this, given the immediacy of their imput. Reason cannot be responsible either: for whatever argument philosophers can produce, since it has not been discov-ered so far, it would have to be very far-fetched and sophisticated, and hence by definition an unlikely candidate for the reason why "children, peasants, and the greatest part of mankind" think of continuous objects in spite of the disconti-nuities of sensation.

The construction of continuous objects turns out to be the result of a syn-thesis operated by the imagination, which attributes continuity when it finds habitual constancy or coherence – the constancy of "my bed and table, my books and papers" with their unchanging appearances (T194-5), and the coherence of the alteration of the fire in the fireplace: "when I return to my chamber after an hour's absence, I find not my fire in the same situation, in which I left it: But then I am accustom'd in other instances to see a like alteration produc'd in a like time, whether I am present or absent, near or remote" (T195). As Hume puts it, the imagination fills the gaps through a kind of mental inertia: "when set into any train of thinking, [it] is apt to continue, even when its object fails it, and like a galley put in motion by the oars, carries on its course without any new impulse" (T198). The supposed continuity is the result of the imagination's inertial movement started by the observed partial uniformity and coherence.

There is here another parallelism with the case of the idea of cause and effect: for the inertial movement of the mind which constructs a continuous object to underpin the constancy or coherence of fleeting sense impressions is of a sort similar to the expectation created by causation, that is, to the easy passage of the mind from the impression of the cause to the lively idea of the effect which usually accompanies it. Such sense impressions as the creaking of a door while I am not looking are unreflectively connected by the imagination to the other sense impressions usually associated with it, regardless of their actual presence in this particular occasion – in this case, say, the sight of the door opening. As a result, "I suppose that the door still remains, and that it was open'd without my perceiving it" (T196-7). Without this, a creaking door heard but not seen would be in contradiction with my common experience, in which doors creaking and being seen moving in certain ways always go together. Indeed, Hume suggests that such a door could put under strain the whole system of maxims on which we base our causal reasonings, by confusing our habitual associations of constantly conjoined events.

So, to summarise with the words of two recent readers of Hume, the whole point of Hume's discussion here is that "the notion of physical objects and processes is built up from sense impressions by means of the imagination".[18] While thus making our experience an experience of objects, our imagination also gives rise to the notion that the objects of our experience are distinct and independent existences – in the manner that in fact we only know to be the case, at least in principle, for our individual perceptions. This is how our experience is an experience of apparently external objects. Each of these objects consists of a number of self-contained sense impressions clustered together by our imagination on the basis of a pattern arising from habit. They are fictions of the imagination.

This is not, of course, the end of the story; and the part played by philosophy in its next instalment is of some interest. The question of the existence of continuous and separately existent objects opens up a fatal dialectic between vulgar thought and philosophy:

> for philosophy informs us, that every thing, which appears to the mind, is nothing but a perception, and is interrupted, and dependent on the mind; whereas the vulgar confound perceptions and objects, and attribute a distinct continu'd existence to the very things they feel or see (T193).

To describe the unreflexive point of view of the "vulgar", Hume suggests, the words "perception" and "object" may be used as synonyms, as both mean "what any common man means by a hat, or shoe, or stone, or any other impression, convey'd to him by his senses" (T202). Note that this is what we all do most of the time, regardless of how philosophically sophisticated we are when we are wearing our philosophical hat: even philosophers are common people when the intellectual tension of their pursuits relaxes, making them snap back to nature so that they "take their perceptions to be their only objects, and suppose, that the very being, which is intimately present to the mind, is the real body or material

existence" (T206). But how can our sensations continue to exist when we are not
having them? This contradiction is exposed by a "philosophy" which Hume
seems to consider as the result of the most basic, and entirely natural use of
reason.

It is again a fact of human nature that our mind feels ill at ease with contradictions, so this one gives rise to a "system of the double existence", something
of a natural, and hence timeless, simplified version of Lockeanism: our perceptions are not all there is; underlying them there is a world of continuous and
independent objects. While the objects of the vulgar are unreflectively identified
with perceptions, within this philosophical system objects are split into perceptions, with which we are familiar and which are dependent on the mind and
therefore interrupted and fleeting, and something else, which we can only suppose to be similar to those perceptions, except that we endow it with continuity
and independence. In other words we try to pacify our mind by satisfying at the
same time our reason – with the attribution of the interruptions to perceptions –
and our imagination – with the attribution of the continuity to independent
objects (T215). This is as much relief as we can get in this case: clearly not a
solution of the contradiction, but an attempt to defuse the mind's unease. In this
sense, to a true philosopher both the vulgar system and the philosophical system
of the double existence are unacceptable, the one for being based on an illusion
and riddled with contradictions, and the other for being in the same position with
the addition of a touch of absurdity.[19]

The conflict can only be reconciled at a metaphilosophical level, and the
solution is the position of the "true philosopher", that is, Hume's own famous
"moderate scepticism". To the eyes of a "moderate sceptic" the spot of ink is
both the obvious object of common sense (this is when he is not philosophising,
that is, most of the time), or a devilishly complicated set of interactions involving
impressions, ideas, habits and operations of the imagination, even acts of judgment. And as far as the existence out there of such continuous and independent
things as "spots of ink" is concerned, "moderate scepticism" means practising
philosophy "in a careless manner" – that is, being able to live with the fact that
our belief in it is at the same time natural, hence absolutely inevitable, and
philosophically incoherent (T218, 273).

                                    III.

Let us now go back to Humean objects and their origin. As I have said, these
"objects" are constructed by our imagination out of sense impressions which are,
in principle, self-contained existences – that is, before we start philosophising
our imagination synthesises the manifold of our scattered sense impressions,
organising and coordinating them into things such as "spots of ink", pens and
sheets of paper, hats and shoes. These are the objects with which we are familiar
when we are living our life without thinking about philosophical issues; when we

start wondering about them, we get entangled in a web of metaphysical inconsistencies and miseries. This situation can only be made tolerable if we regard it as the inevitable result of the features and limitations of human nature and reason.

This way of expressing the situation is deliberately suggestive of a very Prussian Hume. My suggestion is indeed that Humean sense impressions may be regarded, in some important respects, as similar to Kantian phenomena; and that while he is clearly very far from denying the existence of an external world, Hume does, like Kant, work at tracing the limits of our knowledge of it. In this sense I think that to Hume, as to Kant, epistemology is preliminary to metaphysics, because it inevitably determines the scope of metaphysics – accounting for human perception is the first step towards any discussion of what there is, for, as we have seen, before worrying about whether objects are causally connected to each other, or indeed whether they exist out there, we are to decide in what way we come to think of our experience as being experience of "objects". There are also, however, big differences; and I would like to conclude by briefly mentioning one.

In his classic "A Prussian Hume and a Scottish Kant" Lewis White Beck noticed that in treating first causation, then objects, the order Hume is following is the reverse of what Kant was to do in the *Critique of Pure Reason* – as if Kant had put the second analogy of experience, on the irreversibility of time and the objectivity of causal succession, before the first, on duration and the objectivity of permanence.[20] I can think of two reasons for this inversion. The first is Hume's famous commitment to "carelessness and inattention" as a form of wisdom. Unlike Kant, who, perhaps paradoxically, wanted metaphysics to get somewhere, if not to progress and be a science, Hume, convinced as he was that vulgar thought, false metaphysics and true metaphysics are all equally insuppressible expressions of human nature, adopted a drastically non-cumulative approach: every new question is tackled against the background of the whole system of commonly held beliefs left unscathed by former discussions, and, in turn, leaves that system entirely unshaken. With this we are at the first and most immediate problem I signalled above in Hume's passage on the spot of ink, his adoption of Lockean-cum-natural philosophical and medical terminology and approach which is so evidently at odds with much of his stated convictions. And this is why in spite of all that has happened in the course of Book 1 of the *Treatise* at the beginning of Book 2, devoted to the passions, he feels that, after repeating that sense impressions are the ones "which without any introduction make their appearance to the soul", he can explain:

original impressions or impressions of sensation are such as without any antecedent perception arise in the soul, from the constitution of the body, from the animal spirits, or from the application of objects to the external organs (T275).

Here the "application of objects to the external organs", again so reminiscent of Locke's causal theory of perception, figures among the possible "unknown causes" of our sensations, together with the "constitution of the body" and the "animal spirits" of contemporary medical and physiological accounts of perception; and they are all referred to in the typically flat, matter-of-fact manner of a Hume who is focussing his sharp mind somewhere else, a Hume who is preparing to tackle a different question.[21]

My second possible explanation is, in fact, the same, but put in more complex and philosophically attractive terms. Here I can only offer it in the form of a brief suggestion. It has to do with a remark by Hume to the effect that even in the case of the identity of objects the synthesis can only be operated through the relation of cause and effect:

we readily suppose an object may continue individually the same, tho' several times absent from and present to the senses; and ascribe to it identity, notwithstanding the interruption of the perception, whenever we conclude, that if we had kept our eye or hand constantly upon it, it wou'd have convey'd an invariable and interrupted perception. But this conclusion beyond the impressions of the senses can be founded only on the connexion of *cause and effect*; nor can we otherwise have any security, that the object is not chang'd upon us, however much the new object may resemble that which was formerly present to the senses (T74).

This reference to a wider role of causation as the connective tissue of our experience is tantalisingly short, and, I find, not entirely clear. It certainly seems to suggest that there is no way we can first consider and guarantee the existence of objects, then tackle the issue of how they may be connected to each other by causal links – or, as Hume himself puts it, duration is only conceivable on the basis of succession and coexistence: for when we think of an unchangeable object in time this is "only by a fiction of the imagination, by which the unchangeable object is suppos'd to participate of the changes of the co-existent objects" (T200-201). The very way we organise our scattered sense data into those clusters that we call "objects" already shows the natural relation of cause-and-effect at work. Thus, there is no obvious order for us to follow when we try to uncover the foundations of human knowledge: duration, causation and coexistence are so interdependent, that we can only tackle them, in whatever order we wish, one at a time and against the background of the other two firmly taken for granted in the commonsensical way.

NOTES

1. D. Hume, *A Treatise of Human Nature*, ed. L. A. Selby-Bigge, 2nd ed. revised by P. H. Nidditch. Oxford: Clarendon Press 1978, p. 1. From now on cited in the text as T.
2. This discussion presupposes that in calling ideas "images" of impressions Hume is using the term "image" in a metaphorically extended sense. For the standard view that Hume literally equates ideas with mental images see J. Laird, *Hume's Philosophy of Human Nature*. London: Methuen 1932, pp. 26-7, and the sensitive treatment in C. Maund, *Hume's Theory of Knowledge*. London: Macmillan 1937, pp. 169-70; often taken for granted, see for example A. Flew, *Hume's Philosophy of Belief*. London: Routledge and Kegan Paul 1961, p. 22. I discuss this issue in *Space and the Self in Hume's "Treatise"*. Cambridge: Cambridge University Press 1998, ch. 1, pp.15-20.
3. See for example *An Essay on the Origin of Evil. By Dr. William King, late Lord Archbishop of Dublin*, (trans. Edmund Law). Cambridge 1731, ch. 1, King's note A, esp. p. 6, and Law's defence of Locke's notion of "idea", note 3, p. 7.
4. See also D. Hume, *Enquiries concerning the human understanding and concerning the principles of morals*. Ed. L. A. Selby-Bigge, 3rd edition, wed. revised by P. H. Nidditch. Oxford: Clarendon Press 1978, p. 11. From now on cited in the text as E.
5. B. Stroud, *Hume*. London: Routledge and Kegan Paul 1977, pp. 27-9.
6. The "missing shade of blue" is a classic brain-teaser in Hume scholarship. For a recent discussion see D. Garrett, *Cognition and Commitment in Hume's Thought*. Oxford: Oxford University Press 1997, ch. 2, "The copy principle", pp. 43-57.
7. See J. J. Richetti, *Philosophical Writing: Locke, Berkeley, Hume*. Cambridge, MA: Harvard University Press 1983, ch. 4, pp. 183-263, for a detailed and illuminating discussion of rhetoric and philosophy in Hume's writing.
8. See for example the discussions in J. L. Mackie, *Problems from Locke*. Oxford: Clarendon Press 1976, ch. 2, pp. 37-71; and J.W. Yolton, *A Locke Dictionary*. Oxford: Blackwell 1993, entries "Idea", "Knowledge", and "Representation".
9. See also the introductory section of Book 2, T275-6.
10. To mention just one of them as an example, Hume is explicit in pointing out that his distinction and correlation of impressions and ideas involves a version of Locke's famous rejection of innate ideas: see T7 (and E17n).
11. See *Space and the Self*, ch.1, passim, for a full discussion of the difficulties connected with Hume's notion of visual unit, and the relation between that notion and his talk about the nature and properties of "real" space.
12. *Ibid.*, pp. 52-4, for detailed discussions of Hume's natural philosophical talk and its sources.
13. *Ibid.*, pp. 46-55.
14. Laird, *Hume's Philosophy*, p. 68.
15. M. Greene, "The objects of Hume's *Treatise*", in: *Hume Studies* 20, 2, 1994, pp. 163-77.
16. The classics on this are E. Craig, *The Mind of God and the Works of Man*. Oxford: Clarendon Press 1987; John Wright, *The Sceptical Realism of David Hume*. Manchester: Manchester University Press 1983; and G. Strawson, *The Secret Connexion. Causation, Realism, and David Hume*. Oxford: Clarendon Press 1989. For a balance see the recent R. Read and R. J. Richman (Eds.), *The New Hume Debate*. London: Routledge 2000.
17. See K. Winkler, "The New Hume", in: *Philosophical Review*, 100, 1991, pp. 541-579 (reprinted in Read and Richman, *op. cit.*), and the elegant response by P. Kail, "New Humes and Old: Sceptical Realism and the AP Property" (paper presented at the 1999 Hume Society Conference), for the latest stages of the discussion.
18. A. Hausman and D. Hausman, "Idealising Hume", in: *Hume Studies*, 18, 2, 1992, pp. 209-18, esp. p. 211; see also Hausman and Hausman, "Hume's use of illicit substances", in: *Hume Studies*, 15, 1, 1989, pp. 1-38, and the discussion in Stroud, *Hume*, pp. 96 ff., esp. p. 106.
19. Hume's actual position in this section, in particular with respect to "the system of the double existence", is controversial: see J. W. Yolton, *Perceptual Acquaintance from Descartes to Reid*. Oxford: Blackwell 1984, pp. 147-64, and A. Baier, *A Progress of Sentiments: Reflections on*

Hume's "Treatise". Cambridge, MA: Harvard University Press 1991, pp. 101-128; also the debate between F. Wilson, "Is Hume a sceptic with regard to the senses?", in: *Journal of the History of Philosophy*, 27, 1, 1989, pp. 49-73, and D. W. Livingston, "A Sellarsian Hume?", in: *Journal of the History of Philosophy*, 29, 2, 1991, pp. 281-96.

20. "A Prussian Hume and a Scottish Kant", in: L. W. Beck, *Essays on Kant and Hume*. New Haven: Yale University Press 1989; repr. in: B. Logan (Ed.), *Immanuel Kant's Prolegomena in Focus*. London: Routledge 1996, pp. 139-55, esp. n. 39, pp. 154-5.

21. See also Greene, *op. cit.*, pp. 171-2.

Department of History and Philosophy of Science
University of Cambridge
Free School Lane
Cambridge CB2 3RH
U.K.
mfs10@cus.cam.ac.uk

MICHAEL FRIEDMAN

KANT, KUHN, AND THE RATIONALITY OF SCIENCE[*]

In the Introduction to the *Critique of Pure Reason* Kant formulates what he calls "the general problem of pure reason," namely, "How are synthetic a priori judgements possible?" Kant explains that this general problem involves two more specific questions about particular a priori sciences: "How is pure mathematics possible?" and "How is pure natural science possible?"– where the first concerns, above all, the possibility of Euclidean geometry, and the second concerns the possibility of fundamental laws of Newtonian mechanics such as conservation of mass, inertia, and the equality of action and reaction. In answering these questions Kant develops what he calls a "transcendental" philosophical theory of our human cognitive faculties – in terms of "forms of sensible intuition" and "pure concepts" or "categories" of rational thought. These cognitive structures are taken to describe a fixed and absolutely universal rationality – common to all human beings at all times and in all places – and thereby to explain the sense in which mathematical natural science (the mathematical physics of Newton) represents a model or exemplar of such rationality.[1]

In the current state of the sciences, however, we no longer believe that Kant's specific examples of synthetic a priori knowledge are even true, much less that they are a priori and necessarily true. For the Einsteinian revolution in physics has resulted in both an essentially non-Newtonian conception of space, time, and motion, in which the Newtonian laws of mechanics are no longer universally valid, and the application to nature of a non-Euclidean geometry of variable curvature, wherein bodies affected only by gravitation follow straightest possible paths or geodesics. And this has led to a situation, in turn, in which we are no longer convinced that there are any real examples of scientific a priori knowledge at all. If Euclidean geometry, at one time the very model of rational or a priori knowledge of nature, can be empirically revised, so the argument goes, then everything is in principle empirically revisable. Our reasons for adopting one or another system of geometry or mechanics (or, indeed, of mathematics more generally or of logic) are at bottom of the very same kind as the purely empirical considerations that support any other part of our total theory of nature. We are left with a strongly holistic form of empiricism or naturalism in which the very distinction between rational and empirical components of our total system of scientific knowledge must itself be given up.

This kind of strongly holistic picture of knowledge is most closely identified with the work of W.V. Quine. Our system of knowledge, in Quine's well-known figure, should be viewed as a vast web of interconnected beliefs on which ex-

M. Heidelberger and F. Stadler (eds.), History of Philosophy and Science, 25–41.
© 2002 *Kluwer Academic Publishers. Printed in the Netherlands.*

perience or sensory input impinges only along the periphery. When faced with a "recalcitrant experience" standing in conflict with our system of beliefs we then have a choice of where to make revisions. These can be made relatively close to the periphery of the system (in which case we make a change in a relatively low-level part of natural science), but they can also – when the conflict is particularly acute and persistent, for example – affect the most abstract and general parts of science, including even the truths of logic and mathematics, lying at the center of our system of beliefs. To be sure, such high-level beliefs at the center of our system are relatively entrenched, in that we are relatively reluctant to revise them or to give them up (as we once were in the case of Euclidean geometry, for example). Nevertheless, and this is the crucial point, absolutely none of our beliefs is forever "immune to revision" in light of experience:

The totality of our so-called knowledge or beliefs, from the most casual matters of geography and history to the profoundest laws of atomic physics or even of pure mathematics and logic, is a man-made fabric which impinges on experience only along the edges. Or, to change the figure, total science is like a field of force whose boundary conditions are experience. A conflict with experience at the periphery occasions readjustments in the interior of the field. ... But the total field is so underdetermined by its boundary conditions, experience, that there is much latitude of choice as to what statements to reëvaluate in the light of any single contrary experience. ...

If this view is right ... it becomes folly to seek a boundary between synthetic statements, which hold contingently on experience, and analytic statements, which hold come what may. Any statement can be held true come what may, if we make drastic enough adjustments elsewhere in the system. Even a statement very close to the periphery can be held true in the face of recalcitrant experience by pleading hallucination or by amending certain statements of the kind called logical laws. Conversely, by the same token, no statement is immune to revision. Revision even of the logical law of the excluded middle has been proposed as a means of simplifying quantum mechanics; and what difference is there in principle between such a shift and the shift whereby Kepler superseded Ptolemy, or Einstein Newton, or Darwin Aristotle?[2]

As the last sentence makes clear, examples of revolutionary transitions in our scientific knowledge, and, in particular, that of the Einsteinian revolution in geometry and mechanics, constitute a very important part of the motivations for this view.

Yet it is important to see that such a strongly anti-apriorist conception of scientific knowledge was by no means prevalent during the late nineteenth and early twentieth centuries – during the very period, that is, when the great revolutions in geometry and mechanics we now associate with the work of Einstein were actually taking place. If we begin with the key figures in the philosophy of non-Euclidean geometry, for example, whereas it is certainly true that Hermann von Helmholtz viewed the choice between Euclidean and non-Euclidean geometries as an empirical one, he also suggested that the more general structure of space common to both Euclidean and non-Euclidean systems (that of constant curvature or what Helmholtz called "free mobility") was a necessary presupposition of all spatial measurement and thus a "transcendental" form of our spatial intuition in the sense of Kant. And, partly on this basis, Henri Poincaré went even further. Although no particular geometry – neither Euclidean nor non-Euclidean – is an a priori condition of our spatial intuition, it does not follow that the choice between them, as Helmholtz thought, is an empirical one. For there remains an irreducible gulf between our crude and approximate sensory experience and our precise mathematical descriptions of nature. Establishing one or another system of geometry, Poincaré argued, therefore requires a free choice, a *convention* of our own – based, in the end, on the greater mathematical simplicity of the Euclidean system.[3]

Nor was such a strongly anti-apriorist conception of scientific knowledge adopted by the first scientific thinkers enthusiastically to embrace Einstein's new theory. These thinkers, the logical empiricists, of course rejected the synthetic a priori in Kant's original form. They rejected the idea of absolutely fixed and unrevisable a priori principles built, once and for all, into our fundamental cognitive capacities. In place of an holistic empiricism, however, they instead adopted a radically new conception of the a priori. Perhaps the clearest articulation of the logical empiricists's new view was provided by Hans Reichenbach in his first book, *The Theory of Relativity and A Priori Knowledge*, published in 1920.[4] Reichenbach distinguishes two meanings of the Kantian a priori: necessary and unrevisable, fixed for all time, on the one hand, and "constitutive of the concept of the object of [scientific] knowledge," on the other. Reichenbach argues, on this basis, that the great lesson of the theory of relativity is that the former meaning must be dropped while the latter must be retained. Relativity theory involves a priori constitutive principles as necessary presuppositions of its properly empirical claims, just as much as did Newtonian physics, but these principles have essentially changed in the transition from the latter theory to the former: whereas Euclidean geometry is indeed constitutively a priori in the context of Newtonian physics, for example, only *infinitesimally* Euclidean geometry is constitutively a priori in the context of general relativity. What we end up with, in this tradition, is thus a relativized and dynamical conception of a priori mathematical-physical principles, which change and develop along with the development of the mathematical and physical sciences themselves, but which nevertheless retain the characteristically Kantian constitutive function of making the empirical natural knowledge thereby structured and framed by such principles first possible.

Rudolf Carnap's philosophy of formal languages or linguistic frameworks, first developed in his *Logical Syntax of Language* in 1934, was the most mature expression of the logical empiricists's new view.[5] All standards of "correctness," "validity," and "truth," according to Carnap, are relative to the logical rules definitive of one or another formal language or linguistic framework. The rules of classical logic and mathematics, for example, are definitive of certain logical calculi or linguistic frameworks, while the rules of intuitionistic logic and mathematics (wherein the law of excluded middle is no longer universally valid) are definitive of others. Since standards of "validity" and "correctness" are thus relative to the choice of linguistic framework, it makes no sense to ask whether any such choice of framework is itself "valid" or "correct." For the logical rules relative to which alone these notions can be well-defined are not yet in place. Such rules are *constitutive* of the concepts of "validity" and "correctness" – relative to one or another choice of linguistic framework, of course – and are in this sense a priori rather than empirical.

This Carnapian philosophy of linguistic frameworks rests on two closely related distinctions. The first is the distinction between formal or *analytic* sentences of a given framework and empirical or *synthetic* sentences – or, as Carnap puts it in *Logical Syntax*, between *logical rules* ("L-rules") of a linguistic framework and *physical rules* ("P-rules"). The L-rules include laws of logic and mathematics (and may also, at least in spaces of constant curvature, include laws of physical geometry), whereas the P-rules include empirical laws standardly so-called such as Maxwell's equations of electromagnetism. In this way, Carnap's distinction between L-rules and P-rules closely parallels Reichenbach's distinction, developed in his 1920 book, between "axioms of coordination" (constitutive principles) and "axioms of connection" (properly empirical laws). Carnap's differentiation between logical and physical rules (analytic and synthetic sentences) then induces a second fundamental distinction between *internal* and *external* questions.[6] Internal questions are decided within an already adopted framework, in accordance with the logical rules of the framework in question. External questions, by contrast, concern precisely the question of which linguistic framework – and therefore which logical rules – to adopt in the first place. And since no logical rules are as yet in place, external questions, unlike internal questions, are not strictly speaking rationally decidable. Such questions can only be decided conventionally on the basis of broadly pragmatic considerations of convenience or suitability for one or another purpose. An overriding desire for security against the possibility of contradiction, for example, may prompt the choice of the weaker rules of intuitionistic logic and mathematics, whereas an interest in ease of physical application may prompt the choice of the stronger rules of classical logic and mathematics.

Now it was precisely this Carnapian philosophy of linguistic frameworks that formed the background and foil for Quine's articulation of a radically opposed form of epistemological holism according to which no fundamental distinction between a priori and a posteriori, logical and factual, analytic and synthetic can in fact be drawn. As we have seen, it was in Quine's 1951 paper, "Two Dogmas of Empiricism," where his challenge to the analytic/synthetic distinction was first made widely known, that the holistic figure of knowledge as a vast web of inter-connected beliefs also first appeared. But it is important to see here that it is Quine's attack on the analytic/synthetic distinction, and not simply the idea that no belief whatsoever is forever immune to revision, that is basic to Quine's new form of holism. For Carnap's philosophy of linguistic frameworks is wholly predicated on the idea that logical or analytic principles, just as much as empirical or synthetic principles, can be revised in the progress of empirical science.[7] Indeed, as we have seen, Reichenbach's initial formulation of this new view of constitutive a priori principles was developed precisely to accommodate the revolutionary changes in the geometrical and mechanical framework of physical theory wrought by Einstein's development of the theory of relativity. The difference between Quine and Carnap is rather that the latter persists in drawing a sharp distinction between changes of language or linguistic framework, in which constitutive principles definitive of the very notions of "validity" and "correctness" are revised, and changes in ordinary empirical statements formulated against the background of such an already-present constitutive framework. And this distinction, for Carnap, ultimately rests on the difference between analytic statements depending solely on the meanings of the relevant terms and synthetic statements expressing contentful assertions about the empirical world.

Quine's attack on the analytic/synthetic distinction – and thus on Carnap's particular version of the distinction between a priori and empirical principles – is now widely accepted, and I have no desire to defend Carnap's particular way of articulating this distinction here. I do want to question, however, whether Quinean epistemological holism is really our only option, and whether, in particular, it in fact represents our best way of coming to terms with the revolutionary changes in the historical development of the sciences that are now often taken to support it.

Quinean holism pictures our total system of science as a vast web or conjunction of beliefs which face the "tribunal of experience" as a corporate body. Quine grants that some beliefs, such as those of logic and arithmetic, are relatively central, whereas others, such as those of biology, say, are relatively peripheral. But this means only that the former beliefs are less likely to be revised in case of a "recalcitrant experience" at the periphery, whereas the latter are more likely to be revised. A reasonable scientific conservatism prefers to revise less central, less well-entrenched beliefs before it is forced to revise more central and better entrenched beliefs. Strictly speaking, however, empirical evidence – either for or against – spreads over *all* the elements of the vast conjunction that is our total system of science, wherein all elements whatsoever equally face the "tribunal of experience." And it is in this precise sense, for Quine, that all beliefs whatsoever, including those of logic and mathematics, are equally empirical.

But can this beguiling form of epistemological holism really do justice to the revolutionary developments within both mathematics and natural science that have led up to it? Let us first consider the Newtonian revolution that produced the beginnings of mathematical physics as we know it – the very revolution, as we have seen, that Kant's conception of synthetic a priori knowledge was originally intended to address. In constructing his mathematical physics Newton created, virtually simultaneously, three revolutionary advances: a new form of mathematics, the calculus, for dealing with infinite limiting processes and instantaneous rates of change; new conceptions of force and quantity of matter encapsulated in his three laws of motion; and a new universal law of nature, the law of universal gravitation. Each of these three advances was revolutionary in itself, and all were introduced by Newton in the context of the same scientific problem: that of developing a single mathematical theory of motion capable of giving a unified account of both terrestrial and celestial phenomena. Since all three advances were thus inspired, in the end, by the same empirical problem, and since they together amounted to the first known solution to this problem, Quine's holistic picture appears so far correct. All elements in this particular system of scientific knowledge – mathematics, mechanics, gravitational physics – appear equally to face the "tribunal of experience" together.

Nevertheless, there are fundamental asymmetries in the way in which the different elements of this Newtonian synthesis actually function. To begin with the relationship between mathematics and mechanics, Newton's second law of motion says that force equals mass times acceleration, where acceleration is the instantaneous rate of change of velocity (itself the instantaneous rate of change of position). So without the mathematics of the calculus this second law of motion could not even be formulated or written down, let alone function to describe empirical phenomena. The combination of calculus plus the laws of motion is not happily viewed, therefore, as a conjunction of propositions symmetrically contributing to a single total result: the mathematical part of Newton's theory rather supplies elements of the language or conceptual framework, we might say, within which the rest of the theory is then formulated. And an analogous (if also more subtle) point holds with respect to the relationship between Newton's mechanics and gravitational physics. The law of universal gravitation says that there is a force of attraction, directly proportional to the product of the two masses and inversely proportional to the square of the distance between them, between any two pieces of matter in the universe – which therefore experience accelerations towards one another in accordance with this same law. But relative to what frame of reference are the accelerations in question defined? Since these accelerations are, by hypothesis, universal, no particular material body can be taken as actually at rest in this frame, and thus the motions in question are not motions relative to any particular material body. Newton himself understood these motions as defined relative to absolute space, but we now understand them as defined relative to an arbitrary *inertial frame* – where an inertial frame of reference is simply one in which the Newtonian laws of motion actually hold (the center of mass frame of the solar system, for example, is a very close approximation to such a frame). It follows that without the Newtonian laws of motion Newton's theory of gravitation would not even make empirical sense, let alone give a correct account of the empirical phenomena: in the absence of these laws we would simply have no idea what the relevant frame of reference might be in relation to which the universal accelerations due to gravity are defined. Once again, Newton's mechanics and gravitational physics are not happily viewed as symmetrically functioning elements of a larger conjunction: the former is rather a necessary part of the language or conceptual framework within which alone the latter makes empirical sense.

Now the Newtonian theory of gravitation has of course been superseded by Einstein's general theory of relativity, and one might naturally expect Quine's holistic picture of knowledge to describe this latter theory much more accurately. General relativity, like Newtonian theory, can be seen as the outcome of three revolutionary advances: the development of a new field of mathematics, tensor calculus or the general theory of manifolds, by Bernhard Riemann in the late nineteenth century; Einstein's principle of equivalence, which identifies gravitational effects with the inertial effects formerly associated with Newton's laws of motion; and Einstein's equations for the gravitational field, which describe how the curvature of space-time is modified by the presence of matter and energy so as to direct gravitationally affected bodies along straightest possible paths or geodesics. Once again, each of these three advances was revolutionary in itself, and all three were marshalled together by Einstein to solve a single empirical problem: that of developing a new description of gravitation consistent with the special theory of relativity (which is itself incompatible with the instantaneous action at a distance characteristic of Newtonian theory) and also capable, it was hoped, of solving well-known anomalies in Newtonian theory such as that involving the perihelion of Mercury. And the three advances together, as marshalled and synthesized by Einstein, in fact succeeded in solving this empirical problem for the first time.

It does not follow, however, that the combination of mathematical theory of manifolds, geodesic law of motion, and field equations of gravitation can be happily viewed as a symmetrically functioning conjunction, such that each element then equally faces the "tribunal of experience" when confronted with the anomaly in the perihelion of Mercury, for example. To begin again with the relationship between mathematics and mechanics, the principle of equivalence depicts the space-time trajectories of bodies affected only by gravitation as geodesics in a variably curved space-time geometry, just as the Newtonian laws of motion, when viewed from this same space-time perspective, depict the trajectories of bodies affected by no forces at all as geodesics in a flat space-time geometry. But the whole notion of a variably curved geometry itself only makes sense in the context of the revolutionary new theory of manifolds recently created by Riemann. In the context of the mathematics available in the seventeenth and eighteenth centuries, by contrast, the idea of a variably curved space-time geometry could not even be formulated or written down, let alone function to describe empirical phenomena. And, once again, a closely analogous (but also more subtle) point holds for the relationship between mechanics and gravitational physics. Einstein's field equations describe the variation in curvature of space-time geometry as a function of the distribution of matter and energy. Such a variably curved space-time structure would have no empirical meaning or application, however, if we had not first singled out some empirical phenomena as counterparts of its fundamental geometrical notions – here the notion of geodesic or straightest possible path. The principle of equivalence does precisely this, however, and without this principle the intricate space-time geometry described by Einstein's field equations would not even be empirically false, but rather an empty mathematical formalism with no empirical application at all.[8] Just as in the case of Newtonian gravitation theory, therefore, the three advances together comprising Einstein's revolutionary theory should not be viewed as symmetrically functioning elements of a larger conjunction: the first two function rather as necessary parts of the language or conceptual framework within which alone the third makes both mathematical and empirical sense.

It will not do, in either of our two examples, to view what I am calling the constitutively a priori parts of our scientific theories as simply relatively fixed or entrenched elements of science in the sense of Quine, as particularly well-established beliefs which a reasonable scientific conservatism takes to be relatively difficult to revise. When Newton formulated his theory of gravitation, for example, the mathematics of the calculus was still quite controversial – to such an extent, in fact, that Newton disguised his use of it in the *Principia* in favor of traditional synthetic geometry. Nor were Newton's three laws of motion any better entrenched, at the time, than the law of universal gravitation. Similarly, in the case of Einstein's general theory of relativity, neither the mathematical theory of manifolds nor the principle of equivalence was a well-entrenched part of main-stream mathematics or mathematical physics; and this is one of the central reasons, in fact, that Einstein's theory is so profoundly revolutionary. More generally, then, since we are dealing with deep conceptual revolutions in both mathematics and mathematical physics in both cases, entrenchment and relative resistance to revision are not appropriate distinguishing features at all. What characterizes the distinguished elements of our theories is rather their special *constitutive function*: the function of making the precise mathematical formulation and empirical application of the theories in question first possible. In this sense, the relativized and dynamical conception of the a priori developed by the logical empiricists appears to describe these conceptual revolutions far better than does Quinean holism. This is not at all surprising, in the end, for this new conception of the constitutive a priori was inspired, above all, by just these conceptual revolutions.

It is no wonder, then, that in Thomas Kuhn's theory of the nature and character of scientific revolutions we find an informal counterpart of the relativized conception of constitutive a priori principles first developed by the logical empiricists. Indeed, one of Kuhn's central examples of revolutionary scientific change, just as it was for the logical empiricists, is precisely Einstein's theory of relativity.[9] Thus Kuhn's central distinction between change of paradigm or revolutionary science, on the one side, and normal science, on the other, closely parallels the Carnapian distinction between change of language or linguistic framework and rule-governed operations carried out within such a framework. Just as, for Carnap, the logical rules of a linguistic framework are constitutive of the notion of "correctness" or "validity" relative to this framework, so a particular paradigm governing a given episode of normal science, for Kuhn, yields generally-agreed-upon (although perhaps only tacit) rules constitutive of what counts as a "valid" or "correct" solution to a problem within this episode of normal science. Just as, for Carnap, external questions concerning which linguistic framework to adopt are not similarly governed by logical rules, but rather require a much less definite appeal to conventional and/or pragmatic considerations, so changes of paradigm in revolutionary science, for Kuhn, do not proceed in accordance with generally-agreed-upon rules as in normal science, but rather require something more akin to a conversion experience.

Indeed, towards the end of his career, Kuhn himself drew this parallel between his theory of scientific revolutions and the relativized conception of a priori constitutive principles explicitly:

Though it is a more articulated source of constitutive categories, my structured lexicon [= Kuhn's late version of "paradigm"] resembles Kant's a priori when the latter is taken in its second, relativized sense. Both are constitutive of *possible experience* of the world, but neither dictates what that experience must be. Rather, they are constitutive of the infinite range of possible experiences that might conceivably occur in the actual world to which they give access. Which of these conceivable experiences occurs in that actual world is something that must be learned, both from everyday experience and from the more systematic and refined experience that characterizes scientific practice. They are both stern teachers, firmly resisting the promulgation of beliefs unsuited to the form of life the lexicon permits. What results from respectful attention to them is knowledge of nature, and the criteria that serve to evaluate contributions to that knowledge are, correspondingly, epistemic. The fact that experience within another form of life – another time, place, or culture – might have constituted knowledge differently is irrelevant to its status as knowledge.[10]

Thus, although Quine may very well be right that Carnap has failed to give a precise logical characterization of what I am here calling constitutive principles, there is also nonetheless no doubt, I suggest, that careful attention to the actual historical development of science, and, more specifically, to the very conceptual revolutions that have in fact led to our current philosophical predicament, shows that relativized a priori principles of just the kind Carnap was aiming at are central to our scientific theories.

But this close parallel between the relativized yet still constitutive a priori and Kuhn's theory of scientific revolutions implies (as the last sentence of our passage from Kuhn suggests) that the former gives rise to the same problems and questions concerning the ultimate rationality of the scientific enterprise that are all too familiar in the post-Kuhnian literature in history, sociology, and philosophy of science. In particular, since there appear to be no generally-agreed-upon constitutive principles governing the transition to a revolutionary new scientific paradigm or conceptual framework, there would seem to be no sense left in which such a transition can still be viewed as rational, as based on good reasons. And it is for precisely this reason, of course, that Carnap views what he calls external questions as conventional as opposed to rational, and Kuhn likens paradigm shifts rather to conversion experiences. It appears, then, that all we have accomplished by defending the relativized yet still constitutive a priori against

Quinean holism is to land ourselves squarely in the contemporary "relativistic" predicament, wherein the overarching rationality of the scientific enterprise has now been strongly called into in question.

The underlying source of this post-Kuhnian predicament, as we have seen, is the breakdown of the original Kantian conception of the a priori. Kant takes the fundamental constitutive principles framing Newtonian mathematical science as expressing timelessly fixed categories and forms of the human mind. Such categories and forms, for Kant, are definitive of human rationality as such, and thus of an absolutely *universal* rationality governing all human knowledge at all times and in all places. This conception of an absolutely universal human rationality realized in the fundamental constitutive principles of Newtonian science made perfectly good sense in Kant's own time, when the Newtonian conceptual framework was the only paradigm for what we now call mathematical physics the world had yet seen. Now that we have irretrievably lost this position of innocence, however, it would appear that the very notion of a truly universal human rationality must also be given up. It would appear that there is now no escape from the currently fashionable slogan "all knowledge is local."

Yet Kuhn himself rejected such relativistic implications of his views. He continued to hold, in a self-consciously traditional vein, that the evolution of science is a rational and progressive process despite the revolutionary transitions between scientific paradigms which are, as he also claims, absolutely necessary to this process. The scientific enterprise, Kuhn suggests, is essentially an instrument for solving a particular sort of problem or "puzzle" – for maximizing the quantitative match between theoretical predictions and phenomenological results of measurement. Given this, however, there are obvious criteria or "values" – such as accuracy, precision, scope, simplicity, and so on – that are definitive of the scientific enterprise as such. Such values are constant or permanent across scientific revolutions or paradigm-shifts, and this is all we need to secure the (non-paradigm-relative) rationality of scientific progress:

[W]hether or not individual practitioners are aware of it, they are trained to and rewarded for solving intricate puzzles – be they instrumental, theoretical, logical, or mathematical – at the interface between their phenomenal world and their community's beliefs about it. ... If that is the case, however, the rationality of the standard list of criteria for evaluating scientific belief is obvious. Accuracy, precision, scope, simplicity, fruitfulness, consistency, and so on, simply *are* the criteria which puzzle solvers must weigh in deciding whether or not a given puzzle about the match between phenomena and belief has been solved. ... To select a law or theory which exemplified them less fully than an existing competitor would be self-defeating, and self-defeating action is the surest index of irrationality. ... As the developmental process continues, the examples from which practitioners learn to recognize accuracy, scope, simplicity, and so on, change both within and between fields. But the criteria that these examples illustrate are themselves necessarily permanent, for abandoning them would be abandoning science together with the knowledge which scientific development brings. ... Puzzle-solving is one of the families of practices that has arisen during that evolution [of human practices], and what it produces is knowledge of nature. Those who proclaim that no interest-driven practice can properly be identified with the rational pursuit of knowledge make a profound and consequential mistake.[11]

Thus, although the process of scientific development is governed by no single conceptual framework fixed once and for all, science, at *every* stage, still aims at a uniform type of puzzle-solving success, Kuhn suggests, relative to which *all* stages in this process (including transitions between conceptual frameworks) may be judged. And there is then no doubt at all, Kuhn further suggests, that science, throughout its development, has become an increasingly efficient instrument for achieving this end. In this sense, therefore, there is also no doubt at all that science as a whole is a rational enterprise.

This Kuhnian defense of the rationality of scientific knowledge from the threat of conceptual relativism misses the point, I believe, of the real challenge to such rationality arising from Kuhn's own historiographical work. For it is surely uncontroversial that the scientific enterprise as a whole has in fact become an ever more efficient instrument for puzzle-solving in Kuhn's sense – for maximizing quantitative accuracy, precision, simplicity, and so on in adjusting theoretical predictions to phenomenological results of measurement. What is controversial, rather, is the further idea that the scientific enterprise thereby counts as a privileged model or exemplar of rational knowledge of – rational inquiry into – nature. And the reasons for this have nothing to do with doubts about the incontrovertible predictive success of the scientific enterprise – they do not call into question, that is, the *instrumental* rationality of this enterprise. What has been called into question, rather, is what Jürgen Habermas calls *communicative* rationality.[12] Communicative rationality, unlike instrumental rationality, is concerned not so much with choosing efficient means to a given end, but rather with securing mutually agreed upon principles of reasoning whereby a given community of speakers can rationally adjudicate their differences of opinion. It is precisely this kind of rationality that is secured by a shared paradigm or conceptual framework; and it is precisely this kind of rationality that is then profoundly challenged by the Kuhnian theory of scientific revolutions – where it appears that succeeding paradigms, in a scientific revolution, are fundamentally non-intertranslatable and thus share no basis whatsoever for rational mutual communication. Pointing to the obvious fact that science has nonetheless continued to increase its quantitative accuracy, precision, and so on is thus a quite inadequate response to the full force of the post-Kuhnian relativistic challenge to scientific rationality.

Kuhn's notion of normal science, as we have just seen, is itself based on an *intra*-framework notion of communicative rationality – on shared rules of the game, as it were, common to all practitioners of a single given paradigm. What we now need to investigate, then, are the prospects for a comparable notion of *inter*-framework communicative rationality, capable of providing similarly shared principles of reasoning functioning *accross* revolutionary paradigm-shifts.

Let us first remind ourselves that, despite the fact that we radically change our constitutive principles in the revolutionary transition from one conceptual framework to another, there is still an important element of *convergence* in the very same revolutionary process of conceptual change. Special relativistic mechanics approaches classical mechanics in the limit as the velocity of light goes to infinity; variably curved Riemannian geometry approaches flat Euclidean geometry as the regions under consideration become infinitely small; Einstein's general relativistic field equations of gravitation approach the Newtonian equations for gravitation as, once again, the velocity of light goes to infinity.[13] Indeed, even in the transition from Aristotelian terrestrial and celestial mechanics to classical terrestrial and celestial mechanics we find a similar relationship. From an observer fixed on the surface of the earth we can construct a system of lines of sight directed towards the heavenly bodies; this system is spherical, isomorphic to the celestial sphere of ancient astronomy, and the motions of the heavenly bodies therein are indeed described, to a very good approximation, by the geocentric system favored by Aristotle. Moreover, in the sublunary region close to the surface of the earth, where the earth is by far the principal gravitating body, heavy bodies do follow straight paths directed towards the center of the earth, again to an extremely good approximation. In all three revolutionary transitions, therefore, key elements of the preceding paradigm are preserved as approximate special cases in the succeeding paradigm.

This type of convergence between successive paradigms allows us to define a *retrospective* notion of inter-framework rationality based on the constitutive principles of the later conceptual framework: since the constitutive principles of the earlier framework are contained in those of the later as an approximate special case, the constitutive principles of the later framework thus define a common rational basis for mutual communication from the point of view of this latter framework. But this does not yet give us a *prospective* notion of inter-framework rationality accessible from the point of view of the earlier framework, of course, and so it does not yet provide a basis for mutual communication that is truly available to both frameworks.[14] Nevertheless, such a prospective notion of inter-framework communicative rationality also begins to emerge when we observe that, in addition to containing the constitutive principles of the older framework as an approximate special case, the concepts and principles of the revolutionary new constitutive framework evolve continuously, as it were, by a series of natural transformations of the old concepts and principles.

The Aristotelian constitutive framework, for example, is based on Euclidean geometry, a background conception of a hierarchically and teleologically organized universe, and conceptions of natural place and natural motion appropriate to this universe. Thus, in the terrestrial realm heavy bodies naturally move in straight lines towards their natural place at the center of the universe, and in the celestial realm the heavenly bodies naturally move uniformly in circles around this center. The conceptual framework of classical physics then retains Euclidean geometry, but eliminates the hierarchically and teleologically organized universe together with the accompanying conceptions of natural place. We thereby obtain an infinite, homogeneous and isotropic universe in which all bodies naturally move uniformly along straight lines to infinity. But how did we arrive at this conception? An essential intermediate stage was Galileo's celebrated treatment of free fall and projectile motion. For, although Galileo indeed discards the hierarchically and teleologically organized Aristotelian universe, he retains – or better, transforms – key elements of the Aristotelian conception of natural motion. Galileo's analysis is based on a combination of what he calls naturally accelerated motion directed towards the center of the earth and uniform or equable motion directed horizontally. Unlike our modern conception of rectilinear inertial motion, however, this Galilean counterpart is uniformly *circular* – traversing points equidistant from the center at constant speed. But, in relatively small regions near the earth's surface, this circular motion is quite indistinguishable from rectilinear motion, and Galileo can thus treat it as rectilinear to an extremely good approximation. And it is in precisely this way, therefore, that the modern conception of natural (inertial) motion is actually continuous with the preceding Aristotelian conception of natural motion.

An analogous (if also more complex) point can be made concerning the transition from Newtonian mechanics and gravitation theory, through special relativity, to general relativity. The key move in general relativity, as we have seen, is to replace the law of inertia – which, from the space-time perspective inaugurated by special relativity, depicts the trajectories of force-free bodies as geodesics in a flat space-time geometry – with the principle of equivalence, according to which bodies affected only by gravitation follow geodesics in a variably curved space-time geometry. How did Einstein actually make this revolutionary move, which represents the first actual application of a non-Euclidean geometry to nature? Einstein's innovation grows naturally out of the nineteenth century tradition in the foundations of geometry, as Einstein interprets this tradition in the context of the new non-Newtonian mechanics of special relativity. The key transition to a non-Euclidean geometry of variable curvature in fact results from applying the Lorentz contraction arising in special relativity to the geometry of a rotating disk, as Einstein simultaneously delicately positions himself within the debate on the foundations of geometry between Helmholtz and Poincaré. In particular, whereas Einstein had earlier made crucial use of Poincaré's idea of convention in motivating the transition, on the basis of mathematical simplicity, from Newtonian space-time to what we currently call Minkowski space-time, now, in the case of the rotating disk, Einstein rather follows Helmholtz in taking the behavior of rigid measuring rods to furnish us with an empirical determination of the underlying geometry – in this case, a non-Euclidean geometry.[15]

In each of our revolutionary transitions fundamentally philosophical ideas, belonging to what we might call epistemological meta-paradigms or meta-frameworks, play a crucial role in motivating and sustaining the transition to a new first-level or scientific paradigm. Such epistemological meta-frameworks guide the all-important process of conceptual transformation and help us, in particular, to articulate what we now mean, during a given revolutionary transition, by a natural, reasonable, or responsible conceptual transformation. By interacting productively with both older philosophical meta-frameworks and new developments taking place within the sciences themselves, a new epistemological meta-framework thereby makes available a prospective notion (accessible even in the pre-revolutionary conceptual situation) of inter-framework or inter-paradigm rationality.

In the transition from Aristotelian-Scholastic natural philosophy to classical mathematical physics, for example, at the same time that Galileo was subjecting the Aristotelian conception of natural motion to a deep (yet continuous) conceptual transformation, it was necessary to eliminate the hierarchical and teleological elements of the Aristotelian conceptual framework in favor of an exclusively mathematical and geometrical point of view – which was encapsulated, for the mechanical natural philosophy of the time, in the distinction between primary and secondary qualities. Euclidean geometry, as an exemplar of rational inquiry, was of course already a part of the Aristotelian framework, and the problem then was, accordingly, to emphasize this part at the expense of the hylomorphic and teleological conceptual scheme characteristic of Aristotelian metaphysics. This task, however, required a parallel reorganization of the wider concepts of Aristotelian metaphysics (concepts of substance, force, space, time, matter, mind, creation, divinity), and it fell to the philosophy of Descartes to undertake such a reorganization – a philosophy which in turn interacted productively with recent scientific advances such as Copernican astronomy, new results in geometrical optics, and Descartes's own initial formulation of the law of rectilinear inertia. Similarly, in the transition from classical mechanics to relativity theory, at the same time that Einstein was subjecting the classical conceptions of space, time, and motion to a deep (yet continuous) conceptual transformation, philosophical debate on the foundations of geometry between Helmholtz and Poincaré, in which empiricist and conventionalist interpretations of that science opposed one another against the ever-present backdrop of the Kantian philosophy, played a central role – and, in turn, was itself carried out in response to mathematical advances in the foundations of geometry made throughout the nineteenth century.[16]

So what we see here, I finally want to suggest, is that a reconceived version of Kant's original philosophical project – the project of investigating and philosophically contextualizing the most basic constitutive principles defining the fundamental spatio-temporal framework of empirical natural science – plays an indispensable orienting role with respect to conceptual revolutions within the sciences precisely by generating new epistemological meta-frameworks capable of bridging, and thus guiding, the revolutionary transitions to a new scientific framework. This peculiarly philosophical type of investigation thereby makes available prospective notions of inter-framework rationality in the light of which radically new constitutive principles can then appear as rational – as Descartes's appropriation and transformation of the concepts of Aristotelian-Scholastic meta-physics made the new mechanical natural philosophy a reasonable option, for example, or Einstein's appropriation and transformation of the earlier epistemo-logical reflections of Poincaré and Helmholtz did the same for relativity theory.

In place of the Quinean figure of an holistically conceived web of belief, wherein both knowledge traditionally understood as a priori and philosophy as a discipline are supposed to be wholly absorbed into empirical natural science, I am therefore proposing an alternative picture of a thoroughly dynamical yet nonetheless differentiated system of knowledge that can be analyzed, for present purposes, into three main components or levels. At the base level, as it were, are the concepts and principles of empirical natural science properly so-called: empirical laws of nature, such as the Newtonian law of gravitation or Einstein's equations for the gravitational field, which squarely and precisely face the "tribunal of experience" via a rigorous process of empirical testing. At the next or second level are the constitutively a priori principles that define the funda-mental spatio-temporal framework within which alone the rigorous formulation and empirical testing of first or base level principles is then possible. These rela-tivized a priori principles constitute what Kuhn calls paradigms: relatively stable sets of rules of the game, as it were, that make possible the problem solving activities of normal science – including, in particular, the rigorous formulation and testing of properly empirical laws. In periods of deep conceptual revolution it is precisely these constitutively a priori principles which are then subject to change – under intense pressure, no doubt, from new empirical findings and especially anomalies. It does not follow, however, that such second-level constitutive principles are empirical in the same sense as are the first-level principles. On the contrary, since here, by hypothesis, a generally-agreed-upon background framework is necessarily missing, no straightforward process of empirical testing, in periods of deep conceptual revolution, is then possible. And here our third level, that of philosophical meta-paradigms or meta-frameworks, plays an indispensable role, by serving as a source of guidance or orientation in motivating and sustaining the transition from one paradigm or conceptual frame-work to another. Such philosophical meta-frameworks contribute to the rationali-ty of revolutionary scientific change, more specifically, by providing a basis for mutual communication (and thus for communicative rationality in Habermas's sense) between otherwise incommensurable (and therefore non-intertranslatable) scientific paradigms.

None of these three levels are fixed and unrevisable, and the distinctions I am drawing have nothing to do, in particular, with differing degrees of certainty or epistemic security. Indeed, the whole point of the present conception of relativ-ized and dynamical a priori principles is to accommodate the profound con-ceptual revolutions that have repeatedly shaken our knowledge of nature to its very foundations. It is precisely this revolutionary experience, in fact, that has revealed that our knowledge *has* foundations in the present sense: subject-defin-ing or constitutive paradigms whose revision entails a genuine expansion of our space of intellectual possibilities, to such an extent, in periods of radical conceptual revolution, that a straightforward appeal to empirical evidence is then no longer directly relevant. And it is at this point, moreover, that philosophy plays its own distinctive role, not so much in justifying or securing a new paradigm where empirical evidence cannot yet do so, but rather in guiding the articulation of the new space of possibilities and making the serious consideration of the new paradigm a rational and responsible option. The various levels in our total evolving and interacting system of beliefs are thus not distinguished by differing degrees of epistemic security at all (neither by differing degrees of centrality and entrenchment in the sense of Quine nor by differing degrees of certainty in the more traditional sense), but rather by their radically different yet mutually complementary contributions to the total ongoing dialectic of human knowledge – a dialectical process in which mathematical scientific knowledge continues to provide us with the best exemplar we have of human rationality (that is, our very best example of *communicative* rationality) in spite of (and even because of) its profoundly revolutionary character.

NOTES

*   I originally presented, as the Vienna Circle Lecture at HOPOS 2000, a different paper, entitled "The Idea of a Scientific Philosophy." This paper will appear as the First Lecture in my *Dynamics of Reason: The 1999 Kant Lectures at Stanford University* (Stanford: CSLI Press, 2001). The present paper describes some of the main ideas of these Lectures as a whole. I am grateful to Friedrich Stadler and Michael Heidelberger for allowing me to publish the present paper here. I am also indebted to comments from Heidelberger that prompted notes 13 and 14 below.

1.   The "general problem of pure reason," along with its two more specific sub-problems, is formulated in §VI of the Introduction to the *Critique of Pure Reason* at B19-24. Sections V and VI, which culminate in the three questions "How is pure mathematics possible?", "How is pure natural science possible?", and "How is metaphysics as a science possible?", are added to the second (1787) edition of the *Critique* and clearly follow the structure of the 1783 *Prolegomena to Any Future Metaphysics*, which was intended to clarify the first (1781) edition. This way of framing the general problem of pure reason also clearly reflects the increasing emphasis on the question of pure natural science found in the *Metaphysical Foundations of Natural Science* (1786). For an extended discussion of Kant's theory of pure natural science and its relation to Newtonian physics see Friedman, *Kant and the Exact Science* (Cambridge, Mass.: Harvard University Press, 1992), especially chapters 3 and 4.

2.   From the first two paragraphs of §6, entitled "Empiricism without the dogmas," of "Two Dogmas of Empiricism," *Philosophical Review* 60 (1951): 20-43; reprinted in *From a Logical Point of View* (New York: Harper, 1953), pp. 42-3.

3.   For extended discussion of Helmholtz and Poincaré see my "Helmholtz's *Zeichentheorie* and Schlick's *Allgemeine Erkenntnislehre*," *Philosophical Topics* 25 (1997): 19-50; "Geometry, Construction, and Intuition in Kant and His Successors," in G. Scher and R. Tieszen, eds., *Between Logic and Intuition* (Cambridge: Cambridge University Press, 2000); and *Reconsidering Logical Positivism* (Cambridge: Cambridge University Press, 1999), chapter 4.

4.   Reichenbach, *Relativitätstheorie und Erkenntnis Apriori* (Berlin: Springer, 1920); translated as *The Theory of Relativity and A Priori Knowledge* (Los Angeles: University of California Press,

1965). The distinction between the two meanings of the Kantian a priori described in the next sentence occurs in chapter 5.

5. Carnap, *Logische Syntax der Sprache* (Wien: Springer, 1934); translated as *The Logical Syntax of Language* (London: Kegan Paul, 1937).

6. This distinction is first made explicitly in Carnap, "Empiricism, Semantics, and Ontology," *Revue Internationale de Philosophie* 11 (1950): 20-40; reprinted in *Meaning and Necessity*. 2nd ed. (Chicago: University of Chicago Press, 1956).

7. Carnap explicitly embraces this much of epistemological holism (based on the ideas of Poincaré and Pierre Duhem) in §82 of *Logical Syntax*. Quine is therefore extremely misleading when he (in the above-cited passage from §6 of "Two Dogmas") simply equates analyticity with unrevisability. He is similarly misleading in §5 (p. 41) when he asserts that the "dogma of reductionism" (i.e., the denial of Duhemian holism) is "at root identical" with the dogma of analyticity.

8. For an analysis of the principle of equivalence along these lines, including illuminating comparisons with Reichenbach's conception of the need for "coordinating definitions" in physical geometry, see R. DiSalle, "Spacetime Theory as Physical Geometry," *Erkenntnis* 42 (1995): 317-37.

9. Kuhn develops this example in *The Structure of Scientific Revolutions*. 2nd ed. (Chicago: University of Chicago Press, 1970), chapter 9. There is some irony in the circumstance that Kuhn introduces this example as part of a criticism of what he calls "early logical positivism" (p. 98).

10. Kuhn, "Afterwords," in P. Horwich, ed., *World Changes* (Cambridge, Mass.: MIT Press, 1993), pp. 331-2.

11. Kuhn, "Afterwords" (note 10 above), pp. 338-9.

12. See Habermas, *Theorie des Kommunikativen Handelns* (Frankfurt: Suhrkamp, 1981), vol. 1, chapter 1; translated as *The Theory of Communicative Action* (Boston: Beacon, 1984).

13. In Kuhn's own discussion of the theory of relativity (see note 9), he explicitly denies that classical mechanics can be logically derived from relativistic mechanics in the limit of small velocities. His main ground for this denial is that "the physical referents" of the terms of the two theories are different (*op. cit.*, pp. 101-2). Here, however, I am merely pointing to a purely mathematical fact about the corresponding mathematical structures

14. That the convergence in question yields only a purely retrospective *reinterpretation* of the original theory is a second (and related) point Kuhn makes in the discussion cited in note 13 above, where he points out (p. 101) that the laws derived as special cases in the limit within relativity theory "are not [Newton's] unless those laws are reinterpreted in a way that would have been impossible until after Einstein's work." I believe that Kuhn is correct in this and, in fact, that it captures a centrally important aspect of what he has called the non-intertranslatability or "incommensurability" of pre-revolutionary and post-revolutionary theories.

15. For a detailed discussion of this case see my "Geometry as a Branch of Physics," in D. Malament, ed., *Reading Natural Philosophy* (Chicago: Open Court, 2001).

16. See again the reference cited in note 15 above.

Dept. of History and Philosophy of Science
Indiana University
Goodbody Hall
Bloomington, IN 47405
U.S.A.
mlfriedm@indiana.edu

Lothar Schäfer

Neo-Kantian Origins of Modern Empiricism: On the Relation Between Popper and the Vienna Circle

Modern empiricism is usually thought to have emerged in opposition to the then dominant school of neo-Kantianism. True as this may be, it has blinded us to the fact that Kantian and more surprisingly even neo-Kantian elements of philosophy have also had a positive influence upon the development of the new empiricism. One episode in which this influence proves itself in fact dominant and which I will present in the following concerns the philosophical position which Popper adopted vis-à-vis logical empiricism – as advocated by Wittgenstein and Schlick – in Vienna in the late twenties and early thirties of the last century.

From the very beginning, Popper's thought is characterized by a puzzling, if not contradictory, attitude towards the philosophy of the Vienna Circle. First there is a manifest congruence in fundamental tenets of philosophical research: both see their work in the tradition of the philosophy of enlightenment; both promote the transformation of traditional epistemology into philosophy of science; both try to answer to the new situation in physics after the breakdown of Newtonian mechanics; both aim at a logical analysis of the rationality of science, which they see as restricted to the context of justification – yet, at the same time, Popper is highly critical of basic assumptions of logical empiricism which he and others label as "positivistic".

Popper feels so strongly about his opposition to positivism and about the success of his criticism that in his autobiographical retrospective [1] he raised the rhetorical question "Who killed logical positivism?" (UQ 87), only to answer it by confessing, that he himself had to admit responsibility for having killed positivism with his own hands, as it were.

Popper is quite outspoken about the motives and ideas that brought him into opposition to the basic tenets of logical empiricism: it was his own alignment with Kant's epistemological programme which provided arguments for him against the foundations of positivism. In fact, Popper sees his opposition to the Vienna Circle as a repetition of the critique Kant had directed against Hume.

The Vienna Circle and Popper's relation to it appear – in Popper's perspective – as the stage set for the remake of a prominent philosophical story. In much the same way in which Kant's critical philosophy superseded the sceptical empiricism of Hume, Popper sees himself as triumphant over the programme of logical empiricism of the Vienna Circle.

M. Heidelberger and F. Stadler (eds.), History of Philosophy and Science, 43–55.

What I would like to show is that Popper's understanding of Kant's theoretical philosophy is inspired by neo-Kantian revisions and that the development of his own methodology is in many facets indebted to the neo-Kantianism of the Marburg School.

It may look rather odd and implausible to bring Popper's falsificationist methodology in a positive connection with neo-Kantianism. For, although Popper's endorsement of Kant as an enlightenment philosopher and as a representative of critical philosophy is very plausible, the neo-Kantian version of the foundational programme must appear as the very opposite of a Popperian approach to epistemology, which is strictly anti-foundationalist in character.

One of the fundamental ideas of the Marburg School was the interpretation of Kant as a Newtonian. In other words, they saw quite clearly the intimate connection between the *Critique of Pure Reason* and the *Metaphysical Foundations of Natural Science:* the foundational transitions of Kant leading, by means of deduction and construction, from the level of pure transcendental reasoning to the metaphysical principles of natural science and then to empirical science proper provided an account a priori for the rational and mathematical structure of physical theory[2]. The members of the Marburg School realized that Kant's theory of experience in general amounted to a transcendental justification of Newtonian physics, a theory which had lost its fundamental status in physics through the new theories of Einstein and Bohr. Interpreting Kant as a Newtonian meant for the neo-Kantians to emphasise the need for revisions in the conception of transcendental philosophy as presented in Kant's original writings, revisions that were considered necessary in order to accommodate the recent developments in logic, mathematics and especially in physical theory. Paul Natorp's book *Die logischen Grundlagen der exakten Wissenschaften* (1910) was a large-scale attempt to produce a transcendental foundation of the new state of knowledge in mathematics and physics, in the course of which he had taken into account in particular the new logicist conception of mathematics that had been developed by Russell and Frege, on the one hand, and Einstein's special theory of relativity on the other. Natorp's book was highly regarded – even by philosophers outside the neo-Kantian camp – as it dealt primarily with publications of scientists; the discussions of philosophical issues were not confined to the boundaries set by school traditions. Nevertheless, despite its open attitude towards ever more different movements and positions in philosophy of science, and despite its emphasis on the open character of the development of science as such, (Kant's starting point of the 'fact of science' was transformed into the permanent *'fieri* of science') the neo-Kantian philosophy of science remained committed to a strict foundational programme of justification based on "transcendental deductions".[3]

Therefore, *prima facie* it is implausible to see Popper's position in alliance with transcendental foundationalism, be it in Kant's original orientation or in the most advanced version of neo-Kantianism. However, in what follows, I shall present evidence showing that Popper's neo-Kantian adoption of Kant's theoretical philosophy gave shape to his own position and to his critical attitude toward the philosophy of the Vienna Circle.

In Popper's *Logic of Scientific Discovery* [4] this neo-Kantian scenario is no longer prominent. But all the more it is manifest in the manuscript that was its precursor, namely that part of the voluminous manuscript (written between 1930 and 1932, read and commented by Carnap, Feigl et al. for publication in the series of the Vienna Circle) from which in a revision process the drastically abridged text of the *Logic of Scientific Discovery* emerged. The conserved part of the original manuscript was edited by Troels Eggers Hansen, only in 1979, under the title: *Die beiden Grundprobleme der Erkenntnistheorie* [5]. In what follows, I will draw on materials mainly from this manuscript, with occasional additions from Popper's intellectual autobiography.

Although Popper describes his own opposition vis-à-vis the Vienna Circle as a repetition of Kant's criticism of Hume and hence identifies the Vienna Circle with Hume, this is only half of the story. The programme of logical empiricism, to him, is also engaged in solving what he calls the "Kantian problems". Popper regards logical empiricism even "as one of the most interesting attempts to solve the Kantian problems undertaken since Kant" (GE 18), counting among the logical positivists Russell, Schlick, Philip Frank, Wittgenstein, Reichenbach and Carnap, but restricting his subsequent discussion mainly to Schlick and Wittgenstein. The problems Popper is referring to under the label of "Kantian problems" are two: first, the problem of reconciling empiricism and rationalism; second, the problem of demarcation between science and metaphysics. However, as the argumentation develops, Popper loses sight of the fact that both he and his opponents are pursuing rival attempts to solve the Kantian problems. He restates their relation more radically as Kantian versus anti-Kantian in character. This will become evident in Popper's construction of the so-called epistemological antinomy (see below pt. 5) concerning the understandability of the world.

In connection with Popper's tendency to delineate his own position from that of the Vienna Circle we notice a considerable shift in emphasis, such that the position of logical positivism is identified with inductivism, Popper's own position with deductivism. The distinction Kant had made in his *Critique of Pure Reason* – the distinction between the analytic and the dialectic – provides the model for Popper's reorientation. Says he: "Kant's *Critique of Pure Reason* seeks ... mainly to answer the same questions which I have called the basic problems of epistemology: The 'Transcendental Analytic' is concerned with the problem of induction, ... the 'Transcendental Dialectic' with the problem of demarcation." (GE 17) The first problem, i.e. the problem of induction, Popper calls the "Hume problem", whereas the second, the problem of demarcation, is understood as forming the "Kant problem" in the singular.

With respect to the distinction between analytic and dialectic in the *Critique of Pure Reason,* Popper saw the philosophers of the Vienna Circle concerned only with issues of the analytic and taking sides with Hume, and he saw his own approach as an attempt to defend the Kantian conception of theoretical philosophy in a comprehensive way. Popper's defense of Kantianism required far-reaching revisions, no doubt. Yet as he saw it, they were a necessary reaction to the new situation in science and they were conceived in the spirit of Kant.

In this vein Popper says: "But against modern disparagement of Kant, I want to emphasize at this place that in this work I want to adopt Kant's *problem*, advocate his *method* and defend even quite essential parts of his *solutions*." (GE 18)

This is quite a remarkable adoption of Kantianism, indeed.

For Popper's understanding of Kant, the new introduction to *Die beiden Grundprobleme der Erkenntnislehre*, written in 1979, is illuminating. From this concentrated sketch it is obvious that the Kant of the *Prolegomena* and of the Preface to the second edition of the *Critique* forms the essentials of his understanding of Kant. In the sequence of the well known questions, How is mathematics possible?, How is pure natural science possible?, How are synthetic judgements a priori possible?, Popper finds the clue to transcendental reasoning as such. To quote Popper: "The only way how Kant's question can be understood is this: that Kant, starting with Hume's scepticism, felt the existence of Newton's physics to be paradoxical ... . The existence of Newton's mechanics is a paradox for a sceptic and leads directly to the question: ... How is the existence of such a science possible?" (GE XVII)

In the following, I would like to show under five headings how, and to what extent, features of Popper's methodology are informed by Kantian and esp. by neo-Kantian elements, and how these facets determined his opposition to the approach of the Vienna Circle. The headings are: 1. Transcendental method; 2. Transcendence of representation; 3. Fictionalism and hypothetism; 4. Regulative ideas; 5. Epistemological antinomies.

1. TRANSCENDENTAL METHOD

The fact or the existence of knowledge as it is manifest in mathematics and physics is accorded a key rôle in launching the philosophical question for the conditions of the possibility of such knowledge, just as the neo-Kantians had done it. The method according to which we are able to assess claims of knowledge that can neither be identified with logical analysis nor with a mere description of empirical methods, must still be a transcendental one in the Kantian sense. Says Popper: "There is a specifically epistemological method, i.e. a *transcendental method;* ... a method which, if used correctly, is not only completely *harmless,* but plainly *unavoidable;* a method which probably has been used in a more or less conscious manner by *every* epistemologist (since Kant)." (GE 57)

Kant's transcendental method, in spite of the justificatory touch it has, is attractive or unavoidable according to Popper, as it offers a possibility of immanent or internal criticism of science and enforces logical analysis. The specific character of transcendental method consists in its "reference to the fact of *the existence of empirical science*, in particular to the *methods* as they are actually in use in the empirical sciences for testing and justifying their results". (GE 57)

The task of the epistemologist consists in accounting for the methods actually applied in science, in the same sense in which it is the task of the natural scientist to take into account the conditions as they are actually present in nature. Consequently, Popper conceives of the transcendental method as "an analogue to the empirical method; and epistemology stands to the natural sciences in a relation which is similar to that of the latter to the empirical world." (GE 57)

Orthodox Kantians will probably object, that there can be no analogy between empirical and transcendental methods, and that in Popper's reconstruction the transcendental method is losing its specific status, i.e. the status of strictly a priori reasoning. Popper could, however, reply that Kant himself speaks of transcendental philosophy as such and of the *Critique* esp. as an experiment of reason imitating the method of chemists and the natural scientists generally.

Popper, when outlining the transcendental method, is citing as his authority not only Kant's first *Critique*, but also Külpe, Natorp, Cohen, Riehl, Schuppe, Wundt and Rehmke (GE 58). So this is a place where Popper is putting himself in line with quite an impressive selection of neo-Kantians as well.

Transcendental method aims at establishing the possibility and validity of objective knowledge by means of a critical analysis of the methods used in actual research and of the basic concepts in which we describe physical theory. The proofs which Kant had added to the section about the principles of pure understanding in the second edition of the *Critique* form, according to Popper, the very core of the *Critique*, in much the same way in which the Marburg School had seen it.

2. TRANSCENDENCE OF REPRESENTATION

Hume's sceptical position had questioned the possibility of defending universal statements about nature. In opposition to this denial, Kant attempts to show that *any experience whatsoever* implies assumptions of the kind disputed by Hume, in other words, even singular experience presupposes assumptions about universal laws. To put it in Kantian terms: the concept of any possible experience implies the existence of universal laws of nature. According to Popper, Kant had demonstrated "that all scientific objectivity *presupposes* the existence of *laws*, independent of the question whether the cognitive statement represents a *particular* observation or formulates a law of nature, a strictly universal statement about reality." (GE 66)

Experience implies regularity, i.e. accordance with laws. With great emphasis, Popper represents this apparently trivial insight to be Kant's most important discovery, namely the discovery "that any knowledge of reality, that 'the possibility of experience', that the *objectivity of knowledge* rests entirely on the existence of regularities, of laws." (GE 68) Kant's discovery of the priority of the general regularities of nature over the singular experiences is preserved in Popper's conception of the "transcendence of representation".

With this notion he is challenging the doctrine of immanence ("Immanenz-lehre") proposed by Moritz Schlick. The positivistic ideal of immanentism demanded that one restrict epistemic ambitions to the mere description of the given, of the factual, and refrain from moulding hypotheses. Such a restriction to the immediately given is, however, absurd, if we want to remain faithful to the task of reconstructing the form of experience as exposed and practised in the sciences. Laws of nature transcend the realm of singular experience and they are indispensable for any genuine knowledge, esp. where predictions and explanations are concerned. General laws or theories enable us to predict effects of a new kind and that they may have observable consequences in domains that were left unexplored so far, thereby carrying experimental research into new fields.[6] To transcend the given, then, is not a vice, as positivism wants it, but a virtue, provided we adopt a critical attitude with respect to the hypotheses.[7] – The fact that scientific method is in accordance with the "transcendence of represen-tation" forms the basis of Popper's criticism of positivism, which he still calls "transcendental criticism". – So Popper, at that time, not only speaks of a *transcendental method* but of *transcendental criticism* as well.

3. FICTIONALISM AND HYPOTHETICISM

Experience, even singular experience, presupposes the command of general laws or theories, and Kant's programme entertained the ambition to demonstrate, exclusively by means of pure reason, the form of every possible experience, i.e. the general laws of nature. Popper denies the possibility of such a demonstration. Yet, he does not give up Kant's ambition altogether but gives it a fallibilistic twist. For him it is impossible to give a demonstration a priori of the universal laws of nature, but he can concede that there is no need for such a proof either; it is sufficient to assume that events take place in accordance with the laws of nature *as if* they were strictly universal laws. (GE 78)

In connection with this Popper makes use of a typical neo-Kantian concept which was developed by Hans Vaihinger in his well known and influential *Philosophy of the As-If*[8]. Vaihinger's position is often referred to as fictionalism, a position which was adopted by Popper at that time esp., as the term fiction seems to match perfectly well with the hypothetical status which Popper attrib-utes to the general laws of nature.

In the new introduction of 1978 and in a number of footnotes newly added to the text Popper dissociates his own fallibilist position from Vaihinger's fiction-alism and calls it a great mistake in the original text, not to have separated Vaihinger's conception of the epistemic status of hypotheses from his own.

There is an obvious motive for this withdrawal: Vaihinger's fictions can be neither true nor false, and they lack any ontological relevance. A position according to which universal laws of nature could be neither true nor false is unacceptable, as Popper states in retrospective. And it should have been all the more unacceptable in 1930, as in his opposition to logical positivism Popper is questioning the so-called "Scheinsatzposition" of Wittgenstein and Schlick. Yet, the Scheinsatzposition of the positivists has much in common with Vaihinger's fictionalism. According to Wittgenstein and Schlick, universal statements about reality are pseudo-statements; in spite of their grammatical structure, they cannot be considered as propositions at all, but function as *rules of inference* – to pass from one singular statement to another singular statement – and therefore cannot be true or false. In the same manner, Vaihinger, in the *Philosophy of the As-If*, had argued that a law of nature is "only an auxiliary expression for the totality of relations within a group of appearances" and had understood the law as a "summatory fiction".[9] Since in retrospective Popper notices the similarities between Vaihinger's fictionalism and the *Scheinsatz* (pseudo-sentence) position of his opponents, Popper is revoking his own former adoption of it.

However, this revocation should not blur the fact that Popper owes important insights to the as-if consideration. For Popper is freeing himself from the burden of Kant's "apriorism" just by adopting Vaihinger's figure of the as-if. Says he: "If one repudiates the synthetic-apriori turn of the transcendental deduction, then one must use the expression 'regularity' (i.e. 'accordance with a law') in the *whole chain of reasoning* exclusively in the sense of 'as-if-regularity'." (GE 72)

The fundamental orientation of Kant's transcendental idealism, the prescriptive power of reason as expressed in the formula of the Copernican turn, can easily be accounted for in the framework of the as-if-consideration. Vaihinger's concept of the as-if did not accentuate the moment of the unreal and fictitious, but in accordance with the original meaning of the Latin word "fingere" it emphasised the moments of free production, of building, forming, shaping something[10], as the Latin "ars fingendi" signifies the activity of the sculpturer. This aspect, with which Popper wanted to avoid the horns of the dilemma between apriorism and positivism, i.e. to strike the happy mean between "normal-statement-position" and "pseudo-statement-position", Popper tried to catch in the term hypothetism ("Hypothetismus"), a term which did not enter the *Logic of Scientific Discovery* or his later work, *Conjectures and Refutations (1962)*, when he preferred to speak of hypotheses as conjectures, as bold conjectures. Universal statements about reality are but preliminary assumptions, unjustified anticipations, which can be true or false in principle, but which we will never be able to prove to be true (nor can we ever produce a conclusive disproof of a theory!). The attempt of the human mind to imprint a general form of law on nature has no guarantee of success, but remains a risky step and may be frustrated.

4. REGULATIVE IDEAS

Kant's doctrine of the regulative function of the ideas of reason had been the starting point for neo-Kantianism to effect a general shift of theoretical philosophy towards methodology, most forcefully pursued by Ernst Cassirer. Even the principles of pure understanding ("Grundsätze des reinen Verstandes"), which Kant had accounted for as constitutive for both the objects and the form of every possible experience, were stripped of their ontological vestment and reinterpreted in purely methodological terms.[11] This neo-Kantian shift of ontology towards methodology could be endorsed by Popper without restriction.

In order to underline his commitment to Kantianism, Popper quotes a long passage from the section of the *Critique* entitled "Of the regulative employment of the ideas of pure reason" and states that it could serve well as a motto of his own work. And it is worthwhile mentioning that Popper at the same place speaks in the propagandistic tongue of neo-Kantianism: "This quotation shall support my conviction, that the thread of epistemological discussion has to be resumed exactly, where post-Kantian metaphysics had disrupted it: with Kant." (GE 321) In the same manner Otto Liebmann had ended each chapter of his polemic and programmatic book *Kant und die Epigonen* (1865) with the battle-cry: "Therefore, it is necessary to go back to Kant!"

Two features of Popperian philosophy have their origin in the *Dialectic* of the *Critique:* the first concerns the rôle of metaphysical ideas for empirical research as guiding principles; the second concerns the problem of demarcation between science and metaphysics, and leads to the distinction between a criterion of meaning (Sinnkriterium) and the criterion of demarcation (Abgrenzungskriterium). As the connection between the latter and the dialectic of the *Critique* is rather obvious, I shall concentrate my remarks on the first: the progressive, generative, heuristic function of ideas of reason.

Ideas about fundamental constituents of matter, the kinds of interactions or forces that obtain between bodies, generally speaking, cosmological ideas, function as inspirations and motors for concrete research projects. As Popper puts it in the Preface to the *Logic of Scientific Discovery:* "... it is a fact that purely metaphysical ideas ... have been of the greatest importance for cosmology. From Thales to Einstein, from ancient atomism to Descartes' speculation about matter, from the speculations of Gilbert and Newton and Leibniz and Boscovic about forces to those of Faraday and Einstein about fields of forces, metaphysical ideas have shown the way." [12]

Ideas of reason in Kant's sense transcend the scope of experience and no object in the domain of experience can ever correspond to them adequately. Yet, although ideas do not refer directly to objects of experience, they direct and guide the employment of understanding in the field of possible experience. Ideas of reason represent rules for empirical research of a higher order, as it were, in order to promote further investigation and to commit research to the idea of permanent progress. The employment of ideas of reason is problematic or hypothetical in the emphatic sense of the word, as it looks at nature as if nature were structured in such and such a way, say, governed by gravitational forces acting at a distance, or by fields of forces. Popper subscribed to Kant's doctrine of the hypothetical employment of reason, which aimed at the systematic unity of knowledge, taking the ultimate unity to mean the one true physical theory towards which empirical research is converging in the long run. The concept of a unified and true theory, a theory which can explain everything and has no false consequences is but an idea, yet it is indispensable for empirical research insofar as research strives for the goal of complete knowledge. For Popper, truth proves to be a "regulative idea" in Kant's sense. Kant's teachings about the heuristic function of ideas form, no doubt, the basis for Popper's conception of the growth of scientific knowledge, as shown in the *Logic of Scientific Discovery*, and for his conception of the "metaphysical research programme", which he developed during the fifties and which became the nucleus for the contributions of Agassi, Feyerabend and Lakatos to the methodology of modern empiricism.[13]

5. EPISTEMOLOGICAL ANTINOMIES

According to Kant, metaphysical theses about reality, as found in traditional cosmology, cannot be refuted by experience, but their truth cannot be established either. Reason can argue with the same plausibility both in favour of its truth and in favour of its falsehood. Reasoning without the support of experience is bound to run into antinomies, pure reason is dialectical in itself.

Popper not only endorses the section on the cosmological quarrel of reason from Kant's *Critique* but enriches the types of antinomies by a new kind which he calls epistemological antinomies. Examples of epistemological antinomies would be the opposition of idealism vs. realism, of conventionalism vs. empiricism, of epistemological optimism vs. pessimism. Popper gives special attention to one version of the latter under the title "intelligibility or unintelligibility of the world". Popper asserts that all these positions can be developed and maintained consistently and that they are all irrefutable on empirical grounds. Therefore all of them give rise to undecidable antinomies. Since the thesis of the intelligibility of the world rests on the presupposition that there are certain universal regularities (be it in the sense of as-if regularity or of standard regularity), we have to realise that there is no way to establish this condition as fulfilled by necessity; we can imagine, without running into contradictions, that the state of nature changes from that of order into a state of chaos, in which conditions of intersubjective testability or reproduction of test situations would no longer obtain.

Disclosing the antinomy of the intelligibility of the world is helping Popper, as he says, not only to advance his criticism of Kant's apriorism and his criticism of positivism at the same time but is also helping him to arrive at the root of the problem. (GE 77) Kant (and the rationalists) hold the thesis of the intelligibility of the world, i.e. they hold that general states of affairs exist and that we can have knowledge of them. In opposition to it logical empiricists (like Wittgenstein) and the sceptics hold the antithesis i.e. that there are singular states of affairs only and that statements about general states of affairs or regularities of nature have no meaning. According to Popper's construction the positions of Kantianism and of positivism appear as thesis and antithesis of an epistemological antinomy; both positions make claims about reality which cannot be decided on empirical grounds and have to be classified as metaphysical in the old sense of the word. Therefore, both positions have to be excluded from scientific discourse (GE 74).[14]

Along the same line, in which Kant had given a dissolution of the cosmological antinomies, Popper is proposing a solution to the epistemological antinomy of apriorism and positivism. The partial decidability of universal statements about nature corresponding to general regularities functions as an analogue to Kant's notion of the "thing in itself", which Kant used for his dissolution of the cosmological antinomies (GE 315).

```
                         Epistemological Antinomy


          Thesis:                              Antithesis:

            Kant                          Wittgenstein/Schlick
         Apriorism                             Positivism
Normal-statement position             Pseudo-statement position


                           Solution:

                            Popper
                          Hypothetism
                     One-sided decidability
```

6. CONCLUDING REMARKS

As obvious as these influences of neo-Kantianism on Popper's development were, Popper was quite unprepared to accept it when I presented my material to him. This was when I met him in London in 1988 after having sent him a copy of my book[15] published some months before. Proud as he was of his philosophical affiliation to Kant himself, who Popper never ceased to admire, he refused to accept that doctrines of neo-Kantians might have had any influence upon him whatever. It took almost a year for Popper to make up his mind on it and to admit what he had denied so violently before. From the letter he wrote to me on 3 September 1989 I would like to quote the following passage: "Today I only would like to say that much, that you are right to a large extent: three "neo-Kantians" – Vaihinger, Natorp and Cassirer – have had influence upon me. Especially, I guess, the first (Vaihinger). But I had a completely negative view of Hermann Cohen. Since Cohen was considered to be the leader of neo-Kantianism in my early years, I have never felt close to the school of neo-Kantianism. I felt much more attracted to Kant than to the philosophers mentioned. I did admire Kant (which I still do), something I am unable to say of the others. Vaihinger I found conceited and full of vanity; Cassirer I found cold and arrogant (I made his personal acquaintance). For Natorp I felt some sympathy from a distance, but I have not learnt much from him. So I expected much more to learn from Fries and from Nelson, both brilliant Kantians, and that helped me to challenge psychologism ..."

To sum up: Popper's emphatic opposition to basic tenets of logical empiricism and the confidence with which he is holding a strong position against the positivism of Wittgenstein and Schlick rest on the idea that he, Popper, is renewing the arguments Kant had produced against Hume. His attack on positivism culminates in demonstrating what he calls "the basic positivist contradiction", which for Popper boiled down to a demonstration that transcendental philosophy in Kant's sense and positivism are incompatible. Says Popper: "The criticism of logical positivism which plays a prominent role among the critical arguments of this work will prove the incompatibility of positivist and transcendental tendencies: the basic positivist contradiction." (GE 59)

Popper continues to state that Kant was the first to see this contradiction and that this insight led him beyond Hume. "What Kant objects to Hume's positivism is exactly the same what is here brought forward against strict positivism." (GE 59)

In his autobiographical retrospective *Unended Quest*, looking back at the early thirties, Popper says: "At that time I looked upon myself as an unorthodox Kantian ... And I used to think in those days that my criticism of the Vienna Circle was simply the result of having read Kant, and of having understood some of its main points." (UQ 82f.)

So far I have presented a piece of history as experienced and interpreted by one of the prominent protagonists of modern empiricism. I will not question whether or not Popper has given us an adequate description of the situation and how the very same situation is experienced and interpreted from the side of the opponents. (Carnap, for one, could never understand why and how Popper placed himself in opposition to the programme of logical empiricism.)

I will, rather, conclude with a proposal for systematic work to come: if it was correct to attribute shortcomings of logical empiricism (or the breakdown of positivism) to its neglect of transcendental epistemology and if it is true that critical rationalism could advance as a consequence of the renewal and reproduction of arguments as Kant had produced against Hume, it may be a promising project of philosophy of science to make yet another attempt to reread Kant in the light of recent developments in the sciences. Whatever we may gain in terms of particular insights and conceptual instruments from such a rereading, the most promising effect will be the reorientation of work to fundamentals of methodology. Since Kant's own theoretical philosophy had been inspired and shaped by his lifelong attempts to give an adequate account of the science of his time, it could help Popper to give shape to his own fallibilist ideas, and it may again serve us as a prototype[16] in our own attempts to account for the rationality of science as we have it now.

The systematic expectations which were connected with a careful study of Kant's theoretical philosophy did function as the starting point and driving force for neo-Kantianism. Neo-Kantianism, in spite of its common programmatic orientation on Kant, developed with considerable diversity. Attempts to repeat under new conditions in science and philosophy the task Kant had done in his time have produced new systematic contributions to contemporary philosophy – not only in the case of Popper's original conception of methodology and critical rationalism.[17] And, therefore, it may serve us well in our attempts to give shape und coherence to present-day philosophy of science.

NOTES

1. Karl Popper, *Unended Quest: An Intellectual Autobiography*. London & Glasgow: Fontana / Collins 1976. – Originally published as "Intellectual Autobiography" in: Paul A. Schilpp (Ed.), *The Philosophy of Karl Popper*, 2 Vols., La Salle, Ill.: Open Court 1974. – Quotations "UQ" refer to the revised edition by Fontana.
2. Cp. Lothar Schäfer, *Kants Metaphysik der Natur*. Berlin: de Gruyter 1966.
3. The neo-Kantians of the Marburg School tried to enforce the logical form in transcendental argumentation leading them to give up the independent faculty of intuition, which Kant had introduced in the Transcendental Aesthetic of the *Critique*. Yet, their conception of logic is not to be identified with formal logic in our sense, as logic in the neo-Kantian sense aims at the generation of categories and, therefore, is of ontological significance. Cp. Geert Edel, *Von der Vernunftkritik zur Erkenntnislogik. Die Entwicklung der theoretischen Philosophie Hermann Cohens*. Freiburg/München: Alber 1988.
4. Karl Popper, *Logik der Forschung*, (Wien: Springer 1934). Tübingen: Mohr (Siebeck) ³1969.
5. Karl Popper, *Die beiden Grundprobleme der Erkenntnistheorie*, (ed. by T. E. Hanssen). Tübingen: Mohr (Siebeck) 1979. Quotations in the text ("GE") refer to this edition, the translations are mine.
6. In his attack on the doctrine of immanentism Popper is in alliance with Pierre Duhem and gives credit to him. Duhem's *Aim and Structure of Physical Theories* was read and discussed widely among philosophers of science in Vienna after Mach had prefaced the German edition of it (1908).
7. In neo-Kantianism, too, we find a strong criticism of the notion of the "given", which could have been attractive for Popper. Cp. Johannes von Malottki, *Das Problem des Gegebenen*. Berlin: de Gruyter 1929.
8. Hans Vaihinger, *Die Philosophie des Als Ob*. Berlin 1911.
9. Ibid., Part II, § 15.
10. Ibid., ch. XX.

11. As early as 1883 Hermann Cohen, the leading head of the Marburg School of neo-Kantianism, had argued that the second law of Newtonian mechanics, the law of inertia, did not describe facts of nature, but formulate a methodological presupposition of physical research. Hermann Cohen, *Das Prinzip der Infinitesimal-Methode*. Berlin 1883, p.48.

12. Karl Popper, *The Logic of Scientific Discovery*. (London: Hutchinson 1959), New York: Harper & Row 1968, p. 19.

13. Cf. Karl Popper, "Philosophy and Physics", in: *Atti del XII. Congresso Internationale di Filosofia*, Firenze 1961, pp. 367-374. Joseph Agassi, "The Nature of Scientific Problems and their Roots in Metaphysics", in: Mario Bunge (Ed.), *The Critical Approach to Science and Philosophy*, London 1964, pp. 189-211. Paul Feyerabend, "How to be a Good Empiricist", in: Bernard Baumrin (Ed.), *Philosophy of Science*: The Delaware Seminar, Vol. 2, 1962, pp. 3-39. Imre Lakatos, "Falsification and the Methodology of Scientific Research Programmes", in: Lakatos and Musgrave (Eds.), *Criticism and the Growth of Knowledge*, Cambridge 1970, pp. 91-195.

14. In his later work Popper did not rest with this position. Although the epistemological antinomies were construed as undecidable in GE, Popper later opted for one side and claimed to have good reasons for his option (GE 79, n.*2). However this development of Popper is to be assessed. Popper had seen (in GE) the fundamental importance of the antinomies in the *Critique* in its power to demolish the epistemic ambitions of pure reason in experience, which amounted to a transcendental justification of empiricism (here to be understood in a wide sense).

15. Lothar Schäfer, *Karl R. Popper*. München: Beck 1988.

16. "So Kantian thought stands as a model of fruitful philosophical engagement with the sciences." Michael Friedman, *Kant and the Exact Sciences*, Cambridge, Mass.: Harvard UP 1992, p. XII.

17. It may be risky, to name just some prominent examples, as the selection will be subjective and accidental. However, the list of philosophers who have developed their own ideas by way of examining and rereading Kant, incomplete and debatable as it certainly is, looks rather impressive and promising: M. Heidegger, K. Jaspers, N. Hartmann, W. Sellars, P. Strawson, S. Körner. A short estimation of Kant's influence on present-day philosophy gives Wolfgang Stegmüller, *Hauptströmungen der Gegenwartsphilosophie*, Stuttgart: Kröner (5. Aufl.) 1975, pp. XXVII-XXIX.

Institut für Philosophie
Universität Hamburg
Von-Melle-Park 6
D-20146 Hamburg
Germany
schaefer@kassandra.philosophie.uni-hamburg.de

KENNETH SIMONSEN

CONCERNING SOME PHILOSOPHICAL REASONS FOR THE RECOURSE TO MATHEMATICS IN THE STUDY OF PHYSICAL PHENOMENA IN THE THOUGHT OF NEWTON AND LEIBNIZ

Considering the physics of Newton and of Leibniz, we are confronted with two different ways of explaining physical phenomena and with different kinds of concepts. Whether we speak about space, time, force, matter, vortices etc., their conceptions diverge.

In this paper, we are not principally concerned with a comparison of their respective theories of physics considered as a whole or as fully accomplished. Thus, a straight comparison of their respective concepts and of their world systems will not be a main issue. And, if we touch upon some theological or metaphysical arguments in favour of their physics, we will not give an account on how these are discussed in later years when their systems enter in conflict with each other.

Today, our interest in Newton's and Leibniz's physics is more focused on what happens before their theories find their final expression. We will pay attention to how different concepts are articulated during this period preceding their "accomplishment". For, these concepts change their meaning and signification as a result of the authors' reflection on them and/or as a consequence of their meditation on others. And it seems to us that the question of the role and place attributed to mathematics within the theory is of great relevance in this context.

By studying their thought as a theory "en devenir" or as a "work in progress", we believe being not only able to examine the evolution of central concepts in their systems, but also to have a closer look on how Newton and Leibniz consider the problem of mathematisation[1]. For, both inherit the belief from Descartes and Galileo that mathematics are appropriate to explain physical phenomena, and they do both have the intention of elaborating a physical theory where mathematics play an essential role. But in doing that, they are both led to clarify how a mathematical theory is possible or acceptable, i.e. on which grounds the possibility of having recourse to mathematics in the study of physics (mechanics) is laid. And further on, this process sometimes implies conceptual consequences for the theories themselves. "Work in progress" certainly also means work of deepening and clarifying.

M. Heidelberger and F. Stadler (eds.), History of Philosophy and Science, 57–65.
© 2002 *Kluwer Academic Publishers. Printed in the Netherlands.*

TOWARDS AN ABSOLUTE SPACE. NEWTON AND THE PROBLEM OF MOTION

Then, concerning Newton, we can consider that, even though he produced many
of his ideas of physics in the 1660s, it is principally in the period from the *On
Motion*[2] of 1684 to the *Principia*[3] of 1687 that his ideas became a part of a
process of maturation leading to their final expression in a coherent theory. And,
it was particularly in this process that Newton encountered the problem of
mathematisation of physics.

The story is well known. When Edmund Halley came on visit to Newton in
August 1684, he asked him what trajectory a planet should follow if one supposed an inverse square law for the central force acting upon it. According to
John Conduitt, Newton answered that it should be an ellipse[4]. And when Halley
asked him why he thought so, he said that he had calculated it. But as he couldn't
find it, he promised to send him the solution later. Some months later then,
Halley received a short tract entitled *On Motion* which constitutes "the point of
departure" of what was to become the celebrated *Principia*.

In this tract, Newton elaborates a mathematical representation of planetary
motion. Following a synthetic scheme, he presents – like in the *Principia*, a set
of definitions, hypothesis (laws in the *Principia*) and lemmas before demonstrating eleven theorems and problems. However, this set differs slightly from the
one in the *Principia*. In *On Motion*, we don't find any definitions of quantity of
matter, quantity of motion nor is the force defined in exactly the same manner.
And there are no trace – at least not in this part of the tract – of what was to become the second and third laws of motion. Nor does he mention absolute space
and time.

It is interesting to note that, while the initial propositions don't change essentially from manuscript to manuscript, the set of definitions, hypothesis and lemmas as well as the scholia following some of the propositions *do*. These changes
show how Newton is confronted with two kinds of requirements; one concerning
the clarity and the pertinence of fundamental concepts on which the truth of the
following propositions largely depends; the other involves the relation between
theory and real world. However, these two requirements are not necessarily independent. Reflecting on the relation between theory and real world may modify
the set of fundamental concepts, qualitatively or quantitatively. And vice versa, a
modification of these concepts may have a consequence for the apprehension of
the relation theory – real world.

In the first revision of *On Motion* entitled *On Motion of Spherical Bodies in
Fluidis*[5], Newton takes these aspects into account. He largely modifies the
hypothesis and lemmas. The 'hypothesis' become 'laws' (*leges*). He introduces a
first assertion of the second law of motion as well as two new laws inspired by
the principle of inertia. He also makes a first step towards the method of first and
last ratios.

However, we will concentrate on a problem encountered in an addition to the scholium of the fourth theorem.[6] In *On Motion*, he had only considered the motion of a body revolving around a force centre considered – not as a body – but as a simple mathematical point. He never envisaged possible actions of other bodies. In the new scholium, on the contrary, he introduces a "whole space of the planetary heavens"[7] and defines a centre of gravity in it. Leaning on one of the new laws (law 4) he states that this space is either at rest or in uniform rectilinear motion[8]. And from the other new law (law 3) he infers that all motions between planets are the same whether this space is at rest or in uniform rectilinear motion[9]. Consequently, he considers this space to be immobile, and hence the centre of gravity will be immobile as well.

Now, defining a centre of gravity implies that he takes the motion of all planets into account. On the other hand, if the conception of a common space being always equal in relation to the planets' motions is true, then it has important consequences for his mathematical explanation of planetary motion as it appears in the *On Motion*. Newton is well aware of it. In fact, it means that the centre of force doesn't coincide with the centre of gravity, and hence will no longer be immobile as presupposed in his mathematical theory. At this stage, though, Newton accepts that his theory is a simplification of the real world – it would indeed "exceed the force of human wit"[10] to consider all these motions at the same time. Nevertheless, he explains, the mathematical theory still has its power of explanation – for by taking the mean values of observations, one should still obtain an ellipse near the calculated one.

So, in this revision Newton considers the complexity of the real world and evaluates its consequences for his mathematical theory of planetary motions. What allowed him to do this, though, was the extension of the principle of inertia to the system of planets, postulating an immobile space. But, it is not yet an absolute space, and we have to move further to the *On Gravitation*[11] to discover it.

In this manuscript it is not the particular question of planetary motion that is the main problem, even though it constitutes an important element in the argumentation, *but* rather the general question of how we can possibly speak of motion mathematically.

After having defined place, body, rest and motion, he is led to argue against the natural philosophy of Descartes. He attacks his conception of motion arguing that it leads to absurdities and contradictions. He is refuting the relativity of motion the Cartesian conception implies, stating that it is impossible to ever know which body is truly in motion and which is truly at rest. The reason for this is, according to Newton, that Descartes did not distinguish between matter and extension (*étendue*) as independent of body and matter.

Furthermore, Newton states that the Cartesian point of view implies that it is impossible to clearly define places (*locus*) in relation to what the displacement of a body and thus its motion can be identified. In his opinion the Cartesian motion is not a motion at all because we cannot assign any fixed places, and hence no displacement nor any speed. Consequently, he is led to assert the necessity of relating places and local motion to something immobile.

This something immobile is the space itself and Newton characterises it as infinite, eternal and divisible. It is all over uniform and has no capacity whatsoever of resistance or of provoking change in motion. It is not a substance nor an accident, but an emanating effect of God and a certain affection of all being. It is something without which nothing could ever exist. This space is called absolute in the manuscript *On Motion of Bodies in Uniformly Yielding Media*[12]. It has then acquired the necessary characteristics (infinite, eternal, divisible, non-corporeal) in order to constitute a reference frame for a mathematical treatment of motion. It allows Newton to identify real motion and to distinguish it from the relative. Force becomes a causal principle that enables him to that; real motion or rest "are never changed except by force impressed on the body moved or at rest, and are always changed after [the action of] such force, while relative motion can be changed by forces impressed on other bodies"[13].

TOWARDS AN ABSOLUTE FORCE. LEIBNIZ AND THE PROBLEM OF MOTION

The case of Leibniz' physics seems more complicated, because he never wrote a treatise like the *Principia* and because his physics involves assembling of many different manuscripts and articles that cover different sides of it. Nevertheless, we can identify an evolution in his thought with respect to the problem of mathematisation of physics.

In a letter of July 1676 to Edmonde Mariotte, Leibniz expresses his belief that "there exist physical effects of which it is possible to find the last cause"[14]. And this is the case, according to Leibniz, "when a truth of physics depends on a truth of metaphysics or geometry". He is probably referring to his discovery in the *On the Secret of Reducing Motion and Mechanics to pure Geometry*[15] written that same summer. In it, he takes up the problem of mathematisation of physics, stating that the solution should be found in a principle which allows such a "reduction of mechanics to geometry". Being inspired by the equilibrium law of Archimedes which allows a mathematical explanation of statics, Leibniz states that by this principle, one should be able to establish equations of relations in mechanics. However, it is the geometry that offers him the key to a solution:

As the principle of calculation usually is laid down from the equation between the whole and all its parts, the principle in all mechanics depend on the equation between the full cause and the whole effect.[16]

This is, in fact, the first known statement of the principle of equivalence between the full cause and the whole effect, and it implies that something should be conserved. It is a principle of conservation that enables Leibniz to treat physical phenomena mathematically. But the question then is: what is to be conserved? Leibniz speaks about a power. "Of the full cause and whole effect, he says, the power (*potentia*) is the same" [17]. However, what is exactly meant by this "power" or "*potentia*" doesn't become clear until February 1678 when Leibniz accomplishes a genuine "research project" entitled *On Collisions of Bodies* [18]. Intent on finding some fundamental laws of mechanics based on this principle of conservation, he considers the case of collisions. Without going into detail, we should nevertheless mention that he supposes at first this "*potentia*" to be the Cartesian absolute force (or quantity of motion) [19]. After several calculations he obtains some numerical results that he compares with those obtained from an experiment done on a double pendulum system. Now, these results diverge considerably, and Leibniz continues his work. He then restates the cited principle adding that a "force" still has to be conserved. However, he doesn't reconfirm that this "force" is the Cartesian quantity of motion, but rather:

one should not estimate the force in the body by its speed and magnitude, but by the height from which it descends. [20]

This statement along with Galilee's law for falling bodies allow Leibniz to establish a new measure for force: mv^2 – the product of the body's magnitude or quantity and the square of its velocity. By adopting this measure for the absolute force, he is now able to establish what is today known as the conservation principle of quantity of motion which is essentially different from the Cartesian as this last one doesn't take the direction into account, notably because it was considered absolute.

In *On Collisions of Bodies*, Leibniz expresses an occasionalist point of view asserting that a "very wise cause" maintains the bodies in motion. But a year later, he starts revising this conception. In two short manuscripts published by Loemker under the title *On the Elements of Natural Science* [21], he indicates the direction that his thoughts on the absolute force make it necessary to link it more strongly to a metaphysical conception. He states – as before – that the laws of motion "depend upon the mathematical principle of the equality of cause en effect", and that they are among those things that "cannot be explained from the necessity of matter alone" [22]. But he goes even further:

In fact, he says, I have contended that the reasons for physical motion cannot be found in mathematical rules alone but that metaphysical propositions must necessarily be added. [23]

However, although he asserts the necessity of including metaphysical explanations in order to fully understand mechanics, he doesn't yet make a clear step in any specific direction. But, in a letter to the Duke Johann-Friedrich, he writes that he had "re-established demonstratively the substantial forms that the Cartesians pretend to have exterminated" [24].

We then understand the significance that the concept of force takes in the *Discourse on Metaphysics* [25] of 1686 and in the following correspondence with Antoine Arnauld [26].

In article 18, we read:

although all particular phenomena of nature can be explained mathematically or mechanically by those who understand them, nevertheless the general principles of corporeal nature and of mechanics itself are more metaphysical than geometrical, and belong to some indivisible forms or natures as the causes of appearances, rather than to corporeal mass and extension. [27]

In fact, Leibniz is here touching on the same kind of problem as Newton: how to determine true motion. But for Leibniz, change in motion is not sufficient to determine which body is truly in motion and which is truly at rest. Motion considered by itself "is not something entirely real". Separated from the force it is something relative and thus it is impossible "to determine, merely from a consideration of the changes [of positions], to which body we should attribute motion and rest". The force, on the contrary, is absolute and is "something more real". It "is a sufficient basis to attribute motion to one body more than to another" [28].

In fact, as he explains in article 21 of the *Discours*, if mechanical rules did not depend on metaphysics but only on geometry, then it can be shown that the "smallest body would impart its own speed to the largest body without losing any of this speed"[29] – in allusion to his own *Theory of Abstract Motion* of 1671[30].

Now, if the force is "something more real", that is not barely because it provides a better measure for explaining physical phenomena mathematically. It is also because it belongs to "some indivisible forms", i.e. to a "substantial form" or "corporeal substance". For as he puts it in a letter to Arnauld:

[...] we cannot determine which of the changing subjects [the motion] belongs to, unless we have recourse to the force which is the cause of motion and which is in Corporeal substance. [31]

Hence Leibniz goes beyond the question of the identification of which body moves truly and which is truly at rest. He raises the question as to individuation which is central to the mathematical study of mechanics. That is, Leibniz clearly expresses that mechanics is not a question of geometry alone, but that there is something in bodies which cannot be explained in terms of extension and which is necessary for a mechanical explanation. And this something is the force which is always conserved. It is something absolute and real, and is thus connected to the substantial form. In which manner becomes clearer in two treatises published some ten years later, and which we will not treat here, the *A Specimen of Dynamics*[32] and the *A New System of the Nature and of the Communication of Substances*[33], both published in 1695.

CONCLUSION

We have tried to show how the question of the mathematisation of mechanics figures in the thought of Newton and Leibniz. In order to identify this problem, we have found it useful to focus on their theory of mechanics as a "work in progress". By doing this we realise that for both the question of the mathematisation of mechanics is connected to the problem of true motion. Furthermore, they both resolved this problem by claiming something absolute; Newton proposed an absolute space and Leibniz a metaphysical absolute force. However, we have seen that they were confronted with this problem in slightly different ways.

Newton did not seem concerned with the problem of mathematisation before his first mathematical and idealised construct faced the complexity of the real world leading to the introduction of an immobile space in which the motions of the planets could be explained mathematically. However, it was essentially in the confrontation with the Cartesian conception of motion that he realised the importance of being able to distinguish true motion from relative motion in a mathematical theory of mechanics. And it was in this context he was lead to assert the necessity of an absolute space, by introducing a reference frame which enables mathematical treatment of real motion and identification of causal forces.

Leibniz, on the other hand, raised the question of how to reduce mechanics to geometry before successfully accomplishing a mathematical treatment of it. He deduces the principle of equivalence of the full cause and the whole effect which enabled him to find a measure of the conservation force. The problem of motion only appeared with a vengeance after this force was conceptually and mathematically identified. This problem then changed the signification of force in Leibniz' thought. For the important thing for Leibniz, was not barely a problem of identifying positions in an immobile space and thus be capable of mathematical treatment. It was rather a question of individuation, i.e. of finding something in the bodies which was more real and connected to something substantial. According to him, mechanics could not only be a concern of geometry and change of positions. One had to identify the causes, i.e. the forces which belong to metaphysics.

Finally, if they both resolved the problem of relativity of motion by stating the need to find the causes of motions, i.e. the forces, they did not come to the same conclusion. Newton's solution resides in a sort of a general mathematical reference frame. Leibniz finds it in a metaphysical substantial form.

NOTES

1.   Here we are using an expression from M. Fichant: "[…], l'individualisation stylistique des configurations idéelles s'impose d'autant plus à l'attention de l'historien qu'il explore un corpus [le corpus leibnizien] essentiellement en devenir, dépôt d'un *work in progress*, dont les strates portent les tracent d'une genèse singulière." in "De la puissance à l'action: la singularité stylistique de la dynamique" in Fichant *Science et Métaphysique dans Descartes et Leibniz*, Paris, PUF, 1998, 205.

2.   I. Newton *De Motu* (1684), published in both Latin and English in I. Newton *The mathematical papers of Sir I. Newton*, éd. D.T. Whiteside, 8 vols., Cambridge, Cambr. Univ. Press, 1967-1981, vol VI, 30-75 [abbrev. *MP*].

3.   I. Newton *Philosophiae naturalis principia mathematica*, Londres, 1687. The standard English translation has been I. Newton *Mathematical principles of natural philosophy and his System of the world*, Eng. translation of 3e ed. by A. Motte (1729), rev. and edit. by F. Cajori, Berkeley, Univ. of California Press,re-ed. 1962. However, a new and very useful translation of the third edition now exists with some references to the first and second editions, I. Newton *Isaac Newton, The Principia: Mathematical Principles of Natural Philosophy*, new translation by I.B. Cohen and A. Whitman, assisted by J. Budenz, preceded by "A Guide to Newton's *Principia*", by I.B. Cohen, Berkeley, Los Angeles, London, University of California Press, 1999. In our talk we only consider the first edition.

4.   J. Conduitt *Memorandum relating to Sr Isaac Newton given me by Mr Abraham Demoivre in Nor 1727*, extracts published in Cohen [1971] *Introduction to Newton's* Principia, Introd. vol to the ed. by Koyré and Cohen, Cambridge, Mass., Harvard Univ. Press, 1971; Cambridge, Cambridge Univ. Press, 1971; Cambridge, Mass, Harvard Univ. Press, 1978, 297-298.

5.   I. Newton *De motu sphaericorum corporum in Fluidis* (1684) published in Latin and English in *Unpublished Scientific Papers of Isaac Newton. A selection from the Portsmouth Collection in the University Library, Cambridge*, ed. A.R. Hall et M. Boas-Hall, Cambridge Univ. Press, 1962, reed. 1978, 243-301. (abbrev [Hall and Hall]) Parts are published, but in Latin only in *MP*, VI, 74-80.

6.   The fourth theorem concerns the demonstration of Kepler's third law.

7.   I. Newton *De motu sphaericorum corporum in Fluidis* (1684), in [Hall and Hall] 280.

8.   I. Newton *De motu sphaericorum corporum in Fluidis* (1684), Law 4, "The common centre of gravity does not alter its state of motion or rest through the mutual actions of bodies. This follows from law 3" in [Hall and Hall] 267.

9.   I. Newton *De motu sphaericorum corporum in Fluidis* (1684), Law 3, "The relative motions of bodies enclosed in a given space are the same whether that space is at rest or whether it moves perpetually and uniformly in a straight line without circular motion" in [Hall and Hall] 267.

10. I. Newton *De motu sphaericorum corporum in Fluidis* (1684), in [Hall and Hall] 281.

11. I. Newton *De Gravitatione* (1685), published in Latin and English in *The Background to Newton's* Principia. *A Study of Newton's Dynamical Researches in the Years 1664-1684*, Oxford, 1965, 219-235. Abbrev. [Herivel] We follow here the interpretation of B.J.T Dobbs stating that the *On Gravitation* was written by Newton after *On Motion of Spherical Bodies in Fluids* but before another manuscript preparing the *Principia*, the *On Motion of Bodies in uniformly yielding Media*. In his recent edition of Newton's *Principia*, I.B. Cohen adopt this point of view. Dobbs, B.J.T [1991] *The Janus faces og genius. The role of alchemy in Newton's thought*, Cambridge, NY, Melbourne, Cambridge Univ. Press, 1991, 138-146 and Cohen, I.B [1999] in "A Guide to Newton's *Principia*" in *Isaac Newton, The Principia: Mathematical Principles of Natural Philosophy*, new translation by I.B. Cohen and A. Whitman, assisted by J. Budenz, preceded by "A Guide to Newton's *Principia*", by I.B. Cohen, Berkeley, Los Angeles, London, University of California Press, 1999, 47, 56-60.

12. I. Newton *De motu corporum in medijs regulariter cedentibus* (early 1685), parts published in Latin and English in [Herivel] 304-320 and in *MP*, VI, 188-194.

13. I. Newton *De motu corporum in medijs regulariter cedentibus* (early 1685), in [Herivel] 310

14. G.W. Leibniz *Sämtliche Schriften und Briefe*, Darmstadt, Leibzig, Berlin, 1923-, II, i, 270 (abbrev. *A*).

15. G.W. Leibniz *De arcanis motus et mechanica ad puram geometriam reducenda* (1676) published in Latin only as beilage 4 (pp.201-205) in H.-J. Hess [1978] "Die unveröffentlichen naturwissenschaftlichen und technischen Arbeiten von G.W. Leibniz aus der Zeit seines Paris-aufenhaltes. Eine Kurzcharkteristik", *Studia leibnitiana, suppl., vol. XVII Leibniz à Paris: Tome I, les sciences*, Franz Steiner Verlag, Wiesbaden, 1978, 183-217.

16. G.W. Leibniz *De arcanis* (1676), "Quamadmodum in Geometria principium ratiocinandi sumi solet ab aequatione quae est inter totum et omnes partes; ita in Mechanicis cuncta pendent ab aequatione inter causam plenam et effectum integrum." in Hess [1978] 203.

17. G.W. Leibniz *De arcanis* (1676), in Hess [1978] 203.

18. G.W. Leinbiz *De corporum concursu* (1678), published in Latin and French in Leibniz, *La réforme de la dynamique, textes inédits*, edition, presentation, translation into French and commentaries by M. Fichant, Paris, Vrin, 1994 (abbrev. [Fichant]).

19. For an analysis of this manuscript, see in particular the comments of M. Fichant in [Fichant] and of F. Duchesneau in Duchesneau, F. [1994] *La dynamique de Leibniz*, Paris, Vrin (Coll. Mathesis), 1994.

20. G.W. Leibniz *De corporum concursu* (1678), "Nempe vis in corpore non aestimanda est a celeritate et magnitudine corporis, sed ab altitudine ex qua decidit." in [Fichant] 134 (Latin) and 269 (French).

21. G.W. Leibniz *Libellus Elementorum physicae* and *Cogitationes de nova physica instauranda* (1679), published in English in Leibniz *Philosphical Papers and Letters*, L.E. Loemker (ed.), Dordrecht, Boston, D. Reidel publ., 1969, 277-290 (abbrev. [Loemker]). Loemker dated these to 1682-84, but we follow here the datation of M. Fichant which proposes the date 1679 in Fichant [1993] Mécanisme et métaphysique: le rétablissement des formes substantielles (1679), *Philosophie*, 1993, 27-59.

22. G.W. Leibniz *Libellus Elementorum physicae* (1679) in [Loemker] 278.

23. G.W. Leibniz *Cogitationes de nova physica instauranda* (1679), in [Loemker] 289.

24. G.W. Leibniz letter to the duke Johann-Friedrich automn 1679 *A*, I, ii , 225

25. G.W. Leibniz *Discours de métaphysique* (1686), we here use the English translation given in G.W. Leibniz *Philosophical Essays* edited and translated by R. Ariew and D. Garber, Inianpolis and Cambridge, Hackett Publ. Co., 1989, 35-68 (abrrev. *AG*).

26. For the Correspondance Leibniz-Arnauld, we also use the English translation given in *AG*. However, this edition only gives extracts from the correspondance.

27. G.W. Leibniz *Discours de métaphysique* (1686), §18, in *AG* 51-52.

28. G.W. Leibniz *Discours de métaphysique* (1686), §18, in *AG* 51-52.

29. G.W. Leibniz *Discours de métaphysique* (1686), §21, in *AG* 53.

30. G.W. Leibniz *De motus abstracti* (1671) A, VI, ii, 258-276.

31. G.W. Leibniz letter to A. Arnauld on april 30 1687, in *AG* 86-87.

32. G.W. Leibniz *Specimen dynamicum*, published in *Acta Eruditorum*, april 1695.

33. G.W. Leibniz *Système nouveau de la nature et de la communication des substances*, published in *Journal des Savants*, June-July 1695.

Université Paris 7, Denis Diderot
REHSEIS
37, rue Jacob
F-75006 Paris
France
simonsen@paris7.jussieu.fr

R. LANIER ANDERSON

KANT ON THE APRIORITY OF CAUSAL LAWS *

Kant famously rejected an empiricist account of causal claims, because it cannot account for the *necessity* and *universality* of causal laws. He then concludes that causal claims must have an a priori basis:[1]

the concept of cause cannot arise in this [empiricist] way at all, but must either be grounded in the understanding completely *a priori* or else be entirely surrendered as a mere fantasy of the brain. For this concept always requires that something *A* be of such a kind that something else *B* follows from it *necessarily* and *in accordance with an absolutely universal rule*. Appearances may well offer cases from which a rule is possible in accordance with which something usually happens, but never a rule in accordance with which the succession is *necessary;* thus to the synthesis of cause and effect there belongs a dignity that can never be expressed empirically, namely that the effect does not merely come along with the cause, but is posited *through* it and follows *from* it. [A 91/B 123-4]

So much is clearly Kant's view, and as long as we remain with the vague formulation that causal generalizations have *some* a priori ground or other, commentators agree.

But the attempt to specify what the "grounding" of causal claims amounts to immediately generates controversy – even about what is supposed to be a priori. Most scholars take Kant to claim apriority only for the *general principle* that every event has some cause, which then underwrites more particular causal generalizations in some way that falls short of actually implying them.[2] Thus, particular causal laws are merely empirical, justified by standard inductive procedures. In the last decade, this interpretive orthodoxy has been seriously challenged by Michael Friedman (1991, 1992a, 1992b), who contends that Kant meant to assert not only the apriority of the general law of cause, but also the apriority (in some sense) of laws of nature. This position gives clearer content to Kant's anti-empiricism, and Friedman can cite Kant himself insisting that no claim could count as a *law* at all unless it participated in apriority in some way (see A 159/B 198; A 227-8/B 279-80; *Prol.* 4: 312; *MFNS* Pref. 4: 468, 469; and *CJ* 5: 182, 183). Nevertheless, Friedman's view is strikingly counterintuitive, precisely in its core idea that the laws of nature are a priori, which – even with the qualification "in a sense" – has seemed a non-starter to many interpreters. For Kant was clearly aware that natural laws can only be discovered on the basis of experience; he says as much repeatedly,[3] and, the commentators insist, surely Kant was *right* to do so.

M. Heidelberger and F. Stadler (eds.), History of Philosophy and Science, 67–80.
© 2002 *Kluwer Academic Publishers. Printed in the Netherlands.*

The primary textual evidence for the standard reading has been the "Second Analogy" chapter of the first *Critique*, which focuses on the apriority of the general causal principle (see, e.g., Allison 1994). I will argue that the Second Analogy itself supports Friedman's position in its most dramatic and counterintuitive form. I will then deploy the resulting ideas to attempt a specification of the "sense" in which laws of nature are a priori "only in a sense." Before reaching the Second Analogy, however, I need to sketch some features of Friedman's interpretation in outline.

1. FRIEDMAN ON THE APRIORITY OF CAUSAL LAWS IN KANT

As Friedman points out, Kant claims both that particular causal laws are empirical and contingent, and that they are a priori and necessary. The position is apparently impossible, but it is attributable to Kant on highly robust textual grounds.[4] As a solution, Friedman suggests that for Kant causal laws must have "a peculiar kind of mixed status" (1992b, 174). They are empirical in that their articulation and defense depend on inductive regularities. Nevertheless, they are also a priori "in a sense" (1992b, 174), because they admit of a distinctive derivation from the a priori law of cause, which Friedman characterizes in terms of a core example, viz., Kant's grounding for the Newtonian law of universal gravitation.

On Friedman's highly illuminating account, Kant presents Newton's achievement not as an inductive theory of motion, but as a construction procedure that makes a meaningful distinction between true and apparent motion possible in the first place. As such, it functions as a theoretical precondition of the empirical doctrine of motion, and thereby gains a kind of a priori status. The construction appeals to a priori sources in the modal categories (possibility, actuality, necessity), and in the laws of mechanics (including a law of inertia, and a law of the equality of action and reaction), which Kant also claims to be a priori.

Friedman reaches the law of universal gravitation by three steps (see Friedman 1992a, 141-64; 1992b, 177-80). In the first step, we begin with the apparent relative motions of satellites in the solar system with respect to their primary bodies. These can be summarized and systematized by Kepler's laws, which are so far merely empirical regularities. Kant therefore subsumes them under the category of possibility, treating them as *possibly* true motions.

In step two, we assume that these apparent motions approximate to the true ones, provisionally subsuming them under the category of actuality. This allows us to apply the law of inertia, and conclude that because the observed motions are rotary rather than rectilinear, there must be an external cause responsible for their deviation from the straight path. The cause that fits the phenomena is a force of attraction centered on the primary bodies, and it now follows mathematically from Kepler's laws (plus inertia) that the attractive force in question obeys an inverse-square law.

In the third step, we apply the equality of action and reaction (Newton's third law) to reach the law of universal gravitation, and eventually an estimation of the relative masses of bodies in the system. If a body, A, attracts another body, B, the equality of action and reaction implies an equal and opposite attraction by B of A. This introduces the masses of particular bodies into our calculations, because the quantities of action equated are the products of each body's mass and acceleration. As Friedman points out, we can exploit this result to compare the masses of the primary bodies, but only under the additional assumption that inverse-square accelerations obtain directly between primary bodies themselves (and not, for example, only indirectly between primary bodies through some additional body, like an intervening aether). This amounts to saying that the inverse-square attractions are universal and immediate. Given the assumption of *universal* gravitation, we can use the third law to compare the masses of any two primary bodies by treating each as a satellite of the other. The inverse-square attraction of step two depends only on distance, and on the mass of the attracting body. In assessing the equal and opposite attractions between two primary bodies, the distance is of course the same (it is the distance between them), so the acceleration of each body (treated as a satellite) is directly proportional to the mass of the other. Thus, applying the action/reaction law to the actual accelerations constructed at stage two permits estimation of the bodies' relative masses, and ultimately the fixing of the center of mass of the whole system, which now provides a privileged frame of reference relative to which we can determine true, and not merely relative and apparent motions, thereby giving the contrast fully legitimate force for the first time.

We can then discharge the assumption of stage two, that the apparent motions approximate to true motions, by identifying actual motions relative to the center of mass frame we have constructed. These actual motions are just those determined by the laws of mechanics, plus the law of universal gravitation. So, for example, we can decide for the (approximate) actuality of the Keplerian over the Tychonian model of the motions of the solar system by showing which set of relative, apparent, i.e., *possible*, motions, *must be* the actual motions, given the laws of motion and universal gravitation.[5] We have therefore subsumed these motions under the category of necessity, and *only thereby* first guaranteed their actuality.

For our purposes, there are two key points to take away from Friedman's discussion. First, the construction that grounds the causal law of universal gravitation has an essential empirical basis. Without Kepler's laws, it would not be possible to calculate the inverse-square law in the first instance, for even if we presuppose an attractive force as essential to matter, the a priori law of inertia by itself cannot determine that force's actual mathematical form. Second, however, the a priori aspect of the construction is equally important. The main point is that we lack a solid grasp of the distinction between true and apparent motion at all, until we have subsumed motions under the category of necessity by *presupposing a priori* the universality of gravitation, as a condition for a meaningful theory of motion. We do not arrive at the actual motions of bodies through induction, which can yield only more and more apparent motions, never anything distinguishable as a true motion. Instead, we construct one set of motions as true, which, out of all the possible motions, *must have been* the actual ones, given the a priori mechanical laws and the assumption of universal gravitation. As it turns out, Kant's general defense of the causal principle in the Second Analogy will exploit an argumentative strategy strikingly similar to this one, identified by Friedman in the *Metaphysical Foundations*.

2. THE APRIORITY OF CAUSAL LAWS IN THE SECOND ANALOGY

The Second Analogy offers Kant's main defense of the apriority of causal claims, but the argument there turns on considerations from the general theory of cognition, rather than a detailed investigation of scientific laws. Kant's grounding of the concept of cause thus follows the *Critique*'s general strategy of arguing that even the most elementary cognitive achievements in fact *presuppose* the legitimacy of central metaphysical principles like causality. The principles cannot have empirical sources, Kant insists, for they are preconditions of all experience. In the account of causation, as in other arguments from his theory of cognition, Kant's reasoning turns crucially on the mechanism of *synthesis* of representations, which orders experience.

The Second Analogy concerns the representation of events. Where empiricist accounts of causation appeal to induction, inferring from patterns in perceived events to the cognition of causes, Kant claims that judgments of causation are already presupposed by the representation of temporally ordered events in the first place. Remarkably, Kant's solution to this problem of "time-determination" (A 177-8/B 220) shares the general structure Friedman identified in the Kantian grounding of Newton's theory: Kant insists that we arrive at an *actual* time order among our representations *only* by a construction that picks out, from a range of *possible* temporal orderings, the one determined as actual because it *must necessarily be* the order. As Kant puts it,

That something happens, therefore, is a perception that belongs to a *possible* experience, which *becomes actual* if I regard the appearance as determined in time ... in accordance with a rule. This rule, however, ... is that in what precedes, the condition is to be encountered under which the occurrence always (i.e., *necessarily*) follows. [A 200-1/B 245-6, my ital.]

Kant executes the broad strategy indicated in the quotation by starting from a premise that time "in itself" (B 233), or "absolute time" (A 200/B 245), cannot be perceived. That is, no fixed frame of temporal reference is directly given to perception, and we cannot simply read off the temporal order of perceptual contents empirically from temporal markers.[6] Rather, the time order in the content of our representations is something we have to *construct*, on the basis of what is directly given to us – i.e., we must "*show* what sort of combination in time pertains to the manifold in the appearances itself" (A 190/B 236; my ital.).

Of course, as Kant points out, the occurrence of distinct perceptions is already successive, since time is the form of our inner sense.[7] But succession's being the *form* of our representations does not yet bring it into the *content* of any one representation. Succession does contribute to the construction of temporal order, although not in the way one might at first think. It is the indispensable mechanism by which we first explicitly represent parts of the manifold separately, which is a *necessary* condition of the representation of distinct contents in a determinate temporal order (see A 99). But the mere succession of perceptions is *not sufficient* for time-determination of their contents.[8] Kant's thought turns on the distinction between a succession of representations, and a representation of succession. The mere succession of representations cannot yet provide a *representation of succession itself*, precisely because successive representation is our mechanism for expressly representing *all* distinctness, whether due to characteristically temporal diversity or not. Kant makes this point by a famous contrast of two examples: the perception of a house, and the perception of a ship being "driven [hinabtreiben] downstream" (A 192/B 237). In both cases, we represent the phenomena by means of a succession of perceptions, but

I still have to show what sort of combination in time pertains to the manifold in the appearances itself even though the representation of it in apprehension is always successive. Thus, e.g., the apprehension of the manifold in the appearance of a house that stands before me is successive. Now the question is whether the manifold of this house itself is also successive, which certainly no one will concede. [A 190/B 235]

So, for example, I represent the parts of the house successively (the roof, the second floor, the first floor, the front steps, etc.) without ever representing an event. By contrast, in the case of the ship being driven downstream, everyone will concede that we *do* represent succession (ship upstream, in front of me, downstream) as the special "combination in time [that] pertains to the manifold ... itself" – and this despite the fact that the states by which we represent the stages of this event are successive in the same way as the representations of the house.

What we need is some process that will connect, or *synthesize*, the partial contents in a way that determines their temporal order. I start by constructing a complex state that reproduces the successive perceptions together, but since I cannot simply *perceive* time markers, the temporal relations among the reproduced contents remain wholly undetermined. Every permutation ordering them is a *possible* time order. That is, since the contents presenting the ship in different positions are all included in the reproducing representation, I could represent their order as upstream, in front, downstream; *or* upstream, downstream, in front; *or* in front, upstream, downstream; etc. In perceiving an event, we somehow pick out one of these possible time orderings as the actual sequence, whereas in cases like the house, the temporal order remains undetermined among the various possibilities, and we do not perceive an event.

Kant's solution is given in the passage quoted above (A 200-1/B 245-6), which outlines his strategy. We successfully identify the actual order by synthesizing the partial contents *under a rule* that makes that sequence *necessary*: "this rule is always to be found in the perception of that which happens, and it makes the order of perceptions ... *necessary*" (A 192-3/B 237-8). As Kant puts it, in the absence of a direct perception of absolute time, "the appearances themselves must determine their positions in time for each other, and make this determination in the temporal order necessary" (A 200/B 245), and they can do this only in combination with such a rule of succession. Once we deploy the rule, we can straightforwardly construct a complex state that explicitly represents temporal succession. That state reproduces all the partial contents, and given the rule for their synthesis, the character of the contents themselves now determines their temporal ordering; their characteristics bring them under the rule, which mandates one particular order as necessary, and therefore actual.

We are now in a position to see how causality is a condition of the possibility of experience. We do succeed in perceiving events, and therefore in representing temporal succession. But, Kant has shown, it is a necessary condition for an explicit representation of succession that we reproduce prior perceptions of partial stages within the event through a rule governed synthesis imposing temporal order. The schematized concept of cause, according to Kant, is just "the succession of the manifold insofar as it is subject to a rule" (A 144/B 183). Thus, causation provides the general pattern of synthesis that makes event perception possible, and the rule governing such synthesis is precisely a causal rule, which therefore counts as a condition of the possibility of event perception.

This argument speaks immediately to our question about the apriority of particular causal laws. Return to the example of the ship, where our problem was how to pick out, from among all the possible temporal orderings, a unique sequence (viz., 1) ship upstream, 2) ship in front, 3) ship downstream) as the actual time order of the event. The Second Analogy claims that the task is achieved by means of an a priori causal rule. But what rule does the work, the general law of cause, or some particular causal rule? On reflection, it is clear that some particular rule is needed, for the general law that every event has a cause must be compatible with all sorts of events (including the ship's moving downstream, its moving upstream, its not moving at all, its seeming to move because I am moving, etc.). So as far as it can determine, either order must still be possible. Indeed, if matters were otherwise, we could not perceive events in which some cause moved the ship upstream. Thus, the rule that makes one particular time order among my several representations of the ship "*necessary*" (A 193/B 238) must be a special causal law – something like "Drifting boats move downstream."[9] Insofar, then, as Kant's argument shows that causal rules are a priori, because they are conditions of the very event perceptions which would pretend to serve as their inductive support, it demonstrates this of *particular* causal laws in the first instance, and only *thereby* of the general category of cause on which they depend. So Kant must be claiming apriority, at least "in a sense," for particular causal laws, all the way down to ones as specific as "drifting boats go down"!

3. CIRCULARITY, SYNTHESIS, NESTING, AND APRIORITY

Kant's conclusion faces an obvious circularity objection. Knowledge of a particular causal law, like "drifting boats go down," simply cannot be a precondition of the very ability to perceive downstream-moving boats, because the only possible justification for such a law would be empirical, and so would depend on prior and independent perceptions of moving boats.[10] The Second Analogy thus seems to leave us in an insoluble chicken-and-egg problem: obviously we cannot acquire particular causal laws except via experience of actual events, but we cannot perceive the events in the first place without already having the laws. Most scholars address the problem precisely by insisting on some version of the standard, strict separation between the a priori category and the particular laws.[11] They can then argue that only the category itself is a presupposition of event perceptions, and that those perceptions underwrite contingent, empirical causal laws. I want to propose an opposite approach to the puzzle.

Kant's result (viz., that causal rules are preconditions of event perceptions) can seem like a claim that event perceptions are actually indirect – somehow inferred on the basis of prior knowledge of the causal rule. I think this is the wrong way of seeing things. The relation between the causal law and the perception of an event should not be understood as an inference in either direction – either as the inductive inference of the law from perceptions, or the inference of some event perception from prior knowledge of a causal rule. Rather, the causal rule and the cognitive achievement of perceiving an event are related by the act of synthesis that makes them both possible at once. For Kant, my perception of the event "A-then-B" *just is* apprehending an A-state and a B-state under the rule that A's cause B's. As he puts it, if I say "that a sequence is to be encountered in the appearance," then this "*is to say as much* as that I cannot arrange the apprehension otherwise than in exactly this sequence," because it falls under a (causal) rule (A 193/B 238; my ital.). The very same act of synthesis produces at once the representation of the law that A's cause B's, and the representation of A-then-B.

This suggests a parallel between Kant's argument for dynamical categories like cause, and his earlier argument for the mathematical categories, which also depends on the claim that two cognitive tasks (one more basic, the other more intellectual) are accomplished by the same act of synthesis.[12] For example, Kant argues that the categories of quantity are preconditions of perception, because the perceptual synthesis of an extended thing, e.g., an elliptical table top, depends on an underlying synthesis of an elliptical space, according to the categories of quantity. According to Kant, the different levels of synthesis he thereby identifies are not separate acts, but aspects of the "same synthesis" (B 203). The conceptual and mathematical synthesis of an elliptical space is just an abstract structural part, or aspect, of the very empirical synthesis of the table top. Thus, the perceptual synthesis depends on the higher, conceptual synthesis not in the way a claim depends on a separate premise from which it is inferred, but as a concrete image depends on the abstract outline it fills in. Even as the abstract structure is *filled in* by concrete details, it *constrains* their shape. Thus, since the synthesis that produces cognition of this table is the *same one* that produces cognition of an ellipse (only now filled in with sensory content), the mathematical truths we can derive about the ellipse will be literally true of the table, and indeed, *a priori* true of it.[13]

Something analogous is true in causal synthesis. When I perceive an event, like the boat drifting downstream, this *just is* a synthesis ordering some reproduced perceptual contents under a causal rule with the form A-states produce B-states produce C-states. But a highly specific rule, like "Upstream drifting boats produce in-front-of-me drifting boats produce downstream drifting boats" can be understood as a detailed specification of a somewhat more general rule, "Drifting boats go down," which can in turn be seen (it is worth noting) as a special case of the gravity law, and ultimately of the maximally general law that events have causes. Here the higher causal laws encode abstract structural features of the same pattern of synthesis that constitutes the event perception. The specific rule derives distinctive cognitive force from the fact that it has an abstract structure homomorphic to the higher laws. Conversely, the more general laws are presuppositions of the specific one, and thence of event perception, in the sense that such particular laws, and even the perceptions they make possible, are specifications (via detailed empirical content) of the very structure provided by the higher laws. It is in this way, I think, that we should understand Friedman's thought that special causal laws are "nested within" (Friedman 1992b, 185) the general causal principle, as well as Kant's own talk of particular laws' one and all "standing under" the category, even though they "*cannot* be *completely derived*" from it (B 165).

Thus, no problem about circularity arises. We cannot derive causal laws from perceived events, nor event perceptions from causal laws, because the synthesis needed to construct either must simultaneously generate the other. Event perceptions still depend on causal rules, but the dependence is not like that of a conclusion resting on its premise, but like that of an image on the outline drawing it saturates. Kant makes this picture of cognitive synthesis explicit in the Schematism chapter, when he identifies a schema for synthesis as a kind of "monogram" (A 142/B 181) (in the sense of a simple outline drawing), which serves as a "rule of the synthesis" (A 141/B 180). The fully concrete synthesis, in turn, *fills in* the monogram by producing a more specific instantiation of the schematic form it represents. Since Kant opens the final chapter of the Transcendental Analytic claiming to have established that "The principles of pure understanding ... contain nothing but only the pure schema, as it were, for possible experience" (A 236-7/B 296), we can conclude that this story about cognitive synthesis is the basic model for his arguments of the Analytic, including, as we have seen in detail, the argument of the Second Analogy.

It still remains to say how the apriority of the category is related to the relative, or partial, or "in a sense" apriority possessed by the particular causal rule. Here I can offer only suggestions, out of which a more complete view might be built.

First, it is worth noting that many judgments containing empirical elements must be counted as a priori. Consider, a standard example of analytic truth, 'Bachelors are unmarried.' The analyticity of the judgment depends on the structure of its concepts; in Kant's preferred formulation, the predicate concept is "contained in" (A 6/B 10) the subject concept. But of course, one first forms the concepts through experience, and in this sense the judgment is not "pure" (A 20/B 34), or completely free of sensory content, despite its apriority. It is instructive here to compare Kant's synthetic a priori laws of mechanics, which instantiate the relational categories via the *empirical* concept of matter. For example, the law of inertia, "Every change of matter has an external cause" (*MFNS*, 4: 543), plugs the concept of matter into the general principle of causality ("Every alteration has a cause"), and thus directly instantiates the general principle by specifying alterations as changes of matter, and specifying the causes of those changes as external, because "matter, as mere object of outer sense, has no other determinations than those of outer relations in space" (*MFNS* 4: 543). While the concept of matter lends detailed content to the law, the basic connection among the judgment's component representations is still a priori, because it is just the link laid down by the causal principle. Of course, the judgment is not analytic, so the connection is not a matter of concept containment. Instead, the judgment connects its component representations according to a pattern that they must have if a unified experience is to be possible. In Kant's terminology, the judgment's *synthesis* is a priori, because its structure is just that of the category, only now made more specific by empirical detail.

Cognitive synthesis played a parallel role in the Second Analogy. There, actions of synthesis constructed an explicit representation of temporal order among perceptions, according to a pattern determining that some must precede and others follow. The pattern in question was just the schema of causation (succession according to a rule), made specific to fit the instant case. This suggests a general picture of the relation between the a priori and empirical elements of a causal judgment: the structural link that sustains the judgment is a priori, because directly due to the category; experience provides concepts that define a special instance of that structure.

Unfortunately, the simple picture cannot be the end of the story. As soon as we take even one additional step beyond the laws of mechanics, to the law of universal gravitation, we have already seen that empirical information must play a deeper role, since the basic relation established by the law (inverse-square attraction) is derivable only through essentially empirical considerations. The law of inertia would be compatible with attractions that took a variety of mathematical forms; we arrive at the inverse-square law only because the *phenomena* reveal that appearances obey Kepler's laws, which, under the assumption of inertia, mathematically entail an inverse square attraction. Kant is well aware of the point already in the Second Analogy:

Now how in general anything can be altered, how it is possible that upon a state in one point of time an opposite one could follow in the next – of these we have *a priori* not the least concept. For these acquaintance with actual forces is required, which can only be given empirically ... [A 206-7/B 252]

According to this passage, the problem is that without experience we still lack a *concept* of *how* something can be altered. Of course, in one sense we do have an a priori concept of how alteration happens, viz., the category of causation itself. So Kant must have something more specific in mind here: we lack a concept of the particular mechanism of the causal rule in a given case. This is why empirical "acquaintance with actual forces" can help.

Kant's picture, then, must be something like this. As I am starting out in cognitive life, my time-determination rules are actually *too specific*. For example, I have two perceptions, of A and B, and to construct an event representation, I synthesize them under a rule that A's produce B's, i.e., a rule entailing that total states just like A bring about total states just like B. The shapeless character of my concepts for A and B, which might be specific and "gerrymandered" to an arbitrary degree from the point of view of the overall system of experience, seriously limits the rule's power. For example, my concept of A-ness might be so specific as to apply to just this one state before me. As such, my beginners' causal rule fails to distinguish among features of A-type states and pick out the causally important ones. Such distinctions would afford me a better causal rule, applicable to other B-producers, which fail to share causally irrelevant features of A, and therefore do not fall under my rule that A's cause B's. But the discovery of such theoretically powerful causal factors amounts to empirical "acquaintance with actual forces." What I still lack with my beginners' causal rule, then, is precisely an adequate *empirical concept* (cf. A 206-7/B 252, cited above) of the nature or specific source of the causal power of the A-type state before me.

Given this picture, we are in a position to produce a more subtle allocation of responsibility between a priori and empirical elements in causal judgments. Very highly specific laws ("Every B-state has an A-state cause") are actually direct instantiations of the causal principle, much like the laws of mechanics were. The shapeless empirical concepts of the specific laws play same the instantiation role filled in mechanics by the concept of matter, even though the inadequacy of those concepts renders particular causal rules revisable in a way the laws of mechanics are not. Thus, for both the most particular and the most general laws of nature, causal synthesis is a priori "in a sense," because 1) it is homomorphic to the structure prescribed by the category, and 2) in virtue of having that structure, it is a precondition of some empirical cognition (e.g., in the theory of motion, in event perception). When I develop improved empirical concepts, however, they

afford the possibility of intermediate laws, whose precise formulation now depends on the appropriate empirical conceptualization of the actual causal factors.
In these intermediate laws, empirical considerations play a significant part in
specifying the nature of the connection between cause and effect, and thereby in
the essential constitution of the causal judgment. How, then, can these laws still
be a priori, even "in a sense"?

Kant sometimes states his conclusion in the Second Analogy like this: "as
soon as I perceive or anticipate that there is in this sequence a relation to the preceding state, from which the representation follows according to a rule, I represent something as an occurrence" (A 198/B 243). That is, we represent an event
either when we perceive (success verb) the rule of succession, or when we only
"anticipate" the rule. Intermediate laws are anticipated rules, and what we anticipate is precisely their a priori status. So when I make ordinary causal judgments,
deploying laws based on empirical concepts which may turn out not to be fundamental, my procedure is, in effect, to place a *bet* that my causal law suffices for
time-determination, because it (or something similar) will, in the limit, be derived from the category, as a precondition for achieving some part of the total
unification of experience. That is, I am betting that in the limit we will see how
to do for this law something analogous to Friedman's construction deriving the
gravity law, which allowed it to participate in apriority by revealing it as a precondition for applying causal analysis to true motions. The apriority of intermediate laws, then, rests ultimately on anticipated constructions that "reconstruct"
the actuality of the empirical generalizations by showing them to be necessary.

With the appeal to what the progress of empirical concept formation and
scientific reconstruction will achieve "in the limit," we enter the territory of
Kant's distinction between regulative and constitutive. If the category of cause
can prescribe its law to experience only through operations that are parasitic on
successful empirical concept formation, it becomes clear why Kant was forced to
count his dynamical categories as merely regulative, relative to the wholly constitutive mathematical categories. The dynamical categories "cannot ... construct" (A 179/B 222) the *actual content* of the empirical representations they
cover (not even the merely formal features of that content), but only the relations
of "real connection" among them. As it turns out, moreover, the specification of
those relations depends in part on the proper empirical conceptualization of the
content. By contrast, the mathematical categories do directly impose their structure onto intuition itself, regardless of the empirical concepts under which it is
classified.[14] In those cases, the only genuinely relevant concepts are the
mathematical ones, which *can* be *constructed*, rather than sought by empirical
means.

It is important to note, however, that the present version of the regulative /
constitutive distinction has its application *within* the realm of the understanding.
Under this regulative ideal, what we look forward to at the limit of theorizing is
not merely the more systematic organization of contingent judgments, but a set
of derivations that will show exactly how the causal laws we initially anticipated

with the help of induction are in fact *actual laws of nature*, because they are *necessary* as preconditions of the fulfillment of the understanding's own project of unifying experience. The picture sketched in this paper therefore needs completion, by a full discussion of the relation between this appeal to the regulative, and the different notion of the regulative deployed by reason, in order to produce the total system of philosophy, as Kant describes it in the appendix on the regulative use of the ideas of reason. Such an account would be the next step in developing the present interpretation of Kant's conception of a priori synthesis and the unity of experience.[15]

NOTES

* The audience at HOPOS 2000, where this material was presented, provided extremely helpful questions and comments. Thanks are also due to Richard Creath, Michael Friedman, Paul Guyer, Ken Reisman, Alan Richardson, Pat Suppes, and Ken Taylor for useful conversations about the issues discussed.

1. Citations to Kant employ the pagination of the *Akademie* edition, except for those to the *Critique of Pure Reason*, which use the standard A/B format to refer to the pages of the first (A) and second (B) editions. Other Kantian works are cited by abbreviations (for the *Prolegomena*, *Prol.*; for the *Metaphysical Foundations of Natural Science*, *MFNS*; for the *Critique of Judgment*, *CJ*). Except as noted, I follow the Guyer/Wood translation of the *Critique*, and the Hatfield translation of *Prol.* Translations from *MFNS* and *CJ* are mine.

2. See, e.g., Beck 1978, 134; Buchdahl 1969a, 1969b; Allison 1983, 1994; Guyer 1987, 1990. There is a hint of the non-standard reading that will be pressed below at Guyer 1998, 131, but the standard reading is reaffirmed at p. 138.

3. "Particular laws, because they concern empirically determined appearances, *cannot* be *completely derived* from the [a priori] categories, although they one and all stand under them. Experience must be added in order to come to know particular laws *at all* ..." (B 165). See also, "To be sure, empirical laws, as such, can by no means derive their origin from the pure understanding, just as the immeasurable manifold of the appearances cannot be adequately conceived through the pure form of sensible intuition" (A 127-8); and "We must, however, distinguish empirical laws of nature, which always presuppose particular perceptions, from pure or universal laws of nature, which, without having particular perceptions underlying them, contain merely the conditions for the necessary unification of such perceptions in an experience ..." (*Prol.* 4: 320).

4. For Kant's insistence that particular causal laws are empirical and contingent, see the passages quoted in the previous note. At the same time, however, Kant often emphasizes the necessity (and therefore apriority) of causal laws, at least in some sense. For example, "Even laws of nature, if they are considered as principles of the empirical use of the understanding, at the same time carry with them an expression of necessity, thus at least the presumption of determination by grounds that are *a priori* and valid prior to all experience" (A 159/B 198); and again, "Now there is no existence that could be cognized as necessary under the condition of other given appearances except the existence of effects from given causes in accordance with laws of causality. Thus it is not the existence of things (substances) but of their state of which alone we can cognize the necessity, ... in accordance with empirical laws of causality" (A 227-8/B 279-80). To the same general point, see also *Prol.* 4: 312, *MFNS* Pref. 4: 468-9, and *CJ* 5: 182-3. This apparent conflict of passages, and the interpretive puzzle it presents, has been noted by Buchdahl 1969b and Guyer 1987, 1990, in addition to Friedman 1992b.

5.   That is, we use the laws of motion and of gravitation to calculate the center of mass of the system, which turns out to be near the center of the sun, and this decides for the Keplerian model.
6.   As Guyer puts the point (at 1987, 244; 1998, 122), our representations do not include clocks in their content the way television sportscasts often do.
7.   Thus, "the manifold of appearances is always successively generated in the mind" (A 190/B 235).
8.   This formulation of the problem addressed by the Second Analogy takes its starting point from Guyer 1987 (esp. pp. 241-59) and Guyer 1989 and 1998, although I differ from Guyer on some details of the exposition of Kant's argument.
9.   Guyer 1998, 131 appears to appreciate this point. But he does not draw the same conclusions from it that I do (cf. p. 138).
10. For a clear and concise statement of this objection, which is widespread in the literature and has fairly deep historical roots, see Allison 1994, 302.
11. Buchdahl 1969a, 1969b, Allison 1983, 1994, and Guyer 1987, 1990 are good examples of the tendency.
12. The parallel is not surprising, I think, because a similar argumentative strategy is essential to Kant's procedure in the Transcendental Deduction of the categories itself, which applies to all the categories alike. Defending this intuition about the Deduction in detail is beyond the scope of the present paper, but is the subject of some work in progress.
13. I discuss this argument (from the Axioms of Intuition chapter) in my "Synthesis, Cognitive Normativity, and the Meaning of Kant's Question '*How are synthetic cognitions a priori possible?*'" (forthcoming).
14. This is why Kant claims that the mathematical categories are constitutive for intuition itself (and *thereby* for experience), whereas the dynamical categories are constitutive only for experience, but merely regulative for intuition:

> In the Transcendental Analytic we have distinguished among the principles of the understanding the *dynamical* ones, as merely regulative principles of *intuition*, from the *mathematical* ones, which are constitutive in regard to intuition. Despite this, the dynamical laws we are thinking of are still constitutive in regard to *experience*, since they make possible *a priori* the *concepts* without which there is no experience. [A 664/B 692]

Because they are constitutive in this way of experience, and of the "real connections" that build experience out of intuitive perceptions, the understanding's dynamical categories are to be distinguished from the strictly regulative principles of reason, under which interpreters in Buchdahl's line of influence would like to place empirical causal laws (thereby denying them apriority in any sense).
15. The fuller development of this distinction between the regulative operations of understanding and of reason, which is needed for a complete specification of the difference between the approach to causality in Kant presented here, and that of Buchdahl, et al., is the subject of some work in progress.

REFERENCES

Allison, Henry. (1983) *Kant's Transcendental Idealism*. New Haven, CT: Yale University Press.
Allison, Henry. (1994) "Causality and Causal Laws in Kant: A Critique of Michael Friedman." In P. Parrini, ed., *Kant and Contemporary Epistemology* (The Hague: Kluwer), 291-307.
Anderson, R. Lanier. (forthcoming) "Synthesis, Cognitive Normativity, and the Meaning of Kant's Question, '*How are synthetic cognitions a priori possible?*'," *The European Journal of Philosophy.*
Beck, L.W. (1978) "Six Short Pieces on the Second Analogy of Experience." In *Essays on Kant and Hume* (New Haven, CT: Yale University Press), 130-64.
Buchdahl, Gerd. (1969a) *Metaphysics and the Philosophy of Science, the Classical Origins: Descartes to Kant* (Cambridge, MA: MIT Press), ch. 8 (pp. 470-681).

Buchdahl, Gerd. (1969b) "The Kantian 'Dynamic of Reason,' with Special Reference to the Place of Causality in Kant's System." In L.W. Beck, ed., *Kant Studies Today* (La Salle, IL: Open Court), 341-74.
Friedman, Michael. (1991) "Regulative and Constitutive," *The Southern Journal of Philosophy* 30 (supp.): 73-102.
Friedman, Michael. (1992a) *Kant and the Exact Sciences.* Cambridge, MA: Harvard University Press.
Friedman, Michael. (1992b) "Causal Laws and the Foundations of Natural Science." In Guyer, ed. 1992.
Guyer, Paul. (1987) *Kant and the Claims of Knowledge.* Cambridge: Cambridge University Press.
Guyer, Paul. (1989) "Psychology and the Transcendental Deduction." In E. Förster, ed. *Kant's Transcendental Deductions: the Three 'Critiques' and the 'Opus Postumum'.* Stanford, CA: Stanford University Press.
Guyer, Paul. (1990) "Reason and Reflective Judgment: Kant on the Significance of Systematicity," *Nous* 24: 17-43.
Guyer, Paul, ed. (1992) *The Cambridge Companion to Kant.* Cambridge: Cambridge University Press.
Guyer, Paul. (1998) "Kant's Second Analogy: Objects, Events, and Causal Laws," in Kitcher 1998.
Kant, Immanuel. (1997 [1783]; *Prol.*) *Prolegomena to Any Future Metaphysics that may be Able to Come Forward as a Science.* Trans. Gary Hatfield. Cambridge: Cambridge University Press. Cited following the pagination of the *Akademie* edition.
Kant, Immanuel. (1998 [1781/1787]; A/B) *Critique of Pure Reason.* Trans. P. Guyer and A. Wood. Cambridge: Cambridge University Press. Citations are to the pagination of the first (A=1781) and second (B=1787) editions.
Kant, Immanuel. (forthcoming [1786]; *MFNS*) *Metaphysical Foundations of Natural Science.* Trans. M. Friedman. Cambridge: Cambridge University Press. Cited following the pagination of the *Akademie* edition.
Kant, Immanuel. (forthcoming [1790]; *CJ*) *Critique of Judgment.* Trans. P. Guyer. Cambridge: Cambridge University Press. Cited following the pagination of the *Akademie* edition.
Kitcher, Patricia, ed. (1998) *Kant's Critique of Pure Reason: Critical Essays.* Lanham, MD: Rowman and Littlefield.

Department of Philosophy
Stanford University
Stanford, CA 94305-2155
U.S.A.
lanier@csli.stanford.edu

Laura J. Snyder

Whewell and the Scientists: Science and Philosophy of Science in 19ᵀᴴ Century Britain *

Introduction

What is the relation between science and philosophy of science? Specifically, does it matter whether a philosopher of science knows much about science or is actually engaged in scientific research? William Whewell is an obvious person to consider in relation to this question. Whewell was actively engaged in science in several important ways, some of which have not been previously noted. He conducted research in a number of scientific fields, he devised new terminology for the new discoveries made by other scientists, and he frequently attempted to guide the experimental work of other scientists. Moreover, he was a philosopher of science who explicitly claimed to be inferring his methodological injunctions from an extensive study of the history of scientific work (the full title of his major work on methodology is *Philosophy of the Inductive Sciences, Founded Upon their History*). My recent study of his unpublished letters and notebooks indicates that the intimate relation between science and philosophy was always foremost in Whewell's mind; it was not a theoretical stance adopted after his philosophy was well-developed, as some have suggested. In this paper I will present an examination of the relation between Whewell's involvement in science and the writing of his philosophy of science as a springboard to consider the more general question (which, of course, cannot be answered in a paper of this length). Further insight into the relation between science and philosophy of science will be gained by contrasting Whewell's strong involvement in science with the lack of such involvement in the case of his philosophical antagonist, J.S. Mill.

Whewell as Scientist

I will start by examining Whewell's own direct involvement in science. Whewell's promise in mathematics and physics was recognized early on, and confirmed by his placing as second wrangler and second Smith's prizeman upon his

M. Heidelberger and F. Stadler (eds.), History of Philosophy and Science, 81–94.
© 2002 *Kluwer Academic Publishers. Printed in the Netherlands.*

graduation from Trinity in 1816. It was reasonable to expect that he would be an important contributor to science. In 1826 John Herschel even suggests that it is Whewell, rather than himself, who is destined for making important discoveries in science. In explaining why he will decline the Lucasian Professorship, Herschel tells Whewell "I am not destined (like great hard-headed thinkers like yourself) to make great inroads into great branches of human knowledge – but rather to loiter on the shores of the ocean of science and pick up such shells and pebbles as take my fancy for the pleasure of arranging them and seeing them look pretty." [1] These high hopes for his scientific career notwithstanding (or perhaps because of the pressure they exerted upon him) Whewell frequently worries that he will not contribute actively to science; thus he tells Herschel that "when I was admitted into the Royal Society I intended, if possible, to avoid belonging to the class of absolutely inactive members, and I have since been on the look out to find among the speculations that come my way some one which ought possibly be worth presenting to it." [2]

Whewell did ultimately become involved in a number of scientific researches. In 1825 he announced himself a candidate for the chair of mineralogy at Cambridge, and then traveled to Berlin, Freiburg and Vienna to study mineralogy and crystallography with the acknowledged masters of the field, including Möhs (indeed, Möhs invited Whewell to follow him to Vienna). [3] While in Germany he converted to the natural classification system in mineralogy from the artificial system then in vogue in England. [4] He had already published five papers in the area when he was appointed Professor of Mineralogy in 1828, and is still credited with making important contributions to the field of crystallography, by giving it a mathematical foundation. [5]

Whewell also followed up on two scientific programs proposed by his philosophical hero, Francis Bacon. In his *Novum Organum*, Bacon had suggested a pendulum experiment for measuring the mean density of the earth. [6] In June of 1826 Whewell and his friend (the future Royal Astronomer) G.B. Airy went to Cornwall, where they spent time in the Dolcoath copper mine comparing the effect of gravity on pendulums at the surface and at a depth of 1200 feet. [7] The experiment was less than a success. The entry on Whewell in the *Dictionary of National Biography* of 1885-90 rather laconically explains that "accidents to the instruments employed were ... fatal to the success of the experiments." [8] The fuller truth is that they kept dropping the pendulums down the shaft!! [9] Two years later they tried again, and published their findings.

More successfully, Whewell picked up on Bacon's interest in the tides. The major issue in research on the tides, for Bacon, Whewell, and current day researchers, is the behavior of the tides in deep water. [10] In an essay entitled "On the Ebb and Flow of the Sea" Bacon had postulated a northward progressing tide over the whole globe. [11] But he realized that there was not enough empirical data to support this supposition strongly, and he urged mariners to record tidal times along the coast of West Africa, among other places. [12] After Bacon, in the late 17th and 18th centuries, analysis of the tides became mainly mathematical, utilizing models which were overly-idealized, such as that of Laplace which postulated an ocean covering the whole globe. But by the 19th century even the scant empirical data which existed showed the inadequacy of such models. In 1830 John Lubbock, a friend of Whewell's, became interested in this topic when he published a summary of Bernoulli's famous essay on the tides.[13] Whewell's interest was stimulated by Lubbock's papers on the tides of 1831 and 1832. He helped Lubbock get grants from the newly-formed BAAS for his research, and suggested the term "cotidal lines" to him, to designate lines joining High Water Times.[14]

Eventually Whewell took over the task of mapping cotidal lines, presenting 16 papers to the Royal Society and several summary reports to the BAAS on the topic between 1833 and 1850. In a recent book on the history of tidal research, Cartwright points out that Whewell was inclined to support Bacon's theory of progressive waves, because most of the available data pointed to that theory: particularly the data for the British Channel and the German Ocean (as the North Sea was then called). The theory was also suggested by the northward progression of tide times along the Western Coasts of Africa and Europe. But as Whewell recognized that the data were still insufficient, he pushed for a large-scale research project of tidal observations. Aided by Captain (later Admiral) Beaufort, Hydrographer of the Navy, and with the support of the Duke of Wellington, Whewell managed to organize simultaneous observations of tides at 100 British coast guard stations for two weeks in June 1834. In June 1835 he organized three weeks of observations along the entire coast of N.W. Europe and Eastern America, including 101 ports in 7 European countries, 28 in America from the mouth of the Mississippi to Nova Scotia, and 537 in the British Isles, including Ireland.[15] This resulted in over 40,000 readings, as Whewell announces in his June 1835 report in the *Philosophical Transactions of the Royal Society*.[16]

Whewell himself reduced all this data and produced his definitive cotidal map in 1836. As Cartwright notes, "comparison with a properly computed tide map of the region shows that Whewell was right in every respect."[17] However, Whewell ultimately believed his work on the tides to be unsuccessful. By the time of his 1847 Bakerian Lecture, which was the 13th of his tide papers, Whewell had lost faith in the theory of progressive tides (in part because of Fitzroy's discussion on the tides in the appendix to volume 2 of his *Narrative*); and, in fact, the theory was eventually proven false. Whewell was also unsuccessful in his attempt to find a law of the diurnal inequality, i.e., for the fact that "the evening tide is higher than the morning tide at one point of the year, and lower at another."[18] Even after the large-scale observations had been examined, he is forced to admit that this law is "not yet precisely known."[19] Eventually he found that the diurnal effects obey no simple law; instead, they are extremely variable in their relation to the principle tides.[20] Indeed, this variation of diurnal effects is still not easily explained. Although he did not accomplish what he hoped to, Whewell was successful in his organization of a world-wide, international research project of tidal observation. In recognition of this work Whewell was awarded a gold medal by the Royal Society.[21] He was no Newton, nor even ultimately a Kepler, but I think it's fair to say an important Tycho Brahe of tidal research.

I turn now to Whewell's work as a provider of terminology for other scientists. I will concentrate on the work Whewell does in this regard for Michael Faraday.

WHEWELL AS TERMINOLOGICAL CONSULTANT

In April of 1834 Michael Faraday writes to Whewell, "My dear Sir, I am in a trouble which when it occurs at Cambridge is I understand referred by every body in the University to you for removal."[22] The "trouble" to which Faraday refers is the need for new terminology, and Whewell's expertise in this area was already at this point known even outside of Cambridge.[23] Besides coining the term "scientist,"[24] Whewell was famously involved with developing terminology for the new sciences and new developments within science during the middle of the 19th century. In the case of Faraday's initial request, new terminology was needed not because he had made a new discovery, but because of his suspicion that the old theory was false and his belief that the old terms implied this false theory.[25] The metallic plates at which the voltaic current enters or leaves a solution of an acid or salt were called "poles," because they were regarded as analogous to the poles of a magnet: i.e., they were believed to exert attractive or repulsive forces upon the particles of the substances lying between them.[26] But Faraday did not believe that the poles acted upon particles; he thought of them more as "gateways," entrances and exits where the current could enter and leave the solution.[27] In his letter to Whewell Faraday describes the problem and explains, "I want therefore names by which I can refer to [what are called the poles] without involving any theory of the nature of electricity." Of course, what he really wants are terms that do not involve the *old* theory of electricity.

Whewell suggests "anode" and "cathode."[28] Faraday is reluctant to accept these at first: he worries that "anode" will be taken to mean "no way."[29] Whewell points out, correctly, that "no body who has any tinge of Greek" could make this mistake. "Anodos" means "a way up" and "cathodos" "a way down."[30] He writes, "I am so fully persuaded that these terms are from their simplicity preferable to those you have printed, that I shall think it a misfortune to science if you retain the latter."[31] But it is not only the simplicity of the terms that recommends them to Whewell: it is also because the terms more fully express Faraday's theoretical point, as Whewell understands it. Whewell explains to Faraday that anode and cathode are preferable to the other terms Faraday suggests instead – Alphode and Betode; Galvanode and Voltaode – because they express not only difference, but *opposition*.[32] Nine days later Faraday writes that he has decided to accept these terms, noting that "I had some hot objections made to them here ... but when I held up the shield of your authority it was wonderful to observe how the tone of objection melted away." [33] Besides anode and cathode, Whewell invents a number of other terms for Faraday, including anions, cations, ions, and, later, diamagnetism and paramagnetism.

Whewell's contribution to scientific terminology was not limited to Faraday's work. He coined several terms still in use in the science of the tides in addition to cotidal lines, such as lunitidal interval and semi-menstrual inequality; (however his name for this science, "tidology," has not been retained). [34] He invented geological terms for Lyell, Murchison and Sedgewick, most famously "Uniformitarianism" and "Catastrophism" but also pliocene, miocene and eocene; Airy and DeMorgan, among others, also ask him for terminology: Airy for the names of instruments [35] and for certain weights and measures, [36] DeMorgan for logical terms. [37]

It is worth emphasizing that Whewell only provides terminology when he believes that he is fully knowledgeable about what is being named and the science involved. Thus, to one of Faraday's requests, Whewell responds: "I am glad to know that you are working, and come to a point where you want new words: for new words with you imply new things. I fear I am not sufficiently in possession of the bearings on the subject on which you are now employed to make my help of any use to you." [38] And to DeMorgan's request: "Indeed I am not sufficiently master of the subject to attempt it at the present, though I am much interested by the account which you give me of the progress of your views." [39] In his section on "The Language of Science" in the *Philosophy*, Whewell makes clear his methodological position on the creation of scientific terminology: "the only persons who succeed in making great alterations in the language of science, are not those who make names arbitrarily and as an exercise in ingenuity, but those who have much new knowledge to communicate." [40] Whewell's work as a terminological consultant, then, indicates that he believed himself to have some mastery of the scientific areas involved. Faraday and the others who consulted him seem to have concurred in this assessment.

WHEWELL AS EXPERIMENTAL GUIDE

Another interesting aspect of his intercourse with scientists – one which has not yet been noted – is that Whewell frequently attempts to guide their experimental work. Reading through all the correspondence, one constantly sees Faraday, Forbes, Lubbock and others being gently pushed by Whewell to perform certain experiments, make specific observations, and to try to connect their findings in ways interesting to Whewell. This activity of Whewell's also implies a deep understanding of the research being conducted by other scientists. Here again I will concentrate on the relation between Faraday and Whewell.

In his correspondence with Faraday, which consists in some 165 letters from 1833 to 1855, Whewell not only suggests terminology but also suggests lines of research and experiments to perform. In at least one case Faraday did pursue a line of research suggested by Whewell, leading him to important results. In 1835 Whewell tells Faraday "I am solicitous to hear of some connection being traced between voltaic action and crystalline structure or crystallising activity ... Crystalline forces are *polar* in their own way, and must, it would seem, be connected with your polar chemical forces."[41] Faraday responds: "I quite agree with you about the importance of the relation of crystalline and chemical polarity – but do not pretend to know any theory about it at present ..."[42] Whewell may well have been the first to suggest this connection to Faraday. His suggestion was based not only on his own interest in crystallography, but on his more general philosophical views: namely, on the belief that science consists in successive generalization of theories, and thus that theories are the more well-supported the more general they become and the more diverse phenomena they encompass.[43] (Indeed, Faraday's work exemplifies this central element of Whewell's philosophy of science; thus Whewell praises Faraday at one point for taking "another grand stride up the ladder of generalization."[44]) Although in 1836 Faraday can say nothing about the connection between crystalline and chemical polarity, two years later he attempts to answer Whewell's question. In his 14th Series of Researches, Faraday conducts an experiment attempting to discover the relation between crystalline polarity and chemical polarity, expressed in terms of electrical forces. The experiment fails.[45] In the second edition of the *Philosophy*, published in 1847, Whewell describes this unsuccessful experiment and notes (closely paraphrasing Faraday's conclusion) that "Although therefore we may venture to assert that there must be some very close connection between electrical and crystalline forces, we are, as yet, quite ignorant what the nature of the connexion is."[46] By the next year, Faraday has been more successful. He writes to Whewell, "You remember our talk about the connexion which ought to exist between crystalline and electrical forces. Well, it is beginning to appear ... I find that certain *crystals* are subject to Magnetic force..."; he then goes on to describe his experiment leading to this conclusion, noting that "the effect depends altogether upon the crystalline structure."[47] In the third edition of his *History*, Whewell adds a passage discussing the successful experiments of Plucker in 1847 and Faraday in 1848 finding the relation between magnetic force and crystalline form.[48]

This line of research was important to Faraday, because the discovery that the action of a magnet affects the plane of polarization of light in a crystal provided Faraday with more experimental evidence for the connection between light, electricity and magnetism, and reinforced his belief in the unity of all forces.[49] Thus in his Bakerian lecture, read on 7 December 1848, he notes the importance of his recent researches in connecting and unifying forces which had previously seemed to be quite diverse, such as magnetism, electricity, heat, chemical action, light, and crystallization. He expresses the hope that, at some point, the unification of forces will continue, and that these forces will be brought "into a bond of union with gravity itself" (p. 129).[50]

While Faraday was investigating this relation between magnetic and optical polarity, Whewell suggests a particular experiment to perform:

I do not know if you have made any further examination of the *amount* of the deviation of the plane of polarization in your experiment (2152) and especially of its amount for different colours. I believe one difficulty in the way of measuring this amount is the feebleness of the action which rotates the plane of polarization. It occurs to me that perhaps you may find some good suggestions, as to the mode of making such experimental measures in Biot's researches on a similar subject, the circular polarization of fluids ... [51]

Whewell goes on to describe Biot's experiment and give Faraday the reference to his published account. Faraday responds by noting that "My course of study, apparatus, and means do not enable me to supply the careful measures which as you say are so much to be desired but I have fallen upon a method of facilitating their estimation." [52]

The fact that Faraday pays attention to the suggestions of Whewell, and indeed often asks for mathematical and experimental advice, clearly indicates that Faraday does not view Whewell as one of those "amateurs" or "volunteers" he complains about as "embarrassments generally to the experienced philosopher." [53] Indeed, some days after announcing to Whewell his discovery of the connection between crystalline and chemical polarity, Faraday invites him to address the Royal Institution on "The Idea of Polarity," saying "the subject is your own ... and I anticipate much information and help from your discourse on it." [54]

A PHILOSOPHY OF SCIENCE, FOUNDED UPON ITS HISTORY

It is obvious, therefore, that Whewell was quite active in science in several important ways and therefore quite knowledgeable about scientific practice. But what is the relation between this involvement in science and Whewell's writing of his philosophy of science? Interestingly, Whewell does not consider knowledge of current science to be sufficient for developing a philosophy of science. It is also necessary to study the history of science: thus, he writes his *History of Inductive Sciences* (1837) before the *Philosophy* (1840), claiming that any philosophy of the sciences must be "founded upon their history." Some commentators have claimed that, on the contrary, Whewell developed an a priori philosophy of science first and then shaped his *History of the Inductive Sciences* to conform to his own view. [55] To a limited extent this is no doubt true. From his days as an undergraduate at Trinity College Whewell saw as his project the advancement of the inductive method in the sciences. As undergraduates, Whewell, Richard Jones and John Herschel formed a circle of like-minded methodologists devoted to the propagation of a renovated type of Baconianism. [56] So Whewell from the very beginning had an inductive methodological view, which influenced his writing of the *History* in one important sense: by leading him to the view that learning about scientific method must be inductive and therefore historical. As he writes to Jones in 1831, "I do not believe the principles of induction can be either taught or learned without many examples." [57]

Examples, then, are needed to fill out the details of this broadly inductive view, and they are to come both from knowledge of current science and knowledge of the history of science. That these examples are to come from both sources is indicated in numerous letters as well as in Whewell's early induction notebooks. Whewell's earliest attempt at a draft of a work on induction appears in a notebook dated 28 June 1830.[58] In this notebook and in several which follow over the next three to four years there are many notes on recent discoveries in science, as well as citations from contemporary scientific works in which scientists express a view of proper scientific method.[59] Yet there are also numerous entries describing the histories of various scientific fields. These entries are interwoven with Whewell's early thoughts on an inductive philosophy of science. In one of these notebooks alone, Whewell takes reading notes on Davy's *Elements of Chemical Philosophy*, Gilbert's *De Magnete*, Brewster's book on Newton, as well as works by Cuvier, Copernicus, Galileo, da Vinci, and Harvey; he describes recent discoveries in Optics by Biot, Young, Fresnel, Arago, and Airy; and he discusses historical material, giving details about the work of Aristotle, Euclid, Plato, Alhazen, Newton, Roger Bacon, Brahe, Kepler Huyghens, and Fraunhofer. Whewell then uses this examination of the history of Optics and current research in the field to outline the "Steps of the Induction of the Theory of Light." [60] In a notebook dated 1831-32, a discussion of the use of conceptions in induction includes notes on the scientific work of Archimedes, Pascal, Aristotle, Descartes, Mersenne, Galileo and Torricelli.[61] In another notebook, dated December 1833, Whewell mentions Herschel's *Preliminary Discourse*, pointing out one problem with his friend's work: namely, that "we do not here find the view of physical science to which we hope to be led: – that if its history and past progress be rightly studied we shall acquire confidence in truth of all kind ..." [62] Moreover, in several letters to Jones in 1834 Whewell describes himself as eager to get to his philosophy of science but determined to finish the history first.[63]

It is important to note that, while writing both the *History* and the *Philosophy*, Whewell attempts to ensure he has a real understanding of the scientific work he is describing. Although Robert Brown's quip about Whewell ("yes, I suppose that he has read the prefaces of very many books" [64]) is understandable, Whewell does more than simply read: as we have seen, he actively practices science. Further, he consults scientists about their own discoveries or those of others throughout the history of their respective fields, and sends proof sheets of the *History* to them for their approval. For instance, Whewell asks his friend Airy to look over his section on the history of Astronomy and to send references to French works on the polarization of light.[65] He sends queries on physics to Forbes[66] and on anatomical science to another friend, the anatomist Richard Owen.[67] He asks Faraday to tell him whether there are any errors describing his Researches in the first edition of the *History*, while he is revising it for the second edition.[68] (Faraday responds by saying there are no errors that he can see.)

So, while I think it would be an overstatement to say that Whewell had no ideas about philosophy of science until after he completed all three volumes of the *History*, it would also be a vast understatement to suggest that the *History* was written to conform to a fully fleshed-out a priori methodological position. Knowledge of both current scientific practice and the history of science are important to Whewell in developing his philosophy of science. We can better understand the situation by contrasting the case of Whewell with that of John Stuart Mill.

THE CONTRAST WITH MILL

Although Mill's infamous education included some science (he seems to have been particularly attracted to chemistry and botany), Whewell's dismissal of him, in a letter to Herschel soon after the publication of *System of Logic*, is not unjust. "Though acute and able," Whewell wrote of Mill, "he is ignorant of science."[69] Indeed Mill admits to being fairly ignorant of the physical sciences. Nevertheless, he appears to agree with Whewell's claim about the necessity of inferring a philosophy of science from knowledge of its practice and history. For example, in the Preface to *System of Logic*, he explains that "on the subject of Induction, the task to be performed was that of generalising the modes of investigating truth and estimating evidence, by which so many important and recondite laws of nature have, in the various sciences, been aggregated to the stock of human knowledge."[70] He thus implies that he has inferred his theory of induction from the method by which truths have, in the past, been discovered. He makes a similar claim at the start of Book Three ("Of Induction").[71] Mill also suggests the necessity of inferring a philosophy of science from science and its history when, in his *Autobiography*, he expresses a debt to Whewell and Herschel, explaining that their books provided him with something needed before he could complete his work: namely, a comprehensive view of "the generalities and processes of the sciences."[72] Indeed, Mill goes so far as to claim that, because of this, without Whewell's *History* a "portion of this work would probably not have been written."[73]

Such comments notwithstanding, it is clear that Mill did not infer his philosophy of science from knowledge of science and its history. Mill wrote *System of Logic* without knowledge of most of the examples included in the published version. The majority of scientific examples were added to the completed manuscript, on Mill's request, by his friend, the logician and scientist Alexander Bain. Bain describes his reaction to seeing the completed manuscript in the following way:

The main defect of the work ... was in the Experimental Examples. I soon saw, and he felt as much as I did, that these were too few and not unfrequently incorrect. It was on this point that I was able to render the greatest service. Circumstances had made me tolerably familiar with the Experimental Physics, Chemistry and Physiology of the day, and I set to work to gather examples from all available sources.[74]

Thus Mill could not have inferred his philosophy of science from these examples. And if he had inferred his philosophy from the examples given by Herschel and Whewell, surely Mill would have included more of these in his book (besides Herschel's example of Wells' research on dew); he would not then have needed Bain to provide new ones.

Even with Bain's help, Mill finds relatively few examples of his philosophy in the history of science; moreover, many of these examples are, as Whewell points out, either incorrect or inappropriate. Thus, Whewell rejects the Wells example on the grounds that it is not really an original discovery, but rather a deduction of particular phenomena from already established general principles.[75] Many of Mill's other examples are taken from Liebig's researches on physiological chemistry, which, as Whewell notes, are so recent that "the most ... sagacious physiologists and chemists cannot yet tell which of them will stand as real discoveries ..." (so it is not clear that they *are* examples of truths discovered by Mill's methods).[76] Indeed, Whewell's criticisms of Mill do not focus on the question of whether Mill's philosophy of science actually was inferred from science and its history, but rather on the fact that his methods are not applied to a large number of appropriate historical cases. He complains that "if Mr. Mill's four methods had been applied by him ... to a large body of conspicuous and undoubted examples of discovery, well selected and well analysed, extending along the whole history of science, we should have been better able to estimate the value of these methods."[77] On Whewell's view, even though Mill did not infer his philosophy from science and its history, it should be possible, if his methods are valid, to find examples of their use in actual scientific practice throughout the history of science.

I think this sheds some light on what Whewell takes to be the important relation between science and philosophy of science. What is important according to Whewell is not whether a philosophy of science is, in fact, inferred from knowledge of past and present scientific practice, but rather, whether a philosophy of science is *inferable from* such knowledge. Any valid philosophy of science must be shown to be exemplified in actual scientific practice throughout the history of science. Thus, even though Bacon did not infer his philosophy of science from a study of the history of science – and indeed, as Whewell notes, he did not have much history of modern science behind him – his philosophy can still be found to be valid if it is shown to be exemplified in the science that has come since him. (To a great extent this is the project Whewell himself has taken on.)[78] Even if Whewell did not, in fact, develop his philosophy of science only after his study of the history and practice of science was completed, he shows us in his works – through numerous apt examples – that his philosophy has been embodied in the practice of science throughout its history.

So, according to Whewell, a philosopher of science need not actually be active in science, as Whewell was, but must know enough about current scientific practice and the history of science to judge whether his or her precepts have been exemplified in science throughout its history. Mill, as we have seen, lacked this knowledge. Did it matter? I have argued elsewhere (and there is no space for reiteration here) that Whewell's philosophy of science, compared to Mill's, more adequately accounts for the practice of scientists.[79] I think it is fair to conclude that Whewell's greater knowledge of scientific practice and the history of science had something to do with his developing the more adequate philosophy of science.

NOTES

*   Some early thoughts on the last two sections of this paper were presented at a group residency on "The Interpenetration of Science and Philosophy of Science" held at the Villa Serbelloni in Bellagio, Italy July 1998. I thank the other participants – especially David Hull, Jon Hodge and Bob Richards – for their comments and stimulating discussion, and the Rockefeller Foundation for hosting the conference. The U.S. Information Agency and the J. William Fulbright Foreign Scholarship Board provided generous support of research, for which I am very grateful. For help in preparing this paper I am also grateful to Arthur Gianelli and Giovanni Giorgini. Finally, I thank the Master and Fellows of Trinity College for their permission to quote from the Whewell Papers.

1.   17 August 1826; Whewell Papers (hereafter WP) Add.ms.a.207 f.12.
2.   15 October 1823; WP O.15.47 f.122.
3.   See letter to Hugh James Rose, 15 August 1825, WP R.2.99 f.125.
4.   See William Whewell, *An Essay on Mineralogical Classification and Nomenclature*. Cambridge: Cambridge University Press 1828; and Harvey W. Becher, "Voluntary Science in Nineteenth-Century Cambridge University to the 1850's", in: *British Journal for the History of Science*, 19, 1986, pp. 57-87.
5.   See William Whewell, "A General Method of Calculating the Angles Made by Any Planes of Crystals, and the Laws According to Which they are Formed", in: *Philosophical Transactions*, 115, 1825, pp. 87-130; Becher *Ibid*, p. 61; Michael Ruse, "William Whewell: Omniscientist", in: Menachem Fisch and Simon Schaffer (Eds.), *William Whewell: A Composite Portrait*. Oxford: Clarendon Press 1991, p. 99; and H. Deas, "Crystallography and Crystallographers in England in the Early 19th Century: A Preliminary Survey", in: *Centaurus*, 6, 1959, pp. 129-48.
6.   See William Whewell, *On the Philosophy of Discovery*. New York: Burt Franklin 1971 (originally published 1860), p. 141.
7.   For a description of this experiment see William Whewell, *Account of Experiments Made at Dolcoath Mine, in Cornwall, in 1826 and 1828*. Cambridge: Cambridge University Press 1828. See also Richard Yeo, *Defining Science: William Whewell, Natural Knowledge, and Public Debate in Early Victorian Britain*. Cambridge: Cambridge University Press 1993, p. 53.
8.   Leslie Stephen, "Whewell", in: Leslie Stephen and Sidney Lee (Eds.), *Dictionary of National Biography*, in 21 volumes. London: 1885-90, vol. 20, p. 1366.
9.   See Whewell, *Account of the Experiments Made at Dolcoath Mine, op. cit.*, p. 8.
10.   See David E. Cartwright, *Tides: A Scientific History*. Cambridge: Cambridge University Press 1999, pp. 2-3.
11.   Francis Bacon, *The Works of Francis Bacon*. J. Spedding, R.L. Ellis, and D.D. Heath (Eds.), in 14 volumes. London: Longman and Co. 1857-61; see vol. V, pp. 443-58.
12.   See Cartwright, *op. cit.*, pp. 26-8.
13.   See Cartwright, *op. cit.*, p. 46.

14. In his 1832 report to the BAAS Lubbock refers to "a series of points which form the crest of the tide-wave, ... which I have called, at the suggestion of Mr. Whewell, cotidal lines." Cited in Cartwright, *op. cit.*, p. 111. See also William Whewell, "Essay Towards a First Approximation to a Map of Cotidal Lines," in: *Philosophical Transactions*, 123, 1833, pp. 147-236.

15. Whewell notes that the observations were made from June 8-28, and occurred in 28 places in America, 7 in Spain, 7 in Portugal, 16 in France, 5 in Belgium, 18 in the Netherlands, 24 in Denmark, 24 in Norway, 318 in England and Scotland, and 219 in Ireland. See William Whewell, "Researches on the Tides – 6th series. On the Results of an Extensive System of Tide Observations made on the coasts of Europe and America in June 1835", in: *Philosophical Transactions*, 126, 1836, pp. 289-341. See also Cartwright, *op. cit.*, pp. 112-14.

16. Whewell, "Researches on the Tides – 6th Series," *op. cit.*, p. 291.

17. Cartwright, *op. cit.*, p. 116.

18. Whewell, "Essay Towards a First Approximation to a Map of Cotidal Lines," *op. cit.*, p. 221.

19. Whewell, "Researches on the Tides – 6th Series," *op. cit.*, p. 306.

20. For example, Whewell found that diurnal age (the time delay of maximum inequality after maximum lunar declination) varies from nearly zero on the Atlantic coast of North America to 3-6 days on European and North Africa coasts (see Cartwright, *op. cit.*, p. 116).

21. William F. Cannon, "William Whewell: Contributions to Science and Learning", in: *Notes and Records of the Royal Society*, 19, 1964, pp. 176-91; see p. 183.

22. WP O.15.49 f.4.

23. Faraday was aware of Whewell's interest in scientific terminology from at least 1831 when, as editor of the *Journal of the Royal Institution of Great Britain*, Faraday had accepted and published Whewell's article "On the Employment of Notation in Chemistry." See S. Ross, "Faraday Consults the Scholars: The Origin of the Terms of Electrochemistry", in: *Notes and Records of the Royal Society*, 16, 1961, pp. 187-220; p. 196.

24. See S. Ross, "'Scientist': The Story of a Word", in: *Annals of Science*, 18, 1962, pp. 65-85.

25. See Brian Bowers, *Michael Faraday and Electricity*. Sussex, U.K.: Priory Press 1974, p. 64; and Geoffrey Cantor, David Gooding, and Frank A.J.L. James, *Faraday*. London: Macmillan 1991, p. 62.

26. See Ross, "Faraday Consults the Scholars," *op. cit.*, p. 191.

27. Cantor, Gooding and James, *Ibid.*, p. 61.

28. See letter of 25 April 1834, in Isaac Todhunter, *William Whewell, D.D.: An Account of His Writings*, in two volumes. London: Macmillan 1876, vol. II, pp. 178-81.

29. 3 May 1834; WP O.15.49 f.5.

30. On why it is preferable to use Greek or Latin terms as the origin of new technical terms in science, see William Whewell, *Novum Organon Renovatum*. London: John W. Parker 1858, p. 319.

31. 5 May 1834, in Todhunter, *Ibid.*, vol. II, pp. 181-3.

32. See letter from Whewell, 6 May 1834; in Frank A.J.L. James (Ed.), *The Correspondence of Michael Faraday*, in four volumes. London: The Institution of Electrical Engineers 1996, vol. II, p. 184.

33. 15 May 1834; WP O.15.49 f.6.

34. See Whewell, *Novum Organon Renovatum, op. cit.*, p. 284, on the need for new names for new sciences: "The subject of the Tides is ... destitute of any name which designates the science concerned about it. I have ventured to employ the term *Tidology*, having been much engaged in tidological researches."

35. 6 November 1850; WP Add.Ms.a.200 f.86.

36. 3 February 1841; WP Add.Ms.a.200 f.38.

37. See letter from Whewell, 29 July 1857; WP O.15.47 f.21.

38. 12 November 1844; WP O.15.47 f.152; see also 14 October 1837, in James, *Ibid.*, p. 464.

39. 29 July 1857; WP O.15.47 f.21.

40. Whewell, *Novum Organon Renovatum, op. cit.*, p. 293.

41. 11 December 1835; WP O.15.49 f.149. During this week in 1835, Whewell was obviously quite interested in crystalline polarity; he also sends a letter to his friend the physicist J.D. Forbes while Forbes is working on the polarization of heat, admonishing him "You must recollect that

now that you have got polarisation, you must make out the bearing of crystallisation upon it" (8 December 1835; WP O.15.47 f.47).

42. 9 January 1836; WP O.15.49 f.9.

43. For more on this aspect of Whewell's philosophy of science see Laura J. Snyder, "Renovating the *Novum Organum*: Bacon, Whewell and Induction", in: *Studies in History and Philosophy of Science* 30, 4, 1999, pp. 531-557.

44. 20 November 1845; WP O.15.47 f.154.

45. Michael Faraday, *Experimental Researches in Electricity*, in three volumes bound as two. New York: Dover Publications 1965 (originally published 1855), Volume I, pp. 538-41.

46. William Whewell, *Philosophy of the Inductive Sciences, Founded Upon Their History*, 2nd edition, in two volumes. London: John W. Parker 1847, vol. I, p. 368.

47. 7 November 1848; WP O.15.49 f.20; see also 8 December 1848; WP O.15.49 f.24.

48. William Whewell, *History of the Inductive Sciences from the Earliest to the Present Time*, 3rd edition, in three volumes. London: John W. Parker 1857, pp. 534-5.

49. See Cantor, Gooding and James, *Ibid.*, p. 83 and Nancy J. Nersessian, *Faraday to Einstein: Constructing Meaning in Scientific Theories.* Dordrecht: Kluwer 1984, p. 54.

50. Michael Faraday, *Ibid.*, Volume II, p. 129. Importantly, these researches eventually lead Faraday to his endorsement of physical lines of force (see L. Pearce Williams, *Michael Faraday.* New York: Basic Books 1964, pp. 436-49.)

51. 7 August 1846; WP O.15.47 f.156.

52. 10 August 1846; WP O.15.48 f.30. Whewell suggests another experiment to Faraday in January 1846 (see letters of 19 January 1846; WP O.15.47 f.55 and 20 January 1846; WP O.15.47 f.18). He also suggests experiments on the polarization of heat to his friend Forbes (see 8 December 1835; WP O.15.47 f.47). Forbes writes to Whewell asking if Whewell has any suggestions for how to set up an experiment Forbes wishes to perform (see 8 February 1840; WP Add.Ms.a.204 f.39).

53. Cited in R.A.R. Tricker, *The Contributions of Faraday and Maxwell to Electrical Science.* Oxford: Pergamon Press 1966, p. 33. See, for example, 13 December 1836; WP O.15.49 f.10, where Faraday asks Whewell for his assessment of the mathematical reasoning in a paper by Mossotti; moreover, he treats Whewell as a confident (begging him not to divulge what he is working on until his papers appear) in letters of 25 October 1837; WP O.15.49 f.12 and 22 November 1845; WP O.15.49 f.16).

54. 13 November 1848; WP O.15.49 f.21.

55. See, for example, Marion Rush Stoll, *Whewell's Philosophy of Induction.* Lancaster, PA: Lancaster Press 1929 and E.W. Strong, "William Whewell and John Stuart Mill: Their Controversy Over Scientific Knowledge", in: *Journal of the History of Ideas*, 16, 1955, pp. 209-31.

56. See Snyder, "Renovating the *Novum Organum*," *op. cit.*

57. 25 February 1831; WP Add.ms.c.51 f.99.

58. In this notebook Whewell claims that in order to judge the methodology of Bacon, it is necessary "to show how this method has been exhibited and exemplified since it was first delivered." In order to do so it is necessary to discuss the history of science: this may explain why Whewell put aside this draft and began working on his history of science. See WP R.18.17. f.12; for quotation see p. xv.

59. For example, in an entry dated July 1831 there is a quote from Dalton followed by a comment by Whewell: "[this is] an excellent description of induction and a good proof of the difficulty in presenting things inductively" (WP R.18.17 f.15, p. 40). In the induction notebook of 1830 there is an extensive discussion of recent discoveries in Geology (WP R.18. 17 f.12; pp. xxiv-xxv and 1ff).

60. See notebook WP R.18.17 f.13; the notebook is undated but it is headed "Induction IV;" the other induction notebooks are dated from 1826-1832.

61. WP R.18.17. f.15.

62. WP R.18.17 f.8.

63. See 27 July, 5 August and 6 August 1834; WP Add.ms.c.52 f.173, f.174 and f.175.

64. Reported by Charles Darwin in his "Autobiography", in: G. de Beer (Ed.), *Autobiographies of Charles Darwin and Thomas Henry Huxley.* Oxford: Oxford University Press 1974, p. 61.

65. See letters from Airy, 11 October 1856; WP Add.ms.a.200 f.114, and 21 April 1831; WP Add.Ms.a.200 f.9.
66. 19 February 1840; WP O.15.47 f.51a.
67. See letters of Owen, February 11 and 19, 1839; WP Add.Mss.a.210 f.55 and f.56.
68. See 7 August 1846; WP O.15.49 f.56.
69. Todhunter, *Ibid.*, vol. II, p. 315.
70. John Stuart Mill, *A System of Logic Ratiocinative and Inductive: Being a Connected View of the Principles of Evidence and the Methods of Scientific Investigation*, 8ᵗʰ edition, in *Collected Works of John Stuart Mill*, volumes VII and VIII. Toronto: University of Toronto Press 1973 (originally published 1872), p. cxii.
71. Mill, *A System of Logic, op. cit.*, p. 283.
72. John Stuart Mill, *Autobiography*, in *Collected Works of John Stuart Mill*, volume 1. Toronto: University of Toronto Press 1981 (originally published 1873), p. 215.
73. See Mill, *System of Logic, op. cit.*, p. cxiii. This raises an interesting question regarding Mill's use of Whewell's descriptions of science. As David Hull has noted (personal correspondence), it seems strange that Mill could use Whewell's descriptions of science in developing a philosophy of science quite different than Whewell's. However, Mill's somewhat naive belief that he could use Whewell's descriptions while rejecting his theory is actually consistent with Mill's (equally naive) denial of Whewell's claim that description/observation is theory-laden; this is seen in their debate over Kepler's discovery of the Martian orbit, which Mill insists consisted in a mere description requiring no conceptual input (for more on this point, see Laura J. Snyder, "The Mill-Whewell Debate: Much Ado About Induction", in: *Perspectives on Science* 5, 1997, pp. 159-98.
74. Alexander Bain, *John Stuart Mill: A Criticism, with Personal Recollections.* London: Longmans, Green and Co. 1882, p. 66.
75. See Whewell, *Philosophy of Discovery, op. cit.*, p. 267.
76. See Whewell to Jones, 7 April 1843; WP Add.ms.c.51 f.227; and Whewell, *Philosophy of Discovery, op. cit.*, pp. 265-6.
77. Whewell, *Philosophy of Discovery, op. cit.*, pp. 264-5.
78. Whewell sees this as his project from the beginning; see the 1830 notebook (Notebook entry dated 28 June 1830; R.18.17. f.12, p. xv).
79. See Laura J. Snyder, "Discoverers' Induction", in: *Philosophy of Science* 64, 1997, pp. 580-604 and Laura J. Snyder, "The Mill-Whewell Debate: Much Ado About Induction", *op. cit.*

Dept. of Philosophy
St. John's University
8000 Utopia Parkway
Jamaica, NY 11439
U.S.A.
snyderl@stjohns.edu

MARK VAN ATTEN

BROUWER'S ARGUMENT
FOR THE UNITY OF SCIENTIFIC THEORIES[*]

1. INTRODUCTION

The Dutch mathematician and philosopher L.E.J. Brouwer (1881-1966) is well known for his ground-breaking work in topology and his iconoclastic philosophy of mathematics, intuitionism. What is far less well known is that Brouwer mused on the philosophy of the natural sciences as well. Later in life he also taught courses in physics at the University of Amsterdam.

Some of his ideas on the natural sciences found a place in his 1907 dissertation, 'Over de grondslagen der wiskunde' ('On the foundations of mathematics'). Brouwer's original plan for his thesis was to present a highly integrated philosophy of mathematics pure and applied. This can be seen from the chapter titles he proposed in a letter to his thesis adviser, Diederik Korteweg:

I. The construction of mathematics

II. The 'genesis' of mathematics related to experience

III. The philosophical significance of mathematics

IV. The foundation of mathematics on axioms

V. The value of mathematics for society

VI. The value of mathematics for the individual

Korteweg made no secret of the fact that he had seen Brouwer apply his talents to a purely mathematical problem, but recognized how urgent the foundational problems were for his student. Still, at times Korteweg warned Brouwer not to err too far away from mathematics proper, and probably this is the reason why most of Brouwer's thoughts on the natural sciences were not included in the final version of his thesis. (For similar reasons, Korteweg made Brouwer remove passages on mysticism and mathematics; otherwise it would have contained gems from his notebooks such as 'One could see as the goal of one's life: abolition and delivery from all mathematics'.)

While crafting this dissertation, Brouwer had much interaction with Korteweg. They met and corresponded frequently. Part of this correspondence has been preserved, in Korteweg's *Nachlaß*. It is in those letters that we find one

95

M. Heidelberger and F. Stadler (eds.), History of Philosophy and Science, 95–102.
© 2002 *Kluwer Academic Publishers. Printed in the Netherlands.*

of Brouwer's most interesting excursions into the philosophy of science. The issue at hand is also briefly mentioned in the 'Rejected parts of Brouwer's thesis', published in 1979 by Van Stigt, but elaborated upon only in a letter dated November 13, 1906. The letter has been translated by Van Stigt and again, in part, by Van Dalen, but the particular argument has not been commented on yet.[1]

This argument is designed to answer the following question: assuming that natural science derives from many different phenomena affecting our various sensory organs in widely different ways, isn't it a surprising fact that these phenomena can all be brought together in one, or a few, mathematical systems?

Brouwer agrees that this at first strikes one as miraculous. But, according to him, the simple explanation is that theories in the natural sciences are in fact concerned with the projections of these phenomena on our measuring instruments, and these instruments are very similar to one another in so far as they are all constructed from solid bodies. Hence the ways in which these instruments can react to impingements exhibit some strong invariants. Brouwer concludes from this that the unity and similarity of these theories derive from their ultimately being founded on the one theory of movements of solid bodies (these transformations Brouwer calls 'the rigid group').

2. THE ARGUMENT AND ITS CONTEXT

Let us now look at the argument.

Brouwer seems to waver a bit between a stronger and a weaker claim. The weaker claim is that laws in different areas of physics exhibit similarities or are subject to the same constraints. We will see that Brouwer mentions conservation laws as an example of such constraints. The stronger claim is that theories in different areas of physics are not only subject to shared constraints but can always be unified.

In a rejected part of Brouwer's dissertation, we find first the stronger and then the weaker claim:

Is it surprising that not only do we succeed in observing sequences which repeat themselves again and again, but that so many groups of phenomena affecting our naive senses in totally different ways can be brought together under a few general aspects which are covered by simple constructible mathematical systems? This really would be a miracle, were it not for the simple fact that the physicist concerns himself with the projections of the phenomena on his measuring instruments, all constructed by a similar process from rather simple *solid* bodies. It is therefore not surprising that the phenomena are forced to record in this similar medium either similar 'laws' or no laws. [2]

My focus will not be on the instrumentalist nature of Brouwer's philosophy of science – a rather common feature in his days – but the extra element of the explanation why unification is at all possible and has rational grounds.

Brouwer elaborates on this in a letter to Korteweg, written right after a long Sunday afternoon discussion of the part just quoted:[3]

Brouwer to Korteweg, November 13, 1906 (*extract*):
Now I would like to add some words about the main issue, that a similarity of laws in different areas of physics is to be expected on account of similarity of the instruments used, and begin with the following remark:
As projected on our measuring instruments there is no difference between the electromagnetic fields of a Daniell-element and that of a Leclancher-element; but if we look at it without prejudice, we have to expect that between both fields there must exist a difference as large as that between copper sulphate and ammonium chloride; only their effect on our counting and measuring instincts, using certain instruments, is the same; it turns out that one and the same mathematical system can be applied to both, but it is only the lack of suitable instruments that so far has prevented us from finding other mathematical systems that are applicable to the one field, but not to the other.
In every phase of the development of physics, the measuring instruments that 'have been found useful' form a limited totality, compared to the totality of measuring instruments that 'could be found useful to govern all kinds of yet unknown phenomena'; parallel to that, the 'mathematical systems applied to nature so far' form a limited totality compared to the totality of mathematics that 'would be applicable to nature, if only physics had grown sufficiently'. [4]

And as every limited group of mathematical systems has its invariants, it is to be expected that every limited group of physical phenomena, precisely because of that limitation, has its invariants as well, in the form of laws or principles that are valid for all phenomena in that group.
Now one might say: 'But why should we expect invariants for the totality of current physics; given that physics does not pose any *specific* limitation, but at will directs its attention to the most heterogeneous things?'
To which it should be answered: 'Certainly there is a *specific* limitation; for after all the mathematical laws observed in nature do not express anything but the relations between measures, derived from the rigid group; only the *influences*, to which those rigid measures are exposed, are varied indefinitely. The other physical quantities are only auxiliary, appropriately chosen for certain influences of the measures, that through their introduction as coordinates simplify the form of the equations of movement or states. Therefore those physical quantities are never themselves measured, only the rigid measures, in fictitious relation to which they have been introduced; thus one never measured magnetic forces and strength of current, but torsion angles of cocoon strings, and the size of the angle is derived from the rigid group. – And also: if we speak of equivalent things, or of undisturbed circumstances, we always mean: as far as the readings of our measure instruments are concerned. And there is only one thing that can be asserted as an empirical truth in itself, i.e.: the group of movements of solid bodies has approximately such and such properties, and as time goes by they remain more or less unaltered.'
— 'But we certainly do measure other things besides rigid measures; e.g. amounts of electricity; e.g., can't we repeatedly give a conductor equal charges, by twice discharging on it the same charging bulb, that acquired its charge twice in the same way? And don't we then know, that the charge of the conductor after the second charging is twice that after the first?'

— 'No; because in what way only can we speak of *amounts of electricity*, in other words, to what extent can the effects of the successive charges be taken to be equal effects? To the extent, for example, that they have identifiable effects on Coulomb's balance. But to what extent may the forces that result in equal torsions, be identified? To the extent that, for example, they keep equal copper weights in balance. But to what extent may the weights of equal pieces of copper be identified? To the extent that the accelerations that they give to the same body (for example in Atwood's instrument) can be identified. But those accelerations are only detected on solid bodies; for accelerations as well as velocities are detected on the rigid group. And it is the same for weights of fluids, we measure them either by their volume – and that is measured on the basis of the rigid group – or transferred as forces on a solid body, for example a balance or a piston.

In this way, every physically measurable quantity is eventually reduced to a measurement in a rigid group; and it is *the laws of those measurements that are sought after in all kinds of different circumstances*. So a *specific* limitation on the physically applicable mathematical systems is to be expected after all, and the existence of invariant properties should not be surprising. Just as an organ pipe refuses to vibrate with other than specific tones, we can expect that the rigid group refuses to *resonate* with other phenomena than those that fall under the principles of energy, action and thermodynamics. – The unknown more general that lies outside of it could then still manifest itself *in* the physical laws as various 'contingent' constants, such as unexplained atomic weights, dielectrical constants, frequencies, specific gravities and so on, and also the 'contingent' fact that that the laws are the way they are and not otherwise.

Let us now see if the argument Brouwer expounds here can be brought to bear on more recent issues in the philosophy of science. It can, but let me warn you: I will do so in a misleading way, as I will explain at the end.

3. SYSTEMATIC CONSIDERATIONS

There are well-known arguments for scientific realism based on unification. Such arguments have been propounded by Putnam, Boyd, and Friedman, and are of the following general form:

1. Theoretical unification is part of scientific practice, and has several benefits.
2. Only scientific realists have rational grounds for unifying theories.
3. Therefore we should be realists.

I will not discuss the question why unification is *valued* – this seems to be mainly a matter of increasing confirmation levels – but rather the *grounds* for unifying theories.

Let me fix some terminology first. A theory, say T3, is a *unification* of T1 and T2 if T3 entails T1, and T3 entails T2. The simplest case is where we take T3 = T1 & T2 (*conjunctive* unification); if T3 is obtained otherwise, we speak of *grand* unification.[5] Examples of grand unification are, first, the electroweak theory, which unifies the electromagnetic theory and the theory of weak nuclear forces; and second, Maxwell's electromagnetic theory which unified theories of electricity and magnetism.

The argument from unification showed up first in the particular form of Putnam's argument from conjunctive unification, or the *conjunction argument* for short: assume that scientists subscribe to theories T1 and T2, and let E be an empirical claim that does not follow from either T1 or T2, but does from their conjunction T1 & T2. In the practice of science, the argument continues, we see that in such situations E will be believed to occur. Now this belief is readily and rationally accounted for from the realist point of view: you accept a theory because you believe that it is true, and if you believe that two theories are true, then it is only rational to believe in their conjunction and any of its consequences as well. A realist accepts a theory because he thinks it is true, and the conjunction of two truths yields truth. An anti-realist accepts a theory for different reasons, but such reasons need not be preserved by conjunction. If theory acceptance depends not on truth but on empirical adequacy, then conjunction is not rationally explainable, as the conjunction of two empirically adequate theories need not itself be empirically adequate. So, the argument concludes, either science as we know it behaves irrationally when it conjoins theories, or realism is true.

Friedman added a new touch to the Putnam-Boyd argument: 'The point of the present discussion is that this practice plays a central role in the *confirmation of theories;* if we give up this practice, we give up an important source of confirmation.'[6] However, Kukla has shown that unification is epistemically advantageous to both anti-realists and realists, and that the attempt to use the effect of unification on confirmation in favour of realism is actually begging the question.[7] Therefore, I will not discuss the matter of confirmation and limit myself to the ontological issue.

There are two common replies to the conjunction argument. One is to deny that in actual science, theories are ever conjoined in the simple and direct way that the conjunction argument requires. This is Van Fraassen's strategy. Alternatively, one could embrace the idea that the practice of science *is* irrational.

However, Brouwer's theory shows a third option. Even if science proceeds rationally, and theories are simply conjoined, realism is still not the only explanation. Let me elaborate.

For my present purposes, I construe realism as defined by two tenets. First, our theories should be taken literally, and second, our theories are true. Then Van Fraassen's anti-realism agrees with the first tenet and denies the second: that is, on his view our theories should be taken literally and need not be true. Brouwer, on the other hand, disagrees with the realist the other way round: our theories should not be taken literally but they are true. Our theories are true but are not *about* what they seem to be: they are not about external reality but about our measuring apparatus. It seems that Brouwer's argument requires one to be a realist about measuring instruments (which are taken to be all based on the motions of the rigid group) and an instrumentalist about the rest. – As an aside I should add that this does not seem to square with the out-and-out phenomenalism that Brouwer voices in other places. But for the present argument, we need not worry about that.

Now, the conjunction argument would not harm Brouwer, as he agrees with the realist that one should believe theories because they are true. And if you believe, with Brouwer, that T1 and T2 each express – and can only express – truths about our measuring instruments, then it is rational to believe their conjunction.

The second instance of the unification argument I would like to discuss is given by Friedman. He asks: 'Why should we *ever* regard theoretical structure as something more than a mathematical representation of the observable phenomena?'[8] And his own answer is: because if one regards theoretical structure as something more, this gives extra *unifying power*.

For example, Friedman mentions the phenomena of chemical bonding, thermal and electrical conduction, and atomic energy, and then remarks:

In the absence of the theoretical structure supplied by our molecular model – in absence of a literal molecular world – the behavior of gases simply has no connection at all with these other phenomena, and our picture of the world is much less unified.[9]

Of course Friedman realizes that the point is not that purely phenomenological laws cannot have any unifying power, but

The point is that theoretical structure dramatically increases the unifying power of our total picture of the world, thereby dramatically increasing is potential for confirmation.[10]

So it is a matter of the *range* of unification.

Now let me take this last quote as point of departure for a reply from Brouwer's point of view. It is precisely the wide range of unification noticed by Friedman that motivated Brouwer to come up with his anti-realist explanation! After all, Brouwer began his considerations by wondering: 'Isn't it surprising that not only do we succeed in observing sequences which repeat themselves again and again, but that so many groups of phenomena affecting our naive senses in totally different ways can be brought together under a few general aspects which are covered by simple constructible mathematical systems?'

I think Brouwer would reply to Friedman along the following lines. If one wants to establish unity somewhere, this unity has to be founded on something already recognized as a unity. The unity achieved through the unification of theories is founded on the unity of something else. For the realist, the unity of theories is founded on the unity of the outside world: if we take T1 and T2 to be *true*, and *literally* so, we are using the presupposition of the independent existence of the outside world. Brouwer shows an alternative unity to found the unity of theories on: the uniformity of our measuring instruments. On his view, we have grounds just as rational as the realist's to unify these theories.

The *rational* role that the theoretical structure (the world, as realistically construed) plays in Friedman's argument can equally well be played by a simpler structure, i.e., that of our measuring instruments. It is not true that a representative (non-literal) account does not allow for such a rational role, as that account just like the theoretical presupposes our measuring instruments and then Brouwer's argument shows that this presupposition already imports a structure that can play the required rational role.

4. CONCLUSION

Brouwer had an early argument how one can deal with the unification argument even if you're not a realist. But now to accept Brouwer's theory as a live option is a move that will not be welcomed with great éclat. It is a move that would betray a myopic view of the history of the philosophy of science. Putnam, Boyd and Friedman designed their arguments to combat modern empiricism and other views that take our theories literally. The suggestion that we should take our theories non-literally – the suggestion we find in instrumentalism, logical positivism and, indeed, Brouwer – had by that time already fallen victim to assaults that are independent of unification. I mention the fact that a non-literal interpretation contradicts the intuitions of the practitioners of science, and that it is not clear how non-literal interpretations can deal with the fact that actual science also predicts novelties. A literal interpretation refers to a model and thereby makes such predictions explainable.

To summarize. Brouwer's letter shows that already very early there was an alternative to the realist's explanation of unification. To put it this way is perhaps anachronistic. In any case, other, more modern arguments have showed up within frameworks more palatable than Brouwer's. So from a systematical point of view, the interest of Brouwer's 1906 argument to ground unification is only marginal; but what is inside these margins, is history.

NOTES

*   I am grateful to Igor Douven for comments and discussion.

1.   See Van Stigt, W. (1979) The rejected parts of Brouwer's dissertation on the foundations of mathematics. *Historia Mathematica*, 6, 385-404; Van Stigt, W. (1990) *Brouwer's Intuitionism.* Amsterdam: North-Holland; Van Dalen, D. (1990) From a Brouwerian point of view. *Philosophia Mathematica*, 6, 209-226; and Van Dalen, D. (1999) *Mystic, Geometer, and Intuitionist.* Oxford: Clarendon Press.
2.   Van Stigt 1979, p. 399.
3.   The translation is mine. The original can be found in Van Dalen, D. (ed.) (1901) *L. E. J. Brouwer: Over de grondslagen der wiskunde.* Amsterdam: Mathematisch Centrum. Pp. 14-18.

4. Here one is reminded of Gödel's reflection on the relation between actual physics and set
    theory:

    [B]esides mathematical intuition, there exists another (though only probable) criterion of the truth of mathe-
    matical axioms, namely their fruitfulness in mathematics and, one may add, possibly also in physics. This
    criterion, however, though it may become decisive in the future, cannot yet be applied to the specifically set-
    theoretical axioms (such as those referring to great cardinal numbers) because very little is known about their
    consequences in other fields. (Gödel K. (1990) *Collected Works*. Volume II. Oxford: Oxford University Press.
    p. 269.)

    Of course results by Feferman and others show that one needs only a small part of set theory to
    obtain sufficient mathematics for current physics.
5. The terms are taken from Kukla, A. (1995) Scientific realism and theoretical unification.
    Analysis, 55, 230-238.
6. Friedman M. (1983) *Foundations of Space-Time Theories*. Princeton: Princeton University
    Press. p. 246 n. 16.
7. Kukla 1995.
8. Friedman 1983, p. 237.
9. Friedman 1983, p. 243.
10. Friedman 1983, p. 244.

Department of Philosophy
Utrecht University
Heidelberglaan 8
NL-3584 CS Utrecht
The Netherlands
Mark.vanAtten@phil.uu.nl

NADINE DE COURTENAY

THE ROLE OF MODELS IN BOLTZMANN'S
LECTURES ON NATURAL PHILOSOPHY (1903 – 1906)

During the mathematics lesson dealing with imaginary numbers, Törless, the young hero of Robert Musil's novel, *The Confusion of Young Törless*, is truly amazed. Imaginary numbers are *impossible*: numbers which, put to the square, give a negative number *cannot exist*. Still, these imaginary numbers seem to be used to reach quite definite and concrete results. This looks to Törless as if mathematics could make you walk steadily on a bridge which had only a beginning and an end and nothing in between, as if the bridge were complete.

Considering the paradoxes mathematics is able to handle – not only those concerning imaginary numbers, but also irrational numbers, the problem of the infinite –, Törless wonders if mathematics could help him to master the shivering depths he had just encountered in his emotional experiences. "If all this is really meant to prepare us for life, as they say, thinks the young boy, it must also contain some reflection of what I am looking for.[1] " However, turning to his professor of mathematics for some clues, the only answer he gets to his puzzlement is an urge to go 'back to Kant'. And while reading the first pages of the *Critique of pure reason*, poor Törless feels as if his brain got slowly crushed by an "old, bony hand".

It strikes me that Ludwig Boltzmann answered the young pupil's scientific and emotional puzzle more pertinently in the *Lectures on Natural philosophy* he gave at the University of Vienna[2] between 1903 and 1906. In his *Lectures*, Boltzmann seems indeed to have a very similar quest to the one Törless got entangled in. On the one hand, the *Lectures* begin with an inquiry into the nature of integer numbers. From there, they follow the construction of imaginary and irrational numbers, of set theory and non-Euclidean geometries. In all these cases, Boltzmann tries to show how one can come to terms with the fictitious entities constructed by mathematics, and what kind of relation they have to the world. On the other hand, as regards the *purpose* of these investigations, Boltzmann explains, in the preparatory notes for his first lectures, that his aim in tackling such dry and severe subjects as numbers and the like is to search the depths of the human heart and soul (very much as the anatomist searches the human body with his scalpel) in order to cure them of the unanswerable enigmas of philosophy and life. And the search, he announces, will excruciate our most intimate feelings[3].

M. Heidelberger and F. Stadler (eds.), History of Philosophy and Science, 103–119.

It is perhaps here our turn to feel somewhat puzzled. If the problematical entities arising in mathematics were indeed widely discussed at the end of the 19th century, no one else in the scientific community (at least to my knowledge) related the discussion to soul and feelings! What is more, Boltzmann doesn't seem to be exactly the right person to deal with questions related to the philosophy of mathematics. In this respect, the content of Boltzmann's *Lectures* seems quite surprising. Boltzmann is indeed no mathematician; he is a physicist known to be, with James Clerk Maxwell and Willard Gibbs, one of the founders of statistical mechanics. The crisis which followed the emergence, in the course of the 19th century, of new, successful disciplines, such as thermodynamics and electromagnetism, resulted in promoting quite a different research program in physics than the traditional one, based on mechanics. Confronted with such a crisis, Boltzmann turned also into one of those 'scientists-philosophers' who, towards the end of the century, tried to ponder over epistemological questions. He thus came to produce, in his later years, a number of conferences and writings (apparently) devoted to save the old program and defend atomism. They are gathered in his *Popular Writings* published in 1905, one year before his death[4]. How then can one understand that the quite straightforward path driving Boltzmann from physics to the epistemology of physics could happen to lead to the philosophy of mathematics we find in the *Lectures on Natural philosophy*?

In what follows, I will first try to show that the mathematical spirit of the *Lectures* is linked with Boltzmann's defence of the freedom of the scientist to forge hypothetical entities going beyond experience. This defence is indeed, to begin with, centered on atomism, but eventually broadens, and takes a more general, or rather, more fundamental turn in the *Lectures*. In this connection, I will try to show how Boltzmann's line of defence is related to his views on mathematical language, and on how this language relates to the world through *models*. We will then come to see how the *Lectures* succeed in bringing together the realms of science and feelings, and how, in so doing, they may, perhaps, suggest an answer to Törless.

I. A CLUE TO THE TOPIC OF THE *LECTURES ON NATURAL PHILOSOPHY*: BOLTZMANN'S ANALYSIS OF THE MEANING OF DIFFERENTIAL EQUATIONS IN THE *POPULAR WRITINGS*

The mathematical spirit of the *Lectures* will appear more natural if one pays due attention to a particular argument developed in the *Popular Writings*[5]. The argument in question is a critique directed against the privilege assigned by mathematical physics to the representation of phenomena in terms of differential equations. Part of this privilege was supported by a wide-spread epistemological position which Boltzmann names "mathematical phenomenology" – in the following I shall name it simply "phenomenology", without intending any reference to Husserl's philosophy.

According to the phenomenological perspective, science should abandon its metaphysical drive to *explain* phenomena. Instead of explaining, and therefore introducing hypothetical entities such as atoms and forces which defy observation, it should only seek to *describe*. Such a description was to be achieved in terms of differential equations which were believed to contain no hypothetical feature and to reflect the continuous aspects of phenomena.

However, expressing phenomena in terms of differential equations seems to avoid any kind of hypothetical import only if one entertains the thought that the meaning of the differential symbolism is fixed by reference to sensible representations or intuition. Such a belief can indeed be fostered by the way differential equations are often written, formed and dealt with in physics. When writing $dy = f'(x)\, dx$, for instance, one can easily think that dy and dx stand for elements of length that can be taken as small as one wishes – thus faithfully translating the continuous aspect of phenomena. Moreover, it is often through the consideration of elements of length, surface or volume, that differential equations are formed.

Yet, although such representations may seem good enough to back up routine work, argues Boltzmann, they cannot really account for the way the equations are actually used – and this by the phenomenologists themselves!

[G]etting used to the symbols of integral [and also, of course, of differential] calculus resembles getting used to expressions like cm s^{-1}. The convenience thus achieved may however lead to many faulty inferences if one forgets the meaning arbitrarily given to division by a second.[6]

Indeed, if one pays due attention to the concrete aspects involved in the way one handles these equations, one is bound to recognize, Boltzmann contends, that the meaning of the equations of mathematical physics is to be elucidated, not in connection with the vague intuition of indefinitely divisible elements of length or volume (which gives no definite indication as to how the symbolism should be used), but in connection with the *concept of the limit* of a sequence of points or numbers[7]. And this leads one back to *definitions*.

Indeed, considering, in the first place, the rules governing the manipulation of the symbolism of the differential calculus, one is not entitled to consider dx and dy in isolation, as it is the case when one represents to oneself two elements of lengths. One has to consider the *unit* dy/dx which, *by definition*, stands for a *limit*, not for the quotient of two lengths[8]. Such a limit is defined by an arithmetical operation bearing on sequences of points (or rather numbers: x_0, $x_0 + h$) meeting specific arithmetical requirements:

$$\frac{dy}{dx} = f'(x_0) = \lim_{\Delta x \to 0} \frac{\Delta y}{\Delta x} = \lim_{h \to 0} \frac{f(x_0 + h) - f(x_0)}{(x_0 + h) - x_0}$$

This account, grounded on definitions, gives clear instructions on how to use the symbolism. It serves whenever one has to *demonstrate* something relative to differential equations[9]. It is also indispensable, notes Boltzmann, in order to calculate *solutions* when the equations are too complicated to be solved analytically, which is indeed generally the case[10].

In fact, according to Boltzmann, differential equations are really *instructions for calculations*, as becomes clear in the process of computing approximate solutions. In most cases, solving Fourier's equation for the conduction of heat requires replacing, according to the definition, the differential equation by the finite difference equation of which it is the limit[11]:

$$\frac{\partial T(x,t)}{\partial t} = C \frac{\partial^2 T(x,t)}{\partial x^2}$$

becomes:

$$\frac{T(x,t+k)-T(x,t)}{k} = \frac{T(x+h,t)-2T(x,t)+T(x-h,t)}{h^2}$$

(In the simple case where the conduction takes place along the linear space dimension Ox; $T(x, t)$ stands for the temperature at the point x at time t. The constant C has been taken equal to 1 in the finite difference equation.)

When computing a solution, one therefore doesn't work on a continuum but on a discrete net of points, making the distances k between "points" of time and h between points of space become smaller and smaller according to a well defined rule ensuring the convergence of the process, a rule which Boltzmann insists on making specific.

After this clarification pertaining to the symbolism, the way differential equations relate to the phenomena they are supposedly describing is bound to appear in a completely different light. The second concrete aspect Boltzmann points to deals indeed with the way these equations relate to experience. As the above clarification makes clear, differential equations do not deal with indefinitely divisible elements of length, but rather with *points* or, even more accurately, with *numbers*. There are, however, no *points* in physical space; more generally, there are no *numbers* in physical phenomena. As Hermann Weyl will put it some time later: "There are no points in the jelly of the continuum." Still, it is necessary to introduce artificial cuts in such a continuum in order to be able to apply any arithmetical operation. It is indispensable, writes Boltzmann, in order to reach beyond a "hazy" picture of phenomena and construct a *distinct* and, above all, *analysable scientific* representation[12].

If one puts together this line of reasoning with the ideas Boltzmann develops in his article entitled "Models", differential equations never appear to be compared with the phenomena themselves but with, what he calls in the latter article, an "arithmetical analogy" of the phenomena[13]. Since one always substitutes for longitude, distance, temperature a scale of numbers, we are confronted, according to Boltzmann, with a process very similar to the representation by means of *models*. The process is only *similar* because Boltzmann wishes to keep the term 'model' to name concrete objects in three dimensions. Still, he invites us to recognize that differential equations, far from directly 'describing' phenomena, relate to them through an "arithmetical analogy" – we would say today, an 'arithmetical model' – that we have constructed according to our special aims.

To sum up the preceding discussion. Phenomenologists appear to ascribe a purely intuitive (and subjective) meaning to the equations of mathematical physics. However, their inconsistency shows through their own use of the symbolism, which they underrate, but which, in fact, manifests what *can be meant* by these equation. The arguments Boltzmann opposes to the soundness of the phenomenological perspective are twofold:

(i)   First, concerning the mathematical symbolism: the meaning of differential equations is fixed by the arithmetical definitions of the derivative and, hence, of the limit. Boltzmann presents these definitions as *instructions* for the *use* of the symbols, that is, for carrying out definite calculations. Such an account stresses the importance of the technical apparatus and changes profoundly the way one views the relation of the mathematical representations in question to experience.

(ii) Second, concerning the relation to phenomena: Once the arithmetical character of the language used is recognized, one is bound to acknowledge that differential equations have nothing 'naturally' in common with phenomena as they appear. The question of application presents itself, therefore, necessarily in new terms. Paying attention to the concrete features of application, one discovers that the equations cannot be confronted to phenomena but only to an 'arithmetical model' of these phenomena.

The mathematical and technical apparatus is therefore not dispensable with, once the result obtained (as Mach contends for instance). On the contrary, it makes the construction of a model appear as a genuine part of scientific construction. And this model can be very different from the kind of representations one has in mind when one believes the mathematical language to be a mere translation of sense perceptions or intuition.

Indeed, Boltzmann's argument reveals, amongst other things, that, without their knowledge, phenomenologists have resort to an atomistic model. This model is a very rudimentary one, since it reduces to the discrete net of points underlying the definition and the numerical calculations mentioned above. Phenomenologists only add the further hypothesis that the model is experimentally adequate when the distance between the points is taken smaller and smaller – which completely lacks proof, and certainly goes beyond observation!

Phenomenologists are therefore – and this, according to their own epistemological criteria – inconsistant in opposing other, more elaborate atomistic constructions (as the one developed by Boltzmann in the kinetic theory of gases, for instance). Disregarding the importance of the rules governing the mathematical symbolism *makes them blind to what they actually do and mean*, which is settled, *independently of the thinking subject*, by these very rules. Overlooking the conceptual constructions which stand at the basis of the differential calculus' mode of representation, they are unable to see that mathematical equations do not *translate* or *describe* phenomena, but relate to them through a model. They are, consequently, also unable to appreciate the hypothetical features which are part and parcel of the equations of mathematical physics, and hence, unable to recognize the limitations of these features and feel free to refine them – not to mention putting them radically into question.

This argument of the *Popular Writings* is essential to understand Boltzmann's epistemological defence of atomism – as Torsten Wilholt shows in this volume[14]. I will not go further in this direction. My aim is rather to show that this argument gives us also a clue for understanding the topic of Boltzmann's *Lectures on Natural philosophy* – without forgetting its purpose.

II. THE TOPIC OF THE *LECTURES ON NATURAL PHILOSOPHY*: BRINGING TOGETHER THE CRISIS IN PHYSICS AND THE TRANSFORMATION OF MATHEMATICS

The *Lectures* appear indeed to display the background of the argument against phenomenology – a background which remains only implicit in the *Popular Writings*.

Most of the time, the roots of our deepest inconsistencies are nurtured in an objective situation. The *Lectures* reveal that between Boltzmann and what he denounces as the internal inconsistency of the phenomenologists there is all the distance introduced by the transformation of mathematics which took place during the 19th century, and finally resulted in the constitution of 'pure' mathematics, separated from the natural sciences.

When Boltzmann settles the meaning of differential equations by returning to the arithmetical definition of the derivative, he implicitly appeals to the rigorous foundation given to the infinitesimal calculus, and to mathematical analysis as a whole, by the works of Cauchy, Weierstrass, and others. Such a foundation was achieved through an *artificial reconstruction* of the calculus. Within this reconstruction, the mathematical symbolism was disjoined from the geometrical and dynamical representations which had indeed inspired its birth in the 17th century, and used to fix its meaning and support its results[15]. The differential calculus was re-established on the basis of purely arithmetical concepts, through a process the mathematician Felix Klein, a close collegue and friend of Boltzmann's since his Munich professorship (1890-1894), described as the "movement of the arithmetization of mathematics".

In the argument sketched in section I, the phenomenologists seem to have stopped at the traditional way (stamped, intuitive and subjective) of accounting for the meaning of the differential symbolism, typical of a time where mathematics and physics were strongly interrelated. Boltzmann's position seems to fall in with the new arithmetical spirit of mathematics. In this respect, his criticism of the phenomenological point of view can appear as an instance of the well-known battle of the turn of the century against 'psychologism' taking place, this time, not in the field of logic or mathematics but in the field of physics. The contents of the *Lectures on Natural philosophy* confirms this suggestion by making Boltzmann's endorsement of the arithmetical point of view appear explicitly. In fact, a quite unambiguous maxim comes up in the preparatory notes for the *Lectures* and organizes Boltzmann's program of studies: "*Vom Zahlbegriff allein müssen wir ausgehen.*"[16] ("We must proceed only from the concept of number.")

Boltzmann's engagement shows quite deeply and accurately, beyond his partiality for the primacy of arithmetics, in his endorsement of the basic ideas that underlied the rigorization of mathematics. In the *Lectures*, he presents the language of arithmetics as the true means to *clarify* our concepts, to *control our inferences* (replacing appeals to evidence by demonstrations, uncovering hidden hypotheses). Finally – and I will now concentrate on this issue –, he contends that taking the arithmetical symbolism as a guide liberates one from the grip of intuition (regarded as thinking habits), and promotes the construction of entirely new objects as well as the exploration of entirely unforeseen paths of investigation.

The latter issue is essential to my subject since it reveals how the question of meaning is related to the question of the *freedom* of the scientist to fashion hypotheses – which is indeed the real question at stake in the debate between Boltzmann and the phenomenologists. In this respect, Boltzmann's originality is to have seen the connection between, on the one hand, the methodological and epistemological debates taking place in physics, and, on the other, what happens in the mathematics of his time; in other words, to have seen the connection between the contemporary crises in physics and the one in mathematics.

Matters of meaning and of the freedom of the scientist lead directly to Boltzmann's discussion of models in the *Lectures*. If one believes, like phenomenologists or Kantians, that the meaning of a symbol is unproblematically and uniquely determined by reference to sensible or intuitive representations, one also quite naturally believes oneself to be in a position to decide which scientific representations are permissible and which are not. Thus Boltzmann's German colleague physicist, Heinrich Hertz, who openly acknowledged his debt to Kant in his *Mechanics*, could indirectly set restrictions on the construction of scientific representations by claiming that our physical representations should be "permissible", that is, that they should conform to the 'laws of thought' and to intuition. In this respect, phenomenologists and Kantians took most seriously the traditional question "What can I know?", which puts forward the primacy of our *faculties* of representation. The objective turn in mathematics invited one to dismiss this subjective orientation (which relates, moreover, to *fixed* faculties, *invariable* powers of representation). It suggested, instead, to take more seriously matters of *formulation*, putting forward the importance of *signs and language*. Adopting a discursive mode for constructing mathematical entities confronts mathematics and, according to Boltzmann, *also physics* with the possibility of creating entirely new objects not connected with intuitive representation whatsoever. And indeed, in his own book on mechanics, Boltzmann, in tune with Klein, sees no ground upon which the use of the continuous non-derivable functions constructed by Weierstrass could be *a priori* dismissed in physics[17] (as a possible representation of the path of a material point, for instance), although these functions seem to baffle our powers of represention and go against anything we are ready to consider permissible. The analysis of Brownian motion will prove him right.

We have made analytical functions into a representation of the facts of experience. That these functions are differentiable cannot be taken as proof that empirically given functions are equally so, since the number of conceivable undifferentiable functions is just as infinitely great as that of differentiable ones.[18]

Boltzmann's commitment to the spirit of the arithmetization's movement shows that he is far from entertaining classical empiricist views. But he is equally far from adhering to any kind of formalism, or from being inclined to justify mathematical constructions by reference to ideal entities: Boltzmann is neither a formalist nor a platonist, nor any kind of semantic realist. In this respect, Boltzmann's position is deeply linked with his conception of language. Mathematical language is not the language of nature, nor is it mysteriously connected to a world of ghostly essences; it also isn't the language of sensibility or intuition. Mathematical language, but also ordinary language, are constructed by us. And we, that is, humanity in the course of evolution, have set its meaning – a meaning which is therefore not intangible but, on the contrary, adjustable and perfectible. "One shouldn't ask: is matter [a] contin[uum], but rather say that the word contin[uum] is a word coined by us.[19]" It therefore rests *on us* to define it as precisely as possible.

In any event, one cannot decide, just on the face of it, that a system of symbolic constructions is impermissible because it is contrary to intuition or that a system is permissible simply because it seems to entail no contradictions. The *Lectures* show indeed most clearly that the only decisive criterion in such matters of permissibility is whether one can find (or *construct*) a use for the system. That is to say, whether one can find an *interpretation* of the system – we would say today a 'model' –, and *do* something with it, showing thus the *fruitfulness* of the system.

In Boltzmann's argument against phenomenology we saw that differential equations dealt with mathematical points or numbers, and that they were valid only in so far as their relation with a model (or a class of models) was concerned. However, it is in the *Lectures* that Boltzmann really clarifies and develops his views on the subject of models, although he does so not so much in the field of physics but rather in that of mathematics, and, for the reason already stated, without using the term 'model' (Boltzmann uses rather the term "*Versinnlichung*", or "*interpretieren*").

In the following, I will outline Boltzmann's ideas on the role of models through two examples taken from the *Lectures*. The first deals with the classical account of the justification of imaginary numbers. It will clarify the role of models with respect to meaning and validity. The second deals with Cantor's theory of transfinite numbers. It will reveal another function of models which grounds the appeal to hypothesis in scientific construction as well as the claim of the freedom of the scientist. We shall also see how it can suggest an answer to the puzzle concerning the purpose of the *Lectures*.

III. THE TWO FUNCTIONS OF MODELS

The geometrical interpretation of imaginary numbers

The first appeal to models occurs in the lectures dealing with imaginary numbers[20] – the very same numbers Törless found so bewildering in Musil's novel. Boltzmann's account of how imaginary numbers came to be accepted is now classical, but its wording has the merit of bringing out the issue quite clearly: a system of operations remains a mere calculus without meaning or validity unless it is completed by a model.

As long as they hadn't been justified through the exhibition of an interpretation, imaginary numbers remained, in Boltzmann's own words, utterly nonsensical, empty *signs*. And the calculus developed with these signs remained a formal, arbitrary calculus. It is only when a geometrical interpretation was exhibited that one was able to ascertain that calculations with these 'signs' were pertinent – that is to say that they had, at one and the same time, use and validity. Each complex number (sum of a real and an imaginary number) was set in correspondence with a point of the plane, and one could thus establish that, transposed in the geometrical plane, all calculations with complex (and therefore imaginary) numbers gave indeed correct results.

It is to be noted that what Boltzmann places in correspondence are two *discursive systems*: on the one hand, an arbitary system of calculations with signs – complex 'numbers'; on the other, still another system of symbols – points of the plane, or rather couples of numbers. The validation of the first system occurs because the second system is unambiguously correlated with concrete, familiar manipulations, the outcome of which are completely unproblematical.

Boltzmann insists distinctly that the correspondence established between the calculus and its interpretation is entirely our own doing; it is therefore arbitrary: the two systems set in relation have, intrinsically, nothing to do with one another. Imaginary numbers entertain no natural relation with the geometrical plane no more than real numbers naturally correspond to the straight line in geometry[21]. In fact, the calculus with imaginary numbers can have many interpretations. Beside the geometrical interpretation of imaginary numbers, Boltzmann mentions a physical interpretation in terms of the phase of an electromagnetic wave. In other words, imaginary numbers are *signs which can symbolise in several different ways*.

However, if the meaning and validity of a system of signs is revealed through its connection with a field of use, one is bound to draw the conclusion that the validity of such a system is *only relative to one or several domains of objects*. Boltzmann insists indeed that one should never believe a proposition to be absolutely valid: each time that we wish to apply it to a new field, one ought to make sure, by a number of manipulations, that the proposition applies to the new category of objects considered. Such a limitation is not to be regretted. Quite on the contrary, it is a promise of surprises and conquests to come. This will appear in full light when we turn to Boltzmann's discussion of Cantor's set theory.

Evaluating Cantor's set theory

Boltzmann is ready to admire Cantor's eagerness to conquer the "real infinite"[22]. Still, he refuses to consider this conquest as achieved and to name it a "result"[23]. Cantor has indeed succeeded in laying down a set of definitions and rules in order to work with infinite numbers. But it seems to Boltzmann that Cantor has only managed, as yet, to circle round the difficulties connected with the manipulation of the infinite without really clearing them out. After giving an account of the theory in lecture no. 8, Boltzmann observes that a lot in it remains obscure and incomprehensible to him. His reservations seem essentially motivated by the fact that set theory shelters ambiguity and, therefore, cannot provide clear-cut options in order to orient us properly in decision making and action[24]. What then is Boltzmann's overall position regarding the legitimacy and prospect of Cantor's endeavour?

For one thing, Boltzmann refuses Cantor's alleged 'demonstration' of the *existence* of the actual infinite: it is incorrect and psychological[25]. He considers transfinite numbers as new fictions, new senseless signs akin to the imaginary numbers before they had received an interpretation. According to Boltzmann, the theory should be regarded as a purely formal calculus because it can avail itself of no interpretation providing evidence of its consistancy, and, above all, showing that we can *do* something with it, whether in physics or in mathematics.

As a matter of fact, Boltzmann takes some pains to show that physics provides no field of application for Cantor's theory because physics only uses the infinite as a limiting process. Boltzmann is nevertheless ready to think that mathematics may indeed be entitled to "try to determine more precisely" the concept of infinity, and to conceive of the "real" infinite and the "real" continuum[26]. What would seem needed here, in Boltzmann's mind, would rather be the statement of the freedom of the scientist to create new objects – that is, more a bold stroke than fishy demonstrations. It is perhaps worth reminding, in this connection, that in 1925, in his article "On the infinite", David Hilbert also turned to physics to see if one could find there any contents able to secure and justify introducing the notion of the actual infinite. The "net result" of Hilbert's inquiry was that:

> we do not find anywhere in reality, a homogeneous continuum that permits a continued division and hence would realize the infinite in the small. The infinite divisibility of a continuum is an operation that is present only in our thought [...].[27]

Boltzmann could only point to the possibility that the process of division would have to be stopped at a certain scale, whereas Hilbert could, in 1925, rely on the evidence of atoms, electrons and even quanta. As we know today, set theory cannot be saved that way.

For all that, we are still left with the question of Boltzmann's position. I would phrase it this way: instead of taking a position in the 'space of truth', Boltzmann concludes by stating his position in the 'space of time'. No doubt Cantor means to find a use for his theory; it is, moreover, largely rooted in former mathematical constructions. If no interpretation is available yet, an interpretation may still come, may still be *constructed*; maybe not. Only the future will tell. *But Boltzmann himself is not in a position to know.*

Such an answer may seem dull and altogether disappointing. The rest of my paper will try to bring out the philosophical import of this conclusion as it appears in the *Lectures*, and how it clarifies the role of models.

Boltzmann's realism and antifoundationalism

Boltzmann is certainly prone to have mixed feelings towards set theory: the prospect of being able to handle the infinite, and therefore also the continuum, *directly* was liable to put in some difficulty his argument for the conceptual priority of atomism. On the other hand, the point at issue, beyond set theory, was the fate of theory *per se*, the recognition of the indispensability of hypotheses – an issue, the importance of which was fitted to silence any particular controversy. Beyond philosophy of science, it opened indeed on philosophy.

In Boltzmann's mind, the domain of the hypothetical, the power to sustain the tension of the incomprehensible are *consubstantial* to the objective point of view. In physics and everyday life, they are not only essential to the progress of science, but also part and parcel of *realism*[28]. Reality can only be approached by taking the paths of fiction – as did Christophus Colombus, a favorite hero of Boltzmann's.

Indeed, to accept the risk of hypothesis and of the incomprehensible is a way to acknowledge the exteriority of the world – that it is not *my* representation. It is also, at the same time, to acknowledge the limitations of thought. Such a limitation has, however, nothing to do with the idea that there is a limit to knowledge – we saw that Boltzmann considered it perfectly possible that we might be able to think the real infinite. The limitation at issue is that of the *present* configuration of rationality. According to Boltzmann, we have no reason to assert, like Hertz, that the actual structure of our thought is fitted to reflect the world. The so called 'laws of thought' are not necessarily adequate to the structure of the world. To postulate such an adequacy would be to postulate something like a miracle; or to adopt a genuine Kantian perspective where one deals only with objects internal to representations. But in that case, any contradiction, any paradox involving the laws of thought would have to be put on the account of our transcendental constitution and, therefore, would have to be considered as irretrievably unsolvable – which would not only result in limiting knowledge, but, even more unfortunately, promote the most harmful "illness of the intellect" [29].

As for him, Boltzmann appeals to Darwin's theory of evolution. The brain of animals and of humans has evolved through interactions with an environment; and so has the eye. That doesn't mean that they are *perfect*. In any case, they are, at best, adapted to past experience. The limitation of our thoughts is on a par with the limitation of our field of experience to past experience.

This is perfectly in tune with the way models accomplish their logical function of validation: the validation of a system of symbols is only relative to a specific domain of uses. There is therefore no 'universal sentence' that would hold in the absolute, independently of any context. In this connection, Boltzmann refers to the paradox of the liar to illustrate how careful one should be even with the law of excluded middle[30]. Boltzmann's favourite phrase is that there is no absolute truth[31]. And he makes it very clear in his correspondence with Brentano that, according to him, there is no validity by virtue of concepts alone, *even in logic* [32] – in other words, that there is nothing such as a truly analytic proposition.

This means, amongst other things, that facing a theory (such as, for example, set theory) we are not in a position to say that it is impermissible simply because it matches, as yet, no experience, no representation of ours, or because it is contrary to our laws of thought. To make such statements of impermissibility, one would have to believe that the experiences and concepts we are now able to conceive can, in principle, circumscribe the whole of what is possibly thinkable, or the whole of possible experience. Whereas they can only encompass the experiences we are, at present, able to *imagine* (by reproducing and re-arranging familiar representations). Such statements would imply further that meaningfulness – and therefore the possibility for a system to be valid – is a matter decided by perception or the transcendental structure of human thought. And, above all, they would imply that there is only one system, the one we actually have; further developments being simply added or deduced, in full transparency and at no cost.

Yet the objective point of view requires that we step outside the sphere of our representations; outside the sphere of what we are merely able to imagine. One way to do it, and to be confronted with totally new and alien suggestions, is to work with signs so as not to be driven by allegedly obvious representations, only dictated by habits of thought which induce us to "hold mere feeling for truth (*bloß Gefühl für Wahrheit halten*) [33] ". It is, in other words, to take decidedly the "abstract point of view" as Boltzmann calls it, pondering on the construction of non-Euclidean geometries[34]. In this connection, one should be cautious not to identify scientific fictions with mere figments of the *imagination*. Imaginary numbers didn't arise from imagination; they were suggested by the concrete manipulation of mathematical signs: the problem with them was, precisely, that they were impossible to imagine!

If our abstract mathematical constructions, even contrary to intuition, find a field of use, then, says Boltzmann, our present modes of thought and representation will have to undergo transformations:

If one can absolutely not represent this to oneself [Boltzmann is lecturing on non-Euclidean geometries], one must precisely try to modify our representation; if subsequent results show that it is desirable, one must try to get used to represent to oneself [such a thing]. (*Wenn man das sich absolut nicht vorstellen kann, muss man eben die Vorstellung zu modifizieren suchen; man muss sich daran zu gewöhnen suchen, wenn durch anderweitige Resultate konstatiert wird, dass es doch wünschenswert ist, sich das vorzustellen.*) [35]

The new constructions will remain bewildering until our thought adapts, gets transformed by integrating new domains of rationality – as it had been the case with imaginary numbers.

This is to say that the *logical* function of models is completed by a *psychological* function. The way we think cannot be transformed at will. The integration of new features of rationality, the transformation of intuition and laws of thought occur in time through the process of using models: according to Boltzmann, these transformations cannot be the outcome of a purely intellectual decision. In fact, it is a practical process, involving body and action. Talking about the space of colours, Boltzmann states with some irony – and perhaps melancholy too – that only if his food and life depended on it could he achieve to figure intuitively a straight path, a circle etc. in the space of colours[36]. Our intellectual capacities are rooted in the relation of our body with our environment; this we must learn to bear: thought is not free.

It only depends on us to learn to accept (or accept to learn?) what the evolution of mathematics invites us to acknowledge (and, for Boltzmann, mathematicians, when they take "the abstract point of view", are indeed the true philosophers): that we should give up the illusion that we can possess an "*Erkenntnisgrund* (foundation for knowledge)". That we should abandon the search for foundations which reveals our hope to survey (even if only in its principles) the one and unique system; our hope to rule out the passage by those "fictions" which can cause utter bewilderment, the feeling of being lost, and, above all, the necessity for us to *change*.

CONCLUSION

Boltzmann's opposition to foundationalism in all its forms runs throughout the *Lectures*. While presenting the integer numbers in the first lectures, Boltzmann had criticized all the attempts made to define them. He had furthermore very lucidly recognized axiomatization to be a *description* rather than a foundation[37]. Again, if a definition or foundation were possible, we would be in a position to master the entire evolution of arithmetics. This evolution would proceed smoothly and require no transformation of our thinking habits and relation to the world. It would only unfold, make explicit the contents virtually present in the opening definitions, in total transparency (and time would be inessential). How then, asks Boltzmann, could we be confronted to the emergence of 'irrationalities' we are unable to understand, such as the imaginary numbers, as the irrational numbers[38]? How could there be room left for other, entirely new systems to emerge?

Mathematics depends on certain perceptible signs (letters, calculations); but they are completely suspended in the air, writes Boltzmann. We have absolutely no foundations. (*Mathematik haftet an gewissen sinnlichen Zeichen (Buchstaben, Rechnungsoperationen); aber sie schweben rein in der Luft, man hat gar keine Grundlage.*)[39]

As Boltzmann puts it: there is no "*Erkenntnisgrund* (foundation for knowledge)", there is only a "*Beweisgrund* (foundation for proof)". And in order to be able to prove anything we must start with premises that can neither be proved[40], nor be secured by reference to 'definitions' or ideal objects: in addition to Nietzsche's announcement of the "twilight of the gods (*Götzendämmerung*)", Boltzmann calls for a "twilight of ghosts (*Gespensterdämmerung*)"[41]. The starting point cannot be *justified*; we must *accept to adopt* it.

Thus, if arithmetics stands on integer numbers, we must *accept* these numbers as *given* to us: *certainly not a priori*, but partly through genetic inheritance (we have the *ability* to count), and partly through education. For Boltzmann, integer numbers are what we have learned at school by counting, almost at the same time we learned how to speak[42]. And if the system of integer numbers is so rich, can keep such surprises in store, it is because it stems from humanities immemorial practical relation with the world.

In this sense, the goal announced in his preparatory notes seems indeed to have been met. Instead of searching for essences or assured truths to start from, we begin with transmission, education, humanization; we unfold and get transformed in action. The evolution of mathematics while inviting, according to Boltzmann, to renounce the craving for foundations, seems to teach us to accept our heteronomy: the insistence on foundations is only a way to refuse to acknowledge our dependance upon our body and community – to deny our limitations.

To take language seriously – not as an alien language, but as our own – is to accept, at one and the same time, that language can guide us and deeply affect us; that it can require us to abandon the convictions and intuitions that presently organize our relation to the world. Or even worse, it can also make us accept that we are limited in our powers to change.

We have indeed the power to make entirely new possibilities shine through our risky constructions. But these possibilities may be out of the scope of our understanding, of our capacities to find the words that could bring them home to us; may be for others to pick up with their new words and their new doings, as Boltzmann confesses is his case with respect to set theory. So that, perhaps, the ultimate thing to learn would then be, quoting the closing words of Musil's novel, to accept to sense in our present "impossibility of finding words [...] the discrete tugging of the future".

ACKNOWLEDGEMENT

My warmest thanks go to Denis Jolivet for helping me out with most distressing computer problems whilst I was preparing the conference of which this paper is an outcome.

NOTES

1. Robert Musil, *Die Verwirrungen des Zöglings Törless*. Hambourg: Rowohlt Verlag, 1957. Quotation translated from the French edition: Robert Musil, *les Désarrois de l'élève Törless*, translation by Philippe Jaccottet. Paris: Editions du Seuil, 1984, p. 119.
2. Ilse M. Fasol-Boltzmann (Ed.), *Ludwig Boltzmann Principien der Naturfilosofi. Lectures on Natural Philosophy 1903-1906 with Two Essays by Stephen G. Brush and Gerhard Fasol.* Berlin, Heidelberg: Springer-Verlag, 1990. Boltzmann was then replacing Mach, a victim of a stroke, at one of the two chairs of philosophy.
3. Ludwig Boltzmann, "[3. Vorlesung]", in: Ilse M. Fasol-Boltzmann, *loc. cit.*, pp. 83-85. When bracketed the name of the lecture refers to the preparatory notes for the corresponding lecture.
4. Ludwig Boltzmann, *Populäre Schriften*. Leipzig: Johann Ambrosius Barth, 1905. Partly translated in: Ludwig Boltzmann, *Theoretical Physics and Philosophical Problems*, edited by Brian McGuiness, translated by Paul Foulkes, Vienna Circle Collection, vol. 5. Dordrecht: D. Reidel Publishing Compagny, 1974. Whenever possible, citations and references will refer to the English edition.
5. Ludwig Boltzmann, "On the indispensability of atomism in Natural Science (*Über die Unentbehrlichkeit der Atomistik in der Naturwissenschaften*)" and "More on atomism (*Nochmal über Atomistik*)", in: Ludwig Boltzmann, *loc. cit.*, resp. pp. 41-53 and pp. 54-56.
6. *Ibid.*, p. 44.
7. Ludwig Boltzmann, "On the indispensability of atomism in Natural Science", *ibid.*, p. 43 and p. 50; "More on atomism", *ibid.*, p. 55. See also, Ludwig Boltzmann, "On the Development of the Methods of Theoretical Physics in Recent Times (*Über die Entwicklung der Methoden der theoretischen Physik in neuerer Zeit*)", *ibid.*, pp. 77-100, especially p. 97.
8. See Boltzmann's account of the definition of velocity in his *Lectures on the Principles of Mechanics*. Ludwig Boltzmann, *Vorlesungen ueber die Principe der Mechanik*, 2 vols. Leipzig: Johann Ambrosius Barth, 1897 and 1904. Vol. 1, § 3, pp. 10-13. Ludwig Boltzmann, *Lectures on the Principles of Mechanics* (excerpts), in: Ludwig Boltzmann, *Theoretical Physics and Philosophical Problems, loc. cit.*, pp. 223-254, p. 231-233.

9. And this, also in physics, as it appears in one of Boltzmann's first scientific articles. Ludwig Boltzmann, "Über die Integrale lineare Differentialgleichungen mit periodischen Koeffizienten", in: Ludwig Boltzmann, *Wissenschaftliche Abhandlungen*, Fritz Hasenöhrl (Ed.), 3 vols. Leipzig: Johann Ambrosius Barth, 1909. Vol. 1, pp. 43-48.

10. Ludwig Boltzmann, "On the indispensability of atomism in Natural Science", in: Ludwig Boltzmann, *Theoretical Physics and Philosophical Problems.*, *loc. cit.*, p. 43.

11. Ludwig Boltzmann, "More on atomism", *ibid.* The actual finite difference equation doesn't appear explicitly in the text, but has the advantage of condensing in one line Boltzmann's explanations. The same line of thought is already at the basis of Boltzmann's discretization of energy in the famous article where he demonstrates the so called "H Theorem". See Ludwig Boltzmann, "Weitere Studien über das Wärmegleichgewicht unter Gasmolekülen", in: Ludwig Boltzmann, *Wissenschaftliche Abhandlungen*, *loc. cit.*, vol. 2, 1872, pp. 316-402. It is also clearly displayed in the way Boltzmann accounts for his results in: Ludwig Boltzmann, *Lectures on Gas Theory*, translated by Stephen G. Brush. Berkeley, Los Angeles: University of California Press, 1964, vol. 2, § 81. For more detailed comments on these matters, see Nadine de Courtenay, *Science et épistémologie chez Ludwig Boltzmann – La liberté des images par les signes*, Thèse. Paris, 1999. Chap. 5 and épilogue.

12. Ludwig Boltzmann, "On the Indispensability of Atomism in Natural Science", in: Ludwig Boltzmann, *Theoretical Physics and Philosophical Problems, loc. cit.*, p. 43.

13. Ludwig Boltzmann, "Models", in: Ludwig Boltzmann, *Theoretical Physics and Philosophical Problems, loc. cit.*, pp. 213-220, p. 215.

14. See also Nadine de Courtenay, *loc. cit.*, chap. 5.

15. On this subject, see for instance, Morris Kline, *Mathematical Thought from Ancient to Modern Times*, 3 vols. New York, Oxford: Oxford University Press, 1990. Especially vol. 2, chap. 26 and vol. 3, chaps. 40, 43.

16. Ilse M. Fasol-Boltzmann (Ed.), *loc. cit.*, "[12. Vorlesung]", pp. 96-98, p. 97. The phrase appears in the preparatory notes for the lesson n°12, but recurs throughout the *Lectures* in different forms. See in particular lectures n° 11 and 12.

17. Ludwig Boltzmann, *Lectures on the Principles of Mechanics* (excerpts), in: Ludwig Boltzmann, *Theoretical Physics and Philosophical Problems, loc. cit.*, p. 233. Felix Klein, *Lectures on Mathematics*. New York, London: Macmillan and Co., 1894, p. 48.

18. Ludwig Boltzmann, *Lectures on the Principles of Mechanics* (excerpts), in: Ludwig Boltzmann, *Theoretical Physics and Philosophical Problems, loc. cit.*, p. 233. In fact, the continuous differentiable functions constitue the exception!

19. Ilse M. Fasol-Boltzmann (Ed.), *loc. cit.*, p. 77. Note taken on the back of Boltzmann's notebook for the *Lectures*.

20. Ilse M. Fasol-Boltzmann (Ed.), *loc. cit.*, "6. Vorlesung", pp. 174-180, and "7. Vorlesung", pp. 180-186.

21. *Ibid.* p. 180.

22. Ilse M. Fasol-Boltzmann (Ed.), *loc. cit.*, "8. Vorlesung", pp. 187-193; "9. Vorlesung", pp. 194-200; "10. Vorlesung", pp. 201-205.

23. *Ibid.*, p. 202.

24. The lecture was delivered in 1903, but there is no evidence that Boltzmann knew about Russell's disclosure of the paradoxes the very same year. Although his diagnosis seems right, his argumentation doesn't fall in with the now standard account.

25. *Ibid.*, p. 190.

26. *Ibid.*, p. 189.

27. David Hilbert, "On the infinite", English translation by S. Bauer-Mengelber, in: Jean van Heijenoort (Ed.), *From Frege to Gödel. A Source Book in Mathematical Logic, 1879-1931*. Cambridge, London: Cambridge University Press, 1967, pp. 369-392, p. 371.

28. A fuller argumentation of this essential feature of Boltzmann's thought would require, in addition to what follows, a detailed analysis of the article he used to consider as his single genuine "philosophical article": "On the question of the objective existence of processes in inanimate nature (*Über die Frage nach der objectiven Existenz der Vorgänge in der unbelebten Natur*)", in: Ludwig Boltzmann, *Theoretical Physics and Philosophical Problems, loc. cit.*, pp. 57-76. For such an analysis, see Nadine de Courtenay, *loc. cit*, chap. 3 and chap. 4.

29. Ilse M. Fasol-Boltzmann (Ed.), *loc. cit.*, "5. Vorlesung", pp. 168-174, p. 173.
30. *Ibid.*, pp. 169-170.
31. *Ibid.*, "[5. Vorlesung]", pp. 87-89, p. 87.
32. Letters from Boltzmann to Franz Brentano, in: Walter Höflechner, *Ludwig Boltzmann Leben und Briefe*, Publikationen aus dem Archiv der Universität Graz, vol. 30. Graz: Akademische Druck- u. Verlagsanstalt Graz, 1994. Letters dated from Vienna, 26 December 1904, pp. 380-382 and Vienna, 4 January 1905, pp. 383-384.
33. Ilse M. Fasol-Boltzmann (Ed.), *loc. cit.*, notes taken on the 31 October 1904, p. 111.
34. *Ibid.*, "11. Vorlesung", pp. 205-212, p. 209.
35. *Ibid.*, "12. Vorlesung", pp. 212-221, p. 216.
36. *Ibid.*, "14. Vorlesung", pp. 230-240, p. 236.
37. *Ibid.*, "4. Vorlesung", pp. 162-168. Boltzmann's criticism is not exactly directed against the axiomatizations of Peano and Hilbert, but against the 'demonstrations' of the properties of the elementary operations Helmholtz develops, in a kindred spirit, out of 'definitions' in his famous article entitled "Numbering and Measuring".
38. *Ibid.*, "6. Vorlesung", p. 175. In this lecture, Boltzmann argues, more precisely, against the view that integer numbers are *a priori*. But the former lectures allow to direct the argument against any inclination to account for integer numbers by "sharp definitions".
39. *Ibid.*, "17. Vorlesung", pp. 258-273, p. 259.
40. *Ibid.*, notes taken on the 26 October 1904, p. 109.
41. *Ibid.*, "5. Vorlesung", p. 172.
42. *Ibid.*, "4. Vorlesung", p. 164; and also, "[4. Vorlesung]", p. 85.

Conservatoire National des Arts et Métiers
Equipe REHSEIS du CNRS
17, rue Gramme
F-75015 Paris
France
decourtenay@wanadoo.fr

SUSAN G. STERRETT

PHYSICAL PICTURES:
ENGINEERING MODELS CIRCA 1914 AND IN
WITTGENSTEIN'S *TRACTATUS* *

INTRODUCTION

Today I want to talk about an element in the milieu in which Ludwig Wittgen-stein conceived the *Tractatus Logico-Philosophicus* that has not been recognized to date: the generalization of the methodology of experimental scale models that occurred just about the time he was writing it. I find it very helpful to keep in mind how this kind of model portrays when reading the *Tractatus* – in particular, when reading the statements about pictures and models, such as:

- That a picture is a fact (*TLP* 2.141 [1]),
- That a picture is a model of reality (*TLP* 2.12),
- That the "pictorial relationship" that *makes a picture a picture* is *part of that picture* (*TLP* 2.1.5.3), and
- That a picture must have its pictorial form in common with reality in order to able to depict it (*TLP* 2.17).

And, when reading the sections of the *Tractatus* about objects and states of affairs, or atomic facts, such as:

- That the form of an object is the possibility of its occurring in states of affairs, or atomic facts (*TLP* 2.0141);
- That in a state of affairs, or atomic fact, objects fit into one another like links of a chain (*TLP* 2.03), and
- That if all the objects are given, then all *possible* states of affairs, or atomic facts, are also given (*TLP* 2.0124).

The generalization and formalization of the methodology of experimental scale modelling I am referring to was presented in London in 1914 just months before an incident involving a scale model that Wittgenstein said prompted him to think of a proposition as a picture. However, Wittgenstein would certainly have encountered specific applications of the methodology in the experimental work he had done as an engineering research student in Manchester prior to going to Cambridge to study logic with Bertrand Russell. A specific application of physi-cal similarity sometimes used in aeronautical, hydraulic, metereological and even

M. Heidelberger and F. Stadler (eds.), History of Philosophy and Science, 121–135.
© 2002 *Kluwer Academic Publishers. Printed in the Netherlands.*

some pure scientific applications was referred to as "hydrodynamical similarity", though many in the profession remained unconvinced of its validity.

The more general formalization of the method of experimental scale modelling involves a dimensional analysis of the quantities occurring in the equations and appeals to the principle of the dimensional homogeneity of equations, as did the specific application of hydrodynamical similarity. However, the advance to a more general formalization was a shift from examining the form of the specific equations governing a particular field, such as fluid dynamics, to a realization that the most general form of an equation can be specified in terms of the characters (that is, the kinds) of the physical quantities involved in the equation. In the specific as well as the general formulations of the methodology, what is used to establish similarity between two physical situations is knowledge of the *kinds*, as well as the number, of the physical quantities involved in the equation; what are meant by "physical quantities" here are quantities such as velocity, density, temperature, viscosity, surface area, some of which may be expressible in terms of others. However, as we shall see, this shift has somewhat less to do with the ontology of physics than it might at first sound, and somewhat more to do with the symbolism used to express relations among them than one might at first suspect. There is no restriction on the kinds of quantities that can be considered, which is what makes the method completely general. And, though conventions do exist, there is no restriction on which quantities are chosen as the fundamental ones in terms of which others are expressed.

What "dimensional" indicates here is not only the familiar kind of distinction between length and area, where a length would have dimensions of a linear unit while an area would have dimensions of a linear unit squared, i.e., the familiar distinction between one-dimensional and two-dimensional space. Rather, "dimensional" is general enough to cover any kind of quantity one may wish to use, including new kinds that haven't yet been included in physics. (Thus, any two velocities would be of the same dimension, and the product of a velocity and a time would have the same dimension as a length.) From knowledge of the *kinds, as well as the number, of physical quantities involved in the equation*, one can deduce how these quantities can be combined to form different dimensionless parameters that together can be used to classify (though not necessarily determine) the behavior of the physical situation. Once the values of all the dimensionless parameters relevant to a certain phenomenon are determined, so is the behavior of the physical system, if by behavior we mean only the behavior with respect to the particular phenomenon of interest (e.g., the existence of flow eddies, the motions of certain objects such as ships or ballons, certain reaction forces from jet sprays, etc.). However, in spite of having the values of the dimensionless parameters that do characterize the behavior of interest, it may still not be possible to predict the behavior of interest on the basis of the equations alone, due to, for example, intractability of the solution of the governing equations. Yet, one can use experiments in conjunction with things deduced from the equations to predict the behavior, in the following manner: often it is possible to build a scale model of a physical system, such that the scale model and the physical system being modelled have the same values of the dimensionless parameters that characterize the kind of behavior of the system in which one is interested. Then, it is possible to set up correspondences between the quantities in the two systems, so that one can read what would happen in the system being modelled from the behavior of the scale model, using rules of correspondence that tell you how to transform the quantity in the scale model into the value in the system being modelled. For instance, one may deduce that the time elapsed between two events in the model has to be multiplied by a factor of 1000 to obtain the time elapsed between the corresponding events in the situation being modelled; or perhaps that the velocity measured in the scale model has to be divided by a factor of 4 to obtain the corresponding velocity in the situation being modelled, and so on. The scale model by itself really tells you nothing, without knowledge of these rules of correspondence (sometimes misleadingly referred to as "modelling laws").

The generalized formulation of similarity methods presented in 1914 proceeded by way of considering "the most general form of an equation". This is not my paraphrasing, but is quite explicit in the oft-cited landmark paper: Edgar Buckingham opened his paper *On Physically Similar Systems: Illustrations of the Use of Dimensional Equations* with a section entitled: "The Most General Form of Physical Equations." He begins: "Let it be required to describe by an equation, a relation which subsists among a number of physical quantities of n different kinds."[2] Considering equations that describe material systems, he adds a condition that the equations be "complete", which, roughly speaking, is the condition that the only constants that appear in the equation are dimensionless. (This is not so much a restriction on the kinds of things expressed by equations that can be treated by the method as it is on the forms of equations that can be treated by the method, for, Buckingham explains, if a constant with dimensions does appear in an equation, the fact that the constant is not dimensionless may reflect that the constant is dependent on some other quantities in the equation; in cases where the equation can be rewritten to recognize this, doing so yields a complete equation.) Then, Buckingham reasons from a principle he takes to hold for what he calls "complete" equations of any sort whatever: they will be dimensionally homogenous. He attributes the statement of the principle to Fourier, referring to it as "familiar", at least in the form that "all the terms of a physical equation must have the same dimensions."[3] He goes on to show that, if you know all the quantities that are involved in such a "complete equation", and you know the *kind* of quantity each is, that's all you need to know in order to do all of the following:

(i) Figure out the minimum number of dimensionless parameters required to characterize the behavior of the system,

(ii) Actually construct one such set of dimensionless parameters. (The dimensionless parameters are formed by combining the quantities together in ways that yield a set of mutually independent dimensionless parameters; many such sets are possible, but the number of mutually independent ones is determined. Some examples of dimensionless parameters are: Mach number which is a ratio of two velocities (the ratio of the relative velocity of an object and the fluid in which it is immersed to the celerity of the fluid (the velocity of sound in the fluid at the temperature and pressure of the fluid in which the object is immersed), and hence is a pure number with no dimension; Reynolds number, which is the product of a density, a velocity and a length, divided by a viscosity, also a dimensionless ratio.)

(iii) Figure out the correspondences between quantities in physically similar systems, that is, get the (so-called) "modelling laws" used to co-ordinate values of quantities between the two systems, one of which is regarded as the model, the other of which is regarded as the thing modelled. (Examples here would be corresponding (elapsed) times or corresponding velocities.)

Buckingham showed how to do all these things in the paper in which he presented what he called "the pi-theorem" (due to the use of the Greek letter π to indicate dimensionless parameters) in April of 1914 in London. He was systematizing things that had already been proven, putting things that were already in use in special cases into a more general form, and trying to explain their significance. He remarked: "While this theorem appears rather noncommittal, it is in fact a powerful tool and comparable ... to the methods of thermodynamics or Lagrange's method of generalized coordinates." When he presented the "pi-theorem" the following year in the United States, one of the commentators remarked:

... the Π-theorem is closely analogous to thermodynamics and the phase rule. Thermodynamics affords certain rigid connecting links between seemingly isolated experimental results, while the phase rule tells us the number of degrees of freedom of a chemical system. The Π-theorem likewise affords rigid connecting links ... just as the phase rule tells us [the number of degrees of freedom in a chemical system], so also the Π-theorem tells us [the number of degrees of freedom in a physical system]

... The kernel of the paper is a theorem which is merely a restatement of the requirements of dimensional homogeneity, announced by Fourier nearly a hundred years ago, and extensively used by Rayleigh and others. But [...] Gibbs' phase rule, too was new only in form, not in substance, yet it served as the crystallizing influence which caused an immense number of latent ideas to fall into line, and we may expect the Π-theorem to play a similar role.

This inevitable development of technical physics into a unified branch of science ..., can be facilitated if writers on the problems of hydro- and aerodynamics, heat transmission and the like will be as introspective as possible, explicitly calling attention not only to their results, but to their methods of reasoning as well. For in every successful artifice of reasoning, there must be some element which is universal and capable of being generalized ... [4]

The actual statement of the "pi-theorem" in the paper presented in London in early 1914 as follows (here I have used the letter "p" rather than the Greek letter pi and the letter "Y" rather than the Greek capital letter psi):

... any equation which describes completely a relation subsisting among a number of physical quantities of an equal or smaller number of different kinds, is reducible to the form

$$Y(p_1, p_2, p_3, \text{etc.}) = 0$$

in which all the p's are all the independent dimensionless products of the form $Q_1{}^x$, $Q_2{}^y$, ..., etc. that can be made by using the symbols of all the quantities Q. [5]

Notice that in this "most general form of a physical equation", there are no algebraic constants such as signs for addition between terms, and no signs for mathematical relations such as logarithms or trigonometric operations. I remark on this here because of the analogous move in the *Tractatus*, of showing that there are no logical constants such as signs for "and" and "or" in the "general form of a proposition". To properly appreciate the reasoning for and significance of Buckingham's statement, I suggest the reader read and reflect upon the rest of the paper from which the above quote was excerpted. Perhaps it may be helpful here to explain, however, that the exponents of the Q's indicate repeated operations.

Would Wittgenstein have known of this result? It's hard to say what he would have heard of between the time he spent isolated in Norway in 1913 and entering the military in 1914. Lord Rayleigh, who was appointed president of the First British Council of Aeronautics in 1909 (which occurred during the time Wittgenstein was a research student in aeronautical-related engineering at the University of Manchester) had been writing about a less formalized version of the method for a over a decade; he wrote about Manchester engineering professor Osborne Reynolds' remarkable successes using the principle of "dynamic similarity" in hydrodynamics and he explained the method of dimensional analysis used to establish it. But, the main ideas behind the methods of physical similarity, even if they were taking a generalized form in England only that year, may have already been around in Germany. The reason I say this is that another commentator at Buckingham's 1915 presentation in the United States remarked that Buckingham "has struck the keynote of a new development of technical physics ... The importance of technical physics, as a branch of subject matter, is already so clearly recognized in Germany that laboratories are being established devoted exclusively to this field." However, whether this is referring to events in Germany after Wittgenstein's departure in 1908 is difficult to say, for, although Prandtl was brought to Göttingen to establish a laboratory around 1905, it is not clear how accepted model experiments were in Germany at that time – Prandtl's key paper on the importance of model experiments aimed at justifying and promoting widespread use of the methodology appeared in 1910.

Thus, although it is very hard to say just how much of this, how early, and in what form, was around in Wittgenstein's milieu, I think it *is* fair to say that the practice of dimensional analysis, and of efforts to formalize it and generalize it, was probably part of the milieu of anyone studying aeronautics in the years just prior to 1914. And that is what Wittgenstein was doing from 1906 through 1911. The practice of dimensional analysis, though perhaps not treated as a topic in its own right, was to be found popping up here and there in various engineering papers, texts, and research work. The notation used generally took the form of expressing the dimensions of a quantity in terms of three fundamental quantities, denoted thus: [M] for mass, [L] for linear dimension, and [T] for time. Thus in reasoning about dimensions, the dimensions of any term designating a velocity would be expressed by $[L] [T]^{-1}$. The dimensions of any length, whether a distance traversed or a diameter, was expressed by [L], of any surface, $[L]^2$, and of any volume $[L]^3$. The dimensions of the density of a fluid would thus be indicated by $[M] [L]^{-3}$. It is instructive to use the notation to verify that Reynolds number is dimensionless: one would do so by noticing that writing the dimensions for density, velocity, linear dimension, and the inverse of viscosity yields the following "chain": $[M]^1 [L]^{-3} [L]^1 [T]^{-1} [L]^1 [M]^{-1} [L]^1 [T]^1$ which can be rearranged as: $[M]^1 [M]^{-1} [L]^3 [L]^{-3} [T]^{-1} [T]^1$. This notation was widespread; it appears in various technical and scientific papers as well as in Horace Lamb's textbook, *Hydrodynamics*.[6]

That Wittgenstein was reading Russell's *Principles of Mathematics* and was almost certainly aware of Russell's 1908 paper presenting the theory of types[7] around the same time might have made him especially sensitive to the fact that the basis for similarity to be found in dimensional methods was not yet completely formalized. Here is an excerpt from a letter Wittgenstein wrote to Russell in 1912:

What I am *most* certain of is ... the fact that all theory of types must be done away with by a theory of symbolism showing that what seem to be different kinds of things are symbolized by different kinds of symbols which cannot possibly be substituted in one another's places. ... Propositions which I formerly wrote e2(a, R, b) I now write R(a, b).[8]

One thing we do know is that, according to Wittgenstein's own account, it was reading about the use of an experimental scale model – though in a courtroom, rather than in a laboratory – that stimulated him to think of a proposition as a picture.

PROPOSITIONS AS PICTURES

The incident that Wittgenstein said prompted him to think about a proposition as a picture occurred during the first few months of military service, in the fall of 1914, and is recorded in a diary entry. He wrote: "In a proposition a world is as it were put together experimentally. (As when in the law-court in Paris a motor-car accident is represented by means of dolls, etc.)"[9] That insight, that in a proposition a world is put together experimentally, occurs in the context of considering the most general concept of the proposition, and of co-ordination between proposition and situation; an *earlier* entry on the very same day reads thus: "The general concept of the proposition carries with it a quite general concept of the co-ordination of proposition and situation: The solution to all my questions must be extremely simple." His diary indicates that he thinks his insight about a world being put together experimentally does hold the solution to all his questions, for he then notes: "This must yield the nature of truth straight away (if I were not blind)."

The comment to friends that reading the newspaper account of how dolls were used to represent a motor-car accident in a law-court was pivotal in coming to think that propositions represent by being pictures really does appear accurate: with one exception[10], there is no mention of pictures in previous entries in the Notebooks, in any of the manuscripts entitled *Notes on Logic* (1913) nor in the notes dictated to Moore in Norway in April of 1914.[11] But, there is an abundance of entries on the proposition as picture in writings after that. And, the question of how a picture portrays is salient throughout his subsequent writings, although that question undergoes shifts in emphasis as time goes on.

Wittgenstein's biographer, Brian McGuinness, has remarked that this particular development in Wittgenstein's thinking (i.e., that a proposition is a picture or model) is probably not as crucial as it is often taken to be.[12] I think he has a point, in that, although the entry in September of 1914 marks the advent of the discussion of pictures, some of the features of the account in the *Tractatus* of what picturing is already appeared in earlier notes, though they are not formulated in terms of picturing.[13] For example, in the *Notes on Logic*: "Propositions ... are themselves facts: that this inkpot is on this table may express that I sit on this chair,"[14] and (Summary) "that a certain thing is the case in the symbol says that a certain thing is the case in the world."[15] That these ideas appear in his writings over a year before he made the notebook entry about the dolls in the law court suggests that the significance of the insight he had on that occasion, i.e., the insight to think of a proposition as a picture, may be that thinking about the use of a scale model in a courtroom provided a way to put certain thoughts he had already been formulating, rather than marking a totally unprecedented turn of thought.

That he specifically mentions "experimentally" ["probeweise"] here is more significant than might at first be obvious, if we look at what Boltzmann wrote about experimental models. Recall that Boltzmann appears first on the list of those whose work Wittgenstein said: "I have only seized on it immediately with a passionate urge for the work of clarification."[16] In his entry on "Model" written for the *Encyclopedia Britannica*, which was included in an anthology of Boltzmann's popular writings published in 1905 and which Wittgenstein is known to have read, Boltzmann explicitly emphasized that experimental models were of a different sort than the kind with which he was comparing mental models:

A distinction must be observed between the models which have been described and those experimental models which present on a small scale a machine that is subsequently to be completed on a larger, so as to afford a trial of its capabilities. Here it must be noted that a mere alteration in dimensions is often sufficient to cause a material alteration in the action, since the various capabilities depend in various ways on the linear dimensions. Thus the weight varies as the cube of the linear dimensions, the surface of any single part and the phenomena that depend on such surfaces are proportionate to the square, while other effects – such as friction, expansion and condition of heat, &c., vary according to other laws. Hence a flying-machine, which when made on a small scale is able to support its own weight, loses its power when its dimensions are increased. The theory, intiated by Sir Isaac Newton, of the dependence of various effects on the linear dimensions, is treated in the article UNITS, DIMENSIONS OF. [17]

Thus the experimental models represent a challenge; the relationship between model and what is modelled is *not* like the relationship between a mental model and what is modelled by it. This puzzled me for some time, for it's clear that Boltzmann realizes that there *is* a reliable methodology involved in experimental models. Why does he leave the topic in such an unresolved state, then? The answer to this is that the piece really was unfinished: when the *Encyclopedia Britannica* article was commissioned, Boltzmann tried to beg off on it, asking that it be assigned to someone else if possible. He protested that "what I write about technical and machine models will not be complete and will have to be enlarged by an English technician." [18] When transmitting the article, he says that he is sending it only because he has promised to do so, but that he is not at all pleased with it. [19] It is unlikely that Wittgenstein, who is reported to have made scale models of a sewing machine and airplanes in his youth, would not have noticed the glaring difference between mental models and experimental scale models to which Boltzmann is drawing attention. Thus, what was striking about the incident does not fit tidily with the notion of model found in Boltzmann. What it does fit well with is the notion of experimental engineering scale models.

WHAT ABOUT HERTZ?

What about Hertz? A number of people in the audience may be thinking "Doesn't she know the story about Hertz?" That is, there is a by-now-standard view, suggested by von Wright and developed by James Griffin, that "The picture theory comes almost in its entirety from Hertz." [20] Many other commentators have agreed.

Yes, I do know the story about Hertz. And, given such overwhelming agreement by so many commentators, and hardly any dissent, I guess I have the burden of explaining two things. First, why Wittgenstein, who prefaced the *Tractatus* by saying that he wasn't going to identify the sources of his ideas, specifically mentions Hertz's *Principles of Mechanics* in the *Tractatus*. And, secondly, why von Wright's speculative suggestion made in a footnote (i.e., "It would be interesting to know whether Wittgenstein's conception of the proposition as a picture is connected in any way with the Introduction to Heinrich

Hertz's *Die Prinzipen der Mechanik*."[21]) actually does seem to pan out, i.e., that those who investigated it by laying the two works side by side did find striking parallels between Wittgenstein and Hertz?

To answer why Wittgenstein mentioned Hertz in the *Tractatus*: well, consider the intended audience of the *Tractatus*: Russell, Frege, and those familiar with their works. Wittgenstein does not mention Hertz in the sections of the *Tractatus* that contain the statements I've mentioned above, about models and pictures, and about objects and states of affairs, or atomic facts. In the two passages in the *Tractatus* Wittgenstein *does* mention Hertz (in a remark that in Hertz's terminology, only uniform connections are thinkable, and in a parenthetical reference to Hertz's use of degrees of freedom of a dynamical system), I think he is just mentioning Hertz to illustrate a point to the reader, not citing him as a reference for a claim or notion. Still, why Hertz? Well, though it is seldom mentioned, recall (or take my word for it) that Russell himself discussed Hertz's *Principles of Mechanics*, as did many German language authors, including Mach, Helmholtz and Boltzmann.[22] Hertz's book, unlike the technical papers and texts Wittgenstein might instead have referred to, was thus well-known to Russell and to anyone who had read Russell's *Principles of Mathematics*, published in 1903. And Wittgenstein certainly knew the book well: von Wright reports that Wittgenstein "asked someone for advice about literature on the foundations of mathematics and was directed to Bertrand Russell's *Principles of Mathematics* ... It seems clear that the book profoundly affected Wittgenstein's development."[23] Russell's *Principles of Mathematics* closes with a chapter entitled "Hertz's Dynamics"; in this last chapter Russell discusses Hertz's book.[24] In fact, Russell describes Hertz's use of uniform connections, thinkable connections, and possible connections.

If von Wright's account of the role Russell's *Principles of Mathematics* had in leading Wittgenstein to study with Russell is right, Wittgenstein would certainly have remembered Russell's prominent discussion of Hertz's *Mechanics*. I do not mean to deny that Hertz was an important influence on Wittgenstein philosophically, only that what the references to Hertz in the passages that are cited in support of the claim that Wittgenstein's notion of a proposition as a picture is derived from Hertz' discussion of dynamical systems actually reflect is more likely that Wittgenstein recognized that Hertz's *Principles of Mechanics* was a much-discussed book with which Russell and any readers of Russell's *Principles of Mathematics* would be familiar.

To answer the second question – why the suggestion that Hertz was the source of the picture theory seems to pan out upon investigation, in that one can find what seem to be striking parallels between the *Tractatus* and Hertz's book – requires a bit more background. Hertz himself says that much of his book incorporates standard treatments of mechanics that were given in other mechanics textbooks, citing Thomson and Tait's classic *Principles of Mechanics and Dynamics*,[25] among others. These books contain the notions of generalized coordinates, which are mutually independent and together determine the state of a dynamical system, and every single one of these texts mentions that the number of independent generalized coordinates is the number of degrees of freedom of the system. My answer will be, in short, that much of what people see in common between the *Tractatus* and Hertz's book are very basic themes dating to eighteenth century mechanics and that these themes are also common between experimental engineering scale models and Hertz's book. What I will show, in addition, is that there are in fact important differences between the notion of model and picture in the *Tractatus* and in Hertz's book, and that these differences are also differences between experimental scale models and the dynamical models of Hertz's book. As I mentioned earlier, Boltzmann actually identified a key difference between these two notions of model.

But my main point isn't one of criticism. Like many of those who push the story about Hertz, I think that the engineering practice of the time is part of the story of the genesis of the *Tractatus*. I'd like to see my contribution here as carrying this investigation along by providing a more expansive milieu in which there was a more suitable notion of model that, though distinctive in some important ways, nonetheless had some things in common with Hertz's dynamical models.

THE EXPANDED MILIEU

I have a wealth of information about the engineering milieu and the fascinating story of how similarity methods were developed and gained acceptance in a longer work I'm writing on the subject, but, in this short talk, I can only give a very general outline of the development of the methodology of "dynamic similarity." Similarity methods really developed, blossomed, and flourished in England in the nineteenth century, due in large part to Osborne Reynolds's work at the University of Manchester. Reynolds retired from his post as an engineering professor at Manchester in 1905, and, that same year, the mathematician Felix Klein, upon recognizing Ludwig Prandtl's work in air flow as revolutionary, invited him to Göttingen to set up a laboratory there. Germany soon became the premier place for aerodynamical work, in large part due to Prandtl's efforts to ensure the proper methodology for engineering models was understood and accepted.

The mathematician Horace Lamb, the author of the basic compendium on theoretical hydrodynamics, was at Manchester, England, though, before, after, and during the time Wittgenstein studied there. It's recently been discovered that Wittgenstein owned a copy of Lamb's *Hydrodynamics* in German translation – this information is thanks to Brian McGuinness and Peter Spelt. They describe and discuss the marginal notes Wittgenstein made in the book in their paper; they infer that he most likely owned the book while studying in Berlin, and speculate that he may have chosen to go on to Manchester because that is where Lamb was.[26] In fact, Wittgenstein evidently did discuss some equations with Lamb, and attended mathematics lectures in Manchester. Aeronautics at that time was a study being synthesized from practical and theoretical fields. As a historian of hydraulics has put it: "On the one hand, the continued divergence of [experimental hydraulics and theoretical hydrodynamics] had, by the beginning of the 20th century, brought them very far apart. On the other hand, it was in this period that the increasingly stringent demands of aeronautics gave rise to an effectively new branch of science to bridge the apparent gap between theory and fact."[27] Even though Germany was becoming the premier place to study aeronautics around the time Wittgenstsein went to England, England still had a great strength in model experiments at the time: a wind tunnel was built in London in 1903, and an improved one in 1910. Lord Rayleigh's appointment as president of the first British Council on Aeronautics came in 1909, during the time Wittgenstein was doing aeronautical research on propellers and jet engines. Rayleigh used the position to continue his championing of the method of dimensional analysis and the employment of "dynamical similarity" to model experiments.

So there's no question that the notion of scale model, and the notions of dynamical similarity and dimensional analysis, were part of the milieu in which Wittgenstein conceived the *Tractatus*. Now, as I mentioned earlier, the notion of a scale model includes the notion of the scale model *as a transformation* (Abbildung) of another situation that preserves similarity with respect to some behavior. This German word for transformation is the word used in the *Tractatus* that is often translated as "pictorial form", or the "form of representation" [that Wittgenstein says makes a picture and is part of the picture]. To explain how these notions are all of a piece in the method of scale models, it may help to appreciate the problems people had in understanding why and how scale models were to be used. When first arguing for the use of the method of scale models, Froude's defense of the method was at the same time an explanation of how to use data collected from scale models and captures the notion of a transformation in more concrete terms. Froude had to address the concern that "It is observed that models, when towed through the water make proportionately much larger waves than full size ships ..."[28]. His argument against this was that if the full-size ship were travelling through the water at a velocity *corresponding to* that of the small scale ship it, too, would make large waves. Drawing on a similarity principle, Froude explains how the observations on his small-scale model experiments are to be understood:

... the diagram which exhibits to scale the resistance of a model at various successive velocities, will express equally the resistance of a ship similar to it, but of (n) times the dimension, at various successive velocities, if in applying the diagram to the case of the ship we interpret all the velocities as ($\sqrt{n}$) times, and the corresponding resistances as (n^3) times as great as on the diagram.[29]

It certainly seems to me that this is the notion of model involved in the idea of a proposition as a picture, and a picture as a model. And it makes a difference which notion of model one does use: For this kind of model, the picture "reaches right up to reality" as Wittgenstein put it in the *Tractatus*. It is not in a separate realm somewhere and in need of application. It is not ambiguous regarding what it pictures. It needs no interpretation.

Further, the similarity between two systems is established by dimensionless parameters, which are formed by combining quantities in constrained ways: the dimensions of the quantities relevant to the behavior of interest determine how those quantities may be combined into various dimensionless parameters. This is very much like the constraints due to the forms of objects work in objects combining to make states of affairs, or atomic propostions (*Sachverhalte*). Even putting historical details aside, there is much to recommend the claim that the notion of scale model is a much more appropriate notion of model to have in mind when reading the parts of the *Tractatus* about propositions as models or pictures.

Yet I do not want to say that one should keep the notion of a scale model in mind to the exclusion of features of dynamical models mentioned in Hertz's *Principles of Mechanics*. Rather, there are various conceptual themes that can be found both in Buckingham's essay on physically similar systems and in Hertz's short section on dynamical models. Further, I think that these common themes reflect a common ancestry. In a paper entitled "On a Theorem Concerning Geometrically Similar Motions of Fluid Bodies, with an application to the problem of Steering Air Balloons"[30] Hermann Helmholtz explicitly suggested the use of "dynamical similarity" and dimensional analysis to infer the behavior of one physical situation from observations on another. The paper was presented in Berlin in 1873, and given that the application is to air balloons, Wittgenstein may well have known of it in one form or another. Helmholtz, of course, was Hertz's teacher.

But there's an even more general background: Lagrangian mechanics, which contains the notion of generalized coordinates as one of its most basic features. Generalized coordinates are mutually independent, but together determine the state of the dynamical system. The number of generalized coordinates required is the number of degrees of freedom of the system. Transformations of one system to another that is similar in some ways are effected by changes of coordinates. Boltzmann relied upon the generalized coordinate approach to explain thermodynamic processes, Helmholtz to provide an example of a mechanical model in which the phenomena of electrodynamics would arise, and Hertz to provide a new basis for a forceless mechanics. Lamb treats fluid flow problems as dynamical systems for the similar reasons.[31] And, certainly the methods of scale modelling evolved from notions of dynamical similarity. All this is by way of saying that, once one looks at this expanded milieu, it becomes clear that focusing only on Hertz's *Principles of Mechanics* when looking for notions of physical models is ignoring other salient and relevant kinds of models.

THE LABORATORY AND THE COURTROOM

I have one thing to add in closing: given that I am emphasizing how ubiquitous scale models were in laboratories, then why is it the use of a scale model in the courtroom, rather than a laboratory, that kicks off the thought that a proposition portrays by being a picture? Here's my speculation: in a courtroom, there are "questions of fact" and "questions of law". "Questions of fact" would include questions about how things happened, and what the consequences of certain actual and counterfactual events were or would be. "Questions of law" would include questions about responsibility and blame. One can use the scale model only to establish answers to questions of the first sort. In fact, once all the questions of the sort that could be settled by a scale model are settled, questions about responsibility, blame, and regret, are still untouched. In such a context, anyone who thought that empirical propositions might have anything to say about such questions should – and often would – be brought to realize that they don't. In a law court, someone following such a line of thought might be silenced by being told that the question they are attempting to provide evidence for to the jury is a question of law, not of fact. Wittgenstein did say that the most important point of the *Tractatus* was an ethical one, and was made by what he didn't say. Perhaps there is an analogous point about ethics that resonated when reflecting on the limits to what a scale model could portray about a situation in the context of a courtroom, rather than a laboratory. He ended his preface to the work by saying that, if he was correct in believing that he had found the solution of the problems of logic, that this shows "how little is achieved when these problems are solved." So perhaps it was the interest in exploring limits to what can be portrayed that set him to work on working out the idea that a proposition is a picture.

NOTES

*   This paper is the text of a short talk given on July 6, 2000 at the HOPOS 2000 conference held at the University of Vienna. Special thanks to Brian McGuinness and Peter Spelt for providing me with a preprint of their paper, and for answering various questions.

      A longer version of the talk was given under the same title on November 17, 2000 at the University of North Carolina at Chapel Hill. Thanks to numerous members of the audiences at those talks for helpful and stimulating questions and remarks. A book-length work on the subject is also in progress.

1.   Henceforth "*TLP*" refers to Ludwig Wittgenstein, *Tractatus-Logico Philosophicus*. Translated from the German by C. K. Ogden, with an introduction by Bertrand Russell. London and New York: Routledge & Kegan Paul Ltd., 1922. Another translation used here is: Ludwig Wittgenstein, *Tractatus Logico-Philosophicus*. Translated by D. F. Pears & B. F. McGuinness, with the Introduction by Bertrand Russell. London and Henley: Routledge & Kegan Paul, 1974 (revised

version of edition first published in 1961). The numbers refer to the statement numbers of the original German text.

2. Edgar Buckingham, "On Physically Similar Systems; Illustrations of the Use of Dimensional Equations", in: *Physical Review* 4,4,1914, pp. 345-376.
3. *Ibid.*, p. 346.
4. Edgar Buckingham, "Model Experiments and the Forms of Empirical Equations", in *Transactions of the American Society of Mechanical Engineers* 37, 1915, pp. 292.
5. Edgar Buckingham, "On Physically Similar Systems; Illustrations of the Use of Dimensional Equations", *loc. cit.*, p. 376.
6. Horace Lamb. *Hydrodynamics.* New York: Dover Publications, 1945. (Republication of the 6th edition published in 1932 by Cambridge University Press.)
7. Bertrand Russell, "Mathematical Logic as Based on the Theory of Types", in: *American Journal of Mathematics* 30, 1908, pp. 222-262.
8. B. F. McGuinness, *Ludwig Wittgenstein: Letters to Russell, Keynes, and Moore.* Oxford; Basil Blackwell, 1977.
9. Ludwig Wittgenstein, *Notebooks 1914- 1916, 2nd Edition.* Edited by G. H. von Wright and G. E. M. Anscombe. With an English translation by G. E. M. Anscombe. Chicago: The University of Chicago Press, 1979; p. 7e.
10. The exception is a statement "Philosophy gives no pictures of reality." in "Notes on Logic", printed as Appendix I in Ludwig Wittgenstein, *Notebooks 1914-1916, loc. cit.*, p. 106.
11. Ludwig Wittgenstein, "Notes Dictated to G. E. Moore in Norway April 1914", printed as Appendix II to Ludwig Wittgenstein, *Notebooks 1914 - 1916, loc. cit.*, pp. 108-119.
12. Brian McGuinness, "The Grundgedanke of the Tractatus", in Godfrey Vesey (Ed.), *Understanding Wittgenstein.* Ithaca, New York: Cornell University Press 1974, pp. 48-61.
13. Von Wright, too, feels Wittgenstein had the main ideas for the Tractatus before this. G. H. von Wright, "Ludwig Wittgenstein: A Biographical Sketch" in G. H. Von Wright, *Wittgenstein.* Minneapolis: University of Minnesota Press n.d., pp. 13-34.
14. Ludwig Wittgenstein, "Notes on Logic" printed as Appendix I to Ludwig Wittgenstein, *Notebooks 1914 - 1916, loc. cit.*, p.97.
15. *Ibid.*, p. 96
16. Brian McGuinness, *Wittgenstein: A Life. Young Ludwig 1889 - 1921.* London: University of California Press 1988, p. 84.
17. Ludwig Boltzmann, "Model" in: Brian McGuinness (Ed.) *Theoretical Physics and Philosophical Problems: Selected Writings / Ludwig Boltzmann.* Vienna Circle Collection, Vol. 5. Dordrecht; Boston: Reidel Pub. Co. 1974, pp. 220.
18. Ludwig Boltzmann to Joseph Larmor, January 7, 1900 in: John Blackmore (Ed.) *Ludwig Boltzmann / His Later Life and Philosophy, 1900 - 1906. Book One: A Documentary History.* Boston Studies in the Philosophy of Science, Vol. 168. Dordrecht; Boston; London: Kluwer Academic Publishers, p. 57-58.
19. Ludwig Boltzmann to Joseph Larmor, Vienna, undated. Letter #5 in: John Blackmore (Ed.) *op. cit.*, p. 58.
20. James Griffin, *Wittgenstein's Logical Atomism.* (Reprinted from corrected sheets of the first (1964) edition). London: Oxford University Press 1965, pp. 99.
21. G. H. von Wright, *op. cit.*, p. 29n.
22. Bertrand Russell, *The Principles of Mathematics.* Cambridge: Cambridge University Press 1903.
23. G. H. von Wright, *op. cit.*, p. 18.
24. It does seem to me a curious omission that Griffin does not mention Russell's prominent treatment of Hertz in *The Principles of Mathematics.*
25. William Thomson (Lord Kelvin) and Peter Guthrie Tait, *Principles of Mechanics and Dynamics.* (First published in 1879 under title: *Treatise on Natural Philosophy.*) New York: Dover Publications 1962.
26. P. D. M. Spelt and B. F. McGuinness, "Marginalia in Wittgenstein's Copy of Lamb's Hydrodynamics", forthcoming in *Wittgenstein Studien.*
27. Hunter Rouse and Simon Ince, *History of Hydraulics.* State University of Iowa: Iowa Institute of Hydraulic Research 1957, p. 229.

28. Quoted in Rouse and Ince, *ibid.*, p. 184.
29. *Ibid.*
30. Hermann Helmholtz, "Ueber ein Theorem, geometrisch ähnliche Bewegungen flüssiger Körper betreffend, nebst Anwendung auf das Problem, Luftballons zu lenken" delivered to the Berlin Society, in: *Monatsberichte der Königl. Akademie der Wissenschaften zu Berlin, vom 26. Juni 1873*, S.501-504. Reprinted in Helmholtz's *Wissenschaften Abhandlungen*, Vol. I, p.158ff.
31. Horace Lamb, *op. cit.*, p. 187 ff.

Department of Philosophy
Duke University
Durham, North Carolina 27708
U.S.A.
sterrett@duke.edu

GARY HARDCASTLE

THE MODERN HISTORY OF SCIENTIFIC EXPLANATION

To be a philosopher of science means, among other things, to have an account of what scientific explanation is, or, at the very least, to have a response to various accounts of scientific explanation on offer from *other* philosophies of science while earnestly working toward what one hopes will be one's own, original account. One presumption clearly and often lying behind such work is that science provides two kinds of knowledge. There is propositional knowledge, "knowledge that" or "knowledge what," and there is some other kind of knowledge, something *beyond* propositional knowledge, usually called "knowing *why*." We can know *that* the moon will have such a phase at this or that time, *that* home sales will always slump following a rise in interest rates, or *that* probably no two snowflakes are the same shape, without knowing *why* the moon will have that phase, home sales will fall as interest rates rise, or no two snowflakes (probably) have the same shape. But science, so the common contemporary presumption continues, fills in the missing knowledge – it tells us why. How science does this, when (if ever) it can't, and what the nature of this sort of knowledge *is* are precisely the issues that separate theorists of explanation. There are, of course, deflationary views of explanation, which reduce explanation to other properties or eliminate explanation altogether (e.g., van Fraassen's pragmatic account of explanation), but these are a decided minority. The vast majority of the work on scientific explanation takes itself to be addressing a certain, distinct, kind of knowledge. This is, moreover, a familiar and introductory point made in philosophical discussions of explanation. I rehearse it here because it has a role to play later, in my discussion of Carl Hempel and Paul Oppenheim's 1948 article, "Studies in the Logic of Explanation" (hereafter, 'SLE').[1]

Which brings me to a second presumption, just as much in the background of current work in the philosophy or explanation. Almost without exception, this work locates itself – and *must* locate itself – with respect to the "covering-law" account of explanation now associated with Hempel and articulated in detail first in later sections of SLE. Moreover, it is widely understood that at the core of the covering law account is the idea both that to explain something is to display its "nomic expectability" – that is, to show how the thing was to be expected given the true laws and the particular circumstances in place prior to its occurrence – and that to explain something by displaying its nomic expectability is different from, for example, reducing it to more familiar phenomena or showing how it was caused by what preceded it. The presumption to which I want to call attention is not simply that SLE is an important paper, nor even that it is a focal point

137

M. Heidelberger and F. Stadler (eds.), History of Philosophy and Science, 137–145.
© 2002 *Kluwer Academic Publishers. Printed in the Netherlands.*

for the philosophical examination of explanation (nor, of course, that the ideas in it are correct), but that it is regarded as a starting point, a foundation on which the rest is to add. Not all, or many, contemporary accounts of explanation are in agreement with Hempel and Oppenheim's covering law account. But it is one thing to disagree with your intellectual parentage, and another thing to disavow it. When it comes to explanation, contemporary philosophers of explanation do hardly any of the latter, however much they do of the former.

In keeping with the fast-developing enterprise of learning more about (and doing better) philosophy of science by casting an historical eye on the philosophy of science itself, Hempel and Oppenheim's 1948 paper has recently become the subject of historical scrutiny, indeed, scrutiny from two philosophers able to claim participant status in the last five decades of scientific explanation. Wesley Salmon and Nicholas Rescher offer interestingly different accounts of what is significant about, and what went into, SLE.[2] Salmon, for example, presents SLE as a precise distillation and explication of an idea of scientific explanation which was until then only hazily formulated, and which was both inspired by and dependent upon the rise of scientific realism (i.e., the view that a theory's theoretical terms refer and the things it says about the unobservable bits of the world are true, or at least mostly true) among scientists at the turn of the century. According to Salmon, realism among scientists begot a hazy but fruitful notion of scientific explanation, which in turn begot a careful explication in Hempel and Oppenheim's (mainly Hempel's) hands. Philosophers, Salmon suggests, were slow to embrace realism, but after seeing the explication of explanation in SLE three major figures in logical empiricism[3] – Rudolf Carnap, Hans Reichenbach, and Hempel – became realists, at least in good time. Imbued with realism and the model of precision contained in SLE, philosophers of science tackled explanation with vigor, in the spirit of Hempel and Oppenheim, and they continue to do so. This is a good thing, claims Salmon, for only a century ago scientists and philosophers didn't even believe there *was* such a thing as scientific explanation, or knowledge *why*. Here then is how Salmon places SLE in particular with respect to what he refers to as a "remarkable reversal of attitude":

A large proportion of the philosophers and scientists who, at the beginning of the century, denied the possibility of scientific explanation, also denied the existence of such unobservables as molecules, atoms, and electrons.... When Hempel wrote ["Studies in the Logic of Explanation"], I think that the issue [of realism] had already been settled, but that philosophers were slow to perceive the import of certain scientific developments.... [But in SLE] the deductive nomological explanation of particular facts was first presented with an unprecedented degree of precision. [Hempel and Oppenheim] state explicitly that the account they offer is not novel; they cite a number of nineteenth and twentieth century authors, including John Stuart Mill and Karl R. Popper, as anticipators. However, even though the basic idea is not new, this 1948 article is the fountainhead from which practically all subsequent work on scientific explanation flowed.... Philosophers in roughly the first half of the twentieth century who wondered about scientific explanation may have had no clear idea of what it might be. "Explanation" (especially without the qualifier "scientific") seems so vague and ambiguous, with so many subjective overtones, that it is hard to see what sense can be made of it. Hempel and Oppenheim gave a clear model. [4]

Following Salmon, Rescher affirms SLE's status as a "fountainhead" of contemporary work on explanation. But here SLE is presented neither as a reflection of nor as contributing cause toward scientific realism, nor is it described as an attempt to articulate, however precisely, a view of explanation which was widely shared but in need all the same of a formalized treatment. On Rescher's account, SLE originated in the efforts of Oppenheim, Hempel, and Olaf Helmer to systematize and formalize the degree of support a theory has in the light of its evidence. SLE was, on Rescher's account, almost equal parts Hempel's logical expertise and Oppenheim's life-long concern with finding orders, hierarchies, and concepts common among the sciences. The spark to this mixture came from Helmer, and consisted in an almost literal twist on Carnap's then-contemporary attempts to formulate a measure of confirmation, dc(theory/evidence), with the formal properties of a probability. At Helmer's suggestion, according to Rescher, Hempel and Oppenheim considered the converse relation instead, pr(evidence/theory) – a measure of the likelihood of the *evidence* on the basis of the *theory* – as a potential confirmation measure. As Rescher notes,

The step from [pr (evidence/theory)] to a covering law model of explanation – as we have it in the Hempel-Oppenheim paper – is both short and easy. Both approaches have it that the optimal theory is that one which best explains the evidence. And of course if the reasoning is strictly deductive, the probability at issue will be 1.... We are thus led to a deductive model where the step from the theory (plus prevailing boundary conditions) to [observation of] the evidence involves a step of certainty. [5]

The value of this step, on Rescher's view, is the *unification* it provides. As Rescher puts it,

We have here an instance of the confluence of inclinations. But the main themes of the Hempel-Oppenheim paper – namely the so-called nomological-deductive model of scientific explanation and the thesis of explanation-prediction symmetry – reflect at once the interdisciplinarity (or, rather, non-disciplinarity) of Oppenheim's concern for structure and the mathematico-logical formalism that Hempel drew from the Reichenbach-Carnap tradition. [6]

Here then are two very different accounts of SLE. Though there are significant strands of truth in both (more, I think, in Rescher's), neither one, I will argue, gets the story quite right. After revisiting the centerpiece of SLE, the formal account of explanation that Hempel and Oppenheim offer in its Section 7 (in which definitions of law and explanation are provided for the simple formal language L), I will discuss several issues germane to of the covering law account. At least some of these will be familiar ground for philosophers of science, but they appear in a new light from the point of view of Hempel and Oppenheim's formal account. The issues to be raised include, for example, the historical context of their enterprise as Hempel and Oppenheim understood it, the fact that Hempel and Oppenheim identify their (informal) account as causal, the extent to which their project is an exercise in unification, and, relatedly, *explication* as we find it in SLE. I will then assemble these pieces into an account of SLE which has, I believe, advantages over both Rescher's and Salmon's.

Although the account of explanation most often drawn from SLE is the informal one described in its first four sections, that informal account is, for Hempel and Oppenheim, a *preliminary* analysis of explanation; indeed, it is an elementary analysis of a concept of explanation which Hempel and Oppenheim take to be fairly widely received among philosophers and scientists, in spite of the fact that there is, as they put it, "considerable difference of opinion as to the function and essential characteristics of scientific explanation."[7] Hempel and Oppenheim's formal account of scientific explanation, quite often ignored in discussions of the covering law model, is in fact crucial to it, for it is intended to clear up these differences of opinion and provide what Hempel and Oppenheim call a *rigorous* analysis. Here it is important also to recognize that Hempel and Oppenheim's move to formalization, and particularly to an account of *explanation-in-L* for the formal language L, as opposed to explanation for any language, formal or not, is motivated by the need to limit scientific laws to those which are both not finite in scope and which make no reference to individuals. And this, in turn, is motivated by a reflection on universalized sentences which, Hempel and Oppenheim claim, "we would refuse" to recognize as laws.[8] Since it is often almost impossible to determine if the meaning of a natural language predicate avoids reference to an individual, i.e., is "purely qualitative," the account must (at first glance) be tied to a formal language, which is "governed by a well-determined system of logical rules, and in which every term either is characterized as primitive or is introduced by an explicit definition in terms of the primitives."[9] But this, Hempel and Oppenheim point out, "does not by itself suffice to overcome the specific difficulty under discussion," for the primitive predicates of L cannot receive their meanings from within L itself, but must receive their meanings via "semantical laws of interpretation" which ultimately connect them to natural language.[10] If the predicates in the *natural* language have their meanings fixed by reference to individuals, then so will the primitive predicates defined in terms of them; until this problem is solved for natural language "the problem of an adequate definition of purely qualitative predicates remains open."[11] The point worth calling attention to here is that, although Hempel and Oppenheim offer, strictly speaking, a definition not of explanation but of *explanation-in-L*, they recognize from the start that their account depends on no less than an acceptable semantics for natural language, and specifically on the definition of a qualitative predicate in English. There is no suggestion that Hempel and Oppenheim regret or bemoan this; the point to be taken from it is that the notion of explication at work here involves a strong – indeed, inevitable – tie to natural language on two counts: in its somewhat fuzzy, malleable, semantics and in its according significant status to "what we are willing to say" about which sentences count as laws and which sets of sentences count as explanations.

Although Hempel and Oppenheim define not explanation but explanation-in-L, where L is a simple first-order predicate calculus with the usual logical apparatus (although without identity) and with the usual rules of formation and inference, they do not pretend that L is sufficient for science. Indeed, they state that "the question is open at present whether a constitution system can be constructed in which all of the concepts of empirical science are reduced, by chains of explicit definitions, to a basis of primitives of a purely qualitative character." [12] But they consider a definition of explanation-in-L sufficient to their purposes, in part because the analysis of law and language associated with it is "far from trivial." The account on offer, then, is in the form of a definition for the expression, "the ordered couple (T, C) constitutes an explanans for the sentence E," where E is a singular sentence, that is, one without variables. (T, C) constitutes an explanans for E just if (T, C) is a *potential* explanans for E, T is a theory and C is true. These three components of the definition are, of course, defined in turn. (T, C) is a *potential* explanans for E if and only if

1. C is singular (it contains no variables) and T is essentially generalized (it is quantified and not logically equivalent to a singular sentence),
2. E is derivable from (T, C) jointly, and
3. T is compatible with at least one class of basic sentences which has C but not E as a consequence.

The second condition, that E be derivable jointly from (T, C) (it is a consequence of the third condition, in fact, that E does not follow from C alone) reflects Hempel and Oppenheim's requirement that explanation in science make use of *true* scientific laws, for T – what Hempel and Oppenheim call a theory – is any true essentially generalized sentence in L, i.e., a true quantified sentence (either existentially or universally quantified) which is equivalent to no singular sentence. This requirement is perhaps the most well-known feature of Hempel and Oppenheim's theory of scientific explanation. Interestingly, the law in question might be an existential rather than a universal generalization, raising the question of why laws – understood as true generalized statements in L – play such a prominent role in the Hempel-Oppenheim account of explanation. This is a question to which neither Salmon nor Rescher attends, but to which I will offer a conjectural answer. Hempel and Oppenheim's emphasis on laws in explanation is driven by the unificationism at the very heart of SLE, a theme reflected also in Hempel and Oppenheim's familiar thesis that explanation and prediction are logically indistinguishable. "Whatever will be said in this article concerning the logical characteristics of explanation or prediction" they wrote, "will be applicable to either, even if only one of them should be mentioned." [13] Taking the import of prediction to be confirmation, recognizing that confirmation bears on all generalized statements, whether universal or existential, and adding both this symmetry thesis and the emphasis on unification, we can, I believe, grasp the motive for finding in every explanans a general statement. "Explanations" which proceeded without laws would characterize, or take place in, sciences with different logics of confirmation (on pain of violations of the symmetry of explanation and prediction), which is to say, sciences with different logics altogether.

Returning to the formal definition of a potential explanans. The third condition, that T be compatible with at least one class of basic sentences which has C but not E as a consequence, is designed to rule out non-explanations that would otherwise arise if C, the singular sentence in the explanans, were simply a conditional sentence with a singular instance of T as its antecedent and E as its consequent. By this means, note Hempel and Oppenheim, "any given particular fact could be explained by means of any true lawlike sentence whatsoever." [14] The idea behind this third condition is to block such C's by guaranteeing that, given T, C can be verified while E is not, for in the objectionable cases the verification of C depends upon the verification of E. To get at this idea more precisely, Hempel and Oppenheim define the verification of a singular sentence S as "the establishment of the truth of some class of basic sentences which has S as a consequence," [15] where a basic sentence is either a singular sentence without connectives – i.e., an atomic sentence – or the negation of such a sentence. The effect of requiring that there be a class of basic sentences which verify C but not E, given L, is, then, to insulate C from E in the desired fashion. What motivates this third condition fundamentally, though – what Hempel and Oppenheim ultimately find objectionable about the constructed C – is that it is tantamount to self-explanation, in which explanans and explanandum are identical. The countenancing of self-explanation is presumably not a desirable consequence on any theory of explanation, but there are many reasons – even incompatible reasons – for rejecting self-explanation. Hempel and Oppenheim's reason, though, seems to be just the one mentioned earlier – namely, the need to involve a generalization of some sort in the explanans. I therefore trace this third component of the definition of a potential explanans, like the first, to SLE's bedrock theme, unification.

This then is what Hempel and Oppenheim took to be the rigorous explication of a widely shared notion of explanation (recall that a potential explanans for E becomes an honest-to-goodness explanans just when T is a theory and C is true.) Let us return to Rescher and Salmon's respective discussions of SLE, employing some of the points mentioned above in a critical bent against these two papers, and along the way further developing my own account.

There is one thing that Salmon gets right. Recall that one theme Salmon emphasizes is the role of rigorous explication in SLE, almost as an aim in itself. I think this is largely correct, both in the sense that this is what occurs, in SLE, and in that this is a substantial part of Hempel and Oppenheim's conception of their project, a project they, indeed, took themselves to be continuing rather than initiating. Salmon notes that Hempel and Oppenheim cite both Karl Popper and John Stuart Mill as "anticipators" of the covering law account; in fact the list of sympathizers cited by Hempel and Oppenheim includes, besides Mill and Popper, William Jevons, C. J. Ducasse, Clark Hull, Herbert Feigl, Morton White, J. H.

Woodger, Felix Kaufmann, E. C. Tolman, Edgar Zilsel, and Kurt Grelling.[16] As Hempel and Oppenheim understood it, a covering law account of explanation was nothing new. I've also noted above that Hempel and Oppenheim saw their project as indebted to what we are inclined to say about, for example, what does (or does not) counts as a law, which I take to be a mark of an explication project. One further example in this regard: For Hempel and Oppenheim, the matter of whether to restrict laws to *true*, pure and essential generalizations, as opposed to pure and essential generalizations for which we only have very strong evidence, turns wholly on "the meaning customarily assigned to the concept of law in science and in methodological inquiry."[17] Laws are true in SLE for no reason other than the desire to not diverge from how scientists talk.

But, as I've suggested above, explication is not the whole story behind SLE. At a deeper and more fundamental level is the drive for unification, understood here in the rather simple sense of singularity of method. One argument for this claim I've sketched above; the symmetry thesis defended by Hempel and Oppenheim is itself best explained by reference to unificationism. Let me supplement that argument briefly by pointing out one kind of instance in which the aim of explicating what we are inclined to say comes into conflict with some aspect of Hempel and Oppenheim's unificationist project, and in which the latter wins. There are many cases from disparate sciences in which the citation of future goals or aims, or even individual events, count as complete explanans, and indeed Hempel and Oppenheim recognize such explanans as part and parcel of the practice of these various sciences. Readers of SLE will recall the considerable work expended in bringing (and often forcing) these practices into the covering-law mold. What could be behind such effort but the primacy of unity over explication in Hempel and Oppenheim's project?

What about Salmon's second theme, regarding the role of scientific realism as cause, and as effect, in relation to SLE? This is a much broader claim, and one much harder to assess, not only because SLE is mute on the question of scientific realism, but also because one suspects it is a dangerous anachronism to impose what Salmon and others have come to mean by scientific realism in 1998 upon the views Hempel, Oppenheim, and their peers in 1948. And then there is Hempel and Oppenheim's repeated announcement that their's is an account of *causal* explanation. What exactly this meant in 1948, and what relation it may or may not have to realism, I will not try to address here.

What has been said so far, however, has an implication which tells against Salmon's account. I've sketched an argument that various features of SLE incline us to read it as an exercise in unification. Further, we've noted that Hempel and Oppenheim take themselves to be explicating a common and not particularly contested notion of explanation. On Hempel and Oppenheim's own view, what SLE offers is a clear explication of scientific explanation, which, in the end, presents explanation as a flavor of confirmation or, more broadly and accurately, systematization. But if all this is true, in what sense could there have been the "remarkable reversal of attitude" of which Salmon speaks, namely, the acceptance of the view that science can, and does, provide knowledge *why* something happens, in addition to the knowledge *that* it does happen? The answer, I suspect, is that there was no such change of attitude before, around, or even for some years after, 1948, at least among Hempel, Carnap, their peers, and their students. By this I do not mean that the presumption that science provides some new kind of knowledge has been there all along, but just the opposite; in SLE what we have is a characterization of explanation as knowledge *why* which assimilates it fully to knowledge *what*. This is explicit at least one point in SLE, when Hempel and Oppenheim summarize an exemplar explanation by noting that "here again, the question 'Why does the phenomenon occur?' is construed as meaning 'according to what general laws, and by virtue of what antecedent conditions does the phenomenon occur'?" [18] It is a tempting corollary to this suggestion that we have been led by Salmon to retrospectively associate with SLE a conception of scientific knowledge which is in fact foreign to Hempel and Oppenheim's 1948 project.

Given my comments on Salmon, my points of agreement with Rescher with respect to SLE may be clear. In emphasizing the unificationist project in SLE, which Rescher traces to Oppenheim, Rescher has, I think, identified a key aspect of SLE. Relatedly, and as significantly, Rescher does not find in SLE the first articulation of a covering law account or an episode of attitude reversal regarding the legitimacy of scientific explanation. But briefly, it may also be clear in what respect I think Rescher misses an important dimension of SLE, for in his account the role of explication in the Hempel-Oppenheim project, prized by Salmon, is hardly recognized. This is not merely to slight an instance of skillful work in analytic philosophy; it is to miss what I think may be an important motivation and at the same time an important *effect* of SLE. As an exercise in explication SLE deserves recognition not just as a paper but as an attempt to influence the behavior of scientists and philosophers. Any account of SLE which overlooks this obvious point is, I think, a non-starter. Explication had, for Hempel and Oppenheim, a practical point. And in this respect, it is also crucial to recognize, as Salmon does, that SLE was remarkably successful. SLE is installed in the canon of philosophy of science, where it will remain for perhaps decades. And this, despite that it is not, at least not always, correctly understood.

NOTES

1. Carl Hempel and Paul Oppenheim, "Studies in the Logic of Explanation", in: *Philosophy of Science*, 15, 1948, pp. 135-175. Reprinted with a postscript in: Carl Hempel, *Aspects of Scientific Explanation*, New York: The Free Press 1965, pp. 245-290. All page references here to Hempel and Oppenheim's 1948 paper are to the 1965 reprinting.
2. Wesley Salmon, "The Spirit of Logical Empiricism: Carl G. Hempel's Role in Twentieth-Century Philosophy of Science", in: *Philosophy of Science*, 66, 3, 1999, pp. 333-350; and Nicholas Rescher, "H_2O: Hempel-Helmer-Oppenheim, An Episode in the History of Scientific Philosophy in the 20th Century", in: *Philosophy of Science*, 64, 2, 1997, pp. 334-360.
3. Salmon distinguishes logical positivism from logical empiricism, the former (but not the latter) being committed to foundationalism and infallibilism and disinterested in probability. See Salmon, pp. 334ff.
4. *Ibid.*, pp. 339-340.
5. Rescher, "Hempel-Helmer-Oppenheim", *op. cit.*, pp. 351-352
6. *Ibid.*, p. 346.
7. Hempel and Oppenheim, "Studies in the Logic of Explanation", *op. cit.*, p. 245.
8. *Ibid.*, p. 266.
9. *Ibid.*, p. 269.
10. *Ibid.*
11. *Ibid.*, p. 270.
12. *Ibid.*, p. 271.
13. *Ibid.*, p. 249.
14. *Ibid.*, p. 276.
15. *Ibid.*, p. 277.
16. *Ibid.*, pp. 250ff.
17. *Ibid.*, p. 265.
18. *Ibid.*, p. 246.

College of Letters & Science
Department of Philosophy
University of Wisconsin – Stevens Point
Stevens Point, WI 54481-3897
U.S.A.
ghardcas@uwsp.edu

DAVID J. STUMP

FROM THE VALUES OF SCIENTIFIC PHILOSOPHY TO THE VALUE NEUTRALITY OF THE PHILOSOPHY OF SCIENCE

Members of the Vienna Circle played a pivotal role in defining the work that came to be known as the philosophy of science, yet the Vienna Circle itself is now known to have had much broader concerns and to have been more rooted in philosophical tradition (especially neo-Kantianism) than was once thought. Like current and past philosophers of science, members of the Vienna Circle took science as the object of philosophical reflection (whether to provide a foundation for the sciences or simply to clarify scientific terminology and assumptions,) but they also endeavored to render philosophy in general compatible with contemporary science and to define and promote a scientific world view. This latter task seems to continue the work of so-called scientific philosophy, a label embraced by many philosophers in the late nineteenth and early twentieth century, such as Helmholtz, Mach, Avenarius, the neo-Kantians, Husserl, Carus, Peirce, and, of course, Russell during the period when he was applying modern logic to philosophical problems. Russell's program influenced Carnap directly, though the idea of applying modern logic to philosophical problems became a defining feature of analytic philosophy and was applied to many areas of philosophy, not only to the philosophy of science. Scientific philosophy included the promotion of the cultural values of modernity, especially the values embodied in the scientific world conception. By exploring the various meanings ascribed to scientific philosophy in the late nineteenth and early twentieth centuries, I will investigate whether the promotion of scientific philosophy and of the values associated with a scientific world conception is merely part of a transitory social context within which Logical Positivism developed or if it is an enduring part of the philosophy of science. Moreover, the residue of values remaining in the philosophy of science can be brought to light by studying its history.

Let us start by comparing scientific philosophy as it was practiced approximately one hundred years ago and philosophy of science as it is practiced today. Members of the Vienna Circle and their collaborators will be seen as a transitional group between these two types of philosophy, not only because they appear temporally between the other two groups, but also because their views are intermediary. The philosophical categories that I am defining are necessarily broad and imprecise. Scientific philosophy especially has been taken to refer to a large and diverse set of philosophers, but I believe that it is associated with the following: a) A claim that philosophy is objective, genuine knowledge. b) A

M. Heidelberger and F. Stadler (eds.), History of Philosophy and Science, 147–158.

claim of the unity of knowledge, that philosophy and science are continuous. c) An aim to change philosophy in the light of recent scientific advances. d) Advocacy of internationalism in philosophy and in general. e) Promotion of a scientific world conception.

There is no doubt that in some usages, the term 'scientific philosophy' is merely honorific. Since German-speaking philosophers seem particularly susceptible to such usage, it may simply be due to the broadness of the term '*Wissenschaft*', but there may also be more to the story. Hegel represents, perhaps, the most controversial example of 'scientific philosophy' given that many philosophers claiming to be scientific explicitly distanced themselves from Hegel, whose *Science of Logic* would be thought to be neither scientific nor logical by the Logical Positivists and by most current philosophers of science as well. Alan Richardson makes a plausible case for Hegel, arguing that "Hegel's science was science as conceived in his times" (Richardson, 1997, p. 437), yet there is a danger in portraying scientific philosophy broadly enough to include Hegel, since it could disrupt any links that can be found between scientific philosophy and the Logical Positivists. Nevertheless, many who called themselves scientific philosophers had a strong connection to Hegel or more broadly to Idealism. Paul Carus, the editor of *The Monist*, provides a clear example of idealist scientific philosophy, and Russell, who we now know took longer to break with idealism than he avowed later in life, is another (e.g., Griffin, 1991, Hager, 1994). The Logical Positivists have been linked to idealism as well, and not only by Marxists (e.g., Richardson, 1992a, 1992b, Uebel, 1995).

For our purposes here, the religiosity of the idealist wing of scientific philosophers makes it particularly hard to see them as part of the scientific world conception that the Logical Positivists would later champion. Carus's interest in religion seems to be part of his universalism – the unity of knowledge thesis. While it is true that Carus is not considered first-rate, it is not easy to dismiss him from the ranks of scientific philosophers, since he promoted scientists such as Halsted, Mach, and Poincaré and scientific philosophers such as Russell and Peirce by publishing their work in his journal, and he entitled the summation of his life work *Philosophy as a Science* (1909). On the other hand, Carus published as widely on religion and his book on the Devil appears to have been his best seller (1898, 1899, 1900). He was equally committed to the study of religion and the study of science.

Associating Husserl's scientific philosophy with that of the Vienna Circle is problematic as well. His call for philosophy as *streng Wissenschaft* is clearly honorific, that is, a call for philosophy to be objective knowledge, as good as, if different from, natural science. In *The Crisis of European Sciences*, Husserl argues that philosophy has fallen behind the natural sciences and that it must reestablish itself on a new foundation to maintain its traditional place as universal knowledge. The problem with Husserl's version of scientific philosophy, when tied to Logical Positivism and to the philosophy of science, is that Husserl's phenomenology is transcendental and the sciences must be subordinate to it: "Sciences in the plural, all those sciences ever to be established or already under construction, are but dependent branches of the One Philosophy" (1936, 1970) p. 8). This is hardly the perspective of the Logical Positivists, who saw philosophy as continuous with science. Even when he acknowledges the advances of science, Husserl always wants to leave room for transcendental philosophy. There seems to be an echo in Husserl of the project of the Southwest School of neo-Kantians to find a replacement for Kant's transcendental logic in some new transcendental philosophy. By contrast, the Marburg school of neo-Kantians and its followers among the Logical Positivists would attempt to use formal logic to fill that role. Thus, if we follow Michael Friedman's analysis of the continental/analytic split in contemporary philosophy as following from the split within neo-Kantianism (1996), Husserl ends up clearly on the continental, non-scientific side of the divide, despite his professed scientific philosophy.

Calls for philosophy to be counted as genuine knowledge by Carus, Husserl and others is part of a universalism present in scientific philosophy, a vision of the unity of knowledge that claims that philosophy and science are different names for the same discipline (b). At the turn of the century there was a strong sense that science and philosophy had become separate and that philosophy needed to get back in touch with science. Scientific philosophers not only were concerned that philosophy be taken as real knowledge, but they also often demanded that philosophy be given the paramount position as the universal science, as opposed to the merely special sciences. Demands for maintaining philosophy's traditional role as universal knowledge may be a reaction to the academic specialization that was taking place at the end of the nineteenth century, splitting the natural and then the human sciences away from philosophy, leaving philosophy as a study of general (metaphysical) issues that no one can answer scientifically. Husserl often made such claims, and Phillip Frank, in his historical and autobiographical introduction to *Modern Science and Its Philosophy* tells a compatible story as well, even while pointing out that such universalism can turn into idealism or even into Nazi propaganda (1949, p. 193). The problem with the unity of knowledge thesis is that it looks like a return to the traditional role of philosophy as primary, not like an advance to a scientific world view. Therefore, the theme of the unity of knowledge was resisted or modified by the Logical Positivists as we see exemplified by Frank.

A similar but more narrow theme of scientific philosophy was the aim to change philosophy in response to recent scientific advances (c). Starting with his dissertation on geometry, Russell best exemplified this aim among scientific philosophers. Russell's early works took results from contemporary mathematics and science and showed how philosophy must change in order to account for scientific advances. For example, in *Our Knowledge of the External World*, he says that the mathematical theory of the infinite is a triumph of scientific method in philosophy (Russell, 1914, p. 189) and that the new mathematical theory of the infinite surpasses at least two thousand years of philosophical thinking about the infinite (1914, ch. VI and 1917, ch. 5). Similar claims are made about physics in *Our Knowledge of the External World* (1914, ch. IV) and in *Human Knowledge* (1948), Russell explicitly takes scientific knowledge for granted as the proper starting point for philosophical thinking about knowledge. It is striking, however, how much Russell remains focussed on traditional philosophical issues rather than on the content of science (except for his popular books on relativity). Russell is therefore not usually counted as a philosopher of science, though he is, of course, more than anyone else, the model for philosophers of mathematics.

Russell and Moore's dissatisfaction with their idealist teachers is famous and their critics are almost as harsh as those by the Logical Positivists and Empiricists. These attacks on other kinds of philosophy by scientific philosophers seem to be a reaction to the romanticism of the nineteenth century, which had for the first time removed philosophy from science, or perhaps it is better to say that science left philosophy, since scientific philosophy was developed in the context of academic specialization. Scientific philosophers such as Russell, for example, with his degrees in both mathematics and philosophy, exemplify interdisciplinarity, as do, of course, the original members of the Vienna Circle, all of whom had science backgrounds prior to their philosophical work. However, a degree of disciplinary separation must precede the emergence of interdisciplinarity. William James could be a psychologist in a philosophy department without being explicitly interdisciplinary, as Durkheim could be a quantitative sociologist while being trained in and continuing to teach very traditional philosophy. Professionalization and academic specialization throughout the university in the last century explains much, but perhaps not all, of the motivation of scientific philosophers to maintain a unified field of knowledge and of philosophers of science to create an interdisciplinary field. Describing oneself as interdisciplinary is a tacit acknowledgement of the disunity of knowledge.

Scientific philosophers saw themselves as breaking away from idealism and as maintaining a traditional link between philosophy and science. This suggests that there was only a very brief period in the nineteenth century when it would have been unusual for a philosopher to have studied a science, but of course science had changed dramatically by the nineteenth century. Fighting a losing battle against academic specialization, scientific philosophers were trying to maintain the traditional unity of knowledge that had always allowed philosophers to be universalists. Even Russell's more narrow version of the link between science and philosophy is again very traditional, since many earlier philosophers had made similar links, not just nineteenth and twentieth century scientific philosophers. Even though self-described scientific philosophers may have followed science more explicitly than traditional philosophers, it was actually quite common for philosophers to try to integrate their views with the sciences of their day. Aristotle, Descartes, Leibniz, and Kant, for example, were all deeply involved in the science of their time and their philosophies each reflect this. For example, Descartes' dualism provided for the possibility of a unified theory of matter – physics, something that the scholastic hylomorphic theory seemed to rule out (Garber, 1992, pp. 94-5).

The last two themes that I will consider – internationalism and the scientific world conception – explicitly express the values of scientific philosophy. The turn of the century was a time of widespread international scientific cooperation, marked by the scientists' extensive travel to international conferences and World's Fairs, by the development of new organizations, and by work on standards (Lyons, 1963, pp. 223-245). There were numerous international scientific congresses in all fields. For example, the International Mathematical Congresses began with sessions as part of the World's Fair in Chicago in 1893 and continued more formally in a four-year rotation (1897 in Zurich, 1900 in Paris, 1904 in Heidelberg, 1908 in Rome, 1912 in Cambridge, etc. Albers, Alexanderson, and Reid, 1986). The First International Philosophy Congress was organized in 1900 in Paris under the auspices of the *Revue de Métaphysique et de Morale* and Paris was also the site of another World's Fair that year, one of the biggest ever. As is well known, Russell, Couturat, Peano and others attended both the mathematics and philosophy congresses in Paris.

Internationalism at the turn of the century included not only internationalism in science but also in religion, in the peace movement and in the left wing of the labor movement (Bernal, Eijkman, Crawford, Forman, Lyons). Significantly, the two waves of international cooperation – before and after WWI – coincide with the rise of scientific philosophy and with the rise of Logical Positivism, respectively. Internationalism was important to the Logical Positivists as a vehicle to express their opposition to rising German nationalism. One tends to think of science as intrinsically international, thus it would be natural to think of scientific philosophy as intrinsically international as well. However, the temporal parallels between the rise of scientific philosophies and of internationalism in general society require us to consider whether internationalism really is a necessary aspect of scientific philosophy. International cooperation in science went through a difficult period during WWI, with scientists participating in the war effort, especially in Germany, and again in WWII, but science as it is practiced today is surely an international institution. A close comparison of how philosophy has been organized over the last two centuries would be required to see the affect, if any that the rise and fall of international cooperation had on philosophical institutions.

The last theme, the scientific world conception, is not exclusive to scientific philosophers, but rather is also expressed in art and literature and in various social movements, as Peter Galison has shown persuasively in his studies of the connections between the Vienna Circle and the Bauhaus. The two institutions supported each other by expressing a modern, scientific world view and more broadly by developing a modern, scientific way of life (Galison, 1990, p. 716). While striving for a new kind of objectivity, they developed, respectively, an anti-traditional philosophy, and an anti-traditional aesthetic that shared scientism and the use of machine images, and that built from simple elements according to explicit rules in order to avoid intuition and general concepts (Galison, 1990, p. 725).

The scientific world conception is connected to progressive, educated, Enlightenment values, and to the idea of modernity in general. The Logical Positivist's connection to the themes of the Enlightenment is quite explicit in Neurath's work, as has been shown by Thomas Uebel (1996). Enlightenment philosophers such as Condorcet and Kant viewed the rise of modern science as the triumph of reason over dogmatic religion and superstition, claiming that progress in science and technology would have a strongly positive impact on society and that the application of scientific methods to society itself would automatically lead to human happiness and social and political justice: "we shall demonstrate how nature has joined together indissolubly the progress of knowledge and that of liberty, virtue and respect for the natural rights of man" (Condorcet, 1976, p. 215). Neurath and Condorcet share a vision of a social science that will solve human problems by applying the methods and results of science to them. Still, despite some clear successes such as the public health movement in the nineteenth century, the encroachment of technical science on society and the resulting rise of bureaucratic control has been seen as a threat by many. In the late nineteenth and early twentieth centuries, the progressive vision of modern society that arose in the Enlightenment seemed to lead to a society divided by class and gender, to machines of mass destruction, and to alienation and a loss of meaning and value. Technological advances in transportation, food, and medicine have not eliminated hunger and disease and there have been higher civilian casualties with each successive world war – literacy and the dissemination of knowledge did not prevent the genocide of the Holocaust. Nevertheless, in the early twentieth century, artists, architects, philosophers and both left and right wing political movements attempted to apply the modern scientific world view as never before. Is it possible to still believe that advances in science will bring automatic advances in ethics, justice and freedom? Before considering this question, we need to discuss the current state of the philosophy of science.

The philosophy of science as it is practiced today is (a) one of several academic specializations within philosophy, which (b) aims at understanding science and (c) takes science as the subject matter of philosophical reflection. Philosophy of science (d) has no necessary connection to other areas of philosophical study. Comparing the current web site for the journal *Philosophy of Science* with the editorial by W. M. Malisoff from the first issue reveals major changes. The current web site clearly takes for granted that 'philosophy of science' is a well-known and well-understood term, while the inaugural editorial for the journal is entitled "What is Philosophy of Science?" (Malisoff, 1934). The current web site explicitly embraces pluralism, while the inaugural editorial sets out a specific research program, albeit a rather general and open-ended one. The scientific philosophers' claim of the unity of science and philosophy (1b) is explicitly dropped in the inaugural editorial of the journal and is not mentioned at all on the web site. While science is explicitly acknowledged to be important by philosophers of science, the role of science in broader culture is usually not discussed (e). The inaugural editorial comes close in the last item in Malisoff's proposed research agenda, number "VI. Studies in the function and significance of science within various contexts", but even here there is no explicit mention of social issues (Malisoff, 1934, p. 4). There are, of course, occasional works in current philosophy of science that touch on the relation of science to social issues, the debate in the United States over the teaching of evolution or creationism in the schools, for example, or on gender bias in science, but these are not issues that have been considered central to the discipline.

While the more narrow concern for science as an object of philosophical reflection defines the philosophy of science as it is practiced today, many earlier philosophers engaged in philosophical reflection on science as well, almost from the time that science broke away from philosophy, the major difference being that contemporary philosophers of science specialize in one area of the sciences, such as physics or biology, or in scientific method, while for earlier philosophers, Mill and Kant, for example, reflections on science would have been a part of a much more sweeping set of philosophical views. Since specialization is part of a general trend in academia, it alone cannot be used to differentiate between contemporary philosophers of science and earlier scientific philosophers. Thus, there is some question as to how distinctive philosophy of science can really claim to be, outside of the social or institutional fact of academic specialization as it currently exists.

Logical Positivism can be seen to hold an intermediate position between scientific philosophy and the philosophy of science, the Logical Positivists themselves having been at the forefront of establishing philosophy of science as a sub-specialization in philosophy in the United States with a professional association, a journal and biennial conferences. Most of the themes discussed here as forming the basis of both scientific philosophy and of the philosophy of science can be found in the work of the Logical Positivists. From scientific philosophy they took the claim of the unity of knowledge (the continuity between philosophy and science), the aim of changing philosophy in the light of recent scientific advances, and advocacy of Internationalism and the scientific world conception. With contemporary philosophers of science, the Vienna Circle shared the aim of understanding science by taking science as the primary subject matter of philosophical reflection, and they saw themselves as doing something new and different from the work of past philosophers. When members of the Vienna Circle moved to the United States and the movement was renamed Logical Empiricism, the role of science in broader culture was usually not discussed, as it is usually not discussed in contemporary philosophy of science.

What values have been lost in the change from scientific philosophy and philosophy of science? First of all, the aim of making philosophy more scientific has been dropped. The unity of knowledge claim that philosophy and science are continuous, and also the aim of change philosophy in the light of recent scientific advances imply a strong stance on the nature of philosophy that does not allow for pluralism. The Logical Positivists, and especially Alfred Ayer, championed the principle of verification as the tool for removing metaphysical nonsense from philosophy and leaving a scientific core. By contrast, the fact that contemporary philosophy of science is considered one among many sub-specializations in philosophy (2a) almost guarantees a tolerant pluralism. Other elements of contemporary philosophy of science strongly encourage philosophical pluralism as well, since interpretation of science is left exclusively to philosophers of science, who are then excused from having to consider any other philosophical issues. Philosophers can carry on their own studies, independent of each other or only connected to others in their sub-discipline. Thus, while scientific philosophy was at odds with philosophy itself, the philosophy of science is not.

A passage from Reichenbach's *The Rise of Scientific Philosophy* (1951) shows not only the existence of such contention, but also a subtle recognition of the difference between the philosophy of science and scientific philosophy. Reichenbach contrasts the philosopher of science with the traditional philosopher, but then switches labels and calls the hero who is willing to confront the traditionalist a 'scientific philosopher':

The philosopher of the traditional school ... does not realize that philosophical systems have lost their significance and that their function has been taken over by the philosophy of science. The scientific philosopher is not afraid of such antagonism. He leaves it to the old-style philosopher to invent philosophical systems, for which there still may be a place assignable in the philosophical museum called the history of philosophy – and goes to work (1951, 123-4).

There is nothing said about changing philosophy, let alone anything about general social issues, in all but a handful of philosophy of science works. Even Ayer would later apologize for his harsh tone in *Language, Truth and Logic*, calling it "a young man's book." The insular nature of current philosophy of science and the fundamental way that it has lost its value orientation is striking, even if one is glad that pluralism now reigns. Philosophers of science have become so divorced from values that they are not even willing to fight about the nature of philosophy, let alone any broader social issues. Philosophers of science seem less territorial than analytic philosophers in general, who are largely responsible for marginalizing feminist, continental, Asian, and medieval philosophy in the United States.

Second, the Enlightenment optimism about the social impact of science has disappeared in the transition from Logical Positivism to the philosophy of science. There is a long story to be told about how science came to be viewed as value neutral in the nineteenth century, and came to be dominated by machinery rather than virtuosi (Daston and Galison, 1992). Perhaps we are simply witnessing a cultural or social phenomenon in the decline of optimism about science, and philosophy has simply reflected the larger cultural trend, as seemed to be the case with internationalism as well. In the heyday of scientific philosophy, optimism about science was widespread. In the period between the two World Wars when the Vienna Circle was formed, optimism about science again rose, among both the political left and right. We are now in a period of great cynicism, according to many culture critics.

Recently works explore why the members of the Vienna Circle became less political when they came to the United States (e. g. Giere, 1996) and the general chilling effect of McCarthyism on philosophy (McCumber, 2000). Members of the Vienna Circle may have been hiding their past connections to left-wing groups because of the anti-Communist climate in the United States, or they may have felt the need to become respectably academic or, as was noted in a recent email on the HOPOS discussion list, perhaps the political values expressed by the Vienna Circle, with the exception of Neurath, have been rather overstated (Ongley, 2000). However, something like the Enlightenment connection between science and progress is, I want to suggest, built into the Logical Positivists' conception of their work in promoting the scientific world conception. Enlightenment optimism provides one way of understanding how progressive values could be seen to be built into the scientific world conception or even the philosophy of science, even when these seem to be totally disconnected from politics and values. In conclusion, I want to reflect on whether there can be any connection between epistemologies and values and consider the philosophy of science as a possible political force.

Objective value-free knowledge was taken by Enlightenment figures to lead automatically to moral and ethical progress, as well as cognitive and material progress. We will be freer, there will be less war, and there will be more equality between people simply by increasing scientific knowledge, according to Condorcet. If science is necessarily progressive, then the philosophy of science, even if explicitly value neutral, should also be progressive. Reflexively increasing our understanding of scientific knowledge through the philosophy of science is the first step towards being self-consciously scientific and modern. However, how can politics (or values generally) be built into an epistemology or a metaphysics that is explicitly apolitical? Most philosophers of science reject out of hand the idea that science in general could be sexist, for example. However, the same form of argument needs to be made in order to claim that science in general is politically progressive, or that Heidegger's existential phenomenology is Fascist, when both appear on the surface to be value neutral. One reason that it is difficult to prove that a method or an epistemology tends towards particular values is because it often appears to be possible to ground both left-wing and right-wing ideologies on the same epistemology or metaphysics. The philosophy of science as it is practiced today suffers from a further problem in that academic specialization and the general lack of political engagement of academics, especially in the United States, gives philosophers of science very few natural venues for discussing social and political issues.

The issue of whether or not a general epistemology or metaphysics is endowed with a particular value content can be examined with some poignancy by looking at Walter Benjamin's famous paper "The Work of Art in the Age of Its Mechanical Reproducibility" (1936). Benjamin sets out to give an analysis of changes taking place in modern art as a compliment to Marx's analysis of the economic sphere. The central thesis of Benjamin's essay is that in contemporary society, the aura of works of art is destroyed as print media, recordings, and film become widely available. Art that is mechanically reproducible is not unique and can be brought into our homes, thus removing its distance and unapproachability – its status as a special object. This destruction of the aura of the work of art constitutes a fundamental change in our perception of objects that Benjamin sees as an essential part of the creation of mass culture and eventually a classless society, so he claims that his development of new concepts to analyze art will aid revolutionary change. A tension is introduced in the epilogue of his paper, however, where Benjamin raises the specter of the appropriation of mass culture and mechanically reproducible art by Fascists, especially the Italian Futurists. Benjamin calls this appropriation the "aesthetization of politics." The Fascists may attempt to legitimate the Capitalist system by making it aesthetic. He answers the Fascists with a call for the politicization of art (Benjamin, 1936, p. 253).

The right-wing use of technological and scientific imagery that Jeffery Herf analyzed in his book *Reactionary Modernism* (1984) calls into question the Enlightenment link between the scientific world outlook and progressive politics in the same way that the rise of Fascist art called into question Benjamin's link between mechanically reproducible art and progressive politics. There is no guarantee that adopting a scientific world conception will lead to political progress, unless these politically progressive values can somehow be built into this conception. Viewing science and the philosophy of science as value neutral seems to make them apolitical, as do the academic institutional structures under which the philosophy of science has developed. The philosophy of science is no longer a movement with a manifesto, it is a well-integrated part of the academic mainstream.

There are at least four possible ways to recover a modicum of value orientation within the philosophy of science. First, as Warner Wick pointed out long ago, Carnap's explicit recognition of the pragmatic or practical nature of justifications of linguistic frameworks (Carnap's external questions) implies that Logical Positivism is ultimately a political philosophy in the sense that values guide our pragmatic choices. However, Carnap left pragmatics, the realm of practical reasoning, very undeveloped, in comparison with his work on syntax and semantics (Wick, 1951, pp. 50-51). Indeed, Carnap often seems to imply that the choice of a linguistic framework is a matter of indifference. Second, even if there is no direct connection between a particular epistemology and political values, certain epistemologies may limit the range of possible values. This is shown in the philosophy of science itself, to some extent, by those who accept the fact/value distinction, thus eliminating value judgments from science. Third, it may be that some areas of science and the philosophy of science have a strong connection to values and politics, while others have little connection. It is no accident that the human sciences were so strongly emphasized by Enlightenment philosophers like Condorcet and by critics of the Enlightenment like Foucault.

The human sciences have a much more direct impact on society than the exact sciences do, so much so that it is thought to be hard to imagine a value-neutral human science and hard to imagine the social relevance of the exact sciences. Thus, there may be a definite but only partial connection between science and values, and if there is such a connection it needs to be shown in a careful study of a particular science and its historical and social context. Finally, philosophers of science have a role in educating the public about science, either in the classroom or in academic or other writing. We inevitably take stands on many issues in science, and need to be aware of the effect that these stands can have on our audience. Of course, some connections to political topics of the day will be more obvious than others, but by taking science as the object of philosophical reflection, philosophers of science are engaging an institution that plays a major role in contemporary life and are therefore dealing with issues that are often directly related to issues of public concern. Philosophers of science therefore retain their potential to affect public discourse by performing their role as interpreters of science and judges of scientific practice.

REFERENCES

Albers, Donald J., Gerald L. Alexanderson and Constance Reid (1986). *International Mathematical Congresses: an Illustrated History, 1893-1986*. New York: Springer Verlag.

Benjamin, Walter (1936). "The Work of Art in the Age of Mechanical Reproduction." *Zeitschrift für Sozialforschung* translated by H. Zohn, reprinted in *Illuminations*. H. Arendt, ed. New York: Harcourt, Brace and Co, 1968, pp. 219-253.

Bernal, J. D. et al (1975). *Science, Internationalism and War*. New York: Arno Press Reprint of *Science against War*, an excerpt from *Science for Peace and Socialism*, by J. D. Bernal and M. Cornforth, published in 1949 by Birch Books, London; of *The Nazi Attack on International Science*, by J. Needham, published in 1941 by Watts, London; and of *Science in Soviet Russia*, edited by J. Needham and J. S. Davies, published in 1942 by Watts, London.

Carus, Paul (1895). *God: an Enquiry into the Nature of Man's Highest Ideal and a Solution of the Problem from the Standpoint of Science*. Chicago: Open Court.

Carus, Paul (1897). *The Gospel of Buddha according to Old Records*. Chicago: Open Court.

Carus, Paul (1900). *The History of the Devil and the Idea of Evil, from the Earliest Times to the Present Day*. Chicago: Open Court.

Carus, Paul (1909). *Philosophy as a Science: a Synopsis of Writings of Dr. Paul Carus*. Chicago: Open Court.

Condorcet, marquis de (1793). "Sketch for a Historical Picture of the Progress of the Human Mind." *Condorcet: Selected Writings*. Keith Baker, ed. and translator. New York: Macmillan Publishing Co, 1976, pp. 209-282, on p. 215.

Crawford, Elisabeth T. (1992). *Nationalism and Internationalism in Science, 1880-1939: Four Studies of the Nobel Population*. Cambridge and New York: Cambridge University Press.

Daston, Lorraine and Peter Galison (1992). "The Image of Objectivity." *Representations* 40: 81-128.

Eijkman, P. H. and Paul Samuel Reinch (1911). *L'internationalisme Scientifique (sciences pures et lettres)*. La Haye: [W. P. van Stockum et Fils].

Forman, Paul (1973). "Scientific Internationalism and the Weimar Physicists: The Ideology and its Maintenance in Germany after World War I." *Isis* 64: 151-180.

Frank, Phillip (1949). *Modern Science and its Philosophy*. Cambridge: Harvard University Press.

Friedman, Michael (1996). "Overcoming Metaphysics: Carnap and Heidegger." *Origins of Logical Empiricism.* Ronald N. Giere and Alan W. Richardson, eds. Minneapolis: University of Minnesota Press: 45-79.

Galison, Peter (1990). "Aufbau/Bauhaus: Logical Positivism and Architectural Modernism." *Critical Inquiry* 16: 709-52.

Galison, Peter (1996). "Constructing Modernism: The Cultural Location of Aufbau." *Origins of Logical Empiricism.* Ronald N. Giere and Alan W. Richardson, Eds. Minneapolis: University of Minnesota Press: 17-44.

Garber, Daniel (1992). *Descartes' Metaphysical Physics.* Chicago: University of Chicago Press.

Giere, Ronald (1996). "From Wissenschaftliche Philosophie to the Philosophy of Science." *Origins of Logical Empiricism.* Ronald N. Giere and Alan W. Richardson, eds. Minneapolis: University of Minnesota Press.

Griffin, Nicholas (1991). *Russell's Idealist Apprenticeship.* Oxford and New York: Oxford University Press.

Hager, Paul J. (1994). *Continuity and Change in the Development of Russell's Philosophy.* Dordrecht: Kluwer Academic Publishers.

Herf, Jeffery (1984). *Reactionary Modernism .* New York: Cambridge University Press.

Husserl, Edmund (1965). *Phenomenology and the Crisis of Philosophy: Philosophy as Rigorous Science, and Philosophy and the Crisis of European man.* New York: Harper & Row.

Lyons, F. S. L. (1963). *Internationalism in Europe, 1815-1914.* Leyden: A. W. Sythoff.

McCumber, John (2000). *Time in the Ditch: American Philosophy and the McCarthy Era.* Evanston: Northwestern University Press.

Ongley, John (2000). "Re: Kuhn, Fuller, Giere, et al." *HOPOS-L: A Forum for Discussion of the History of the Philosophy of Science.* 11:48 AM June 14, 2000.

Reichenbach, 1951, *The Rise of Scientific Philosophy* Berkeley: University of California Press.

Richardson, Alan (1992). "Logical Idealism and Carnap's Construction of the World." *Synthese* 93: 59-92.

Richardson, Alan (1992). "Metaphysics and Idealism in the Aufbau." *Grazer Philosophische Studien* 43: 45-72.

Richardson, Alan (1997). "Scientific Philosophy." *Perspectives on Science* 5(3): 418-451.

Russell, Bertrand (1914, 1993). *Our Knowledge of the External World as a Field for Scientific Method in Philosophy.* London: Routledge.

Russell, Bertrand (1897). *An Essay on the Foundations of Geometry.* Cambridge: Cambridge University Press.

Russell, Bertrand (1948). *Human Knowledge, Its Scope and Limits.* New York: Simon and Schuster.

Uebel, Thomas E. (1995). "Otto Neurath's Idealist Inheritance: The Socio-economic Theories of Wilhelm Neurath." *Synthese* 103: 87-121.

Uebel, Thomas E. (1996). "The Enlightenment Ambition of Epistemic Utopianism: Otto Neurath's Theory of Science in Historical Perspective." *Origins of Logical Empiricism.* Ronald N. Giere and Alan W. Richardson, eds. Minneapolis: University of Minnesota Press.

Wick, Warner A. (1951). "The 'Political' Philosophy of Logical Empiricism." *Philosophical Studies* II(4): 49-57.

Department of Philosophy
University of San Francisco
2130 Fulton Street
San Francisco, CA 94117
U.S.A.
stumpd@usfca.edu

PATRICK J. MCDONALD

HELMHOLTZ'S METHODOLOGY OF SENSORY SCIENCE, THE *ZEICHENTHEORIE,* AND PHYSICAL MODELS OF HEARING MECHANISMS

1. INTRODUCTION

Hermann von Helmholtz (1821–1894) was one of the most productive scientists to contribute to the understanding of perception. His treatises on vision and hearing continue to exert a notable influence on the sciences of perception.[1] For Helmholtz and his contemporaries, the science of perception was an independent enterprise, distinct from the philosophy of perception, just as the two are distinct today. However the two areas of inquiry were more closely in contact than they are today. In fact, Helmholtz developed a general framework for conceiving of perception that went beyond a purely naturalistic explanation of how humans perceive. The framework outlined a set of normative conditions for perceptual knowledge in general. His general theory has been called the empiricist theory of perception, and it includes as proper parts the sign-theory of perception (or *Zeichentheorie*) and the theory of unconscious inference. In this paper I explore an interpretation of the *Zeichentheorie* and argue that Helmholtz maintains a strong residue of a causal realist theory of perception throughout his life. The argument will include the discussion of a case study from Helmholtz's research on hearing, specifically, his influential resonance theory of hearing.

Several recent discussions of Helmholtz's philosophy have highlighted the unique quality of his sign-theory of perception, or *Zeichentheorie*.[2] There has been some debate regarding the precise status of perceptual reference to physical objects and properties in the course of the evolution of Helmholtz's theory of perception. It is generally agreed that the *Zeichentheorie* postulates a causal connection between the perceiver and the perceived, and that early in his career Helmholtz grants this causal connection strong epistemic status. The causal connection between world and agent, particularly the inference of a causal connection by the perceiver, allows for the perceiver to cross the veil of perception. However, later in his career Helmholtz qualifies the claim regarding access to the external world via the principle of causality. Some have argued that the qualifications regarding perceptual access to the external world amount to a rejection of Helmholtz's early causal realism. The debate on which I will focus concerns the meaning and extent of this qualification. Some suggest that he ends up defending

M. Heidelberger and F. Stadler (eds.), History of Philosophy and Science, 159–183.

some form of pragmatism, while others suggest that he ends up defending a form of idealism.[3]

The case study from his research on hearing, aims to show that despite qualifications that moderate his causal realism, he does not give it up. By causal realism I just mean a theory of perception whereby the act of perception consists in a causal relation between mind-independent objects or events and the perceiver. Perceptual knowledge consists in some specific reference to the properties of physical objects that play a causal role in the perceptual event. For Helmholtz to give up on reference to causal interactions instantiating purely physical properties, not themselves directly sensible, would be giving up on his basic research strategy in two ways. It would have required a major modification in his understanding of the role of one-to-one coordination in the account of perceptual knowledge offered by the *Zeichentheorie*. Abandoning this notion of causality further would have required a major revision in his reductionistic explanatory strategy in sensory science. In both the philosophical sphere and the scientific, he continued throughout his career to attribute an important causal role in perceptual processes to purely physical properties.

Michael Friedman has presented a sophisticated argument that Helmholtz abandons (say post 1866–70) a principle of causality that would transport a perceiver, at least figuratively, across the veil of perception. Friedman overly emphasizes Helmholtz's caution regarding how causal inferences bridge the subject/object gap and thus misinterprets aspects of the *Zeichentheorie* that focus upon the psychogenesis of perceptual abilities and the nature of perceptual objects. Friedman recognizes an important distinction in causality but concludes that Helmholtz abandons the aspect of causal inference that links the agent with the external world.[4] From the 1850s onward, Helmholtz's empiricist theory of perception holds perceptual objects to be mental representations composed of complex associations of subjective sensations. Physical objects are inferred to be the causes of sensations, and perceptions are interpretations of sensations through unconscious inferences that implicitly appeal to law-like correlations joining sensations with acts of willful intervention in the world. But the inferred physical objects have their own physical properties that are correlated with but distinct from sensational properties. From the early 1850s onwards, Helmholtz insists that there is no meaningful resemblance between physical properties such as light frequency and sensational properties such as colors. The realms are correlated by causal inferences proceeding from sensational properties to inferred physical causes. This is what happens in normal perception when one experiences a set of color sensations as a single physical object, localized somewhere in a three-dimensional visual field. In a more advanced stage of inquiry it becomes possible (through physical research) to determine and test the physical properties of such inferred objects and the laws that unite physical interactions as such. Then one can postulate laws connecting the physical properties with sensational properties and perceptual objects. This process assumes a second type of causal relationship among sensations and perceptions.[5]

As part of this general explanatory strategy, Helmholtz applied his considerable skills as a mathematical physicist to develop sophisticated models of
hearing mechanisms. His models have exerted a considerable influence on the
subsequent history of physiological acoustics.[6] The models invoke unobservable
functional properties, justified by their heuristic and explanatory power. However, he argues at length that the models' success in accounting for observed data
warrants the conclusion that the models correctly represent the function of the
inner ear. This says much about the conditions for confirming scientific hypotheses revealing a commitment to explanatory reduction of sensory properties to
their underlying causal, but unobservable physical properties. Despite the
importance of Helmholtz's well-known qualifications concerning causal inferences as the subject's connection to the physical *per se*, he persistently grants
such inferences considerable epistemic weight in his hearing theory in particular,
and in the *Zeichentheorie* in general. His claims about pragmatism and idealism
notwithstanding, he remained a moderate realist. Though true both of his views
on the nature of scientific knowledge as well as perceptual knowledge, I will
focus mostly on the manner in which he remains a realist in his philosophy of
perception.

Helmholtz's realism rests upon his faith in the truth of physical laws. Well-
tested law-like connections among sensations that yield perceptual objects and
general features of our perceptual world are not an infallible source of knowledge. Since perception is pragmatic and goal-oriented it functions not as a mirror
for the world but as a feature detector for those properties that are deemed
essential for the interest of the agent. Perceivers are constantly exposed to a blitz
of information. Because of the unconscious and relatively unsystematic manner
in which it is represented and tested, he argued that perceptual experience is
considerably more fallible than well-tested physical hypotheses. Further, as
Helmholtz notes in the 1866 installment of the *Handbuch der physiologischen
Optik* (hereafter, *Handbuch*), the laws of physics are more certain than the claim
that all men are mortal. The reason is that the causal story is richer in the former
case, largely because it is subject to systematic analysis and testing. Thus determining the genuine status of perception as knowledge requires linking the
structure and content of subjective perceptual experience to the underlying
physical and physiological systems that support perception. The case study in
this paper shows how Helmholtz attempts to do this with the basilar membrane
and its role in pitch perception.

2. THE NATURE OF THE *ZEICHENTHEORIE*

In his 1997 paper, Michael Friedman presents an exceptionally original and
interesting account of Helmholtz's *Zeichentheorie* and its relation to Moritz
Schlick's epistemology in the *Allgemeine Erkenntnislehre*. I will discuss and
contest his claim that Helmholtz's mature position represents a rejection of

causal realism. As noted, the *Zeichentheorie* embraces both the psychogenesis of subjective perceptual experience as well as the attempts to explain this via the physics and physiology of the unobserved causes of sensational content. Basic perceptual objects are constructed, in Helmholtz's view, through the cognitive association of sensations into law-like connections. But sensations are, by hypothesis, assumed to have a causal origin in the physical world, independent of the world as perceived. That is, Helmholtz seems to assume, both in his presentations of the *Zeichentheorie* in lectures and in his development of it in practice, that there are actual physical objects in the world which carry properties that are not directly sensible.

For example, think of the property of 'mass' that a block of iron bears. As a perceptual object, the block of iron as an object is a complicated collection of colors, textures, tastes, smells, and sounds that are associated with it. Identifying such properties with the object "iron block" is the sort of association of sensations that fits a law-like pattern and thus constitutes the iron block as perceptual object. But Helmholtz held that we know more about the iron block. We can assign it a mass, describe its behavior in free-fall, its chemical properties that cause it to have a specific color, or its tendency to vibrate when struck. These propositions require that there be a relation between the so-called purely physical properties and the perceived sensible effects.

The *Zeichentheorie* was outlined in a series of addresses and publications. These "programmatic" texts were given or published throughout his career, and range from his *Habilitationsvortrag* ("Ueber die Natur der menschlichen Sinnesempfindungen"), given in Königsberg in 1852, to his famous "Die Thatsachen in der Wahrnehmung" delivered in Berlin during 1878.[7] His *Zeichentheorie* has been equated with his general theory of perception.[8] While this may be acceptable in rough terms, I would like to limit the *Zeichentheorie* to the following more limited claims:

a.) Sensations are mere *signs* of their objects, resembling the latter in no meaningful way.

b.) Interpretation of sensory signs requires coordinating them into systems of law-like relations.

c.) Law-like relations consist of *intra*-level relations (perception to perception, sensation to sensation, physical property to physical property) as well as *inter*-level relations (physical properties to sensory signs; sensory signs to perceptual objects, etc.)

A simple way to characterize the status of knowledge, perceptual and scientific, within this scheme is that knowledge refers to exceptionless law-like regularities. Correspondence can mean little more than functional dependence. I do not intend to imply by this that any level must be constructed out of any other. In discussions of the epistemology of geometry Helmholtz emphasizes the primacy of the subjective perceptual properties of spatial intuition and the consequences of his epistemology of perception for understanding the nature of geometrical axioms.

His discussion of the causal law in later writings emphasizes the tentative nature of its link between subjective sensory properties and physical properties. Thus he became quite cautious of making any claim to definitive truth regarding unobservable properties in physical theories and models. In 1878 he even calls the existence of an external world, a fruitful hypothesis, concluding that without certainty of its existence, we must simply trust and act. However, at the same time his approach to sensory science was unabashedly physicalistic, and he built his entire methodology on telling the detailed history of a physical signal, following it through physiological processing and terminating in psychic perceptual experience. The interpretation of his *Zeichentheorie* as a theory of perception, as an epistemology, as a methodology of sensory science ought to reflect his appeal to each distinct level of event. Each level bears unique properties, and each has its own status as a node within the total scheme of law-like relations.

The argument for the *Zeichentheorie* relies upon the fact that there are numerous possible causal arrows between various levels. There can be different causes in the object realm (a live orchestra, or a high-fidelity reproduction). A tuning-fork, an electronic synthesizer, or an electrical stimulus of the auditory nerve may each cause the sensation of a simple tone. In short, the same kinds of sensations can be caused by non-identical physical or physiological sources. On the other hand, the same type of physical energies, when processed through different sense modalities cause different subjective experiences. A pressure wave in the air may be perceived as a tone via the hearing system, or as a tactile vibration via the tactile sensory system. Single sensations appear to have multiple relations with objective causes. Thus, starting from sensations and inferring their unobserved causes is a tenuous endeavor.

It is possible, he argues, to coordinate objective causes with subjective experiences. Given a pressure wave of specified frequency acting upon a normal auditory system, there will always be a tonal sensation of a certain pitch. Yet given a tone of a certain pitch, it is not possible unequivocally to determine a pressure wave of a certain frequency. Such tones could be the product of electrical stimulation of a certain auditory nerve, or even of spontaneous brain activity. Right there is a clue that he would never attempt to locate knowledge of objects in laws restricted to coordinations among sensations. His very own *Zeichentheorie* seems to forbid it. Thus, he must always help himself to physical (and physiological) properties to guarantee one-to-one correlations.

As I indicated, the property mismatches among levels suggests another challenge. A major premise supporting the *Zeichentheorie* is that physical and physiological causes of sensations bear properties entirely incommensurable with subjective sensational and perceptual properties. Tuning forks have mass, resonance frequencies, and dispositions to reflect certain wavelengths of light. Air molecules bear similar physical properties, as do the mechanical parts of the auditory system. The auditory nerves have physical properties as well, but they also bear electrochemical properties too. The relevant physical properties are functionally related to subjectively experienced tones, but a frequency of vibration could only resemble a tonal pitch by some extended analogy. Only the subjective properties are observable, strictly speaking, on Helmholtz's view of perception. The physical properties are inferred, essentially resting on the assumption of the causal law.

Now what is the argument provided by Friedman to suggest that Helmholtz abandoned causal realism by the early 1870s and certainly had done so by 1878? He correctly describes the appeal, made by Helmholtz in his 1847 paper on energy conservation, to the principle of causality. There the principle functions as a strongly epistemic bridging concept between the observable appearances and unobservable causes. This connection becomes officially sanctioned as a principle in his theory of perception as presented in the 1855 *Ueber das Sehen des Menschen.*[9] To this point, Helmholtz remains a causal realist. But by 1878 Helmholtz has, so the argument goes, abandoned causal realism, calling both realism and idealism metaphysical hypotheses, and arguing that neither position can be definitively proven. He calls realism a very good hypothesis that accords with common sense and scientific practice much more simply and comprehensively than idealism. He is quite clear about which option he prefers, but notes that it is in the end a preference, not an absolute truth. As a result, Friedman interprets this to mean that Helmholtz has concluded that the causal principle can no longer transport us behind the veil of appearances. It can only allow agents to connect appearances in law-like relations. And the appearances should be understood as sensations such as tones with specific pitch, colors with specific tone, smells with specific qualities and so on. Or they could be understood as well as the perceptual objects composed from such elements and located in a subjective spatial framework.

Now why is Helmholtz thought to have abandoned the possible *external* reach of the causal principle? One of the first reasons given by Friedman is the basic claim of the *Zeichentheorie* itself, that sensations are signs for consciousness and that learning to interpret them is a function of our understanding, a conceptual accomplishment just as much as rational thought. Further, Friedman appeals to the fact that Helmholtz, in the 1866 installment of the *Handbuch*, described this interpretive procedure on "our" side of consciousness as relying on the causal principle primarily to grasp the regularity or law-likeness of subjective sensations. And because it is so described as doing this important "internal" work, it can no longer be thought of as taking us outside to the physical objects assumed to be the causes of the subjective sensations. The emphasis upon the important role of the law-like within the level of subjective sensations is cited to support an exclusion of the external reach of causal inferences. Friedman then concludes that Helmholtz adopted a view of perception such that the important relation between perceptual objects and real objects is cashed out completely at the level of law-like relations among sensations. The only remaining relata then are sensations, laws connecting sensations, and perceptual representations that result. Helmholtz supposedly endorses this in the following statement from his

1878 lecture, "what we can find unambiguously and as fact, however, without hypothetical interpolation, is the law-like in the appearance."[10]

But this statement can be read numerous ways. Helmholtz was frustrated by philosophical discussions as well as by more specific scientific controversies. Of immediate relevance to this issue at hand was his on-going debate with Ewald Hering. Roughly speaking, Hering defended a number of views of depth perception and color vision that together comprise what Helmholtz called the "nativist" view in the *Handbuch*. In numerous publications Hering had criticized particular claims of Helmholtz through the 1860s and 1870s.[11] It is clear that Helmholtz carries on the debate in the 1878 lecture and it seems likely to have led him to qualify his claims about perception in two related but distinct ways. One, he saw in the course of his controversy with Hering that it was exceedingly difficult to prove the soundness of any general theory of perception. So he prudently admitted that his or anyone's claims to isolate a particular causal explanation should be appropriately qualified. Two, he could defend significant claims concerning the laws of sensory experiences without having to debate the further merits of more ambitious causal hypotheses. This quote about the law-like among sensations might very well just be an attempt to isolate some neutral territory among disputing parties.

Further, it is important to note the surrounding textual context. In the prior paragraph of *Thatsachen*, Helmholtz explains why it is important to realize that realism and idealism are both possible as metaphysical hypotheses. The more abstract claims of each metaphysical view are to be seen as hypotheses, and when so recognized are justified to be entertained in scientific discourse. "Science must discuss all admissible hypotheses in order to retain a full overview of all possible attempts at explanation."[12] But the scientific investigator must never forget the hypothetical status of such extra-empirical investigations. When participants in scientific debates defend such hypotheses with zeal and vehemence, he suggests that this is the "customary consequences of the unsatisfying feeling which their defender shelters in the hidden depths of his conscience, about the justness of his cause."[13] It would be just as metaphysical to claim that the physical as such is constituted by law-like relations of sensations.

By suggesting that we find the law-like in the phenomenon, unambiguously and without hypotheses, Helmholtz makes a plea for his empiricist theory of perception. More specifically he wants to defend the claim that we learn to see in depth and that this proceeds through processes of unconscious inference and willful intervention with phenomena. He mentions once again in the lecture that from the earliest moments forward, in perceiving enduring and distinct objects laid out in space, that, "this perception is the acknowledgement of a law-like connection between our movements and the therewith occurring sensations. Thus even the first elementary representations (*Vorstellungen*) contain intrinsically some thinking, and proceed according to the laws of thought."[14] He concludes the paragraph arguing that a sufficiently broad concept of thought allows that everything added in the act of perceptual intuition, which is not there in the raw material of sensations, "can be resolved in thought." And he means by thought, not native modules or Kantian categories, but more abstract law-like associative processes. It is most plausible that the target here is the nativist theory of perception, and the topic is the psychogenesis of perceptual activity.

That Helmholtz's discussion of law-like correlations is posed as a way of defending himself from nativists and not as an abandonment of his old causal realism, is supported by his next topic of discussion. He discusses the concept of substance. Quite apparently he views the concept as epistemically problematic, in that substances can only be conceived concretely indirectly, through exhaustive testing, and even this can not rule out a revision of the concept upon further testing. But he says substance is that which, without dependence on other things remains the same over time. "The relationship which remains alike between altering magnitudes, we call the *law* connecting them. What we perceive directly is only this law." [15] His concern with substance suggests a much moderated but abiding realism.

What else supports this reading? Friedman suggests that Helmholtz presented his theory of perception only schematically in the 1850s. He worked out the empiricist theory in more detail as he published the *Handbuch* and *Tonempfindungen*. "Although he had declared his allegiance to such a theory, and his opposition to "nativism" in 1855, the theory did not acquire clear articulation until 1865-66." [16] This suggests that one is likely to find details in the later work that will differ importantly from the earlier view. There are indeed important differences about the epistemic weight carried by inferences to causal origins in the indirectly perceived external world. That Helmholtz later developed important details in the empiricist theory is taken by Friedman to show that the *Zeichentheorie* means something new and different after 1866.

However, examination of Helmholtz's early lectures on his theory of perception (those delivered in 1852 and 1855) shows that the major themes of the *Zeichentheorie* were already in place, as were the general forms of argument elaborated later. Certainly the view is considerably more developed by 1866. However, if one means by clear articulation a statement of the basic principles, then arguably the basic structure of the theory was in place by 1855. This is borne out by Friedman's summary of what is meant by the empiricist theory. He lists a number of key properties:

1. The ability to see objects in space, i.e. localization, is not an innate capacity of consciousness or of nerve physiological structures. It is learned like a native language or musical training.

2. Learning proceeds via unconscious inferences, based on regularities in sensations. And this process of finding huge numbers of regularities in sensations makes possible the generation of an articulated three-dimensional space of perception, and the ability to localize objects therein.

3. Thus perception of 'objects' as complex constructs of simpler parts, whether visually as three dimensional (or aurally as unified complexes of tones such

as in the perception of tone quality) is an affair of the understanding and interpretation.

4. Thus the role of experiment in perception is essential as a key to interpretation and this is in turn likened to experiment in scientific inquiry.

From these premises, Helmholtz concludes that, like induction in science, induction in perception is used to unify data. Friedman then concludes: "hence, since the primary role for the causal or inductive principle here is now precisely to secure our grasp of regularity or law-likeness on the side of our perceptions, it no longer functions as a bridge to another realm existing *behind* our perceptions."[17] Even if he was right, his conclusions do not follow from this argument. But further, each of these four claims about the nature of perception was articulated by Helmholtz in the 1850s along with his endorsement of a causal realist theory of perception. This is possible because the set of claims is fully consistent with Helmholtz's brand of causal realism, whether somewhat dogmatic (early in his career) or more cautious (later in his career).

Notably Friedman grants the fact that Helmholtz is somewhat inconsistent, in that he sounds like a causal realist at times in later work, for example, at times in the *Handbuch*. But then he quotes the *Handbuch* to suggest that Helmholtz explicitly excludes causal realism. He makes much of Helmholtz's claims that there is no possible sense in speaking of the truth of our representations except a practical truth. Representations can be nothing but symbols or naturally given signs that we use for regulation of motions and actions. Learning to read them correctly achieves the desired results of regulating behavior via making the right predictions of expected new sensations. "Another comparison between representations and things not only fails to exist in actuality – here all schools agree – but any other kind of comparison is in no way thinkable and has no sense at all."[18] This appears to be a dramatic statement suggesting a clear rejection of causal realism. But in reality, it is nothing of the sort. It appears rather to be nothing other than a reiteration of the basic sign character of sensations and the representational nature of perceptions. His point is that, it is an unavoidable fact that the nature of sensations and perceptions reflect not only the properties of the external objects, like tables and chairs. They also are colored by the peculiar properties of our sensory apparatus and the cognitive processes by which we interpret our sensations. Once these facts are considered, he thinks it apparent that the question of comparison in the sense of resemblance is out of the question. Nevertheless, comparison is one thing, functional coordination is another. He thinks it is possible to coordinate representations with independently determined properties, and thus infer properties of the physical causes understood as outside the veil of perception.

More might be said about Friedman's arguments, but I think the point is clear. A complete discussion of the issue would have to address Helmholtz's work on the foundations of geometry. It is on the topic of the foundations of geometry that Friedman finds the most convincing evidence of the abandonment

of causal realism. It should be admitted that Helmholtz does not, to my knowledge, rule out a construction of all physical properties from purely sensational
properties. But as far as I know he never attempted to effect such a construction
systematically. His *Zeichentheorie* suggests that it would be a non-starter
because of problems with the uniqueness of inferences of physical causes from
sensational components alone. Further, his research practice reveals just the
opposite type of endeavor: he attempted to re-construct complex sensational
experiences from their causal origins in basic physical properties and processes.
His attempts to do so reveal the nature of inter-level law-like relations. It also
shows the creative ways he used such a schema to pursue specific questions in
hearing research. And we learn much about the types of constraints he advocated
for the admittedly tentative but genuine claim to knowledge concerning the
unobservable physical causes of subjective tonal phenomena. This can be seen in
his model of the function of the basilar membrane within the inner ear's cochlear
structure.

3. MODELING THE BASILAR MEMBRANE

Helmholtz published the first edition of his *Tonempfindungen* in 1863. The aim
of that work was to explain scientifically consonant and dissonant tone sensations. He had published a number of important papers on physical and physiological aspects of acoustics in the 1850s and early 1860s. However, the published work in 1863 represents a major step into new territory. This major
synthesis unified Helmholtz's prior studies but it also brought together the major
works of his immediate predecessors, such as Georg S. Ohm, August Seebeck,
Alfonso Corti, and Gustav Hällström, among others. But just as importantly he
extended the field of physiological acoustics into new areas, such as the origins
of the theory of musical harmony (structure of scale, chords, and key relationships). Along the way he defended a modified version of Ohm's definition of
simple tone as a pure sinusoidal form, he proposed a theory of tone quality or
timbre, and he proposed his famous resonance theory of hearing. The resonance
theory of hearing suggested an explanation for the dependence of pitch perception on the frequency of acoustic stimuli. It proposed that the mechanisms of the
inner ear respond selectively to various sound frequencies through sympathetic
vibrations. That is, specific vibratory signatures in acoustic energies cause certain structures in the inner ear to vibrate. The theory offered a mechanism for the
ear's abililty to act as a Fourier analyzer, resolving complex wave patterns into
different series of sinusoidal vibrations.

The *Tonempfindungen* was quickly received as a major event in the understanding of hearing as well as musical harmony. Ernst Mach described it in
lectures from 1865 as the successful culmination of centuries of efforts to
explain harmony, though he later proposed important criticisms of the ultimate
success of the resonance theory and the explanation of harmony.[19] Helmholtz's

study proved to be a wealth of stimulating ideas for further work, both for adherents and detractors, largely because it provided a broad framework for pursuing detailed questions. Researchers could see how a number of previously independent areas of study, such as physical acoustics, nerve physiology, and the anatomy of the ear, converged in one discipline. Further Helmholtz initiated conceptual innovations correlating all levels of the hearing processes among the regions of the physical, physiological and psychological. He developed a range of experimental and instrumental innovations as well.

A detailed examination of his model of the basilar membrane, its role in his resonance theory, and his argument for its plausibility illuminates the philosophical significance of his *Zeichentheorie* and his methodological stance towards modeling the unobservable in scientific research. For the purposes of this essay, the more general questions of methodology that arise when one considers the epistemic status of hypotheses regarding unobservable (or at least unobserved) processes, though relevant, will be put to the side. I am concerned to show that Helmholtz was engaged in such theory construction, and that he had a principled philosophical stance to defend such practices. However, at this point, I want to appeal to this case study to illuminate the philosophical structure of his theory of perception. This episode in his practice suggests that, within his scientific study of perceptual processes, the explanation of perceptual acts referred to processes and properties that were not directly observable. And this was not an accidental feature of his research, but reflected an important principle of the *Zeichentheorie*: namely, one-to-one correlations were achievable only from non-perceived physical properties to perceived sensible properties, not vice versa.

In order to understand Helmholtz's model of the basilar membrane and its role in the resonance theory, it will be necessary to discuss some basic ear anatomy and auditory physiology. *Figure 1* shows the basic parts of the ear. It includes the outer ear from the pinna to the external auditory canal. These parts have important though not dramatic effects on the gathering and transmission of sonic energies. The middle ear begins with the tympanic membrane. Its vibrations transmit and amplify the sound energy. Its exact vibratory patterns are still not known, though Helmholtz contributed much to specifying its vibration patterns, as have researchers since, particularly Georg von Békésy. The tympanic membrane is connected to the *malleus* (or hammer) that is connected to the *incus* (anvil), which in turn joins the very important *stapes* (stirrup). The three tiny ossicles transmit and amplify sound energy. Amplification occurs partly from the lever action of the three bones, but mostly because of the ratio of surface area between the tympanic membrane and the stapes is on the order of 22 to one. This partly makes up for the significant impedance mis-match between the vibratory properties of the air and the fluid of the inner-ear (endolymph and perilymph). That is, it takes much more energy per unit area to cause motion in the fluid, than in the air, because of their respective densities.[20]

The transmission between the stapes and the fluid of the cochlea at the oval window marks a very important transition from middle to inner ear. At this point in the process, the physical energies are converted into fluid pressure waves that alter the forms of membranes that extend through the spiral formations of the cochlea. The cochlea is a bony formation, resembling a snail-shell, and contains within it a number of very finely articulated structures. It is here that physical energies are converted into electrochemical energies by the action of the hair cells. We can see in *Figure 2*, that Helmholtz had detailed knowledge of the anatomy of the cochlea. He owed much of this to Alfonso Corti's work in the 1850s. He was aware of inner structure of the cochlea, its three chambers or galleries, as well as the fine details of the so-called organ of Corti (see *Figure 3*). This is extremely important, because the basilar membrane is located here.[21]

The resonance theory focused upon the functions of the mechanisms within the cochlea and particularly on the properties of the structures in the organ of Corti. The important parts included the rods of Corti, the basilar membrane, the inner and outer hair cells, and the tectorial membrane. They are important for a number of reasons, the primary one being that they are the key parts that carry out the functions postulated by the resonance theory of hearing. In the first two editions of the *Tonempfindungen* Helmholtz postulated Corti's arches as being the critical structures that vibrated sympathetically with the frequency of pressure waves in the cochlear fluid. In the third edition of 1870, following publication of a paper in 1869, Helmholtz revises the theory and proposes that the basilar membrane is the critical structure that vibrates sympathetically in certain locations to specific sound frequencies.

Corti's arches run the length of the basilar membrane side by side, bridging the tunnel of Corti and separating the tectorial membrane and the basilar membrane, while running parallel with and separating the inner and outer hair cells. Helmholtz was keenly interested in the nature of Corti's arches and as noted, in the first edition had proposed that they were the key sympathetically resonating structures. Shortly thereafter, C. Hasse found that numerous birds and amphibians lack Corti's arches, yet hear quite well, with excellent discrimination of different pitch. They do possess a basilar membrane and a tectorial membrane. Such animals possess fine hair cells and hairs that cross through the tectorial membrane and are connected to auditory nerve fibers (see *Figure 3* for a sense of how the hairs protruding from the inner and outer hair cells connect to the tectorial membrane). Helmholtz observed that comparative anatomy (likely that of Hasse and Hensen) had shown these hair cells were placed such that they are sheared laterally by Corti's membrane when the basilar membrane oscillates vertically.

Before discussing the details of Helmholtz's argument for his mechanical model of the basilar membrane, let me say a few words about the range of data that constrain his proposed explanation. First and foremost, he had to consider the range of tonal pitch that spans the range of audibility. There had been a number of studies correlating sound frequency ranges and intensities with the range of audibility. One of the main tasks of the theory was to account for the ability to discriminate changes in pitch. He cites Wilhelm Preyer's findings showing that practiced musicians can distinguish a difference of half a vibration per second in the "doubly accented octave" (approximately 500 to 1000 cycles per second [cps]). That is, musicians could tell that there is a difference in pitch between a tone of 500 cycles per second compared to one of 500.5 cps, an obviously impressive capacity of discrimination. Thus in that octave alone there were 1000 distinct pitch levels (500 cps, 500.5, 501, 501.5, etc.).

Further, Helmholtz had conducted numerous experiments using tuned resonators to show the audibility of upper partial or harmonics. The theory had to account for why these tones occur given certain specifiable stimulus frequencies. Specifically he had defended Ohm's definition of audible tones as exclusively sinusoidal periodic patterns. Further, he had done extensive work on the physical properties of sympathetic resonance. This work was of central importance to the resonance theory. This work made plain a number of important facts about vibrating objects. It showed that bodies resonated at characteristic frequencies and that they responded maximally to ambient sound energies at these frequencies. That is, the induced vibrations of objects from ambient sound would be maximally energetic, with the greatest amplitude, if the air vibrations matched the frequency of a body's signature resonance frequency. This meant that the physical properties of bodies that could be correlated with resonance signatures also needed to be specified, specifically their relative damping properties.

There is a distinction to be made here. Different objects have different resonance frequencies. Consider a range of tuning forks. If one strikes a tuning fork with a resonance frequency of 440 cps, then it will vibrate at that frequency, and it produces a physical tone at that frequency. Further, if some other sound source produces a tone at that frequency near the tuning fork, it will set the tuning fork into sympathetic vibrations, largely because the vibrations couple in such a way that the sound reinforces the vibrations of the fork without repeated cancellation. There is also the property of damping to consider. Objects of the same resonance frequency, say a string and a tuning fork, have different physical properties. These properties affect the ease with which the object is set into motion and the duration through which they remain in motion. Strings are easier to set into motion, but cease vibrating relatively quickly. Tuning forks are more difficult, i.e. require greater force, to set into vibration, but remain vibrating for a relatively long time.

Helmholtz formulated a table that assigns numbers to relative damping power and its effects on sympathetic vibration relations. For example, if an object tends to remain vibrating like a tuning fork – say it takes 38 vibrations to lose 90% of its original intensity – then its disposition to be set into sympathetic vibration will be highly sensitive to changes in frequencies. That is, a neighboring sounding body out of tune by an eighth of a tone will only bring about 10% of the vibration intensity as a body tuned to the primary object's resonance frequency. On the other hand, objects that are easier to set in motion will respond to a wider range of frequencies, even if they respond maximally to their signature frequency.

Because tuning forks are hard to set in motion, and remain vibrating for a long time once in motion, they require nearly an exact match to be set into vibration sympathetically. Stretched membranes or light strings are easily set in motion, but they die away quickly. They respond to much greater ranges of frequencies. Helmholtz first estimates the rough parameters of the damping and response character of sympathetically vibrating fibers in the inner ear by analysis of the response to rapidly alternating tones (what he calls "Triller", and what Ellis translates as "shakes"). For the hearer to distinguish different tones played within fractions of a second, the fibers responding to the first tone must damp down considerably, so that the succeeding tone can be registered. Based on these considerations he estimates that the majority of the sympathetically resonating parts would damp down to 1/10 of their original intensity somewhere between 9.5 and 22 vibrations.

Hence when we hereafter speak of individual parts of the ear vibrating sympathetically with a determinate tone, we mean that they are set into strongest motion by that tone, but are also set into vibration less strongly by tones of nearly the same pitch, and that this sympathetic vibration is still sensible for the interval of a semitone.[22]

This has two important consequences. It allows for the plausibility that there would be physiological mechanisms that would respond selectively, based upon sympathetic resonance. Further, it allows for wiggle room in saving the appearance of continuous pitch transformations. There need be no missing pitches in principle. If a stimulus matches the resonance frequency of no specific fiber, it will always fall between some two fibers and will set each in some motion. The mind need only learn how to interpret this comparative response to locate a pitch uniquely.

Helmholtz admitted that there was no way to know precisely which parts of the ear actually vibrate sympathetically with individual tones. This is just the kind of inference to a causal source of an experienced event or process that attempts to bridge the veil of perception in a radical way. It's possible to see how Helmholtz might conceive of an everyday perceptual object as nothing but a construction out of sensations and their law-like relations. In a sense that is how he conceives of such objects qua perceptual objects, or *Vorstellungen*. But the basilar membrane inside the cochlea was not only not observed in this everyday way, its motion was not observable, he thought, in principle. But he still conceives of it as a cause of sensible effects. In such a case he is going beyond an induction from sensations to a hypothetical inference about what physical/physiological structures and functions would cause an observed set of properties.

Helmholtz justifies the resonance scheme with the following argument. The anatomical formations in the organ of Corti do in fact distinguish differently pitched tones, and these in turn are equally well-perceived throughout the range of the musical scale. Further, the elastic formations of the cochlea are connected to different nerve fibers, that is, different locations have very specific, local nerve connections. Consequently, the differential response of the nerves to different acoustic frequencies must be caused by differential tuning of the elastic formations within the cochlea. The proper tones of the individual elastic formations must form a regular series (more or less) through the extent of the musical scale (and somewhat beyond).

He thought the best candidate, roughly speaking, for a location of such functions was the area of the organ of Corti. As mentioned, he had proposed in 1863 that Corti's arches were the likeliest candidates. They might be he thought differently tuned by having varying stiffness and tension. But he revised this in light of Hasse's discovery that birds and amphibians lack Corti's arches. Further, V. Hensen's work published in 1863, demonstrated that the basilar membrane might be a better candidate since it grew in width by approximately 12 times from the base of the cochlea to the apex. Thus this change in breadth was a property that he thought could explain the variable tuning.

There were other important properties of the basilar membrane making it a likely candidate to carry out the appropriate resonating action. Its fibers that ran in the direction of its length appeared weak, since it could be broken easily in this direction. In contrast, the fibers in the transverse or lateral direction were quite strong. This was a crucial clue to the mechanical analogy that he was to develop and propose. Given these features, the membrane could be tightly stretched in the transverse direction, from the modiolus to the outer wall of the cochlea (roughly the direction of right to left along the cross-section of the basilar membrane as seen in *Figure 3*). But it could sustain little tension in the direction of the length of the cochlea. Further, the physical theory of stretched membranes suggested that a membrane with differential tension would behave significantly differently than one with uniform tension in all directions. If a uniformly tense membrane is set in motion, the vibrations spread equally in all directions.

However, with an elastic object constructed like the basilar membrane, the response to induced vibrations would resemble that of a system of stretched strings. That is, if the tension in the direction of the length is infinitesimally small compared to the tension in the direction of the width, then the basilar membrane would respond selectively to stimuli like tuned strings do. Helmholtz speculates that the membranous connection between each fiber would provide a mere fulcrum for the pressure of the fluid on each fiber. The relevant laws of motion would be as if each individual string moved independently of the others, following the periodically alternating pressure of the fluid of the labyrinth.

The exciting tone would set that part (or parts) of the membrane into sympathetic vibration based upon the match with the proper tone of the fibers in that region. He conceived of the fibers as being fairly limited as to the range of response. That is, each fiber would be set in motion by only a limited range of different frequencies. The maximum response would be to a tone exactly matching the "proper" tone of the basilar membrane fiber. Though if it was approximately a semitone as he suggested, that allowed for a significant range of frequencies to have some detectible effect. The degree of vibration would be limited by the damping properties of the membrane's fibers and the damping properties of adjacent parts of the cochlear tissue as well as the fluid in the labyrinth. By the third edition he revised the role of Corti's arches, conceiving of them as playing a more structural role in transmitting the vibrations of the basilar membrane through the surrounding tissue, including the inner and outer hair cells, up through to the tectorial membrane. "According to this view Corti's arches, in the last resort, will be a means of transmitting the vibrations received from the basilar membrane to the terminal appendages of the conducting nerve."[23] This is surprisingly close to the view held presently in its grosser features.

In *Figure 4*, I wish to draw attention to a few of the details of his proposed model of the basilar membrane. It is a physical model in the same sense that celestial mechanics models the motions of the planets, or the equations of fluid mechanics model fluid flow. He first isolates the observed physical properties of the membrane, constrains the models with both anatomical and psychophysical data (e.g. pitch range, temporal characteristics of tones, inter-tonal discrimination data, even the characteristics of timbre). Then he constructs an equation that will allow him to calculate the relevant properties. These would be relevant vibratory physical properties, thus expressed in terms of a periodic function. And this is what he developed in a paper published in 1869, "Ueber die Schallschwingungen in der Schnecke des Ohres", which was developed in more detail in an added appendix to the *Tonempfindungen* (added in the 3rd edition).

Let me first discuss the fruits of the model, then say a few words about the methodological issues. The first point that he stresses is that the number of fibers in the basilar membrane can efficiently account for the range of discriminable pitch values, both in their limits, and the transitions between pitch levels. The crucial property is that sympathetic resonance of the fibers is not restricted to exact matching, rather it falls off quickly but gradually (like a steep bell curve). The central point, he says, is that the locations of maximum resonance are connected with specific nerves. If a stimulus tone falls between the resonance frequency of two distinct fibers, then it would resonate at a level in each neighbor reflecting where it fell between them. That is, the stimulus tone would cause vibrations in both, but the closer to one fiber's resonance frequency, the stronger its effect would be. The effectiveness of such a process would depend upon how well the mind (or other higher central processing location) can analyze or compare the various vibrations excited. The overlapping, though distinct responses postulated here, would explain further why pitch alterations are heard as continuous and not as patchy, that is, occurring in jumps.

The resonance theory in general and the basilar membrane model in particular could account for the nature of complex tonal sensations and their complex relations to acoustic stimuli. If a complex tone, or a chord was sounded then according to the resonance theory, it should stimulate a range of fibers corresponding to the range of tones contained, and the range of response regions postulated by the theory. His experiments with sympathetically resonating bodies, years of reports from musicians, and his own observations with ear-fitted resonators had shown that all musical tones contain a range of harmonics in addition to the fundamental tone. His hypothesis implied that if such ranges of physical tones stimulate a range of fibers, then properly directed attention should or at least could make one aware of the upper harmonics. He reported this to be the case for an impressive range of partials (for some tones going up to something like the 12th partial). He contended further that this could explain why non-sinsusoidal vibrations were distinguished by a characteristic range of upper harmonics they seemed to cause. Further, this shows why an originally simple periodic vibration can produce a sum of different sensations (i.e. harmonics) and thus appears compound in perception. It also easily accommodates his account of timbre or tone quality. In summary, he concludes that one can then think of sensations as reduced to this physiological hypothesis.

The sensations of different pitch would consequently be a sensation in different nerve fibers. ... Physiologically it should be observed that the present assumption reduces sensations which differ qualitatively according to pitch and quality of tone, to a difference in the nerve fibers which are excited.[24]

He likens his strategy here to that found in Johannes Müller's doctrine of specific sense energies. The peculiar quality of sensations of different senses is not so much determined by properties of stimuli (though they are part of the equation), but decisively by the nerves that receive and transmit the stimuli.

He also likens his strategy, which it resembles to a great degree, to that of Thomas Young's three receptor hypothesis for color vision.

The laws of the mixture of colors led Thomas Young to the hypothesis that there were three kinds of nerve fibers in the eye, with different powers of sensation, for feeling red, for feeling green, and for feeling violet. In reality this assumption gives a very simple and perfectly consistent explanation of all the optical phenomena depending on color. And by this means the qualitative differences of the sensations of sight are reduced to differences in the nerves which receive the sensations. For the sensations of each individual fiber of the optic nerve there remains only the quantitative differences of greater or less irritation.[25]

And the same goes for hearing. First, one must say, the qualitative differences of tonal phenomena are explained by their location in the auditory nerves. Further, the qualitative difference of pitch and timbre are reduced to which particular fibers process the sensation. And as for the eye, so for the auditory nerves, the only difference within each nerve is the quantitative difference in the amount of excitement.

4. LESSONS FOR UNDERSTANDING HELMHOLTZ'S THEORY OF PERCEPTION

We are acquainted first in the process of development with sensational events and their attendant properties. That is, long before humans learn to attribute physical properties to sounds, or colors, or tactile sensations, we experience the sensational properties. As our psychological ability to perceive develops, it is these sensational properties that are coordinated into unified perceptual objects such as physical objects, spoken words of language and the like. In this scenario of the psychogenesis of perception, which Helmholtz offers, perceptual / sensational properties play an important role in designating physical properties. Discussion of forces, frequencies, intensity, and the like presupposes measurement, which requires a notion of spatial dimension and temporal sequence, not to mention measurements of such quantities. Helmholtz discusses this point at length in the *Thatsachen* as well as in the essays on geometry, such as "The Origin and Significance of Geometrical Axioms."

Further, it is the case that according to the *Zeichentheorie*, we are not acquainted with physical objects as they are in themselves. They are interpreted as being thus and so via the sensory signs they cause in us. The advance of physics had made possible a more precise language of physical properties, such as frequencies, intensities, and the like, for reference to the stimuli associated with sensory experiences. However, the very notion of sensory content as signs for their inferred causes rests upon a conviction that sensations bear very different properties than do physical stimuli. Thus, it is unlikely that Helmholtz would ever attempt to limit a conception of the physical to laws among sensations. He thought of spatial intuition in such terms, but the objects picked out by physics carry properties and enter relations beyond their spatial properties as schematized by canons of measurement and geometrical representation.

The case study shows as well that Helmholtz attempted to explain the nature of subjective sensory experience by reference to unobserved and possibly unobservable physical events. The term "unobservable" applies here in principle because he thought that attempts to observe the function of the intricate physiological and anatomical structures of the human sensory system was very likely to disturb its processing. He eagerly took up the results V. Hensen achieved in studying the hearing organs of crustaceans, which he thought gave strong confirmation to his theory. Why was such comparative study so crucial? Because the "concealed position and ready destructibility of the corresponding organs of the human ear give little hope of our ever being able to make such a direct experiment on the intonation of its individual parts."[26] Not until Békésy simulated the action of a basilar membrane did hearing researchers begin to come closer to an experimental realization of this unapproachable ideal. But his simulation had limitations similar to Helmholtz's more indirect methods. Mössbauer techniques of inserting tiny radioactive probes to detect movement in the basilar membrane have more closely approached what Helmholtz thought impossible.[27]

The argument for the model is highly analogical. It does not even have a set of data points to "fit" with an equation, much less a set of laws, to explain. Rather there is a more or less well-coordinated set of results from numerous regions and fields that Helmholtz attempted to unify and explain. Part of the allure of the model of the basilar membrane was that if true, it would have allowed him to unify a number of other more limited explanations. It attempted to explain in the form of a causal model. It was causal in both the old fashioned sense of a mechanical model, and in Helmholtz's mature sense of an exceptionless and highly general law or set of laws.[28] Further, it accorded well with his aim to reduce differences in sensational quality to differences in physical quantity. His model showed how it might be plausible. Further, and perhaps of some significance, the model made specific predictions regarding the response of the basilar membrane to highly specific ranges of frequencies. This made it highly testable and in turn helped spur further work on psychoacoustics as well as physiological acoustics.[29]

Thus from a methodological perspective, the direct connection to observable properties is one factor in the epistemic evaluation of a theoretical model. However, there were a number of other factors as well. From the perspective of his theory of perception there are a number of consequences derivable from this episode. The path of explaining perception from laws among sensations to inferred objects may be more secure from a philosophical perspective than inferring unobservable properties that supposedly cause sensations to occur. It is also one that Helmholtz certainly develops in his writings on perception and epistemology. This path though remains more squarely in the realm of describing the psychogenesis of perception, an important task for the empiricist theory of perception and the *Zeichentheorie* in particular.

However, the *Zeichentheorie* offers an account of perceptual knowledge that defines perceptual knowledge as realized in one-to-one correlations among properties. This correlation occurs within the realm of sensations. But it also occurs starting from the inferred and non-sensational physical properties of objects going to the directly perceived sensational properties of perceptual objects. The path from unobserved physical processes to experienced sensory phenomena was very important to explore and explain. In a complete understanding of the *Zeichentheorie*, each path of approach plays distinct epistemic functions, thus neither should ultimately be privileged.

Figure 1 – Anatomy of the Ear. The middle ear begins with the tympanic membrane. Its vibrations transmit and amplify the sound energy. The tympanic membrane is connected to the *malleus* (or hammer) that is connected to the *incus* (anvil), which in turn joins the very important *stapes* (stirrup). The three tiny ossicles transmit and amplify sound energy (S.S. Stevens, et al., *Sound and Hearing*, New York: Time-Life Books, 1965).

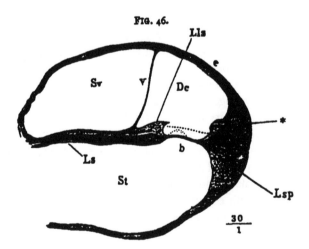

Figure 2 – Cross-section of the Cochlea. If one were to take a slice of one of the spirals of the cochlea as pictured in *Figure 1*, lay it flat, then one would have the object pictured in *Figure 2*. This shows both the three galleries of the cochlea as well as the location where the organ of Corti rests. The stapes forces fluid pressure waves to flow back and forth through the galleries, and this in turn causes a deformation of the basilar membrane which lies within the organ of Corti (*Tonempfindungen*, 138).

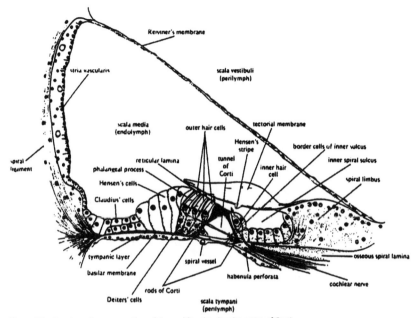

Figure 6.3 Drawing of a cross section of the cochlea showing the organ of Corti situated in the scala media on the basilar membrane.
Drawing by Sarah Crenshaw McQueen, Henry Ford Hospital, Detroit

Figure 3 – Organ of Corti inside the Cochlea. It is important to note that the cross-section of the cochlea pictured in *Figure 2* and the detail of the organ of Corti in *Figure 3*, run the length of the cochlea (William Yost and Donald Nielsen, *Fundamentals of Hearing.* Second Edition. Chicago: Holt, Rinehart, and Winston, 1985, 55).

Figure 4 – Analytical Model of the Basilar Membrane.

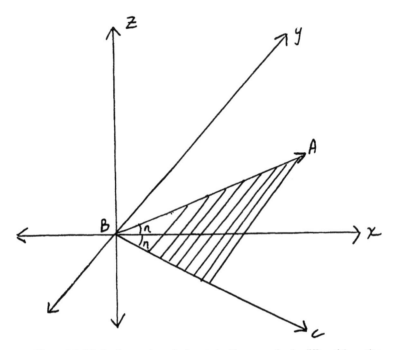

The model of the basilar membrane is characterized by a second order differential equation with the following conditions. Assume the membrane to be stretched between the legs of angle ABC of magnitude 2η, in the plane of the x and y axes. The x-axis bisects the angle; the y-axis bisects the vertex of the angle.

P = the tension of the membrane parallel to the x-axis; Q = the tension parallel to the y-axis. The forces as defined upon a unit square with sides parallel to the x and y axes respectively, would bring the membrane into equilibrium.

Let μ be the mass of the unit square, t the time, and z the displacement of a point in the membrane from equilibrium position. Let Z be an external force acting on the membrane in the direction of positive z, setting it into vibration.

The equation (deduced from Hamilton's principle by Kirchhoff's process is then:

$$Z + P \cdot \frac{d_2 z}{dx_2} + Q \cdot \frac{d_2 z}{dy_2} = \mu \cdot \frac{d_2 z}{dt_2}$$

The equation is restricted by the following boundary conditions:

1) z = 0 along the legs of the angle, when y = ± x · tanη.
2) z = 0, when x = y = 0, or at the vertex of the angle (the origin of the axes).
3) z is finite, when x is infinite.

NOTES

1. The two works are the following: Hermann von Helmholtz, *Die Lehre von den Tonempfindungen als physiologische Grundlage fur die Theorie der Musik.* Braunschweig: Friedrich Vieweg. 1863/77. Translated and Edited by Alexander J. Ellis. 1885. *On the Sensations of Tone: As a Physiological Basis for the Theory of Music.* 2nd ed. London:Longman's and Co. (Reprint. 1954. New York: Dover). The other is: *Handbuch der physiologischen Optik.* Leipzig: Voss. Translated by James P.C. Southall as *Treatise on Physiological Optics,* 3 vols. Milwaukee: Optical Society of America, 1924-25.

2. Primary examples are include the following. Michael Friedman, "Helmholtz's *Zeichentheorie* and Schlick's *Allgemeine Erkenntnislehre*: Early Logical Empiricism and its Nineteenth Century Background." *Philosophical Topics* 25: 1997, 19-50; Hatfield, Gary. *The Natural and the Normative: Theories of Spatial-Perception from Kant to Helmholtz.* Cambridge, MA: MIT Press. 1990; Heidelberger, Michael. "Force, Law, and Experiment: The Evolution of Helmholtz's Philosophy of Science." In *Hermann von Helmholtz and the Foundations of Nineteenth Century Science.* Edited by David Cahan. Berkeley: University of California Press, 1993, 461–97; Gregor Schiemann, *Wahrheitsgewissheitsverlust: Hermann von Helmholtz' Mechanismus im Anbruch der Moderne.* Darmstadt: Wissenschaftliche Buchgesellschaft, 1997.

3. On Helmholtz as a pragmatist, see Hatfield *Natural and Normative.* Heidelberger's "Force, Law, and Experiment" and Friedman's "Helmholtz's *Zeichentheorie*" present two distinct ways of reading Helmholtz as an idealist regarding the nature of perceptual knowledge and experience.

4. This point is articulated at length by Schiemann in *Wahrheitsgewissheitsverlust.*

5. I have tried to clarify this distinction between the two very different directions of causality in *Figure 1.*

6. Georg von Békésy *Experiments in Hearing.* Translated and Edited by E.G. Wever. New York: McGraw Hill. 1960, 471; Edwin G. Boring *Sensation and Perception in the History of Experimental Psychology.* New York: Appleton-Century Crofts. 1942, 404-411; E.G. Wever and Merle Lawrence, *Physiological Acoustics.* Princeton, Princeton U. Press, 1954.

7. The *Habilitationsvortrag* delivered in 1852 was published as "Ueber die Natur der menschlichen Sinnesempfindungen." *Königbsberger naturwissenschaftlicher Unterhaltungen.* Band III, 1-20. It was reprinted in Helmholtz's *Wissenschaftliche Abhandlungen. (Vol. I-III.* Leipzig: J.A. Barth. 1882-95), volume 2: 591-609. The 1878 address "Die Thatsachen in der Wahrnehmung," was printed in Helmholtz's *Vorträge und Reden.* Fifth edition. Braunschweig: Friedrich Vieweg. Translated by Malcolm Lowe in Helmholtz, *Epistemological Writings,* edited by Robert S. Cohen and Yehuda Elkana, Boston: D. Reidel.

8. One can see this tendency in Friedman, "Helmholtz's *Zeichentheorie*," and Schiemann, *Wahrheitsgewissheitsverlust.*

9. Helmholtz, "Ueber das Sehen des Menschen," Lecture delivered in Königsberg, Leipzig: L.Voss. Reprinted in Helmholtz, *Vorträge und Reden:* 87–117.

10. Helmholtz, "Thatsachen," 141.

11. For an extensive discussion of the so-called nativist-empiricist dispute, see R. Steven Turner, "Consensus and Controversy: Helmholtz on the Visual Perception of Space," in Cahan (ed.) *Helmholtz and the Foundations,* 154–203; see also Turner, *In the Eye's Mind: Vision and the Helmholtz-Hering Controversy.* Princeton, Princeton U. Press, 1994, and Hatfield, *Natural and Normative.*

12. Helmholtz, "Thatsachen," 138.

13. Helmholtz, "Thatsachen," 138.

14. Ibid, 138.

15. Ibid, 139.

16. Friedman, "Helmholtz's *Zeichentheorie*," 15.

17. Ibid, 17.

18. Helmholtz, *Handbuch III*, 19)

19. Ernst Mach, "The Fibres of Corti," and "On the Causes of Harmony," *in Popular Scientific Lectures,* Chicago Open Court, 17–31 and 32–47. See also Mach, *The Analysis of Sensations,* New York: Dover.
20. Impedance is a technical term that describes the relative reactivity and resistance of different media to the motion of sonic energy. I pass over the technical details here because they are not directly relevant.
21. *Figure 3* is from a recent text, but it differs little (mostly in clarity and labeling) from the drawing contained in the 4th edition of the *Tonempfindungen.* For a more detailed description, see the captions with Figures 1, 2, and 3.
22. *Tonempfindungen,* 144.
23. *Tonempfindungen,* 147.
24. Ibid, 148.
25. Ibid, 148.
26. Ibid, 150.
27. See William Yost and Donald Nielsen, *Fundamentals of Hearing.* Second Edition. Chicago: Holt, Rinehart, and Winston, 1985.
28. See *"Thatsachen"* for a mature articulation of his notion of how causality and laws are unified.
29. See Wever and Lawrence, *Physiological Acoustics,* for a helpful discussion of early twentieth century attempts to model the function of the basilar membrane. Many theories distinct from Helmholtz's were proposed, though quite similar in general conception.

REFERENCES

Békésy, Georg von. 1960. *Experiments in Hearing.* Translated and Edited by E.G. Wever. New York: McGraw Hill.
Boring, Edwin G. 1942. *Sensation and Perception in the History of Experimental Psychology.* New York: Appleton- Century Crofts.
Cahan, David (ed.) 1993. *Hermann von Helmholtz and the Foundations of Nineteenth Century Science.* Berkeley: University of California Press.
Erdmann, Benno. 1921. "Die philosophischen Grundlagen von Helmholtz' Wahrnehmungstheorie." *Abhandlungen der preussischen Akademie der Wissenschaften, philosophisch-historische Klasse* Nr. 1: 1-45.
Friedman, Michael. 1997. "Helmholtz's *Zeichentheorie* and Schlick's *Allgemeine Erkenntnislehre*: Early Logical Empiricism and its Nineteenth Century Background." *Philosophical Topics* 25: 19-50.
Hatfield, Gary. 1990. *The Natural and the Normative: Theories of Spatial-Perception from Kant to Helmholtz.* Cambridge, MA: MIT Press.
Hatfield, Gary. 1993. "Helmholtz and Classicism: The Science of Aesthetics and the Aesthetics of Science." In Cahan 1993, 522–58.
Heidelberger, Michael. 1993. "Force, Law, and Experiment: The Evolution of Helmholtz's Philosophy of Science. In Cahan, David (ed.) 1993. *Hermann von Helmholtz and the Foundations of Nineteenth Century Science.* Berkeley: University of California Press , 461–97.
Heidelberger, Michael. 1998. "Die Erweiterung der Wirklichkeit im Experiment." In Heidelberger and Steinle (eds) 1998, 71-92.
Heidelberger, Michael and Steinle, Friedrich (eds). 1998. *Experimental Essays – Versuche zum Experiment.* Baden-Baden: Nomos.
Helmholtz, Hermann von. 1852. "Ueber die Natur der menschlichen Sinnesempfindungen." *Königsberger naturwissenschaftliche Unterhaltungen.* Band III. 1-20. Reprinted in Helmholtz 1882-95, 2: 591-609.
Helmholtz, Hermann von. 1855. "Ueber das Sehen des Menschen." Lecture delivered in Königsberg and published in Helmholtz 1903, 87–117.
Helmholtz, Hermann von. 1856. "Ueber Combinationstöne." *Annalen der Physik und Chemie* 99: 497–540.

Helmholtz, Hermann von. 859. "Ueber die Klangfarbe der Vocale." *Poggendorff's Annalen der Physik und Chemie.* Band 106: 280-90. Reprinted in Helmholtz 1882-95, 1: 397-407.

Helmholtz, Hermann von. 1863/77. *Die Lehre von den Tonempfindungen als physiologische Grundlage für dieTheorie der Musik.* Translated and Edited by Alexander J. Ellis. 1885. *On the Sensations of Tone: As a Physiological Basis for the Theory of Music.* 2nd ed. London: Longman's and Co. (Reprint. 1954. New York: Dover).

Helmholtz, Hermann von. 1867/1909-11. *Handbuch der physiologischen Optik.* Leipzig: Voss. Translated by James P.C. Southall as *Treatise on Physiological Optics,* 3 vols. Milwaukee: Optical Society of America, 1924-25.

Helmholtz, Hermann von. 1878. "Die Thatsachen in der Wahrnehmung." Reprinted in Helmholtz 1995 as "The Facts in Perception."

Helmholtz, Hermann von. 1882-95. *Wissenschaftliche Abhandlungen. Vol. I-III.* Leipzig: J.A. Barth.

Helmholtz, Hermann von. 1894. "The Origin and Correct Interpretation of our Sense Impressons." *Zeitschrift für Psychologie und Physiologie der Sinnesorgane.* VII: 81-96. Reprinted in Helmholtz 1971, 501-12.

Helmholtz, Hermann von. 1894. "Introduction to the Lectures on Theoretical Physics (Introduction and Part I)." Reprinted in Helmholtz 1971, 513–29.

Helmholtz, Hermann von. 1903. *Vorträge und Reden.* 5th ed. Braunschweig: Friedrich Viewig und Sohn.

Helmholtz, Hermann von. 1971. *Selected Writings.* Edited by Russell Kahl. Middletown, CN: Wesleyan University Press.

Helmholtz, Hermann von. 1977. *Epistemological Writings.* Edited by Robert S. Cohen and Yehuda Elkana. Boston: Reidel.

Helmholtz, Hermann von. 1995. *Science and Culture.* Edited by David Cahan. Chicago: Chicago University Press.

Lenoir, Timothy. 1994. "Helmholtz and the Materialities of Communication." *Osiris 1994,* 9. Edited by Thomas Hankins. 185-207.

Müller, Johannes. 1837, 1840. *Handbuch der Physiologie des Menschen für Vorlesungen.* 2 vols. Coblenz: J. Hülscher.

Schiemann, Gregor. 1997. Wahrheitsgewissheitsverlust: Hermann von Helmholtz' Mechanismus im Anbruch der Moderne. Darmstadt: Wissenschaftliche Buchgesellschaft.

Turner, R. Steven. 1977. "The Ohm-Seebeck Dispute, Hermann von Helmholtz, and the Origins of Physiological Acoustics." The British Journal for the History of Science. 10: 1–24.

Turner, R. Steven. 1993. "Consensus and Controversy: Helmholtz on the Visual Perception of Space." In Cahan 1993: 154–203.

Turner, R. Steven. 1994. In the Eye's Mind: Vision and the Helmholtz-Hering Controversy. Princeton, Princeton U. Press.

Vogel, Stephan. 1993. "Sensation of Tone, Perception of Sound, and Empiricism: Helmholtz's Physiological Acoustics." In Cahan 1993: 259–87.

Wever, E.G. and Merle Lawrence. 1954. Physiological Acoustics. Princeton, Princeton U. Press.

Yost, William and Donald Nielsen. 1985. Fundamentals of Hearing. Second Edition. Chicago: Holt, Rinehart, and Winston.

Department of Philosophy
Seattle Pacific University
Seattle, WA 98119
U.S.A.
mcdonp@spu.edu

MATTHIAS NEUBER

PHYSICS WITHOUT PICTURES?
THE OSTWALD-BOLTZMANN CONTROVERSY,
AND MACH'S (UNNOTICED) MIDDLE-WAY

1. INTRODUCTION

It is a common view in cognitive psychology that there is a fundamental difference between what may be called de*scriptive* information, on the one hand, and
de*pictive* information, on the other. While the first kind of information is –
ideally spoken – non-pictorial and usually equated with the content of a proposition, the second kind of information is pictorial by definition and accordingly
equated with the content of a mental image. Granting the correctness of this
distinction, cognitive scientists differ on the role played by mental images in
epistemic processes. One faction, represented by the writings of Zenon Pylyshyn,
asserts that mental images are merely an epiphenomenon of a more general and
abstract processing system, and that they are in principle dispensable since they
do not have any truth value.[1] Another faction, represented by the writings of
Stephen Kosslyn, favors the opposite view, maintaining that mental images are
indispensable since they play, in spite of their lack of truth value, a fundamental
role in human thinking.[2] As Arthur I. Miller has proposed in his book *Imagery in
Scientific Thought* (1986), these two factions may be labeled as the "anti-
imagist" and the "pro-imagist" view, respectively.[3]

Now the roots of this rather recent controversy between pro-imagists and
anti-imagists are traceable to the late nineteenth- and early twentieth-century
philosophy of science. During that period, Central European physicists, especially in Austria and Germany, became increasingly dissatisfied with the prevailing mechanical view of nature.[4] Several methodological disputes were the
outcome of this mood, and the perhaps most controversial issue in these disputes
was the role of mental images in scientific theory construction. A good example
in this respect is the late nineteenth-century energetics controversy.[5] Typically
enough, this whole controversy was perceived by its protagonists as a debate
about the usefulness of what they liked to refer to as *pictures* (*Bilder*). To be
sure, the methodological significance of mental images or pictures was not the
only point at issue in that controversy. Other, at first sight physically more
precarious topics, like atomism, irreversibility or the energeticist program itself,

185

M. Heidelberger and F. Stadler (eds.), History of Philosophy and Science, 185–198.
© 2002 *Kluwer Academic Publishers. Printed in the Netherlands.*

formed part of discussion as well. But if we rely on what is reported by contemporary observers, then it appears that all these other topics were nothing else but *special aspects* of the more general debate about the usefulness of pictures.[6]

Why were pictures so important in the energetics controversy? And what were the major positions concerning their use in physics? While the first of these two questions is rather intricate and difficult to answer, the second question seems to permit a concise disquisition. Thus in the present paper, only the second question will be investigated further.[7] I will begin, in Section 2, with the debate between the two main figures in the energetics controversy, Wilhelm Ostwald (1853-1932) and Ludwig Boltzmann (1844-1906), pointing out that Boltzmann argued for a radical "pro-imagist" account of theory construction, which in turn was directed against the likewise radical, but "anti-imagist", account proposed by Ostwald. Section 3 will have its focus on Ernst Mach (1838-1916) and his attempt to reconcile these two positions. It will be shown that Mach was not at all the kind of inexorable iconoclast as which he often is described. Instead, I will suggest taking him seriously in his attempt to figure out a practicable middle-way. This middle-way, however, is not without its own difficulties in its own, as I shall briefly demonstrate in Section 4.

2. "No-Picture View" versus "All-Picture View": Ostwald and Boltzmann on Picturing in Physics

Let us begin with Ostwald and his arguments against the usefulness of pictures. Most of these arguments can be found in his famous lecture "On the Conquest of Scientific Materialism". Given at the *Naturforscherversammlung* in Lübeck in 1895, this lecture embraced two parts: a "destructive" part, where it was shown that the traditional mechanical view of nature was no longer tenable; and a "constructive" part, where energetics was outlined as the alternative by which the mechanical view should be replaced.[8] For our concerns, it will suffice to concentrate on the destructive part of Ostwald's lecture.

Given the programmatic intent of the Lübeck lecture, it is clear from the outset what kind of theoretical constructions Ostwald had in mind when he referred to "pictures". Aiming at a devastating critique of the mechanist approach in science, he confined his statement to "mechanical pictures or analogies, which one is accustomed to calling mechanical theories of the phenomena in question."[9] Pictures, thus, are analogical constructions, based on mechanics, and designed for a mechanical interpretation of non-mechanical phenomena. They are *pictures* insofar as they merely provide an indirect and arbitrarily "enriched" account of what can be immediately observed. Consider, for example, the conception of heat as disordered motion. On Ostwald's view, this would be a paradigm case of a mechanics-based pictorial construction. Heat as such, he would say, cannot be conceived as disordered motion. Something has to be added if we want to conceive it thus. And this additional content is delivered,

Ostwald would continue, by picturing an imaginary mechanism not being identical with what can be immediately observed.[10]

In order to make his approach to mechanical pictures more precise, Ostwald, in his Lübeck lecture, introduced some sort of measurability criterion.[11] According to this criterion pictures must be sharply discriminated from scientific laws and theories. For Ostwald maintained that laws and theories are in principle confined to measurable quantities, whereas the main feature of pictures is that they in principle transcend the realm of such quantities by way of referring to atoms, forces and other not directly measurable entities. The motive for the application of such pictorial constructions is, Ostwald observes, to achieve a *causal explanation* of the phenomena in question.[12] Measurable quantities thus are supposed to be reducible to underlying mechanisms, whereby these mechanisms themselves are not subjected to experimental measurement. This, however, would – says Ostwald – amount to a transgression of what should be acknowledged as the only realistic aim of science, namely "to coordinate realities, i.e., definite and measurable quantities, so that when certain of them are given the others can be deduced."[13] Everything that goes beyond this phenomenalistic task of science must consequently be considered *and refused* as metaphysical in character.[14]

It might be objected that Ostwald at this point clearly goes astray since the (allegedly) metaphysical status of mechanical pictures is not at all an argument against their inner-scientific use. That is, one might concede their metaphysical character and yet make use of them for reasons of heuristics. Still, for Ostwald this would be an absurd strategy. For in his view the use of mechanical pictures inevitably leads to contradictions. The reason for this is that mechanical pictures always add some supplementary content to the immediately observed phenomena, and that this supplementary content sooner or later gets in conflict with established empirical laws.

Ostwald illustrates this point by offering a reconstruction of the history of optics.[15] The corpuscular hypothesis of light, he tells us, had some success throughout the eighteenth century, but it was overthrown at the beginning of the nineteenth century because the picture of mechanically governed particles did not account for the phenomena of interference and polarization. That is, instead of causally explaining these phenomena, the picture of mechanically governed particles entailed a fatal contradiction with what was evidenced by direct experimentation. As a consequence, the corpuscular hypothesis was given up in favor of the wave hypothesis of light. But this hypothesis also suffered from the general disease of mechanistic physics: it turned out as contradictory because of its pictorial constituents. In particular, it was the impossibility of constructing an adequate mechanical picture of the light-bearing ether that eventually caused the physicists to drop the wave hypothesis. From there on, several attempts had been made to conceptualize the various phenomena of light without the help of any pictures. As a result, the electromagnetic *theory* of light emerged from these endeavors. That is, optics had reached the level of methodological maturity in that it now (as "the unforgettable Hertz" had demonstrated) allowed to correlate the experimentally obtained phenomena by means of a coherent set of differential equations. "That the evolution of the theory should end in this point", Ostwald comments, "is a far more convincing argument than any I could adduce against the fruitfulness of the theoretical methods previously followed on mechanical lines."[16]

The general lesson to be drawn from this example is obvious enough: According to Ostwald, physicists should henceforth *completely dispense with pictures* and all other forms of visualization. Although what then remains is nothing else but "bare descriptions", physicists should recognize that these are much more closer to the facts than the imaginary atomist and otherwise pictorial constructions used so far by the adherents of the prevailing mechanical view of nature. Or as Ostwald puts it himself:

We must ... give up all hope of getting a clear idea of the physical world by referring phenomena to an atomistic mechanics. But, perhaps, one of you will say, what means shall we have left of making for ourselves a picture of reality if the conception of atoms in motion is abolished? To such a question I would answer: Thou shalt not make unto thyself any picture or likeness. Our task is not to view the world in a more or less bedimmed and crooked mirror, but as directly as the nature of our mind will permit. [17]

It is often said that Mach also advocated such a descriptionist, or radical positivist, "no-picture-view". But this, I claim, is a misleading allegation. In fact, Mach did consider Ostwald's view as too restrictive. A more liberal account, he urged, should be conceded. And this more liberal account Mach found prepared, though not accomplished, in what Boltzmann had replied to Ostwald in his 1896 essay "A Word from Mathematics to Energetics".

The central aim of Boltzmann's essay was to demonstrate that energetics as a program of research had failed in almost all important respects. It had failed in its attempt to base mechanics on a general "energy principle"; and it no less had failed in its pretended derivation of the entropy law.[18] This is what Boltzmann objected to at the descriptive level. But he had also something to say against Ostwald's "destruction" of the mechanical approach in methodology. It is this part of Boltzmann's critique on which I shall focus now.[19]

Directly referring to the Lübeck lecture, Boltzmann, in the final section of his essay, attempted to show that Ostwald's view of theories was basically misguided. To be sure, Boltzmann did not deny that the aim of a direct, picture-free, description of phenomena might be conceived as a regulative ideal to be strived after.[20] But he refused to accept Ostwald's assertion that pictures are dispensable and hence to be excluded from the area of physics.

What, then, was the alternative Boltzmann could offer? As is well known: the *in*dispensability of pictures.[21] But what exactly does this mean? Two arguments, presented in the 1896 essay, are worth discussing here. The first argument (though actually the second) is related to heuristics and says that Ostwald is simply wrong when he assumes that pictures are an obstacle for the development of theories. In fact, Boltzmann returns, they are "extremely useful". A more adequate account of their historical significance would show, Boltzmann contends, that pictures frequently have served as an effective means for "making new discoveries, for ordering our ideas, for presenting them clearly, and for remembering them more easily".[22] Boltzmann admits that some of these pictures might, like the corpuscular hypothesis of light, eventually turn out to be inappropriate or even false. But he can quote a whole bunch of examples where this was evidently not the case.[23] Since all these successful applications were, inadmissibly, suppressed by Ostwald, his entire case against the usefulness of pictures must be regarded as historically warped. Or to put it in more positive terms: For Boltzmann, the "age-old pictures of mechanical physics"[24] are for the most part still progressive and should therefore not, as Ostwald claims, be given up, but cultivated further.

Now on the face of it, this argument seems rather limited. Ostwald could immediately reply that the successful applications Boltzmann appears to have been successful only because of the descriptive mathematical equations on which pictorial constructions are usually "pinned up". If disentangled, it would become clear that all the virtues Boltzmann ascribes to pictures actually are due to those equations.[25] Thus, in order to convince us of the *indispensability* of pictures, Boltzmann must prove more: He must prove that apart from their heuristic usefulness, pictures, are a *prerequisite* for a mathematized description of phenomena.

Boltzmann was surely aware that this is the way to proceed. Before asserting the heuristic usefulness of pictures, he presented in his essay quite another, stronger, argument by which he hoped the indispensability of pictures to be demonstrated once for all. According to this second argument a picture-based account of theories is unavoidable because "*all* human thoughts are *nothing else* but pictures of reality".[26] Obviously this claim amounts to setting up some sort of radical "all-picture view". Still, in the 1896 essay it does not become sufficiently clear what is implied by such a view. But this should not bother us here. For there are several passages in other writings, where Boltzmann is more explicit about this point. Some remarks on these passages might help to outline further his position.

To begin with, in the 1897 paper on the indispensability of atomism, Boltzmann emphasizes that "[of a comprehensive area of fact] we can never have a direct description but always only a mental picture. Therefore we must not say, with Ostwald, 'do not form a picture', but merely, 'include in it as few arbitrary elements as possible'."[27] This passage clearly indicates that Boltzmann accepted the aim of picture-free description only as a regulative ideal. In truth, Boltzmann is sure, physicists cannot do without pictures. As he sees it, pictures are already indispensable when it comes to conceptualizing our everyday perceptions. Thus he states in his 1899 lecture on the fundamental principles and equations of mechanics:

When Goethe says that only one half of our experience is experience, what he intends to convey by this seemingly paradoxical dictum is surely that in every conceptual grasp of experience or verbal representation of it we must already go beyond experience. The frequently uttered requirement that natural science must never go beyond experience should therefore in my view be reformulated thus: never go too far beyond experience and introduce only such abstractions as can soon be tested by experience.[28]

Posed in this way, Boltzmann's view comes very close to present scientific realist conceptions. Stathis Psillos, for instance, has recently attacked van Fraassen's anti-realist position by making the following move: "Theoretical beliefs are formed by means of abductive reasoning. But so are most of our everyday commonsense beliefs ... The pattern of reasoning, as well as the justification, are the same in both cases".[29] Boltzmann no doubt would have agreed. Like Psillos he believed it would be far from realistic to restrict the scope of science to the immediately perceived, since some inferential (non-perceptual) element must already be allowed as underlying our everyday cognition. Given that science necessarily is based upon the evidence of everyday cognition, we should not subscribe to any theory like Ostwald's. For then, Boltzmann maintains, we unavoidably would end up with some sort of solipsism, which in turn would undermine the physicists' most basic theoretical ambition, namely, to *predict* not yet perceived phenomena.

A final quote may illustrate how Boltzmann is trying to drive this point home. Again in the 1897 atomism-paper, he remarks that "differential equations ... are evidently nothing but rules for forming and combining numbers and geometric concepts, and these in turn are nothing but mental pictures from which appearances can be predicted."[30] This passage is important because it contains an implicit refutation of Ostwald's measurability criterion. As discussed above, this criterion entailed a sharp distinction between theories and pictures, whereby theories were conceived as axiomatic sets of differential equations. Consider, for instance, Fourier's theory of heat conduction on the one hand and the kinetic theory of gases on the other. On Ostwald's view, only the first of these two theories would be a "real" theory. Describing the transfer of heat between the parts of a body (usually an iron bar) at different temperatures, Fourier's theory comprises only such quantities as can be directly measured in the laboratory. Thus, at the very bottom of that theory we have a (partial) differential equation of the form:

$$\mathrm{d}u/\mathrm{d}t = k\left(\partial^2 u/\partial x^2 + \partial^2 u/\partial y^2 + \partial^2 u/\partial z^2\right), \tag{1}$$

where u is the temperature at the point (x, y, z), t the time, and k a constant equal to the quotient of thermal conductivity and the product of density and specific heat. Compare with this, for example, the law of Boyle-Mariotte in one of its kinetic formulations[31]:

$$pV = 1/2\ mnv^2, \tag{2}$$

where only p and V are referring to directly measurable quantities (pressure and volume of a gas, respectively), whereas the symbols on the right side are supposed to represent atomic magnitudes (namely mass, number, and velocity of gaseous *particles*, respectively).

Now, the point to emphasize is the following: While Ostwald would argue that all we need for deriving predictions in physics are equations of type (1), Boltzmann would insist that such equations cannot adequately be applied unless we base them on atomic magnitudes as occurring in equations of type (2). The reason Boltzmann thinks so is that differential equations call for a finite number of material particles, without which the operation of *taking the limit* would remain opaque.[32] Thus, in order to describe phenomena like heat conduction we inevitably must proceed from the assumption that the investigated system is made up by finite quantities, and that it is these quantities by which the corresponding differential equations actually require their mathematical significance. From this Boltzmann concludes that differential equations are not – as Ostwald would have it – an *alternative* to atomist pictorial constructions but rather their most salient mathematical *expression*.[33]

All this no doubt sounds like a strong commitment to a scientific realist position. The assumption of unobservable entities like atoms is characterized as an intrinsic feature of picturing processes, and these processes themselves are seen as a prerequisite for the description and prediction of phenomena. But here one must be careful. For in the last analysis, Boltzmann goes not so far as to assert that atoms must be recognized as ontologically real causes.[34] Rather he conceives of them as mathematically indispensable within the "describe-and-predict approach" itself. That is, like Ostwald he maintains that the attempt to causally explain phenomena would overshoot the mark.[35] Still, the central case against Ostwald's no-picture view remains untouched by this restriction: *Physicists cannot do without pictures, for otherwise they could not justify their application of the calculus.* As we shall see below, Mach did agree with Boltzmann on that issue.

3. MACH'S MIDDLE-WAY:
TAKE PICTURES SERIOUSLY, BUT DO NOT "ATOMIZE" THEM ...

Before investigating Mach's position in the debate over pictures, let me briefly comment on the "received view", according to which Mach was an ally of Ostwald in the energetics controversy. Among Mach's younger physicist contemporaries this view apparently was pretty widespread. Arnold Sommerfeld, for instance, wrote that at the *Naturforscherversammlung* in Lübeck in 1895 – where the energetics controversy had come to a head – "Helm from Dresden gave the report on energetics; behind him stood Ostwald and behind both stood the *Naturphilosophie* [!] of Ernst Mach".[36] Much the same is told by Mach's "archenemy" Max Planck, who retrospectively complained that at the end of the nineteenth century "it was impossible to make any headway against the authority of men like Ostwald, Helm, and Mach"[37], and by Robert Millikan, who found it "amazing" that "outstanding men like Ostwald and Helms [sic!], and even the brilliant philosopher Ernst Mack [sic!], could at that epoch be proponents of the continuum theory".[38] Even nowadays the received view has its supporters. Carlo Cercignani, for example, writes in his 1998 book *Ludwig Boltzmann: The Man Who Trusted Atoms* that Ostwald's rejection of the "atomic picture of reality" was an (admittedly rather naive) "application of Mach's ideas".[39] A similar but more explicit interpretation is delivered by Henk Visser, who in a 1999 paper (on Boltzmann and Wittgenstein) confronts "the English tradition of *picture theories*" with "the German tradition of *radical simplification*", thereby presenting Boltzmann as a follower of the first and both Ostwald and Mach as proponents of the second tradition.[40]

As regards Ostwald and Boltzmann, Visser's approach is surely correct. But is it also adequate with respect to Mach? Admittedly, there is large evidence that Mach approved of all attempts to realize the aim of pure description. But he was not a physical iconoclast like Ostwald. Especially his *Principles of the Theory of Heat*, published first in 1896, can hardly be squared with the received view. A careful reading of this book reveals that Mach's intention was not at all to play off Ostwald against Boltzmann. Rather his intention was to *mediate* between their seemingly extreme positions.

Why do I think so? Well, because Mach explicitly outlined such a mediating view in a remarkable but obviously not very well-known chapter of his *Principles of Heat*. Titled "The Opposition Between Mechanical and Phenomenological Physics", the chapter in question was written in order to effect a "reconciliation".[41] The two extremes to reconcile, Mach pointed out, were traceable to Hooke and Newton, but more forcibly and – as it seemed – implacably defended in the current energetics controversy. Still, for Mach a middle-way, by which especially the views of Boltzmann and Ostwald could be reconciled, was "quite attainable".[42] I shall call this the *reconciliation thesis*, or RT for short.

Next we have to scrutinize how Mach tried to substantiate RT. Let us call the corresponding line of reasoning the *reconciling strategy*, or RS for short. In its most distinct form RS is expressed by Mach's assessment that a theoretical picture "can have a great heuristic value as a working hypothesis and, at the same time, a very slight value as regards its epistemic content".[43] Accordingly, RS requires us to make a clear distinction between two levels of evaluation: on the one hand the *use* of theoretical pictures, on the other hand their epistemological *interpretation*. Furthermore, RS appreciates pictures on the first level, while it apparently depreciates them on the second. Thus, at its face value RS amounts to saying that (a) someone like Boltzmann is right on the use level, whereas (b) someone like Ostwald is right on the interpretation level. However, what I want to suggest is that RS can be made stronger, and that in favor of Boltzmann's point of view.

The chief evidence for my suggestion is the following passage from the above-mentioned chapter of the *Principles of Heat*:

I cannot agree ... that the marvelous forces which people ascribe to the notions used in mechanical physics are now merely transferred to algebraic formulae, and that in place of the mechanical mythology we can simply substitute an algebraic one. The validity of a formula in like manner denotes an analogy between the operation of a calculus and a physical process. Whether this analogy holds or not, in each particular instance, has to be tested. [44]

This passage is revealing because it shows that Mach in fact acknowledged the pictorial character of mathematical equations. A person who attempted to set up an "algebraic mythology" would be someone who, by contrast, supposed that physical reality is so to speak *implied* by treating observed facts mathematically. This, as we have seen, was done by Ostwald. On the latter's view, the differential calculus enabled physicists to correlate phenomena directly. For that no pictures were required: "Just write down your equations, and then you know what can be known at all." For Mach, this attitude was ill-advised, for it could not account for cases where the application of the calculus had led to false results. Moreover, it ignored that differential equations, when used as means to the predictive end, did necessarily transcend the observed facts. Finally and most importantly, it failed to recognize that the respective magnitudes to be related by the use of differential equations were in fact the outcome of such epistemologically in-direct procedures like the employment of units of measurement *instead* of the immediately perceived objects and processes. That the "immediate contact with the system is lost by th[ese] procedure[s]"[45] Mach willingly admitted, adding (at another place) that "we need the unit of measurement because we cannot directly transfer our notion of magnitude – indeed, no thought can be so transferred".[46] Accordingly, for Mach there was no other way than to assume that even our most exact mathematical descriptions ultimately are resulting from manipulating natural phenomena in such a way that they come out as more or less idealized "*constructions* of facts in thought".[47] Whether these idealized constructions were compatible with other, known or yet to be discovered facts had to be tested in each case. Mach's oft quoted device "analogy is not identity"[48], as a consequence, applied to differential equations as well: far from being an incorporation of immediate cognitive security, those equations were as much in need of being constantly *adapted* to the facts as any other kind of theoretical representation. That is, for Mach there was no difference in kind between the abstract mathematical equations of phenomenological physics and the more intuitive conceptions of mechanical physics. Both types of "thought construction" were – though at different degrees – pictorial in character in that they both provided numerically specifiable but at the same time observationally underdetermined *models* of (certain aspects of) reality.[49]

Understood this way, RS is supporting Boltzmann, not only on the use level but also on the interpretation level.[50] Mach apparently felt that Ostwald had gone too far in his iconoclast restrictions. In particular, he recognized that Ostwald's view neglected the constructive element in treating observed facts mathematically. Nevertheless, Mach thought of Boltzmann's counter-thrust as too restrictive in another sense. For even if the analogical, or pictorial, status of the calculus was conceded, its interpretation along *atomist* lines was not compelling altogether. It rather seemed that physics then again would end up at the old "*mechanical* mythology", which Mach thought could not be the intended effect of RT.

Mach's own proposal, then, was to supplant the atomic hypothesis, especially because a mathematically more convenient and epistemologically less "unnatural" conception was at hand. Thus, in a note added to the second edition of the *Principles of Heat* (which appeared in 1900) Mach briefly indicated that Kirchhoff's theory of *volume elements* provided the desired scientific basis for RS. Says Mach:

> The volume elements, with their falls of temperature, behave exactly like finitely extended bodies under similar circumstances; but I have this advantage, that I can build up, out of such small volume elements, any case however complicated, with whatever exactness I desire. I cannot understand, therefore, why every differential equation must necessarily be based upon atomistic views. [51]

I think Mach was completely justified in drawing this conclusion. Boltzmann's peculiar attempt to "atomize" the differential calculus indeed was anything but convincing, let alone compelling. Rather, it ran the risk of trivializing atomism and with it Boltzmann's position on the whole. Just imagine what would happen if any theory which availed itself of differential equations would be interpreted the way Boltzmann demanded. Then any such theory would automatically be an atomistic one. Mach recognized that this can hardly be accepted as a feasible conception. For this reason, he espoused the theory of volume elements, with the help of which he was in a position to account for the fact that differential equations are commonly interpreted over continuous quantities. A further advantage of this approach was – according to Mach – that the assumption of volume elements did not require a transgression of what at least in principle can be regarded as an object of perception. Yet this is not to say that Mach in fact was physically right in favoring volume elements instead of atoms. I just want to suggest that his attempt to block Boltzmann's restrictive atomist conception was appropriate in view of questions of applied mathematics. Whether Mach's continuum-based alternative as such had anything to recommend it, is quite another issue not to be dealt with here.

To summarize: Mach proposed a reconciling middle-way in order to avoid the extreme consequences, both in Ostwald's and in Boltzmann's view of picturing in physics. To that effect he outlined a strategy which I referred to as RS, and which can be interpreted in a weaker and in a stronger sense. Interpreted in the weaker sense, RS gives rise to a purely pragmatic account of picturing in physics. Understood this way, it sides with Boltzmann only insofar as it concedes the heuristic usefulness of pictures. Interpreted in the stronger sense, RS is conceding more insofar as it approves of Boltzmann's view that pictures are to be conceived of as an unavoidable ingredient of treating observed facts mathematically. But it sides with Ostwald by refuting atomism as a scientifically inconvenient position. Thus, according to the stronger version of RS, Mach was a picture-theorist like Boltzmann and, at the same time, an anti-atomist like Ostwald. Whether or not convincing in the end, this mediating point of view, I think, is worth investigating further and hence a challenge for researchers in the history of the philosophy of science.

4. EPILOGUE

The most precarious point of my approach to Mach is its restricted textual basis. There are several other passages in Mach's writings by which the above-propounded reading can easily be invalidated. But this is rather a problem of Mach than of my admittedly selective interpretation. Mach was, it must be emphasized, not very consistent in advocating his "third way". In fact, he was wavering between the weaker and the stronger version of RS, whereby the weaker version eventually prevailed. A good example for this is the following quote from the *Principles of Heat:*

Fourier's theory of the conduction of heat may be characterized as an ideal theory. It is founded, not upon a hypothesis but upon an observable fact according to which the velocity of equalization of small differences of temperature is proportional to these differences themselves. Such a fact can be more precisely established or corrected by finer observations; but it can, as such, enter neither directly nor in its correct mathematical deductions into conflict with other facts. This foundation of the theory, with the entire structure supported by it, remains secure – while a hypothesis like that of the kinetic theory of gases … must be prepared at any moment for contradiction by new facts, no matter how it may have contributed to the survey of the properties of gases up to that time. [52]

Here, we have a clear endorsement of the weaker reading of RS. The only reconciling aspect to observe is Mach's concession that a pictorial construction like the kinetic theory might at best be helpful as an auxiliary hypothesis. But in the long run, Mach confirms, it no doubt will be supplanted by a theory identical in structure with the one Fourier set up. Even if this would be so (which appears unlikely), Mach – on the stronger reading of RS – would have to admit that such an "ideal theory" *as well* must be prepared for "contradiction by new facts". However, the above-quoted passage clearly indicates that Mach eventually refused to do so. And that's the reason, I conclude, why he so often is characterized as a proponent of the anti-imagist tradition in the philosophy of science.

ACKNOWLEDGEMENTS

I would like to thank Michael Heidelberger and David Hyder who supported me in my investigations. I also wish to thank John Blackmore, Robert Deltete, and Patrick McDonald for helpful comments on earlier drafts of this paper.

NOTES

1. See, for example, Zenon Pylyshyn, "What the Mind's Eye Tells the Mind's Brain: A Critique of Mental Imagery", in: *Psychological Bulletin* 80, 1973, pp.1-24.
2. See, for example, Stephen Kosslyn, *Image and Mind*. Cambridge, Mass.: Harvard University Press 1981.
3. See Arthur I. Miller, *Imagery in Scientific Thought: Creating 20th-Century Physics*. Cambridge, Mass.: MIT Press 1986, p.224.
4. For details, see Martin J. Klein, "Mechanical Explanation at the End of the Nineteenth Century", in: *Centaurus* 17, 1973, pp.58-82.
5. For the details of that controversy, see Robert J. Deltete, *The Energetics Controversy in Late Nineteenth Century Germany: Helm, Ostwald, and their Critics* (Yale University doctoral dissertation, 2 volumes). Ann Arbor: UMI 1983. For a brief summary, see Erwin N. Hiebert, "The Energetics Controversy and the New Thermodynamics", in: D.H.D. Roller (Ed.), *Perspectives in the History of Science and Technology*. Norman: University of Oklahoma Press 1971, pp.67-86.
6. Thus writes Georg Helm, himself an energeticist, in 1898: "And so, in the [energetics] controversy ... it is not really a question of atomism or of matter continuously filling space, not a matter of the inequality sign in thermodynamics, or of the energetic foundations of mechanics. All of these are only details. In the final analysis, what is at stake are the principles of our knowledge of nature". And directly referring to the picture debate, Helm asserts: "What was opposed and defended about energetics [at the 1895 *Naturforscherversammlung* in Lübeck] was *the method of being able to talk about natural processes in a language free of pictures*." See Georg Helm, *The Historical Development of Energetics*. Translated by Robert J. Deltete. Dordrecht: Kluwer 2000, pp. 404 and 402.
7. This is not to say that the first question is not worth discussing. But a satisfactory treatment of it would require a thorough consideration of the very complex history that lead to the introduction of the picture concept in the late nineteenth-century physical community. Since this would be a task of its own, I must ignore it here.
8. See Wilhelm Ostwald, "Die Ueberwindung des wissenschaftlichen Materialismus", in: *Verhandlungen der Gesellschaft deutscher Naturforscher und Ärzte* I,1. Leipzig: Vogel 1895, pp.155-156.
9. *Ibid.*, p.160.
10. This train of thought is elaborated at greater length in Wilhelm Ostwald, *Vorlesungen über Naturphilosophie*. Leipzig: Veit & Comp. 1902, pp.206-208.
11. See Ostwald, "Die Ueberwindung des wissenschaftlichen Materialismus", *op.cit.*, p.159. See also Ostwald, *Vorlesungen über Naturphilosophie*, *op.cit.*, pp.211-214, where Ostwald is more explicit about this point.
12. For details, see Ostwald, *Vorlesungen über Naturphilosophie*, *op.cit.*, pp.207-208.
13. Ostwald, "Die Ueberwindung des wissenschaftlichen Materialismus", *op.cit.*, p.162.
14. By the way, it is not at all clear whether this conception of the metaphysical as the not directly measurable is held consistently throughout Ostwald's writings. To some extent it seems to conflict with his substantial interpretation of the energy concept. But this is a large issue, which I cannot address adequately here. For further information, see Wilhelm Burkamp, *Die Entwick-*

lung des Substanzbegriffs bei Ostwald. Leizig: Emanuel Reinicke 1913, esp. pp.84-89 and 104-109.

15. See Ostwald, "Die Ueberwindung des wissenschaftlichen Materialismus", *op.cit.*, p.160.
16. *Ibid.*, p.161.
17. *Ibid.*, p.162.
18. See Ludwig Boltzmann, *Populäre Schriften*. Leipzig: Johann Ambrosius Barth 1905, pp.105-128.
19. For a fair evaluation of Boltzmann's physical objections, see Deltete, *The Energetics Controversy, op.cit.*, pp.462-477.
20. See Boltzmann, *Populäre Schriften, op.cit.*, p.104.
21. A lot has been written on that topic. For more recent contributions, see the references in Henk W. De Regt, "Ludwig Boltzmann's *Bildtheorie* and Scientific Understanding", in *Synthese*, 119, 1999, pp.113-134.
22. Boltzmann, *Populäre Schriften, op.cit.*, pp.136.
23. Among the examples cited by Boltzmann are the mechanical "pictures" of sound, heat and crystallography, which on his view had been confirmed "almost up to the point of certainty". See Boltzmann, *Populäre Schriften, op.cit.*, p.135. A revealing passage, by the way, if one bears in mind that Boltzmann's own work in theoretical physics was mainly devoted to what he here characterized as the mechanical picture of *heat*.
24. *Ibid.*, p.136.
25. Without explicitly referring to Boltzmann, Ostwald in fact espouses this view in his 1902 lectures on natural philosophy. See Ostwald, *Vorlesungen über Naturphilosophie, op.cit.*, pp.213-215. For a very similar contemporary account, see Pierre Duhem, *La Théorie Physique, son objet et sa structure*. Paris: Chevalier et Rivière 1906, p.32.
26. Boltzmann, *Populäre Schriften, op.cit.*, p.130; italics added.
27. Ludwig Boltzmann, *Theoretical Physics and Philosophical Problems*. Edited by Brian McGuinness and translated by Paul Foulkes. Dordrecht: D. Reidel 1974, p.42.
28. *Ibid.*, p.126.
29. Stathis Psillos, *Scientific Realism: How Science Tracks Truth*. London: Routledge 1999, p.211.
30. Boltzmann, *Theoretical Physics, op.cit.*, p.42.
31. The chosen notation is somewhat bizarre, but for the sake of the argument I rely on Ostwald's original formulation. See Ostwald, *Vorlesungen über Naturphilosophie, op.cit.*, p.214.
32. See Boltzmann, *Theoretical Physics, op.cit.*, p.43: "On closer scrutiny the differential equation is merely the expression for the fact that one must first imagine a finite number; this is the first prerequisite, only then is the number to grow until its further growth has no further influence."
33. For a more detailed comment on Boltzmann's (finitist) interpretation of the calculus, see Martin Curd, *Ludwig Boltzmann's Philosophy of Science: Theories, Pictures and Analogies* (University of Pittsburgh dissertation). Ann Arbor: UMI 1978, pp.241-254.
34. This, by the way, is the point where Psillos would dissociate himself from Boltzmann. For on Psillos' view, "in the case of the reference of theoretical terms, 'the existentially given thing' is nothing but a *causal agent*, i.e. an agent which is posited to have the causal power to produce certain effects". See Psillos, *Scientific Realism, op.cit.*, p.290. Boltzmann did, as far as I see, never think of the existence of such "hidden" causes as implied by his all-picture-view (see next footnote).
35. Or in his own words: "What might be the true cause for [the phenomenal world] to run as it does, what is as it were concealed behind it and acts as its motor, these things we do not regard as the business of natural science to explore". See Boltzmann, *Theoretical Physics, op.cit.*, p.109. There are several other passages where Boltzmann says pretty much the same (see, for example, *ibid.*, pp.16, 88, 111, 265n). I can therefore not approve of Yehuda Elkana's claim that "Boltzmann never gave up the idea that the task of science is to explain and not only to predict the phenomena". See Yehuda Elkana, "Boltzmann's Scientific Research Program and its Alternatives", in Y. Elkana (Ed.): *The Interaction between Science and Philosophy*. Atlantic Highlands, N.J.: Humanities Press 1974, p.268n. To be sure, it is likely that Boltzmann *as a working physicist* indeed was driven by the hope to find out something about the causal structure of the world. But *as a philosopher of science* he, in fact, refrained from articulating such a hope.
36. Arnold Sommerfeld, "Das Werk Boltzmanns", in: Wiener Chemiker Zeitung 47, 1944, p.25.

37. Max Planck, *Vorträge und Erinnerungen*, Stuttgart: Hirzel 1949, p.12.
38. Robert A. Millikan, *The Autobiography of Robert A. Millikan*. New York: Prentice-Hall, Inc. 1950, p.85.
39. Carlo Cercignani, *Ludwig Boltzmann: The Man Who Trusted Atoms*. Oxford: Oxford University Press 1998, pp.26-27.
40. Henk Visser, "Boltzmann and Wittgenstein or How Pictures Became Linguistic", in: *Synthese* 119, 1999, p.152.
41. Ernst Mach, *Principles of the Theory of Heat*. Edited by Brian McGuinness and translated by Martin J. Klein. Dordrecht: D. Reidel 1986, p.333.
42. *Ibid.*
43. *Ibid.*, p.445n; translation slightly altered.
44. *Ibid.*, p.334.
45. *Ibid.*, p.77.
46. *Ibid.*, p.370.
47. *Ibid.*; italics added.
48. *Ibid.*, p.333.
49. It is astonishing (but not untypical) that one of the chief proponents of the so-called semantic view, Frederick Suppe, blames Mach for ignoring the fact "that scientific principles contain mathematical relationships not reducible to sensations alone". See Frederick Suppe, "The Search of Philosophic Understanding of Scientific Theories", in: F. Suppe (Ed.): *The Structure of Scientific Theories*. Urbana: University of Illinois Press 1977, p.11. Mach, as we have seen, conceded that mathematical equations have the status of analogical *constructions* being, as such, applicable but not at all reducible to what can be immediately perceived by our senses.
50. This agreement between Mach and Boltzmann is nicely reflected by the latter's confession that Mach's critique of Ostwald's tendency toward a mathematical mythology had "greatly helped in clarifying my own world view". See Boltzmann, *Theoretical Physics, op.cit.*, p.51n.
51. Mach, *Principles of the Theory of Heat, op.cit.*, p.445n.
52. *Ibid.*, p.113.

Institut für Philosophie
Humboldt-Universität
Antwerpener Str. 49
D-13353 Berlin
Germany
matthias.neuber@t-online.de

TORSTEN WILHOLT

LUDWIG BOLTZMANN'S MATHEMATICAL ARGUMENT
FOR ATOMISM

In recent years, the philosophy of Ludwig Boltzmann has become a point of interest within the field of history of philosophy of science. Attention has centred
around Boltzmann's philosophical considerations connected to his defense of
atomism in physics. In analysing these considerations, several scholars have
attributed a pragmatist stance to Boltzmann. In this paper, I want to argue that,
whatever pragmatist traits may be found in Boltzmann's diverse writings, his
defense of atomism in physics can not be analysed this way. In other words, I
wish to show that he did not defend atomism as "preferable for its practical virtues", as has been alleged.[1] On the contrary, Boltzmann considered the atomist
picture to be indispensable – more precisely, an indispensable prerequisite for
making the application of continuous differential equations an understandable
enterprise.

Boltzmann's ideas related to this line of reasoning are scattered across his
writings. Thus, what I label his 'mathematical argument' for atomism is a rather
loosely arranged set of views. In addition to the aim formulated above, it is a
second objective of this paper to bring together various historical details relating
to the mathematical argument and to find a reading that is coherent in itself and
with Boltzmann's other philosophical opinions. My third aim is (at the end of the
paper) to identify an important limitation to Boltzmann's case for atomism as reconstructed here.

1. THE VIRTUE OF ATOMISM

At the beginning, let me remind you of one of Boltzmann's better known essays
on this subject, his 'On the Indispensability of Atomism in Natural Science' (first
published in 1897). My thesis that Boltzmann argued for the indispensability of
atomism and not merely for its pragmatic superiority is on a very superficial
level reinforced by the title of the essay. But on a deeper level, the argument in
this central essay is not so unambiguous. Let me, without further quotations,
sketch a very obvious way to read that argument.

The argument is mainly directed against the mathematical phenomenologists
who demand the abandonment of atomist principles and their replacement by the
idea of a physical continuum. The continuum idea in question is the mathematical continuum concept as it is manifested in the application of real analysis in

M. Heidelberger and F. Stadler (eds.), History of Philosophy and Science, 199–211.
© 2002 *Kluwer Academic Publishers. Printed in the Netherlands.*

theoretical physics. However, Boltzmann claims, the phenomenologists thereby obscure the discontinuous foundations of the continuum concept underlying the differential calculus, which derives from the mathematical concept of limit. This concept, if applied to physical reality, must from Boltzmann's point of view be understood as follows: the limit of a function is established by increasing a certain number (of particles, of sections into which an interval is divided) until further increment of that number would not have any "noticeable" influence on the result any more. So, for Boltzmann, to forget about these foundations and to take the differential equations themselves as the most straightforward representation of reality (as the mathematical phenomenologists do) means only to make an additional assumption: the assumption, that is, that however much our means of observation improves and however more subtle the differences "noticeable" for us become, we will never chance upon a difference between the measured facts and the magnitudes given by mathematical (infinitary) limits. For Boltzmann, atomism is the position which does not make that additional assumption, and since he holds the assumption to be totally unwarranted, the mathematical phenomenologists go beyond the observable facts than the atomists do. Thus, atomism is the most natural stance for a scientist applying the differential calculus to nature.

So far, Boltzmann's reasoning provides an argument that atomism is preferable with respect to the background assumption that the scientific image should contain as few arbitrary elements as possible. This rhetoric is, of course, designed to defeat his scientific adversaries Mach, Ostwald *et al.* with their own weapons. However, I must concede, it does not add up to an argument for the *indispensability* of atomism, because 'making fewer arbitrary assumptions' is a feature that may make a theoretical picture preferable but not automatically indispensable.

2. THE INDISPENSABILITY OF ATOMISM

Fortunately, Boltzmann's argument contains a further, less obvious and less clear element. The essay discussed so far also contains claims to the effect that the idea of a continuum in nature would be incomprehensible without an atomistic foundation:

Atomism seems inseparable from the concept of the continuum. The reason why Laplace, Poisson, Cauchy and others started from atomistic considerations is evidently that in those days scientists were as yet more clearly conscious that differential equations are merely symbols for atomistic conceptions so that they felt a stronger need to make the latter simple.[2]

Here it is no longer 'atomism vs. the continuum' but 'no continuum without atomism'. Contrary to the aforementioned line of reasoning, where the continuum picture was just like the atomist picture but with one more arbitrary and

unnecessary assumption, it is now claimed to be in need of an atomistic interpretation.

Boltzmann returned to this leitmotif many times. The continuum itself is
thereby taken to represent the idea of actually infinite divisibility. But in physics,
infinity can only be viewed as a "Grenzübergang", a transition to the limit.[3] This
is how he represented this central idea in his 1904 address to the Scientific
Congress in St. Louis:

[W]e cannot define infinity in any other way than as the limit of ever-growing finite
magnitudes; at least nobody so far has been able to establish an intelligible concept of
infinity in any other way.[4]

For Boltzmann, this implied that it is indispensable to understand not only the
infinite in terms of the finite, but also the continuum in terms of the discrete.
Let's turn again to the 'Indispensability' essay of 1897, where he states the following:

Do not imagine that by means of the word continuum or the writing down of a differential
equation, you have acquired a clear concept of the continuum. On closer scrutiny the differential equation is merely the expression for the fact that one must first imagine a finite
number; this is the first prerequisite, only then is the number to grow until its further
growth has no further influence.[5]

Obviously, this view requires clarification. Analysis as applied in physics
evidently presupposes the actual continuum of real numbers. It is essential that
for any monotone and bounded sequence, the set of real numbers contains the
actual limit towards which the sequence converges, and *not* just enormously
many elements of the sequence until further growth is not "noticeable" anymore.
(Otherwise, real analysis wouldn't be Cauchy complete.) It is therefore even
characteristic for real analysis that it can*not* be reduced to the finite in the way
Boltzmann envisages.

Is he thus suggesting that the mathematical continuum is somehow an unjustified concept and that science should distance itself from real analysis? Fortunately not. Nowhere does he make this suggestion, and he goes on to apply real
analysis like any other physicist. How then are we to understand his assertion
that in physics, we have to conceive of the continuous differential equations as
"symbols for atomistic conceptions"?

To get further elucidation on this, one has to turn to the lecture course on
Natural Philosophy that Boltzmann gave in the winter term of 1903/04.[6] There, a
difference between Boltzmann's philosophy of pure and applied mathematics
becomes apparent. He naturally introduces infinitary concepts of pure mathematics and even calls it a "legitimate claim" that mathematics be in principle in a
position to grasp the infinite and the "actually continuous".[7] But when he turns to
a closer inspection of infinitary mathematics as explicated by Cantor's set
theory, he makes it clear that this amounts to no more than a "game with concepts".[8] As I will lay out in a little more detail in section 4, Boltzmann thought

that such merely conceptual constructions, not directly modelled on experience, were inadvertently fallible and ambiguous if applied to reality. Thus, to preserve certitude of scientific inference, applied analysis has to be given a finite, discontinuous interpretation. (As Boltzmann wrote in 1897, the "practical utility" of infinitary methods is not to be called into question, but they are "epistemologically inferior to atomistic conceptions".[9])

While a more extensive treatment of Boltzmann's reasons for this claim will have to wait for a moment, we may for now retain the following moral as a tentative interpretation of the claim that "differential equations are merely symbols for atomistic conceptions": there is nothing wrong with the actual mathematical continuum as such, it is only in its application to natural phenomena that it has to be accompanied by an atomistic conception of the underlying reality (in order to prevent ambiguity). Such an atomistically understood reality cannot, of course, be accurately reflected by continuous differential equations. But descriptions in terms of differential equations can be seen as symbols for atomistic conceptions insofar as they can serve to give us good *approximations* to the truths of an indefinitely (but not infinitely) fine-grained discontinuous reality. In fact, Boltzmann suggested just this picture when he spoke of a possible discontinuity of time during his exposition of the general laws of motion in his lectures on mechanics:

Perhaps our formulae are only very closely approximate expressions for average values that can be constructed from much finer elements and are not strictly speaking differentiable. As to that, however, there are so far no indications from experience.[10]

This leads us to another point of clarification. The atomism at stake in the debate around the turn of the century was, of course, an atomism of matter. Yet Boltzmann's argument is directed against continuous concepts as such and thus promotes a more thoroughgoing atomism, including discontinuous time. Boltzmann mostly buries this aspect in the footnotes. In an extensive footnote to the 'Indispensability' essay, he imagines a world of an enormous number of packed spheres that evolve through discontinuous time. He concludes:

If it were possible to find such a picture that showed more comprehensive agreement than ordinary atomism, the picture would thereby be justified. Thus the view of atoms as material points and of forces as functions of their distance is no doubt provisional but must at present be retained failing a better one.[11]

In short: even atomistic physics as practised by Boltzmann is not only approximative (because it still applies continuous equations to a discontinuous reality) but also provisional (because it falls short of a thoroughly atomistic conception and remains restricted to an atomism of matter).

Let me summarise the picture that emerges from the interpretation I have sketched in this section: without an atomistic conception of the underlying reality, continuous differential equations are rooted in mere games with concepts and remain fallible and ambiguous. Therefore, atomism is indispensable for physical science if it is to be saved from ambiguity. There still remains a lot to say about why Boltzmann held these convictions. Before I turn to that question, I wish to respond to a possible objection.

3. ATOMISM: PREFERABLE OR INDISPENSABLE?

My claim that Boltzmann really wanted to argue for the *indispensability* of atomism depreciates the other argument that can be inferred from the 1897 essay, i.e. that atomism is *preferable* on the ground of its making fewer arbitrary assumptions, as explained in section 1, to a mere rhetorical device. An obvious objection to my analysis arises: Why should one accept the suggestion that it's the argument for preferability which has only rhetorical character, while the indispensability of atomism for unambiguous reasoning about the physical continuum constitutes a genuine and fundamental argument of Boltzmann's? Could it not equally well be just the other way round, so that Boltzmann's indispensability considerations turn out to be a rhetorical over-intensification of his plea for preferability?

This objection would in fact fit in well with a very common account of Boltzmann's philosophical activity in the 1890's. This account stresses the fact that during that time, Boltzmann was facing the increasingly growing influence of anti-metaphysical opposition to the (trans-conscious) reality of atoms, brought up by Mach, Ostwald and others. This threatened the core of his physical work and thinking: kinetic theory and physical discontinuity. Therefore, it is often said, he adopted a series of diverse and sometimes incompatible philosophical positions, with the one and only aim of being in the strongest possible position to defend the core of his scientific work: atomism.[12] This account is sometimes applied to his picture theory of scientific representation and to his proposals (sounding like philosophy of language) to cleanse science of 'meaningless' metaphysical questions. Extending this to his indispensability idea seems to add up to a nice objection to my reading: the talk of the indispensability of atomism for our understanding of the physical continuum would then merely be the strongest possible rhetorical device, which Boltzmann pragmatically adopted for the purpose of defending his research programme.

My belief that this is not the case and that there is more to his indispensability considerations is based on my account of how Boltzmann came to think about the indispensability of atomism in the first place. The idea that in physics, infinity can only be viewed as a transition to the limit, presented itself to Boltzmann at least 20 years before the 'Indispensability' essay was published. As I will now explain, I believe that it emerged directly from the concrete conceptual problems of his own research in theoretical physics.

Whereas usually Boltzmann is said to have ignored philosophy before the mid-1880's, the indispensability idea surfaces as early as 1877, in his famous paper on the relation between the second theorem of thermodynamics and the theory of probability. It contains a proof of the theorem which we today call Boltzmann's relation, $S = k \log W$. This important result relates the concept of entropy (S) of a gas in a certain macrostate to the probability (W) of that macrostate. This probability is in turn defined via the relative number of equiprobable microstates corresponding to each respective macrostate. The microstates in question are energy distributions among the molecules of the gas. However, since the kinetic energy that each molecule can assume is a continuous quantity, there are of course infinitely many possible energy distributions corresponding to each macrostate. Therefore, the rough and slightly incorrect characterisation of the concepts involved which I have just given, though it captures the intuitive idea behind the relation between thermodynamics and probability, strictly speaking involves an impossible operation: the quantitative comparison between infinite numbers.

This is how Boltzmann actually proceeded in order to circumvent this difficulty: he first assumed that there is only a finite set of discrete energy values that each molecule can adopt, and only afterwards calculated the mathematical limits which result from infinitely increasing the number and "concentration" of discrete energy values. While this may be quite an ordinary mathematical procedure, it is important to see how Boltzmann himself commented on it in the paper:

> Even if this way of dealing with the problem seems to be very abstract at first sight, with most suchlike problems it leads the quickest way to the goal, and if one considers that *all infinite in nature never means anything other than a transition to the limit*, then one cannot at all conceive of the infinite variety of velocities that each molecule is capable of adopting in any other way, unless as the limiting case which occurs when each molecule can adopt more and more velocities.[13]

As already mentioned above, the italicised expression is a phrase that he keeps on using into his late writings, where it is his slogan for the view that continuous equations presuppose atomistic conceptions which can then be conceived as indefinitely, but not infinitely fine-grained. For two reasons, I take the 1877 paper to confirm my reading that Boltzmann entertained the thought that atomism is indispensable to our understanding of the continuum as a serious element of his physical *Weltbild*, and not only as a rhetorical device. Firstly, it is obvious that the core of the idea, viz. that "all infinite in nature never means anything other than a transition to the limit" is directly related to indispensable problem-solving techniques in his own concrete work as a theoretical physicist. Secondly, and more importantly, he had explicitly stated it long before he first engaged in philosophical and rhetorical battles with energeticists, mathematical phenomenologists and other anti-metaphysical opponents.[14]

4. WHY DOES THE PHYSICAL CONTINUUM PRESUPPOSE A DISCONTINUOUS CONCEPTION?

I shall therefore now turn to a major problem *within* Boltzmann's indispensability considerations. As we have seen, he bases them on his conviction that continuous equations presuppose finite and discrete interpretations; in his words: that all infinite in nature can only be understood as a transition to the limit. But this principle is itself not explicitly justified in the 1877 paper, nor in most of the other places where it is invoked. Boltzmann almost seems to have thought it self-evident. Nevertheless, important motivating ideas for the principle can be found in Boltzmann's philosophy, even if only in his later philosophical prose. In my opinion, two different lines of thought capable of serving as explanations for the aforementioned principle can be extracted from Boltzmann's writings. For both these lines, the only place where they are elaborated is the above-mentioned course on Natural Philosophy.

The *first of the two lines of reasoning* starts in the fourth lecture with the following remark:

It is self-evident that we cannot define the simplest concept, the concept of number, because we cannot resolve it into simpler concepts.[15]

As Boltzmann goes on to explain, the concept in question is, more precisely, that of the positive whole numbers. He sees it as basic and grounded in the practice of counting. How much of it is inherited and how much learned, he finds hard to tell. In any case, we have acquired familiarity with the operations on positive whole numbers. It is only by our deliberate intent to extend the scope of these operations that we construct the other kinds of numbers: fractions as well as negative, irrational and imaginary numbers.[16] Positive whole numbers, on the contrary, have through constant experience made their way into our "forms of thought", which Boltzmann characterises as follows:

... forms of thought, which have through outside influence developed in our brain, in our psyche, and these forms of thought have always led us to correct results; thereby they have become fixed forms of thought.[17]

From this psychologistic position, it is only a very small step to the conclusion that since the experience of counting which informs our laws of thought is always a finite experience, only finite concepts can be completely understood by our thus conditioned minds. Boltzmann nowhere makes this step explicit. He does however, with reference to Bolzano, contend that the infinite is a paradoxical notion. The paradox has, according to Boltzmann, not been solved by Cantor and the set theorists, who have only

... got used to it, they have only shown how the computational operations are to be arranged, so that one does not stumble over it, how one has to calculate in order to get around the paradoxes. [18]

I find it not too bold to round out this line of thought and give the following explanation: The concept of number, whether finite or infinite, is ultimately grounded in experience and in our laws of thought. And since our laws of thought have, through evolutionary aeons and throughout our individual experience, been informed by exclusively *finite* experiences, we can only understand the infinite in nature as a limit of increasing finite magnitudes. To our minds, the infinite as such remains an unfathomable paradox.

The *second line of reasoning* in question is highly original. It is based on an old problem in theoretical mechanics. Boltzmann does not refer to any specific texts in connection with this problem, but only mentions the name of the Slovenian mathematician Georg von Vega. But it can be shown that the problem goes back at least to Leonard Euler's *Mechanica* of 1736.[19] It arises with a very simple mechanical setting: A point mass A is, out of relative rest, attracted to a centre of force S. But Euler calculated the path of A as a limiting case of the general form of motion governed by a single centre of force, elliptical motion. The small axis of the ellipse in Euler's calculation becomes infinitely small, and the ellipse infinitely eccentric; therefore, one apsis of the ellipse moves infinitely close to the focus at S. In the limiting case, A's starting point and S form the endpoints of a straight line. The result is that A will move toward S in a straight line, but upon arrival will be instantaneously reflected and move backwards toward the starting point. Euler accepted the result despite its counter-intuitive character.

Just as an aside, note that also Euler himself saw the reason for the conflict between common sense and his calculation in our limited understanding of infinitary reasoning. Yet, in contrast to Boltzmann, he trusted in mathematics to compensate for a lack of intuitive understanding:

Whatever may be, here one must trust in the calculation rather than in our judgement and declare that we do not deeply comprehend a leap if it occurs from the infinite to the finite.[20]

(He mentions a leap from the infinite to the finite because, according to his calculation, A would momentarily acquire an infinite velocity at S.)

Georg von Vega was not content with Euler's result. His treatment of the case can be found in the third volume of his *Lectures on Mathematics*, first published in 1788.[21] He rejects Euler's calculation and simply calculates the motion of A on a straight line right through S by means of integration, starting from the inverse square force law. Vega himself is not really bothered that this law implies division by zero for the case when A passes S, but Boltzmann tacitly corrects Vega's alternative treatment and introduces a tiny sphere around S where the force is suspended.

Boltzmann describes this as a dilemma: If, in Euler's calculation, the small axis becomes *infinitely* small, and, in Vega's calculation, the sphere around S becomes *infinitely* small, then they are both treatments of the same setting, but with directly opposite results, and it becomes impossible to decide which is the correct one. For Boltzmann, the dilemma is due to the infinitary reasoning involved. By restricting oneself to finitary reasoning, one can resolve the dilemma into a matter of differing cases: If we assume the small axis to be extremely, but not infinitely small, then A will reverse its direction near S, describing an extremely eccentric elliptical motion *around S*. And if we assume a motion on a straight line and a very, but not infinitely small sphere around S where the force is suspended, then A will traverse S and move on in the same direction.

In fact, if we never adopt anything infinite, if we only calculate with finite magnitudes that can be arbitrarily large, we never get to a contradiction. [...] But if A coincides with S and if Newton's law is valid till coincidence, [...] then I enter into set theory, into the really infinite, into the ultimate number ω, and I encounter contradictions.[22]

This whole argument is concerned with such a specialised mechanical puzzle that one naturally doubts whether it can be intended as a serious general argument against infinitary reasoning about nature. But in Boltzmann's own fragmentary notes for the lecture, he suggests just this:

Vega. Collision [thrust]: That is my proof that nature must be constructed atomistically. Not that nature must, but that we must think it thus, if we do not want to apply such absurd concepts as the ones from set theory, which lead to ambiguities.[23]

Thus, for Boltzmann, as I understand him, the Euler Vega dilemma shows nothing but the following: Our ability to reach secure and unambiguous conclusions is limited to reasoning about finite magnitudes. But this is at the same time a limitation of our understanding, because

... the purpose of thinking is to be able to unambiguously draw conclusions everywhere; therefore, we must seek to form our signs of speaking, writing and thinking in such a way, that we express ourselves unambiguously and understand ourselves unambiguously.[24]

So, since we cannot unambiguously draw conclusions about the infinite as such, we cannot understand it. Therefore, we can not understand the infinite in nature, unless we conceive of it as a transition to the limit.

Is there a common idea behind the two lines of reasoning I have just been describing? I think so. Let me try to give an interpretation of how the diverse ideas of Boltzmann that I have presented so far belong together:

While, for Boltzmann, the Euler Vega dilemma seems to constitute exemplary evidence that we cannot unambiguously apply infinitary reasoning to nature, his psychologistic considerations about numbers may provide an explanation why this is the case. Numbers, being indefinable and basic forms of thought, have their justification in our and our ancestors' regular experience with physical objects. This is why we can securely apply finitary mathematics to nature. Infinitary mathematics, on the contrary, is the extension of the laws of thought to a realm far beyond the experience that first established them. And while this may be tolerable as a mere game with concepts, by reapplying this extended mathematics to nature, our judgement becomes ambiguous and fallible. To apply a phrase which Boltzmann often used: The laws of thought overshoot the mark. And since our ability to unambiguously draw conclusions is a prerequisite for understanding, we cannot truly understand the concept of an infinitely divisible continuum. Thus, an atomist interpretation of the continuum becomes indispensable to our understanding.

This is, of course, just an interpretation, though I have tried to stick as closely as possible to Boltzmann's own statements. Boltzmann himself often only hints at the connections between his diverse ideas. Furthermore, some central parts of these ideas, like e.g. the border between finitary and infinitary reasoning, are far from clear.

But these uncertainties notwithstanding, one thing should have become obvious: for Boltzmann, whether or not to adopt an atomistic understanding of nature is not a matter of free choice, to be decided by criteria of usefulness. There are serious constraints to our choice. Against the Kantian spirit prevailing among his contemporaries, Boltzmann conceives of these constraints as grounded in naturalistically understood laws of thought.

5. Conclusion: Limits of Psychologism

The reader may have already stumbled over a caveat in one of the citations given above: Boltzmann claims to demonstrate not that nature must be constituted atomistically, but "that we must think it thus". This, I think, reflects Boltzmann's acknowledgement (at least in 1903), that his mathematical argument falls short of a case for realism about atoms.

As I have reconstructed the argument, the fact that our capacity to unambiguously draw conclusions is limited to the finite is rooted in our laws of thought. But laws of thought, even if conceived as conditioned by our evolutionary history, need not reflect the true constitution of nature. Thus the claim that an atomistic conception of nature is an indispensable precondition for an unambiguously founded science does not imply that nature must *be* discontinuous.

Boltzmann was well aware that the laws of thought he imagined need not be perfectly adapted to promoting our cognition of the truths of nature. He thought of them as a kind of naturalist version of the *a priori*, but without the necessity, as the following remark indicates:

One can call these laws of thought a priori because through many thousands of years of our species' experience they have become innate to the individual, but it seems to be no more than a logical howler of Kant's to infer their infallibility in all cases.[25]

On the contrary, Boltzmann suspected that evolution often favours features that are useful in some cases and then become so deeply entrenched in the organism's biological constitution that they cannot be given up if they turn out not to be appropriate in all circumstances.

This happens especially often with mental habits and becomes a source of apparent contradictions between the laws of thought and the world, and between those laws themselves.[26]

This shows that according to Boltzmann's own conception of laws of thought, they cannot be seen as reliably mirroring the nature that selected them; and the mathematical argument for atomism, in so far as it rests on the requirements posed by our laws of thought, cannot establish that this nature must really be constituted atomistically.

To be sure, the mathematical argument was not Boltzmann's only strategy for defending atomism. The main alternative is, of course, to present independent predictive and explanatory successes of the atomistic hypothesis. Today, this is often seen as the most promising argumentative strategy for scientific realism,[27] and there are places where Boltzmann rests his case solely on this line of reasoning.[28] However, in his later writings, he seems to have given preference to the mathematical argument. In the St. Louis address, he does indeed specify some explanatory successes of atomism, but then goes on to claim:

However, it is not all these facts and the consequences drawn from them that I wish to put forward here, for they cannot resolve the question as to the limited or infinite divisibility of matter. [...] Rather, we will [...] examine the formation of concepts itself in as un-prejudiced a way as possible [...].[29]

The statement is then followed by a short version of the mathematical argument. The citation shows that, though Boltzmann did realise that the mathematical argument was not the only possible case for atomism in physics, he considered it the strongest one, at least in his later years. And to the extent that the mathematical argument was Boltzmann's strategy for propagating atomism, his campaign cannot have been one for the reality of atoms. The argument as I have reconstructed it suggests that he was concerned with the secure and unambiguous foundations of physics rather than with scientific realism.

As a conclusion, I would like to stress once more that the textual evidence shows clearly that for Boltzmann, the constraints of our laws of thought were serious and inevitable. Therefore, the limitations of his naturalistic (or more precisely, psychologistic) reasoning as presented in this section must not be confused with an alleged problem-shift from "Do atoms exist?" to "Is atomism as a mental picture fruitful as the hard core of a scientific research programme?" as diagnosed by Yehuda Elkana.[30] On the contrary, the atomistic conception is presented as indispensable (albeit in a psychological way) and not a question of pragmatic choice. Though naturalism and pragmatism have sometimes gone hand in hand in the history of philosophy, they did not do so in the philosophy of

Ludwig Boltzmann – at least not in his defence of atomism, as I have intended to show in this paper.

NOTES

1. See, e.g., Henk de Regt, from whom the quotation is taken. 'Ludwig Boltzmann's *Bildtheorie* and Scientific Understanding', in: John Blackmore (Ed.), *Ludwig Boltzmann: Troubled Genius as Philosopher*. Dordrecht: Kluwer 1999 (= *Synthese* **119** (1-2)), pp. 113-134, p. 128. Cf. also Yehuda Elkana as cited below in note 30.
2. 'Über die Unentbehrlichkeit der Atomistik in der Naturwissenschaft' (1897), quoted from the translation in Brian McGuinness (Ed.), Ludwig Boltzmann, *Theoretical Physics and Philosophical Problems: Selected Writings*. Dordrecht: Reidel 1974, p. 44.
3. This is a formulation he chose in his lecture course on Natural Philosophy in 1903. Ilse M. Fasol-Boltzmann (Ed.), Ludwig Boltzmann, *Principien der Naturfilosofi. Lectures on Natural Philosophy 1903–1906*. Berlin etc.: Springer 1990, p. 189.
4. 'Über statistische Mechanik' (1904), quoted from the translation in McGuinness (Ed.), *op. cit.*, p. 169. He had already used a very similar formulation in his *Lectures on Gas Theory* in 1896, see Roman U. Sexl (Ed.), *Ludwig Boltzmann Gesamtausgabe*, Vol. 1: *Vorlesungen über Gastheorie*. Graz: Akademische Druck- u. Verlagsanstalt & Braunschweig: Vieweg 1981, p. 5.
5. 'Über die Unentbehrlichkeit der Atomistik in der Naturwissenschaft', *loc. cit.*, p. 43.
6. There is a transcript of Boltzmann's own fragmentary and for the most part hardly intelligible notes for this lecture course in existence, as well as an unknown student's detailed elaboration of it (both of which were published only fairly recently). In the following, I will mostly refer to the student's elaboration. Cf. Fasol-Boltzmann (Ed.), *op. cit.*
7. *Ibid.*, p. 189.
8. *Ibid.*, p. 194.
9. 'Nochmals über die Atomistik' (1897), quoted from the translation in McGuinness (Ed.), *op. cit.*, p. 54.
10. *Vorlesungen über die Prinzipe der Mechanik, Teil I*, quoted from the partial translation in McGuinness (Ed.), *op. cit.*, p. 245 f.
11. 'Über die Unentbehrlichkeit der Atomistik in der Naturwissenschaft', *loc. cit.*, footnote 4, p. 52.
12. For such an account of Boltzmann's philosophical activity, see e.g. John Blackmore: 'Boltzmann and Epistemology', in: Blackmore (Ed.), *op. cit.*, pp. 157-189.
13. "Wenn auch diese Behandlungsweise des Problems auf den ersten Anblick sehr abstrakt zu sein scheint, so führt sie doch bei derartigen Problemen meistens am raschesten zum Ziele, und wenn man bedenkt, daß alles Unendliche in der Natur niemals etwas anderes als einen Grenzübergang bedeutet, so kann man die unendliche Mannigfaltigkeit von Geschwindigkeiten, welche jedes Molekül anzunehmen imstande ist, gar nicht anders auffassen, es sei denn als den Grenzfall, welcher eintritt, wenn jedes Molekül immer mehr und mehr Geschwindigkeiten annehmen kann." 'Über die Beziehung zwischen dem zweiten Hauptsatze der mechanischen Wärmetheorie und der Wahrscheinlichkeitsrechnung resp. den Sätzen über das Wärmegleichgewicht', in: Fritz Hasenöhrl (Ed.), *Wissenschaftliche Abhandlungen von Ludwig Boltzmann*, Vol. 2. Leipzig: Barth 1909, pp. 167 f., my translation, my italics.
14. Apparently, some of Boltzmann's contemporaries did think that his argument suddenly appeared only in the 1890's. Thus in 1898, Georg Helm calls it "Boltzmann's *new* defence of atomism ..." (my italics, "Boltzmann's neue Verteidigung des Atomismus durch den Differentialbegriff"). Georg Helm, *Die Energetik nach ihrer geschichtlichen Entwickelung*. Leipzig: Veit 1898, p. 362.
15. "Es versteht sich von selbst, daß wir den einfachsten Begriff, den der Zahl nicht definieren können, denn wir können ihn nicht in einfachere Begriffe zerlegen." Fasol-Boltzmann (Ed.), *op. cit.*, p. 163, my translation.
16. Cf. *ibid.*, 174 f.

17. "[...] Denkformen, welche sich durch die äußeren Einflüsse in unserem Gehirne, in unserer Psyche herausentwickelt haben, und diese Denkformen haben uns immer zu richtigen Resultaten geführt; dadurch sind sie zu feststehenden Denkformen geworden." *Ibid.*, p. 168, my translation.

18. "[...] sie haben sich nur daran gewöhnt, sie haben nur gezeigt, wie die Rechnungsoperationen einzurichten sind, daß man nicht darüber stolpert, wie man zu rechnen hat, um über die Paradoxien herumzukommen." *Ibid.*, pp. 189 f, my translation.

19. See *Propositio 80*, especially § 655, and confer also *Propositio 32*. Paul Stäckel (Ed.), *Leonardi Euleri Opera Omnia*, Ser. 2, Vol. 1: *Mechanica sive motus scientia analytice exposita*. Leipzig & Berlin: Teubner 1912.

20. "Quicquid autem sit, hic calculo potius quam nostro iudicio est fidendum atque statuendum, nos saltum, si fit ex infinito in finitum, penitus non comprehendere." Stäckel (Ed.), *op. cit.*, § 272, p. 88, my translation. This remark can be found in Euler's discussion of a different setting (assuming a 1/r force law), but immediately followed by a reference to the case I am dealing with here, by which, as Euler claims, the quoted remark gets even more confirmed.

21. See the *Anmerk* for § 230, pp. 505 ff. in Georg von Vega, *Vorlesungen über die Mathematik. Dritter Band, welcher die Mechanik der festen Körper enthält*. 4th edition, Vienna: Tendler 1818.

22. "Faktisch kommen wir, wenn wir nie etwas Unendliches aufnehmen, wenn wir nur mit endlichen Größen rechnen, welche beliebig groß sein können, nie zu einem Widerspruch [...] Wenn aber A mit S zusammen fällt und bis zum Zusammenfallen das Newton'sche Gesetz gilt [...] komme ich in die Mengenlehre hinein, in das wirklich Unendliche, in die letzte Zahl w [*sic*], und ich komme auf Widersprüche." Fasol-Boltzmann (Ed.), *op. cit.*, p. 200, my translation.

23. "Vega. Stoß: Das ist mein Beweis, daß die Natur atomistisch konstruiert sein muß. Nicht, daß Natur muß, daß wir sie so denken müssen, wenn wir nicht so absurde Begriffe, wie die der Mengenlehre anwenden wollen, die zu Mehrdeutigkeiten führen." *Ibid.*, p. 92, my translation.

24. "[...] der Zweck des Denkens ist ja, überall eindeutig schließen zu können; daher müssen wir unsere Sprach- Schrift- und Denkzeichen so zu bilden suchen, daß wir uns selbst eindeutig ausdrücken und uns selbst eindeutig verstehen." *Ibid.*, p. 200, my translation.

25. 'Über eine These Schopenhauers' (1905), quoted from the translation in McGuinness (Ed.), *op. cit.*, p. 195.

26. 'Über statistische Mechanik', *loc. cit.*, p. 166.

27. Especially since Richard Boyd's 'On the Current Status of Scientific Realism', in: *Erkenntnis* 19 (1983), pp. 45-90.

28. See, for a very nice example, his 'Über die Entwicklung der Methoden der theoretischen Physik in neuerer Zeit' (1899), translated in McGuinness (Ed.), *op. cit.*, esp. pp. 98 f.

29. 'Über statistische Mechanik', *loc. cit.*, p. 169.

30. 'Boltzmann's Scientific Research Program and its Alternatives', in: Yehuda Elkana (Ed.), *The Interaction Between Science and Philosophy*. Atlantic Highlands, N.J.: Humanities Press 1974, pp. 243-279, esp. p. 268.

Institut für Wissenschafts- und Technikforschung
Universität Bielefeld
Postfach 100131
D-33501 Bielefeld
Germany
torsten.wilholt@uni-bielefeld.de

ULRICH MAJER

HILBERT'S PROGRAM TO AXIOMATIZE PHYSICS (IN ANALOGY TO GEOMETRY) AND ITS IMPACT ON SCHLICK, CARNAP AND OTHER MEMBERS OF THE VIENNA CIRCLE

INTRODUCTION

In recent years the works of Friedman, Howard and many others[1] have made obvious what perhaps was always self-evident. Namely, that the philosophy of the logical empiricists was shaped primarily by Einstein and his invention of the theory of relativity, whereas Hilbert and his *axiomatic approach* to the exact sciences had comparatively little impact on the logical empiricists and their understanding of science – if they had any effect at all. This is in one respect quite astonishing, insofar as Einstein himself confessed 1921 in his famous lecture before the Prussian Academy of Science that "without it [the axiomatic point of view] it would have been impossible for me to propound the theory of relativity".[2] Hence the simple question arises: why didn't the logical empiricists pay more attention to Hilbert and his axiomatic point of view, than they in fact did? It is an aim of this paper to answer this question and in part to *correct* this one-sided view in the hypothetical or contra-factual sense that the logical empiricists would have done better, if they had paid somewhat more attention to Hilbert and his axiomatic approach to science.

In order to avoid a serious misunderstanding, I should add that I don't blame the logical empiricists for this failure, at least not primarily, but first and foremost Hilbert himself. He, in the first place, failed to make his position sufficiently clear, and did not make much effort to promote his views beyond the narrow circle of mathematical physics in Göttingen. Still a further remark before I get started. If I speak of the logical empiricists, I have primarily Schlick, Reichenbach and (the early) Carnap in mind, and not so much the other members of the Vienna Circle and the Berlin school.

First I'll point out what the logical empiricists, starting with Schlick, assimilated from Hilbert's work, both the early in geometry as well as the later one on the new foundations of mathematics, and how they understood this. Next I'll come to the more important point, which ideas and crucial aspects of Hilbert's work they missed [more or less involuntarily] and, most important, which misapprehensions and wrong conclusions resulted from this failure. Once we have

M. Heidelberger and F. Stadler (eds.), History of Philosophy and Science, 213–224.
© 2002 *Kluwer Academic Publishers. Printed in the Netherlands.*

clarified the misapprehensions, in particular, *why* they are misapprehensions, the way is free to recognise, what a more reasonable philosophy of science should look like and how it could have developed, if one had taken Hilbert's axiomatic point of view more seriously and carefully. The philosophical view that emerges from this I'll call "recursive epistemology".

SCHLICK'S RECEPTION OF HILBERT'S *FOUNDATIONS OF GEOMETRY*

Basically there were only two topics in Hilbert's extensive work on the foundations of mathematics and physics, which received, at least, a certain minimum of attention in the philosophical considerations of the logical empiricists regarding the development of a new brand of "logical empiricism". The first is connected with Hilbert's book on the "Foundations of Geometry" from 1899, and the second with Hilbert's so called "formalism" with respect to a proper foundation of mathematics developed during the twentieth of the last century. (One can already see: In the twenty years between 1900 and 1920 there is nothing in the work of Hilbert and his disciples, which the logical empiricists regarded as interesting or worthwhile to consider. I'll come back to this crucial point.) Yet first let me consider the *Foundations of Geometry* and inquire, which topic or aspect was estimated as important by the logical empiricists?

If we were not accustomed with the answer, we could hardly believe it. Instead of taking the true mathematical *achievements* of this work as important (such as the proof of the *independence* of the Archimedean Axiom from the remaining axioms of Euclid or the invention of the 'axiom of completeness'[3] in Euclidean geometry) Schlick, in the same vein as Frege, picked up a side issue, the so called "implicit definitions". Of course, from Schlick's epistemological point of view this was an important issue. It enabled him to *separate* Hilbert's representation of geometry not only from the suspicious 'pure *spatial intuition*' of Kant, but also from *'physical space'* and its empirical treatment by the theory of relativity. Instead, geometry became an object of mere logical stipulations of formal concepts and relations without any intuitive or physical content.[4] Furthermore, once the separation of 'abstract geometry' from intuition and physical space was established, Schlick was able to explain why (in his view) Poincaré was right and Reichenbach wrong. Poincaré insisted, in conscious opposition to Reichenbach's empiricism regarding physical geometry, on the *conventional* character of physical geometry in principle. This means, we are free to choose a certain geometry not only with respect to the space-time of special relativity, but also and in particular with respect to the geometry of space-time in general relativity[5]. Here comes a rather long quotation from Schlick's *Allgemeine Erkenntnislehre*, which corroborates my claim. Notice, that the AE was published 1918, three years after Einstein had announced his general theory of relativity.

Euclidean geometry has served as the geometry of everyday life, and until a short time ago it seemed to provide the proper foundation for all the purposes of natural science. The

new physics, however, in one of its boldest and most beautiful moves, has concluded from
the Einsteinian Theory of Gravitation that we cannot make do with the Euclidean metrical
determinations if we wish to describe nature with the greatest accuracy and ba means of
the simple laws. According to the theory, a different geometry must be used at each place
in the world, a geometry that depends on the physical state (the gravitational potential) at
that place. On the basis of Einstein's latest work it is likely that world space as a whole
can best be viewed as endowed with approximately "spherical" properties (thus a finite,
although of course also unbounded).

It cannot be emphasized too much that we are not *compelled* to conceive of space in
accordance with a theory of this kind. No experience can prevent us from retaining
Euclidean geometry if we insist on doing so. But then we do not obtain the simplest formulations of the laws of nature, and the system of physics as such becomes less satisfactory. ... The physical description of nature is not tied to any particular geometry and no
intuition dictates that we must base such a description on the Euclidean axiom system as
the only correct one, nor, of course, on any of the non-Euclidean systems either. We select
– in the beginning instinctively, in more recent times deliberately – those axioms that lead
to the simplest physical laws. In principle, however, we could have chosen other axioms if
we were willing to pay the price of more complicated formulations of the laws of nature.
Thus fundamentally the choice of axioms is left to our discretion.

And this means that they are *definitions*.

Our finding then is that geometry, not only as a pure conceptual science but also as
the science of space, does not proceed from synthetic *a priori* propositions. Instead, it
proceeds from conventions, that is, from implicit definitions.[6]

Of course, all this sounds quite familiar. Nonetheless, it would be very interest-
ing to know how Schlick arrived at this peculiar result, which is, as Friedman has
pointed out rightly[7], quite distinct from the usual brand of English empiricism.
Although I have not the time to go into any details, one point seems obvious:
Schlick's position is much more influenced by Einstein and Poincaré than by
Hilbert and a careful reading of his texts. This is indirectly corroborated by
Schlick himself, who, at the end of the section just quoted, praises Einstein for
stating precisely the same insight into the *conventional* character of geometry[8],
whereas Schlick never refers to Hilbert's "Foundations of Physics". This is, by
all means, no accident, because Hilbert criticises Poincaré's conventionalism
vehemently – and even includes in his critique the much-admired Einstein for his
concordance with Poincaré.[9]

But more important than the question of philosophical alliances is in the
present context the question whether Schlick's position, seen as a whole, was
consistent. It was obviously not in agreement with Hilbert's views about geometry, which is immediately clear from Hilbert's uncompromising and repeated
critique[10] of conventionalism. But it was also not consistent in itself. This can be
recognised, if one analyses chapter 7 on 'Implicit Definitions' in AE carefully.
Schlick begins this chapter with the usual story of modern 'post-Euclidean'
geometry: the rejection of spatial intuition as a reliable source for the meaning of
geometrical terms such as 'point', 'line', and 'plane' and the subsequent transformation of geometry into a purely *deductive* science – without any recursion to

intuition. This brings him to Hilbert and his axiomatic presentation of geometry, whose real advantage he localises in that, what he calls – not Hilbert – the method of implicit definitions. What is the logical essence of this method? Schlick gives the following answer.

Inference can proceed *only* from *judgments* or statements. Hence when we utilize a concept in the business of thought, we employ none of its properties save the property that certain judgments *hold* with respect to the concept – for example, that the axioms hold for the primitive concepts of geometry. It follows that for a rigorous science, which engages in series of inferences, a concept is indeed *nothing more* that that concerning which certain judgments can be expressed. Consequently, this is also how the concept is to be defined.[11]

If we ignore for a moment some minor inaccuracies in the statement, Schlick's answer seems to be the following: Logical deductions have to start exclusively from *judgements*, that is from *true* propositions. At the same time, the basic concepts and relations of geometry have to be *defined* (that is endowed with a meaning) solely by means of some axioms (entailing the respective terms), but without any recursion to intuition or some other kind of presupposed meaning. It remains Schlick's secret how the axioms can be judged to be 'true', i.e. to be *valid* propositions of the concepts and relations in question, if *these* concepts and relations have no meaning prior to the definitions by the axioms. Truth of those axioms is, however, an indispensable presupposition – as Schlick himself confesses – in order to use the 'modus ponens' properly in logical deductions.[12] I would say, Schlick simply didn't understand Hilbert's *axiomatic* approach to geometry[13] and, consequently, also not to the natural sciences[14]. The latter point is the more important one and brings me promptly to the main question of this essay. Which of the significant ideas and aspects in Hilbert's foundational work were *not* recognised (or accepted) by the logical empiricists, and which misapprehensions resulted from this ignorance? Let me begin with a particular point and then continue with more general aspects. I presume the following.

SOME OBVIOUS MISAPPREHENSIONS

It never occurred to the minds of the logical empiricists that Hilbert might take geometry as a "natural science", indeed as the "simplest and most perfect" of all natural sciences, as he had always stressed from his very first lecture in geometry in 1893 until his last lecture in 1927.[15] Unfortunately, the only place, where he did not stress that aspect, is his [published] book *Grundlagen der Geometrie*. At least to the extent, the logical empiricists have an excuse for their formalistic reading of Hilbert's *Foundation of Geometry*. However, I bet that even *if* Hilbert had published his opinion, the logical empiricist would have ignored it, just as they ignored Hilbert's second conviction regarding geometry, namely, I quote: "The designated task (the specification of the axioms of geometry) comes up to the logical analysis of our spatial intuition". This claim could impossibly be true according to the epistemological point of view the logical empiricists held with respect to the development of geometry during 19th century. Consequently, they ignored Hilbert's assertion completely faithful to the rule: what's not permitted cannot be possible either! I presume that both assertions, taken together, would have confused the logical empiricist entirely, because they were convinced that both assertions would contradict each other, that geometry as a *natural* science had to be *empirical* and that for this reason its axioms could impossibly rest on a logical analysis of spatial *intuition*. But Hilbert's point of view with respect to geometry is not as absurd [or plainly inconsistent] as it looks at first glance. On the contrary, it makes pretty much sense, if one distinguishes carefully, as Hilbert did, the two categories *natural* and *empirical* sciences, because this opens a solution to the *epistemological* problem that haunted the logical empiricists from the very beginning. How can we *know* the laws of the different kinds of geometry quasi *'a priori'*, i.e. before they have been testified by experiments? After all, the logical empiricists had missed a big opportunity to come to grips with the old problem of physical geometry along the ideas that Hilbert had offered. Instead, they made common cause with Poincaré and his – at their time already outdated – geometrical conventionalism.

Once on the wrong track the logical empiricists missed another, much more important aspect of Hilbert's work: his tireless efforts to extend the axiomatic method beyond the domain of geometry on the totality of physical sciences. Again the logical empiricists can be partially excused for this failure for two reasons. First, Hilbert didn't publish very much in this area, to say the least, and second because the few papers, which he in fact published, did not look very much like an axiomatic treatment of the physical field under investigation. But the logical empiricists cannot be excused completely, because, as far as I know, they never referred to any of Hilbert's papers in physics,[16] although two of the papers dealt with the Foundations of Physics, more properly speaking with the generalised field equations in the frame of general relativity. This was a topic, in which the logical empiricists usually demonstrated great interest. Consequently, the question arises: Why didn't they do just the same in Hilbert's case? I have no direct answer, only an indirect one. But before presenting it, let me quickly outline the main bulk of Hilbert's work in physics as one can find it in the Nachlass in Göttingen.

OUTLINE OF HILBERT'S WORK IN PHYSICS

Hilbert's engagement in physics can be roughly distinguished into three periods. The first began in 1898 with a lecture on the principles of mechanics and lasted roughly until 1910. Note that the first lecture was held before *the Foundations of Geometry* were finished. The second period started about 1911, when Hilbert for the first time remoulded his lecture on continuum mechanics in order to account for the special theory of relativity and its relation to electrodynamics and, astonishingly enough, its consequences for thermodynamics. This period lasted, with a remarkable interlude, to which I'll come to, vaguely until 1918. Then Hilbert's concern with the new foundations of mathematics forced a certain interruption in his occupation with physics. The third and last period is relatively short due to Hilbert's illness. It begins in the winter term 1922/3 with Hilbert's first lecture on quantum-mechanics and ends about 1927, when he – for the last time – read on the mathematical methods of quantum-mechanics. You recognise, the periods match roughly with the three principal domains of physics: classical mechanics, theory of relativity (with electrodynamics), and quantum-mechanics.

Of course, the correspondence is not quite exact. There is the already mentioned interlude of three years between 1911 and 1914, in which Hilbert was mainly occupied with the application of his new theory of integral-equations first to the kinetic theory of gases and then to the theory of radiation. Furthermore, in connection with the kinetic theory of gases Hilbert displayed a long-standing interest in the foundations of statistical mechanics and the (quantum-mechanical) theory of matter. However, much more important than this mere catalogue of research fields and lectures is the observation that in his work in physics Hilbert tried to combine two apparently incongruent points of view: the axiomatic method with an interest in the true foundations of physics. It is crucial to make clear that this was no contradiction for Hilbert, not even a conflict, but the expression of a deep conviction, namely the conviction that physics, as a science about nature, rests on two fundamental procedures: (a) the *collection* of all available facts into an appropriate 'conceptual framework'[17], (b) a *logical analysis* of the dependence as well as the independence of all the facts, collected together. The first is the duty of the experimental physicist, the second the task of the theoretical physicist, or the mathematician. This task not only can but also has to be performed according to the rules of the *axiomatic method*. Consequently, it is absolutely mandatory to specify these rules, and how they can and have to be distinguished from the question of the empirical correctness or truth, if you like, of a certain "axiomatic presentation" of a physical theory, as I'll call the aim of traditional axiomatic.

The Crucial Difference between the Axiomatic Presentation of a Theory and its Logical Analysis by Means of the Axiomatic Method

The main thesis, which I want to put forward in this section, is the following. The logical empiricists – in particular Carnap – never came to grips with the fundamental difference between the 'axiomatic presentation' of a physical theory in the first place and the application of the 'axiomatic method' in the second place. Instead, they jumped immediately from Hilbert's 'axiomatic treatment'[18] of geometry to the *New Foundation of Mathematics* in the twentieth, completely ignoring what had happened in between. In this vein they not only lumped together two quite distinct research programs – namely the logical analysis of geometry on the one hand and the efforts to prove the consistency of arithmetic *once and for all* by meta-mathematical means on the other. But they also failed to distinguish what has to be distinguished extremely carefully: the logical analysis of an axiomatic presentation of a field of facts by means of the axiomatic method and the axiomatic presentation itself, in particular the question as to, whether the axiomatic presentation is true. Of course, both are connected. One can't reasonably perform a logical analysis of a field of facts – whether geometry or arithmetic is irrelevant – without an axiomatic presentation of the field in question and vice versa, one cannot bring a field into an axiomatic form without a certain degree of logical analysis. But being 'connected' doesn't mean being one and the *same;* both have to be distinguished precisely; otherwise senseless questions arise, as has happened time and again; take, for example, the battle about the significance of implicit definitions.

Hence the next thing I should point out is the fundamental differences between both procedures and explain, what their logical interrelationship is, not only in geometry but also in more advanced, full-fledged fields of physics, so to speak, such as the theory of general relativity or quantum-mechanics. But, for lack of space I have to restrict my considerations to one [dangerous] short remark. The axiomatic presentation of a field of facts has to be in the end absolutely *loyal* to the facts. So far, traditional empiricism regarding natural sciences is correct; he is only mistaken about the intricate question of how we approach that end. On the other hand, in the coarse of logical analysis we *intentionally* vary the axioms of a field and inquire among other things whether in the universe of mathematics a structure still exists, which is a model of the resulting system of axioms. This means that the "truth" of the modified axiom system is not judged with respect to the real world, but reduced to the question of *consistency* with respect to the universe of numbers, whose consistency in turn has to be proved in an *absolute* sense, i.e. by formal, finite means. (Precisely to this end Hilbert invented in the twenties his proof theory.) The logical analysis (of a field of facts) by means of the axiomatic method is not just a *useless* playground for jobless mathematicians, as often supposed, but of the highest *significance* for the progress of science. Thus we are not only able to recognise which facts are logically *independent* from other facts of a field under investigation, but also which theories, which "world-descriptions" are logically possible and which not. In this way the mathematical physicists can hurry ahead future experience and tell the experimental physicist how he can test a certain theory and, what is often even more important, how he can't do it.

Geometry itself offers the best example for what I have just stated. For a long time the logical analysis of geometry (including its axiomatic presentation in a corresponding form) was far ahead the *precision* of its application to the real world in physics. Gauß, for example, could not experimentally decide, whether the sum of the angles in the geophysical triangle he had measured was equal or larger than two right angles. He knew already that it could not be smaller than two right ones, because this was according to Legendre's sentences inconsistent with the measurement of parallaxes. The lead of theoretical over experimental geometry (as I'll call it) remained for almost one hundred years until the experimental failure of the addition-theorem of velocities opened for the first time the possibility to apply non-Euclidean geometry to the physics of space-time.

I can't repeat the long and intricate history of the logical analysis of geometry, which culminated in Hilbert's *Foundations of Geometry,* but one point I will mention, because it is connected with a persistent confusion regarding the necessity of "rigid bodies" for distance-measurements – a confusion first created by Poincaré and then repeated by the logical empiricists. According to a theorem by Helmholtz *free mobility of rigid bodies* characterises the spaces *of constant curvature.* Consequently, these spaces *cannot* be discriminated observationally by the free mobility of rigid bodies in space. A choice has to be made. So far Poincaré is right in regarding the inevitability of geometrical conventionalism. But he is mislead if he thinks that the application of geometry to nature *presupposes* the existence of rigid bodies, because otherwise the notion of congruence is not defined. This thought forces itself upon the mind, if one takes *motions,* i.e. the translations and rotations of bodies, as the prime *sources* of our knowledge of space and, hence, of the laws of geometry. But this is neither necessary nor desirable, as Hilbert has shown in an important paper "Über die Grundlagen der Geometrie", published first 1902, three years after the appearance of his book. Unfortunately, this paper is little known and almost totally ignored as a significant contribution to the philosophy of geometry as a natural science.

CARNAP'S LOGICISTIC INTERPRETATION
OF HILBERT'S AXIOMATIC POINT OF VIEW

If we finally turn to Carnap and his recently published script *Untersuchungen zur allgemeinen Axiomatik* [19], which was written shortly after the *Aufbau* had been published, one could have the impression Carnap tried to investigate and to improve the old axiomatic approach of Hilbert's geometry from the new perspective of meta-mathematics. And to some extent he, indeed, does this. But a closer inspection reveals that he is wandering on the same old logicistic trails as Frege and Russell did with respect to the foundations of arithmetic. The main difference between them and Carnap is that he tries to generalise the concept of an axiomatic system in such a way that geometry (or any other theory), if axiomatised according to his proposal, turns out to be nothing but a logical theory. This is, however, definitively not Hilbert's point of view! Let me just mention two aspects. First, Carnap maintains that the descriptive expressions like point, straight line, between, etc. have to be taken as *variables,* which have to be substituted by names for proper concepts and relations, in order to turn the axioms into proper sentences. Although this interpretation of Hilbert's axiomatic treatment of geometry is repeated again and again, it is not Hilbert's point of view. Not only does Hilbert not say such things; it is also not his intention to give the impression as if they were "variables". On the contrary, they have a meaning, which, however, by mere logical means cannot be made more precise than up to isomorphism. Second, Carnap defends once more the view that mathematics is nothing but a branch of logic. Although Hilbert had flirted with logicism for a short period, this is definitively not his point of view in the twentieth. On the contrary, in that period he had stressed more than once that mathematics is in need of intuition, not only on the level of its axiomatic presentation but primarily in the meta-mathematical investigation of its logical structure.

In order to avoid a close misunderstanding I should stress that Carnap's investigation of axiom systems, taken in it self, is nonetheless of considerable value. In fact, it entails some interesting model-theoretic results regarding such meta-theoretical concepts as 'monomorphism', 'categoricity' and 'deductive completeness', as D. Scott has shown.[20] Its principal defect in the present context is just this that it is not in agreement with Hilbert's views neither on the foundations of geometry nor that of arithmetic.

HILBERT'S RECURSIVE EPISTEMOLOGY

Let me close by explaining very briefly what Hilbert's epistemological position is in regard to the historical development of science, in particular to geometry. At the beginning I stressed that Hilbert held two views regarding geometry that seemed to be mutually incompatible: first, geometry is a natural science and second, spatial intuition provides us with the geometrical facts, that the mathematician is going to analyse. Hence, the question arises: How is this *possible?* The answer, very sketchily, runs like this. In the *pre*-critical period, spatial intuition was naively[21] equated with physical space, like primary qualities were identified with physical properties. Then in the critical period of 19th century logical analysis took over the leadership. The result is well known: The possibility of non-Euclidean geometries was recognised but – this is important to note – without Euclidean geometry being experimentally disproved at that time. So far, most people agree. The usual *empiricist* story then continues in the following way:

The recognition of non-Euclidean geometries led to an epistemological distinction between spatial intuition and physical space. For the first time, the possibility occurred that both might have different structures. Hence, from now on it had become an *empirical* question, which kind of geometry was the correct or true geometry of physical space. Accordingly, it was just a matter of time when the question would be decided. And indeed, it took only a century, until Einstein detected that Riemann was right: that the metrical structure of physical space is not Euclidean, but instead varies from place to place, depending on the distribution of masses. Consequently, spatial intuition had to be removed from physics, like other sense qualities, as an *'anthropomorphic'* element.

Now, it is important to note that this was *not* Hilbert's point of view. Although the empiricist story entails a number of truths, it overlooks a crucial point; this point is of decisive significance for the answer to the question, which epistemology is the right one. The development of physics from classical mechanics and electrodynamics first to the special, and then to the general theory of relativity, would not have been possible, according to Hilbert, if physicists had not relied upon the practical correctness of our spatial intuition. Consequently, they built their measurement devices according to the "standards" of Euclidean geometry. Only in this way was it possible for the physicists to improve the precision of time and length measurements to such an unimaginable degree that the incredible small deviations of physical space from Euclidean geometry became experimentally observable. In other words, the progress of physics did in this case depend not only on a *new* geometry, which was known long before the transition took place, but also on the practical reliability of the *old* Euclidean geometry up to the point at which significant deviations from it became observable.

It may be that other beings with other cognitive abilities could have recognised the Riemannian structure of physical space immediately, but we *human* beings were forced for contingent reasons, connected with the evolution of our senses, to go this *recursive* way from primitive sense impressions to space as a form of our sensual cognition, and from here to the cognition of physical space, as a system of relations among material objects, that is completely independent, not only of our intuition of things, but also of their mathematical representations.

NOTES

1. See for example the proceedings of the Carnap-Reichenbach Centennial *Logic, Language, and the Structure of Scientific Theories*, edited by W. Salmon and G. Wolters, University of Pittsburgh Press (1992).
2. A. Einstein "Geometrie und Erfahrung", erweiterte Fassung des Festvortrages, Springer, Berlin 1921, page 6. The relative pronoun "it" refers quite literally to Einstein's conviction that geometry, taken as an axiomatized theory, only together with an interpretation relating the mathematical concept of body to real objects (solid bodies) becomes a physical theory. Insofar such a view presupposes an axiomatic point of view with respect to geometry, the "it" refers indirectly also to Hilbert's approach, which Einstein had discussed earlier.
3. Completeness of a theory here means the same as "being categorical", which in turn is precisely what Hilbert's completeness axiom says: The domain of objects cannot be extended (beyond what has already been introduced) without T becoming inconsistent. See my "Husserl and Hilbert on Completeness"; Synthese 110 (1997).
4. Compare §7 of Schlick's *Allgemeine Erkenntnislehre*, second edition of 1925; henceforth (AE).
5. Poincaré, of course, could not know general relativity, but his arguments regarding the conventional character of the choice of a geometry for characterising physical space were quite general.
6. Schlick, M. *General Theory of Knowledge*, transl., 2nd edition 1925, p. 354f. (Reprint: Springer-Verlag, Wien-New York 1974.)

7. Friedman, M. "Geometry, Convention, and the Relativized A Priori: Reichenbach, Schlick, and Carnap" in *Logic, Language, and the Structure of Scientific Theories*, edited by W. Salmon and G. Wolters.

8. See Einstein, A. "Geometrie und Erfahrung" (1921), where he states: "*Sub specie aeterni* hat Poincaré mit dieser Auffassung nach meiner Meinung Recht. Die Geometrie (G) sagt nichts über das Verhalten der wirklichen Dinge aus, sondern nur die Geometrie zusammen mit dem Inbegriff (P) der physikalischen Gesetze. Symbolisch können wir sagen, daß nur die Summe (G)+(P) der Kontrolle der Erfahrung unterliegt. Es kann also (G) willkürlich gewählt werden, ebenso Teile von P; alle diese Gesetze sind Konventionen." p. 8.

9. See Hilbert's lectures in Hamburg and Zürich "Die Weltgleichungen" (1923), in which he discusses and refutes explicitly Poincaré's conventionalistic position with respect to geometry.

10. For the first time Hilbert articulates his critique in the lectures on "Continuum Mechanics" in 1912 and then repeats it in 1921 and 1923. See my "Hilbert's Criticism of Poincare's Conventionalism", in *Henri Poincaré: Wissenschaft und Philosophie*, edited by Jean-louis Greffe et al., Akademie Verlag Berlin, 1994.

11. Schlick, M. *General Theory of Knowledge*, transl., 2nd edition 1925, p. 33f. (Reprint: Springer-Verlag, Wien-New York 1974.) (Italics are mine.)

12. Otherwise, i.e. without the truth of the "defining" axioms, we cannot use the 'modus ponens' to deduce true consequences from these axioms; instead we wind up in longer and longer sequences of nested *conditionals*.

13. Unfortunately, given the limited space I cannot present the correct interpretation of Hilbert's axiomatic point of view. For the latter see my paper "Hilbert's Axiomatic Method and the Foundations of Science" in *John von Neumann and the Foundations of Quantum Physics*; edited by M. Rédei and M. Stöltzner. Kluwer Ac. Publ. 2001..

14. Likewise, I can't tell the further story of the logical empiricists fight with the *dragon* of 'implicit definitions'. One can find, however, a significant part of this story in D. Howard's essay "Einstein, Kant, and the Origins of Logical Empiricism" (1991) as well as in A. W. Richardson's book "Carnap's Construction of the World". An interesting essay about Carnap's effort to come to grips with the implicit definitions and their notorious ambiguities appeared recently in *Erkenntnis* by S. Awodey & A. W. Carus "Carnap. Completeness, and Categoricity: The Gabelbarkeitssatz of 1928" The paper deals with Carnap's unpublished manuscript of 1928 entitled "Untersuchungen zur allgemeinen Axiomatik". A different story is presented in D. Howard's "Relativity, Eindeutigkeit and Monomorphism: Rudolf Carnap and the Development of the Categoricity Concept in Formal Semantics" (1996) in *Origins of Logical Empiricism*, edited by R. Giere & A. W. Richardson.

15. Notice that 'natural science' should not be equated with 'empirical science'. It could turn out that geometry, although it is a natural science, is not an empirical one. In Hilbert's view these are different notions, belonging to radically different spheres of concepts. The first designates the ontological sphere which we are talking about, the second belongs to the epistemological sphere of how we come to know what we are talking about.

16. There is one exception, which I noticed first recently (when the paper was in print): In the early essay "Über die Aufgabe der Physik" (1923) Carnap mentions in a rather long list of axiom systems in physics also "Mie-Hilbert" – but without any reference to a particular publication of Hilbert. Presumably Carnap had Hilbert's two papers on the "Grundlagen der Physik" from 1915/16 in mind. He never came back to that work.

17. This is my translation of Hilbert's metaphorical expression "Fachwerk der Begriffe"; Hilbert liked to compare science in early 20th century with a timber-framed house that has to be remodeled, because it became too large.

18. By this expression I mean both, the axiomatic presentation of a certain field of facts and the application of the axiomatic method to the axiomatic presentation.

19. Carnap, R. *Untersuchungen zur allgemeinen Axiomatik*, edited by T. Bonk & J. Mosterin, Wissenschaftliche Buchgesellschaft, Darmstadt ([1928] 1999). A short "Bericht über Untersuchungen zur allgemeinen Axiomatik" appeared in *Erkenntnis*, 1 (1930-31). However, more revealing, beside the *Aufbau* (1928), is Carnap's essay "Eigentliche und uneigentliche Begriffe" published in "Symposion: philosophische Zeitschrift für Forschung und Aussprache", Vol. I, Berlin (1927).

20. See Awodey, S. and Carus, A.W, "Carnap, Completeness and Categoricity", first footnote on the first page.
21. By 'naive' I mean: It was not taken into account that there might exist the possibility that space, as we see it, and space, as it exists independent from our perception, could be different. Neither Leibniz nor Newton considered this possibility, although they disagreed about another question, namely whether space is *'absolute'*, that is independent of the existence of material bodies, or only *'relational'*, that is dependent on the existence of bodies. First, Kant introduced the distinction between space, as the form of external intuition, and space as the sphere, in which bodies interact. But even for Kant both could not be different. This was the "clou" of his epistemology.

Institut für Wissenschaftsgeschichte
Humboldtallee 11
D-37073 Göttingen
Germany
umajer@gwdg.de

TILMAN SAUER

HOPES AND DISAPPOINTMENTS IN HILBERT'S
AXIOMATIC "FOUNDATIONS OF PHYSICS"

1. HILBERT'S FIRST NOTE ON THE "FOUNDATIONS OF PHYSICS"

Sixteen years after his "Foundations of Geometry," Hilbert published a communication that bears a similar and, by use of the definite article, even less mistakable title: "The Foundations of Physics." In the opening paragraph of this article, Hilbert announced his intention self-confidently:

In the following, I should like to set up – following the axiomatic method – a new system of fundamental equations of physics, constructed essentially from two simple axioms; equations that are of ideal beauty and in which, as I believe, is contained the solution of both Einstein's and Mie's problems.[1]

The mention of Einstein and Mie refers to the first, immediately preceding sentence where Einstein was credited with a "tremendous formulation of problems" as well as with "sharp-witted methods invented for their solution," and Mie was associated with "profound thoughts and an original formation of concepts." Einstein's all-but-covariant-field-equations theory of general relativity and gravitation and Mie's non-linear generalization of special relativistic Maxwellian electrodynamics to a field theory of matter thus provide the conceptual stage on which Hilbert here puts his own work into the limelight. The central idea of Hilbert's communication is to combine both Einstein's and Mie's theories in a variational framework with an invariant action integral. In the paper, Hilbert showed how to generalize Lorentz-covariant electrodynamics to a generally covariant theory in such a way that also generally covariant gravitational field equations follow from the variation of the action integral.

Hilbert's first note has been a topic of extensive historical discussion in the past few years which I shall not take up again.[2] For the purpose of this paper, let me only quote those two axioms announced by Hilbert in his note. The first axiom reads

> Axiom I: The laws of physics are determined by a world function
> that depends on the components of the metric tensor and its first and
> second derivatives and on the electromagnetic potential and its first
> derivatives. Specifically, the variation of the integral

M. Heidelberger and F. Stadler (eds.), History of Philosophy and Science, 225–237.
© 2002 *Kluwer Academic Publishers. Printed in the Netherlands.*

$$\int H\sqrt{g}d\omega, \quad (g = |g_{\mu\nu}|, d\omega = dw_1, dw_2, dw_3, dw_4) \qquad (1)$$

must vanish for each of the 14 potentials $g_{\mu\nu}, q_s$.[3]

The mere mention of the metric tensor $g_{\mu\nu}$ and the electromagnetic potential q_s invokes both the conceptual framework of Einstein's tensorial theory of gravitation in which the metric tensor is the crucial technical and conceptual ingredient as well as Mie's attempt to formulate the basis for a theory of matter by suggesting how non-linear generalizations of Maxwell's equations could be derived starting from a Lorentz-covariant variational formulation. The axiom also invokes the calculus of variations as the mathematical subdiscipline that provides the resources for the physical theory.

The second axiom postulates general covariance and brings in the theory of invariants as the second mathematical subdiscipline relevant here. It reads

> Axiom II: The world function is an invariant with respect to an arbitrary transformation of the world parameters, i.e. it is an invariant under general coordinate transformations.[4]

The two axioms do not specify the actual form of what Hilbert here calls the Hamiltonian function or world function H. In particular, they do not imply any separation of gravitational and matter parts, nor do they exclude the possibility that the world function would explicitly depend on the electromagnetic potential and hence violate gauge invariance. Only later on in the paper, the world function is split up into a purely gravitational part $K = K(g^{\mu\nu}, g_l^{\mu\nu}, g_{lk}^{\mu\nu})$ and an electromagnetic part $L = L(g^{\mu\nu}, q^\mu, q_l^\mu)$ (with subscripts here denoting derivatives with respect to the coordinates). In fact, in a 1924 reprint of his "Foundations of Physics" in a somewhat revised version Hilbert introduced the splitting of the world function $H = K + L$ into a purely gravitational and an electromagnetic part as an independent third axiom.

On the level of technical representation and as far as identification of the core of the theory is concerned, Hilbert, I should like to suggest, is not so far away from our modern understanding of classical general relativity, a fact that is due to a great extent to the prominent role of variational principles in modern theoretical physics. In fact, if you would go to a theoretical physicist nowadays and ask him to briefly characterize what he would understand under classical general relativity – classical in contrast to non-classical, i.e. quantum field theories, quantum electrodynamics, quantum chromodynamics and the like – he might answer: "That's a theory governed by the Hilbert-Einstein action." The term "Hilbert-Einstein action" is not a patented term, you might also hear him call it "Einstein action" or "Hilbert action" only. Also it doesn't have a canonical meaning, so you would ask him to specify what exactly he means. He would then go to his blackboard and write down something like

$$S_{\text{HE}} = \int d^4x\sqrt{-g}R. \qquad (2)$$

Writing down this formula, he would explain: "You have a $3 + 1$ space-time manifold with a metric field $g_{\mu\nu}$ of Lorentz signature defined on it whose dynamics is governed by a variational principle, and more precisely by a Lagrangian given by the Riemann curvature scalar R." "Is this all?" "No," he might say, "that's the vacuum case, you can also add a Maxwell field to your theory by adding the Maxwell stress-energy tensor to the Lagrangian like this

$$S_{\text{HE}} = \int d^4x \sqrt{-g}(R + \kappa T), \tag{3}$$

where $T = \sum g^{\mu\nu}T_{\mu\nu}$ is the trace of the Maxwell stress-energy tensor, and κ is a coupling constant." "And what do you mean by theory?" "Well, the usual thing. You can now do the variation with respect to the metric components and you'll get the gravitational field equations, or you can do the variation with respect to the electromagnetic potential and get Maxwell's equations. You then look at special situations of the gravitational field equations and try to find solutions, the most important one is Schwarzschild's solution, that'll give you light deflection and also the anomaly of Mercury's perihelion advance. But, of course, you can look at all other kinds of things, except that we don't have so many exact solutions for the field equations. But that's essentially classical general relativity."

"And why do you call it Hilbert-Einstein action?" – "Well, I don't know, that's just how it is usually called. You tell me, you're the historian."

This brief hypothetical interview with a working theoretical physicist was, of course, feigned to suggest that even today we could make sense of Hilbert's claim that we can capture the facts of a field of knowledge by recognizing some particular propositions that suffice to construct and build up on their basis the whole domain of science under investigation. And if we take the physicist's point of view, identifying the theory with its variational formulation, and more specifically with its Lagrangian, we might conclude that Hilbert's name would justly be associated with this account, and not only so because Hilbert would have been the first who suggested the Riemann scalar as the relevant Lagrangian.[5] It is also the particular spirit of this kind of characterizing the theory that we may associate with Hilbert.

Of course, there are differences. Most significantly, we would certainly no longer take classical general relativity with electromagnetic fields as the ultimate and comprehensive basis of all of physics. We would at least allow for other kinds of matter fields. And we would no longer declare this theory the basis of all we know in physics. It would be, at best, one field of knowledge in physical science, relevant on a cosmological scale but more or less irrelevant for the physics of the microscopic realm. Indeed, as we know, there is not even a peaceful coexistence between the relevant theories of these different realms of physics but there are explicit inconsistencies, as the still largely unsuccessful attempts to integrate classical relativistic gravitation theory into a quantum theory testify.

This was clearly different when Hilbert published his note. He believed that the achievements of his axiomatic foundation would amount to much more. In a often quoted, again rather emphatic concluding passage of his paper he confirmed his belief that his two axioms suffice to construct the full theory, and he asserted his conviction

that by means of the fundamental equations set up here, the most intimate and up to now hidden processes within the atom can be explained and that, in particular, a reduction of all physical constants to mathematical constants will be possible – as, on the whole, the possibility comes near that physics turns into a science of the kind of geometry: certainly the most magnificent glory of the axiomatic method which here, as we see, puts the most powerful instruments of calculus, i.e. the calculus of variations and invariant theory, to its service.[6]

While we may take Hilbert's two axioms as a not-too-bad axiomatic approximation of classical general relativity even from a modern point of view, we would no longer follow him in his claims of this concluding passage. We now know that the idea of a unified field theory of gravitation and electricity along the lines suggested in Hilbert's note is not a viable way to arrive at a satisfactory theory that would also account for a quantum structure of matter. It has intrinsic difficulties and it turned out to be utterly inadequate to account for the phenomenological wealth of later elementary particle physics. But what is problematic about this passage from today's point of view – Hilbert's hope that the field equations could eventually be shown to account for phenomena on the molecular level and that also the fundamental physical constants could be deduced from them – was perhaps not so fantastic at the time. Those hopes were simply speculative, and no one at the time, I should like to say, could have decided to what extent those hopes were realistic or not.

It is a central claim of this paper that, from a historical point of view, if we really want to understand what Hilbert was about at the time, we have to realize that both the acceptable aspects of Hilbert's paper as well as the unrealistic, outdated and old-fashioned belief in having solved the problems of physics belong together inseparably as two sides of one and the same coin.

Hilbert was aware that he was formulating an ambitious program, and in his paper he said that this first communication would only contain the basic ideas and that the details had to be left for further communications:

The detailed carrying out as well as above all the specific application of my basic equations to the fundamental questions of electricity theory I shall reserve for subsequent communications.[7]

Hilbert indeed believed that he had essentially solved the sixth problem of his famous Paris list of mathematical problems to be dealt with by twentieth century mathematics. He believed that he had solved the sixth problem of that list, the problem of axiomatizing physics by formulating two axioms that would be at the foundation of all of physics.[8] And he claimed that it was the axiomatic method that had put him in a position to achieve this. If Hilbert's high hopes had been realistic we would now have a different view of the history of theoretical physics. And we might expect that philosophy of science, too, would have taken a different turn in reaction to Hilbert's work. So what went wrong?

2. HILBERT'S SECOND NOTE ON THE "FOUNDATIONS OF PHYSICS"

Already in his first paper Hilbert had announced a follow-up communication presenting details of his theory. From archival evidence we know that he had, in fact, presented such a second communication already two weeks after submission of the first one.[9] It was, however, withdrawn from publication twice, and finally presented to the Göttingen Academy for publication in its proceedings only a year after, in December 1916. This second communication contains an investigation of what Hilbert called some general questions of a physical and logical nature, and also presented some considerations about the integration of the gravitational field equations.

A detailed analysis of Hilbert's second note cannot be presented here.[10] Let me only briefly indicate some of the topics discussed by Hilbert in this note. A central concern of Hilbert is the need to reconsider the notion of causality in a generally covariant theory. He discusses the causality problem both on the level of the causal order of space-time events and on the level of the field equations. In the first case, the causality problem arises since only those space-time events can be causally related that may be connected by a continuous time-like curve, and Hilbert introduces coordinate conditions, now known as Hilbert conditions, that are sufficient to ensure a time-like separation between causally related space-time events. On the level of the field equation, Hilbert identifies the causality problem with the Cauchy initial value problem of the theory of partial differential equations. The problem here is that in contrast to non-generally covariant physics, the number of independent equations is less than the number of independent variables as a consequence of diffeomorphism invariance. Hilbert suggests a sharpening of the concept of causality by introducing the notion of "physically meaningful" statements which are those that are independent of the use of specific coordinate systems. With this notion, Hilbert reformulates the causality principle saying that given a physically meaningful statement about a physical system at some initial time, the state of the system at a later time is uniquely determined by the field equations.

In a second part of the paper, Hilbert turns to a discussion of possible solutions to the field equations. He rederives in a simpler manner Schwarzschild's solution for static, spherically symmetric space-times and also discusses possible trajectories in Schwarzschild space-time. Along the way, he presents one of the first definitions of singularities in general relativity, albeit according to his definition the coordinate singularity at the Schwarzschild radius would be a proper singularity.

Despite the significance of these investigations and insights, one has observes that Hilbert in his second communication did not present the results that had been announced in the first note. From what can be inferred from various manuscripts in the Hilbert archives – both material from his pen and papers written by his assistants – the second communication also did not present all the research that had been done in the meantime. It appears that Hilbert's original program at the time was to find the electromagnetic part L of the Hamiltonian function that would satisfy a number of conditions. It had to satisfy in the first place his axiomatic characterization, i.e. it would depend on the metric components as well as the electromagnetic potential and its first derivatives and be generally invariant. It should then allow him to derive a non-linear generalization of Maxwell's equations that would be both explicitly solvable and the solution of which at small distances should allow for an interpretation in terms of an electron's charge distribution. The viability of this program was indicated by a particular Hamiltonian function given by Mie that provided such a solution mathematically which, however, had already been found by Mie to be unacceptable for physical reasons.

It seems that Hilbert had given up for the time his hopes of finding a presentable solution to the electron problem when he decided to finally issue his second communication in December 1916, and he did not come back to this topic in published work during his life-time. Before proceeding to comment on some of the immediate reactions to Hilbert's work and on his own self-assessment let me briefly comment on the role of the axiomatic approach in Hilbert's second communication.

In light of the emphatic remarks on the role of the axiomatic method for the achievements of his first communication, one may ask what role the axiomatic approach would play in his second communication. We find comments related to the axiomatic approach at two places but the use of axioms is notably less prominent and much more matter-of-fact and of a technical nature. For one, Hilbert comments briefly on a possible "axiomatic construction" of what he calls the "pseudogeometry" of Riemannian space-time with a metric of Lorentz signature. He indicates that one would need two axioms for such a construction – one that would entail that the length of space- and time-like curves are integrals whose integrand would depend only on the coordinates and their first derivatives, and another one that would ensure that pseudo-Euclidean geometry is valid at small distances, i.e. that space-time would be locally Minkowskian.

While the function of these axioms would be to allow the construction of Riemannian space-time from a set of axioms less directly associated with the original formulation of general relativity, the second use of axioms in the paper is of a different nature. Here Hilbert starts his discussion of possible trajectories in space-times characterized by the Schwarzschild line-element by formulating two axioms according to which the motion of a material particle is a time-like geodesic, and light rays are geodesic null lines. Hilbert explicitly asserts that in a fully elaborated theory those two axioms would be deduced from the fundamental equations but for the time being he would only postulate them as independent axioms in order to be able to proceed with the elaboration of their consequences.

3. THE RECEPTION OF HILBERT'S WORK BY HIS CONTEMPORARIES

Let me now look at some contemporary assessments of Hilbert's work. Einstein, Hilbert's direct rival in this field, soon realized that the variational formulation in terms of the Riemann scalar provided a very superior representation of the gravitational field equations compared to the explicit one in terms of the Ricci tensor and its trace that he himself had given. He carefully studied Hilbert's paper, and from his correspondence we know that he had difficulties coming to grips with some mathematical details of Hilbert's work. Twice he had to contact Hilbert and ask him to elaborate on details of his argument. In addition to these difficulties he also disliked the style of Hilbert's work. To his friend Ehrenfest he wrote:

Hilbert's presentation doesn't appeal to me. It is unnecessarily specific, as far as "matter" is concerned, unnecessarily involved, not honest (=Gaussian) in its construction (pretense of the superman by camouflaging the methods).[11]

In this brief statement, we find three reasons why Einstein disliked Hilbert's work, and these three reasons come up again and again. In the first place, Einstein did not agree with Hilbert's electromagnetic specification of the matter tensor, secondly, he found the paper technically unneccessarily complicated, and, last but not least, he despised the fact that the axiomatic method, in his view, did not give the justification that had actually led Hilbert to consider his axioms in the first place.

The first reason seems to be the most harmless one. It amounted to a statement of belief as to what the right physics might eventually turn out to be. In any case, in contrast to an opinion expressed in the literature,[12] I do not think that the electromagnetic world view was unambiguously outdated at the time and that you could not have had well-founded reasons to believe that speculations along the lines of Mie's theory would give you a reasonable theory of matter. It is true, though, that Einstein did not believe that Mie's theory of matter would be the right track. Some months later he wrote to Hermann Weyl:

Hilbert's *Ansatz* for matter appears childish to me, in the sense of a child that does not know of any tricks of outer reality. I am looking in vain for a physical clue that the Hamiltonian function may be formed from the φ_ν, and in fact without differentiation [of the $g^{\mu\nu}$].[13]

While Einstein rejected Hilbert's *Ansatz* as being too restrictive, we would now also immediately reject it as being too general on the grounds that it allows for Lagrangians that are not gauge invariant. But as far as I know the violation of gauge invariance was only put forward as an argument against Hilbert's theory as late as 1921 in Pauli's famous encyclopedia article on relativity.[14] Einstein conceded that the search for a suitable Lagrangian that would allow to account for the electron was part of the agenda but he flatly denied that the axiomatic method would be of any help here. The letter to Weyl goes on:

In any case, it is not to be approved when the solid considerations arising from the relativity postulate are mingled with such daring, unfounded hypotheses about the construction of the electron. I readily concede that finding a suitable hypothesis resp. Hamiltonian function for the construction of the electron is one of the most important tasks of today's theory. But the "axiomatic method" is of little use here.[15]

Einstein here introduces a theme that we find with other authors as well: the association of the axiomatic method with a moment of speculation, with daring and unfounded hypotheses. We find the same association, but with a positive undertone, in Mie's writings. In the introduction of a paper from 1917 on Einstein's gravitation theory and the problem of matter, Mie wrote about the problem of gravitation:

... one has to move on to speculation, a method of research that is rather new and unwonted in physics. One proceeds in such a way that one invents principles that are, at the same time, of great generality and of great fecundity, and from which, above all, the experimentally found laws can be deduced easily. Such principles we have known for a long time in geometry which at bottom also is only a part of theoretical physics, as the so-called axioms, and Hilbert calls the way of research characterized here the 'axiomatic method.'[16]

Mie's characterization, positive as it may be, strikes me as expressing a subtle misunderstanding of Hilbert's own intentions. While the formulation of a set of axioms for a given field of knowledge expressly allows for some freedom as to what axioms may be taken, it is certainly intended to properly represent the body of existing positive knowledge about nature accumulated in the domain of knowledge under investigation, and these axioms are not "invented" in the first place. Reading Mie benevolently, what may be at stake here is a reflexion on the evolving gap between what may be characterized as mathematical and theoretical physics, the former being guided by considerations of mathematical generality and consistency, the latter being oriented towards a theoretical account of observed phenomena.

An expert on both the practical and the theoretical issues of specialization in the mathematical sciences at the time was Felix Klein. He started in 1917 to study closely both Einstein's and Hilbert's works on General Relativity in the course of his lectures on the history of mathematics in the 19th century. Starting with a systematic comparison of the energy expressions put forward by both authors, Klein in 1918 eventually succeeded in substantially contributing to the conceptual clarification of the rather involved question of energy-momentum conservation in General Relativity. When his 1918 papers were republished in his Collected Works, he added some historical footnotes to these. In one of them he commented on the question of priority, expressing his assessment that Hilbert and Einstein had pursued rather independent lines of research. He also commented on the difference in their methodology. Klein wrote:

Einstein proceeds *inductively* and immediately has in mind arbitrary material systems. Hilbert *deduces*, invoking the [...] above-mentioned restriction to electrodynamics, from highest principles, chosen beforehand. Hilbert here in particular goes back to Mie.[17]

More explicitly Klein expressed himself in correspondence with Wolfgang Pauli who at the age of 21 was assigned to write a major review article on the theory of relativity for the Encyclopedia of the Mathematical Sciences. In a series of letters

Klein commented on a first draft of Pauli's review. Quite generally, he praised it for its soberness, and for Pauli's not chiming in to what he called the "idolization of Einstein".[18] But commenting on Pauli's presentation of the field equations of General Relativity, Klein protested against Pauli's slighting Hilbert's achievements as an independent discoverer. About his own assessment of Hilbert, Klein wrote in that letter:

Also with Hilbert there is (in his first note) much subjective construction: the fanatic belief that through pure mathematical reflection the essence of nature could be explained. Plus, mathematically a most disorderly presentation ... But instead of thus setting forth his own way of thought, he chose an axiomatic presentation that nobody understands who has not already mastered the whole thing.[19]

Nonetheless, Klein urged Pauli to acknowledge Hilbert's achievements. Pauli did follow Klein's advice and added a footnote to his presentation. However, this footnote turned out to read like this:

At the same time as Einstein, and independently, Hilbert, formulated the generally covariant field equations (...). His presentation, though, would not seem to be acceptable to physicists, for two reasons. First, the existence of a variational principle is introduced as an axiom. Secondly, of more importance, the field equations are not derived for an arbitrary system of matter, but are specifically based on Mie's theory of matter.[20]

In summary, it is to be observed that Hilbert's work on the "foundations of physics" at least as far as its published account in his two communications is concerned was received rather critically by contemporary physicists and mathematicians. In any case, it was certainly not received as a brilliant example of a new, superior method for investigations in theoretical physics.

4. HILBERT'S OWN ASSESSMENT OF HIS WORK IN HIS 1917 ZURICH LECTURE

Some of Hilbert's substantial results of his two communications, notably the introduction of the Riemann curvature scalar, were quickly taken up and incorporated into the accepted body of knowledge of the developing theory of general relativity.[21] Some other insights were ignored and remained largely neglected in the literature. The role of his axiomatic approach that Hilbert had put forward so prominently in his first note was frowned upon or simply ignored at best. On the other hand, in the context of contemporary discussions in philosophy of science, as discussed by Ulrich Majer in this volume,[22] Hilbert's axiomatic work in geometry played a much more important role than that in physics, and in philosophy of science Hilbert's contributions to general relativity and unified field theory remained largely ignored.

Hilbert himself gave an explicit account of his axiomatic method in a well-known programmatic lecture with the title "Axiomatic Thought" delivered to the

Swiss Mathematical Society in September 1917.[23] It is in this lecture, that the notion of "deepening the foundations" is put forward as an important function of the axiomatic approach which plays a significant role for Michael Stöltzner's discussion of the reception of Hilbert's work by Frank and Hahn.[24] Hilbert starts by observing that the facts of a field of knowledge are always ordered by a half-timbering ("Fachwerk") of concepts which he identifies with the theory of the field, and that in each such theory we always recognize certain propositions which suffice for the construction of its conceptual half-timbering. Attempts to prove those basic propositions leads to a "deepening of the foundations" by the introduction of other axioms from which the basic propositions may be deduced. The lecture then proceeds largely by discussing the two aspects of independence and consistency of such axiomatic formulations. What is remarkable about the Zurich lecture is the extensive use of examples for the illustration of Hilbert's methodological and epistemological concerns. Indeed, Hilbert mentions quite a number of "fields of knowledge" that he has in mind when elaborating on his general philosophical concerns, and about half of these examples are taken from the domain of physics proper, e.g. mechanics, electrodynamics, gas theory, thermodynamics, radiation theory, heat conduction, and the like. The use of these examples emphasizes Hilbert's claim that the axiomatic method not only advances our insight into the essence of scientific thought as such but also into the unity of our human knowledge.

Biographically, the lecture was given immediately after several years of intense work in theoretical physics culminating in the two communications on the "Foundations of Physics" discussed above. At that time, Hilbert intended to turn his attention more to foundational issues in mathematical logic and to take up his investigations of mathematical proofs. It is therefore a central text for Hilbert's own assessment of the role of the axiomatic method in science.

Surprisingly enough, his work in general relativity and field theory plays almost no role in this lecture. This neglect is astonishing not only since he was deeply engaged in this work in the period immediately preceding the lecture. One could also argue that Hilbert's "Foundations of Physics" in particular could have served as prime examples for his general considerations which, indeed, are illustrated extensively by many examples taken from mathematics and physics. His sketch of an axiomatic construction of Riemannian pseudogeometry could have been taken as an example of a "deepening of the foundations" of the axiomatic formulation of a field of knowledge, and the use of axioms for the investigation of the trajectories of material particles and of light rays in Schwarzschild space-time could be taken as a good example of a provisional identification of a domain of facts in a field of knowledge that is to be reduced to deeper axioms in the course of scientific progress. Nevertheless, we only find a mere hint at these investigations in a rather evasive manner:

Here I do not want to dwell upon the most recent ways of laying foundations of physics, in particular of electrodynamics, which are continuum theories through and through and postulate the axiom of continuity to the greatest extent, because these investigations have not come to an end yet.[25]

Strangely enough, in a lecture that recommends the axiomatic method as a means to realize the inherent unity of the mathematical sciences, Hilbert does not even mention the two axioms of his first note in which, as he believed, was contained the solution of both Einstein's and Mie's problems. He had not given up this belief, or else he would not have republished his two notes several years later in a slightly revised version that, however, did not alter much of its contents or add anything to it. Rather the neglect of his own work in his programmatic lecture of 1917 somewhat paradoxically bears witness to the seriousness of Hilbert's intentions and the appropriateness of the axiomatic method for them because he did not ignore the difficulties inherent in the whole endeavor.

NOTES

1. "Ich möchte im Folgenden – im Sinne der axiomatischen Methode – wesentlich aus zwei einfachen Axiomen ein neues System von Grundgleichungen der Physik aufstellen, die von idealer Schönheit sind, und in denen, wie ich glaube, die Lösung der Probleme von Einstein und Mie gleichzeitig enthalten ist." David Hilbert, "Die Grundlagen der Physik. (Erste Mitteilung.)", in: *Königliche Gesellschaft der Wissenschaften zu Göttingen. Mathematisch-physikalische Klasse. Nachrichten* 1916, p. 395. (All quotations translated by the author, if not otherwise indicated.)
2. See, e.g., Leo Corry, "From Mie's Electromagnetic Theory of Matter to Hilbert's Unified Foundations of Physics", in: *Stud. Hist. Mod. Phys.* 30, 2, 1999, pp. 159–183; Jürgen Renn and John Stachel, *Hilbert's Foundation of Physics: From A Theory of Everything to a Constituent of General Relativity*, Max Planck Institute for the History of Science Preprint 118, 1999; Tilman Sauer, "The Relativity of Discovery: Hilbert's First Note on the Foundations of Physics", in: *Arch. Hist. Exact Sci.* 53, 1999, pp. 529–575, and further references cited therein.
3. Hilbert, "Erste Mitteilung", *loc. cit.*, p. 396.
4. *Ibid.*
5. See Sauer, "The Relativity of Discovery", *loc.cit.* for an extensive discussion of the achievements of Hilbert's first note.
6. "Überzeugung, daß durch die hier aufgestellten Grundgleichungen die intimsten bisher verborgenen Vorgänge innerhalb des Atoms Aufklärung erhalten werden und insbesondere allgemein eine Zurückführung aller physikalischen Konstanten auf mathematische Konstanten möglich sein muß – wie denn überhaupt damit die Möglichkeit naherückt, daß aus der Physik im Prinzip eine Wissenschaft von der Art der Geometrie werde: gewiß der herrlichste Ruhm der axiomatischen Methode, die hier wie wir sehen die mächtigsten Instrumente der Analysis, nämlich Variationsrechnung und Invariantentheorie, in ihre Dienste nimmt." Hilbert, "Erste Mitteilung", *loc. cit.*, p. 407.
7. "Die genauere Ausführung sowie vor Allem die spezielle Anwendung meiner Grundgleichungen auf die fundamentalen Fragen der Elektrizitätslehre behalte ich späteren Mitteilungen vor." *Ibid.*, p. 395.
8. For a general discussion of Hilbert's work in physics, see, e.g., Leo Corry, "David Hilbert and the Axiomatization of Physics (1894–1905)", in: *Arch. Hist. Exact Sci.* 51, 1997, 83–198, and Ulrich Majer, "The Axiomatic Method and the Foundations of Science: Historical Roots of Mathematical Physics in Göttingen (1900-1930)" in: Miklos Rédei and Michael Stöltzner (Eds.), *John von Neumann and the Foundations of Quantum Physics* (Institute Vienna Circle Yearbook 8/2000) Dordrecht et al.: Kluwer, 2000, pp. 11–34.
9. For details see, Sauer, "The Relativity of Discovery", *loc. cit.*, p. 560.
10. For further discussion, see Renn and Stachel, "Hilbert's Foundation of Physics", *op. cit.* and Tilman Sauer, "The Relativity of Elaboration: Hilbert's Second Note on the Foundations of Physics", to be published.

11. "Hilberts Darstellung gefällt mir nicht. Sie ist unnötig speziell, was die "Materie" anbelangt, unnötig kompliziert, nicht ehrlich (=Gaussisch) im Aufbau (Vorspiegelung des Übermenschen durch Verschleierung der Methoden)." Einstein to Ehrenfest, 24 May 1916, in: *The Collected Papers of Albert Einstein*, Vol. 8, Princeton: Princeton University Press, 1998, p. 366.

12. Renn and Stachel, "Hilbert's Foundations of Physics", *op. cit.*

13. "Der Hilbertsche Ansatz für die Materie erscheint mir kindlich, im Sinne des Kindes, das keine Tücken der Aussenwelt kennt. Ich suche vergeblich nach einem physikalischen Anhaltspunkte dafür, dass die Hamilton'sche Funktion für die Materie sich aus den φ_ν, und zwar ohne Differentiation [der $g^{\mu\nu}$], bilden lasse." Einstein to Weyl, 23 November 1916, in: *The Collected Papers of Albert Einstein*, Vol. 8, Princeton: Princeton University Press, 1998, p. 366.

14. Wolfgang Pauli, *Relativitätstheorie*, (Enc. d. Math. Wiss. V.19), Leipzig: Teubner 1921; English translation: "Theory of Relativity", Dover: New York, 1981.

15. "Jedenfalls ist es nicht zu billigen, wenn die soliden Überlegungen, die aus dem Relativitätspostulat stammen, mit so gewagten, unbegründeten Hypothesen über den Bau des Elektrons bezw. der Materie verquickt werden. Gerne gestehe ich, dass das Aufsuchen der *geeigneten* Hypothese bezw. Hamilton'schen Funktion für die Konstruktion des Elektrons eine der wichtigsten heutigen Aufgaben der Theorie bildet. Aber die "axiomatische Methode" kann dabei wenig nützen." Einstein to Weyl, 23 Nov. 1916, *loc. cit.*. In a letter to Study, dated 17 November 1918, Einstein praises the latter's criticque of the "Auswüchse der Axiomatik", *loc. cit.*, p. 877.

16. "... man muß zur Spekulation übergehen, einer Forschungsmethode, die in der Physik ziemlich neu und ungewohnt ist. Man geht so vor, daß man Prinzipien ersinnt, die gleichzeitig von großer Allgemeinheit und von großer Fruchtbarkeit sind, und aus denen sich vor allem die experimentell ermittelten Gesetze zwanglos als Konsequenzen herleiten lassen. Solche Prinzipien kennen wir ja längst in der Geometrie, die ja im Grunde genommen auch nur ein Teil der theoretischen Physik ist, als die sogenannten Axiome, und Hilbert nennt den hier geschilderten Forschungsweg "die axiomatische Methode". Gustav Mie, "Die Einsteinsche Gravitationstheorie und das Problem der Materie", in: *Physikalische Zeitschrift* 18, 24, 1917, p. 551.

17. "Einstein geht *induktiv* vor und denkt gleich an beliebige materielle Systeme. Hilbert *deduziert*, indem er übrigens die [...] genannte Beschränkung auf Elektrodynamik eintreten läßt, aus voraufgestellten obersten Variationsprinzipien. Hilbert hat dabei insbesondere auch an Mie angeknüpft." Felix Klein, *Gesammelte Mathematische Abhandlungen* I. Band, Berlin: Springer 1921, p. 566.

18. "Daß Sie die ganze populäre Literatur mit ihren Torheiten bei Seite lassen, überhaupt nicht in die Verhimmelung Einsteins einstimmen, scheint mir ein großer Vorteil Ihres Referates." Klein to Pauli, 8 May 1921, in: Wolfgang Pauli, *Scientific Correspondence with Bohr, Einstein, Heisenberg, a.o. Volume I: 1919–1929*, New York: Springer, 1979, p. 31.

19. "Auch bei Hilbert ist (in seiner 1. Note) viel subjektive Konstruktion: der fanatische Glauben an die Variationsprinzipien, die Meinung, daß man durch bloßes math[ematisches] Nachdenken das Wesen der Natur erklären könne. Dazu mathematisch eine ganz ungeordnete Darstellung ... Statt aber so seinen eigentlichen Gedankengang klarzulegen, wählt er eine axiomatische Darstellung, die Niemand versteht, der nicht das ganze Ding schon gemeistert hat." Klein to Pauli, 8 May 1921, *ibid.*

20. Wolfgang Pauli, "Theory of Relativity", *op. cit.*, p. 145, n. 277.

21. See Renn and Stachel, *op. cit.*

22. Ulrich Majer, "Hilbert's program to axiomatize physics (in analogy to geometry) and its impact on Schlick, Carnap and other members of the Vienna Circle", this volume.

23. David Hilbert, "Axiomatisches Denken", in: *Mathematische Annalen* 78, 3/4, 1918, pp. 405–415.

24. Michael Stöltzner, "How Metaphysical is 'Deepening the Foundations'? – Hahn and Frank on Hilbert's Axiomatic Method" in this volume.

25. "Auf die neuesten Begründungsarten der Physik, insbesondere der Elektrodynamik, die ganz und gar Kontinuumstheorien sind und dem gemäß die Stetigkeitsforderung in weitestem Maße erheben, möchte ich hier nicht eingehen, weil diese Forschungen noch nicht genügend abgeschlossen sind." Hilbert, "Axiomatisches Denken", *loc. cit.*, p. 409.

History and Philosophy of Science
University of Bern
Sidlerstrasse 5
CH-3012 Bern
Switzerland
tilman.sauer@philo.unibe.ch

MIKLÓS RÉDEI

MATHEMATICAL PHYSICS AND PHILOSOPHY OF PHYSICS (WITH SPECIAL CONSIDERATION OF J. VON NEUMANN'S WORK)

1. MAIN CLAIM

The main claim of this talk is that mathematical physics and philosophy of physics are not different. This claim, so formulated, is obviously false because it is over-stated; however, since no non-tautological statement is likely to be completely true, it is a meaningful question whether the overstated claim expresses some truth. I hope it does, or so I'll argue. The argument consists of two parts: First I'll recall some characteristic features of von Neumann's work on mathematical foundations of quantum mechanics and will claim that von Neumann's motivation and results are essentially philosophical in their nature; hence, to the extent von Neumann's work exemplifies what is considered to be mathematical physics, mathematical physics appears as formally explicit philosophy of physics. The second argument is based on a rather trivial interpretation of what mathematical physics is. That interpretation implies that mathematical physics shares some key characteristic features with philosophy of physics which make the two almost indistinguishable.

2. VON NEUMANN'S WORK ON QUANTUM MECHANICS, 1927-1932

Von Neumann's aims and results obtained by his creating abstract Hilbert space quantum mechanics (QM) can be summarized as follows:

- Eliminating mathematical nonsense
  Dirac δ function, eigenvalue problem

- Getting rid of intrinsically unnecessary elements in the theory
  Concrete L^2, ℓ^2 representation of Hilbert space $\mathcal{H}$, fixed basis in $\mathcal{H}$

- Giving a unified, coherent theory of QM
  "axiomatic" presentation of QM

- Solving interpretational problems
  Analysis of the status of causality and QM, completeness of QM measurement problem

M. Heidelberger and F. Stadler (eds.), History of Philosophy and Science, 239–243.

- Investigating intertheory relations
  Relation of QM to classical statistical mechanics and to statistics

It is remarkable that von Neumann did *not* aim at any of the following:

1. Giving any application of the theory

2. Making any physical prediction

3. Discovering anything physically new

Let me support the statement that the above lists are indeed representative of von Neumann's intentions and results by quoting von Neumann himself:

"The object of this book is to present the new quantum mechanics in a unified representation which, so far as it is possible and useful, is mathematically rigorous. ... In particular, the difficult problems of interpretation, many of which are even now not fully resolved, will be investigated in detail. In this context the relation of quantum mechanics to statistics and to the classical statistical mechanics is of special importance. However, we shall as a rule omit any discussion of the application of quantum mechanical methods to particular problems ..."
Mathematical Foundations of Quantum Mechanics, Princeton University Press, 1955, p. vii

"A common defect of all these methods [methods of matrix mechanics and wave mechanics – M.R.] is, however, that they in principle introduce unobservable and physically meaningless elements into the calculation: for the eigenfunctions have to be calculated which virtue of their unit norm remain undetermined up to a constant of norm 1 (the 'phase' $e^{i\epsilon}$), and in the case of a χ-fold degeneration (that is, of a χ-fold eigenvalue) even up to a χ-dimensional orthogonal transformation. Although the probabilities obtained as final results are invariant, it is nevertheless unsatisfactory and it is unclear why the detour by way of unobvervable and non-invariant ones is necessary."
Mathematische Begründung der Quantenmechanik, Collected Works I. p. 153

"We will have to deal solely with such facts which are well-known mathematically and which *also physically offer nothing new* [my emphasis – M.R.], because in essence they appear in the work of Schrödinger cited in note 3, p. 2, and in several works by Dirac."
Mathematische Begründung der Quantenmechanik, Collected Works I. p. 154

3. ABANDONING HILBERT SPACE QUANTUM MECHANICS: VON NEUMANN'S WORK ON QUANTUM THEORY AFTER 1932

In the years 1935-36 von Neumann criticized Hilbert space QM explicitly and advocated an algebraic approach to QM; in particular, he set high hopes on a specific algebraic structure known today as the type II_1 factor von Neumann algebra he discovered with Murray in 1935. The reasons why von Neumann criticized and abandoned Hilbert space QM can be summarized as follows:

- Interpretational inconsistencies in Hilbert space QM
  No von Mises type relative frequency interpretation of quantum probability can be given

- Presence of operationally meaningless mathematical operations and entities in Hilbert space QM
  Usual (composition) product of observables, phase in state vector

- Mathematical pathologies in Hilbert space QM
  All selfadjoint unbounded operators do not form an algebra

- New mathematical discoveries
  Existence of finite, continuous von Neumann algebras

- Perceived greater conceptual coherence of operator algebraic QM
  Unique a priori quantum probability determined by quantum logic

Again it is most remarkable that von Neumann's move beyond Hilbert space QM was *not* motivated by any of the following:

1. Empirical inadequacy of Hilbert space QM

2. Any new empirical/physical discovery

3. Mathematical imprecision/nonsense in Hilbert space QM

To prove that von Neumann had indeed abandoned his brainchild (the abstract Hilbert space QM) for the reasons listed above would require a detailed reconstruction of the development of von Neumann's views on QM, especially in connection with his research on "operator rings", known today as von Neumann algebras. Such a proof cannot be attempted here (see the works [1], [2], [3] and [5] for a detailed analysis of this story). What I wish to point out here is that von Neumann's aims, motivations and results in his work on QM, both in its first phase 1927-1932 as well as after 1932, are explicitly meta-physical, or philosophical – as opposed to being simply physical or purely mathematical. So to the extent von Neumann's work on the mathematical foundations of quantum mechanics is representative of what mathematical physics is, mathematical physics does not seem to be different from what philosophy of physics is about.

This conclusion can be further supported by having a look at the history of 20th Century philosophy of physics: It seems impossible to draw a line between philosophy of physics and mathematical physics in discussions of topics such as foundations of classical statistical mechanics (ergodic theory, theory of dynamical systems), locality and QM, quantum logic, cosmological models, singularity theorems in general relativity, etc.

Blurring of the demarcation between mathematical physics and philosophy of physics is due to the fact that mathematical physics and philosophy of physics have some characteristic features in common; specifically both are *reflective* and *normative*.

If we take mathematical physics as the discipline that creates and investigates mathematical models of physical phenomena, where a mathematical model is viewed as (mathematics + physical interpretation), then mathematical physics is *normative* in the sense that it forces

- conceptual precision:
  mathematics is mathematics – it is not permissible to let the theory contain mathematical nonsense;

- conceptual clarity:
  interpretation = relating explicitly mathematics to physical reality / observation

Also, on this interpretation of mathematical physics, mathematical physics is *flexible rigidity* because

- a given interpretation/model is a binding committment; but

- one has considerable freedom in designing different models – possibly even of the same phenomena.

The reflective character of mathematical physics is nicely expressed by von Neumann:

"I think that in theoretical physics the main emphasis is on the connection with experimental physics and those methodological processes which lead to new theories and new formulations, whereas mathematical physics deals with the actual solution and mathematical execution of a theory which is assumed to be correct per se, or assumed to be correct for the sake of the discussion. In other words, I would say that theoretical physics deals rather with the formation and mathematical physics rather with the exploitation of physical theories."
John von Neumann's letter to R.O. Fornaguerra, December 10, 1947 (unpublished), LOC, Washington D.C., quoted in [6]

Let me conclude by submitting for discussion the following provocatively formulated summary: Mathematical physics is the formally explicit, rational philosophy of physics.

REFERENCES

[1] M. Rédei: "Why John von Neumann did not like the Hilbert space formalism of quantum mechanics (and what he liked instead)", *Studies in the History and Philosophy of Modern Physics* **27** (1996) 493-510.

[2] M. Rédei: *Quantum Logic in Algebraic Approach* (Kluwer Academic Publishers, Dordrecht, 1998).

[3] M. Rédei: "Unsolved problems in mathematics" J. von Neumann's address to the International Congress of Mathematicians, Amsterdam, September 2-9 1954, *The Mathematical Intelligencer*, vol. 21, 1999 p. 7-12.

[4] M. Rédei: "John von Neumann's concept of quantum logic and quantum probability" in [5] 153-172.

[5] M. Rédei, M. Stöltzner (eds.): *John von Neumann and the Foundations of Quantum Physics* (Kluwer Academic Publishers, Dordrecht, 2001).

[6] M. Stöltzner: "Opportunistic axiomatics: John von Neumann on the methodology of mathematical physics" in [5] 35-62.

Department of History and Philosophy of Science
Loránd Eötvös University
H-1518 Budapest 112
Hungary
redei@hps.elte.hu

MICHAEL STÖLTZNER

HOW METAPHYSICAL IS "DEEPENING THE FOUNDATIONS"? – HAHN AND FRANK ON HILBERT'S AXIOMATIC METHOD [*]

Only recently has David Hilbert's program to axiomatize the sciences according to the pattern of geometry left the shade of his formalist program in the foundations of mathematics.[1] This relative neglect – which is surprising in view of the enormous efforts Hilbert himself had devoted to it – was certainly influenced by Logical Empiricists' almost exclusively focusing on his contributions to the foundational debates. Ulrich Majer puts part of the blame for this neglect on Hilbert himself because "he failed to make his position sufficiently clear, and he did not take much effort to promote his views beyond the narrow circle of mathematical physics in Göttingen."[2]

This might excuse Schlick and Carnap who are the protagonists of Majer's paper. But two other core members of the Vienna Circle had studied with Hilbert and continued to work on topics close to his circle, such as variational calculus and relativity theory. The mathematician Hans Hahn spent the winter semester 1903/4 at Göttingen, and the theoretical physicist Philipp Frank went there in the summer of 1906. For the winter term 1903/4, the catalogue of the university lists lectures on "The concept of number and the quadrature of the circle", the "Theory of partial differential equations" and "Exercises on algebra and arithmetic" and "Exercises on the Theory of Differential Equations" (together with Minkowski).[3] For Frank's Göttingen semester there are a "Seminar on the theory of functions" (together with Klein and Minkowski) and lectures on "Continuum Mechanics" and "Differential and Integral Calculus I" (together with Carathéodory). It seems safe to assume that Frank attended Hilbert's lecture on continuum mechanics.[4]

After 1908, Frank would quickly become one of the leading early researchers on Einstein's relativity theory, such that he certainly was aware of Hilbert's 1915 "The Foundations of Physics" – a text of which Majer does not find any substantial mention in Schlick's work. And at least from the days of his Ph.D. dissertation Frank knew enough about variational calculus to understand the technical side of Hilbert's reasoning. Hahn even was one of the leading figures in variational calculus; together with Ernst Zermelo he was entrusted with the respective entry for the *Encyclopedia of Mathematical Sciences.*[5]

Thus for Hahn and Frank, more substantial factors than sheer neglect must have prevented a due appreciation of Hilbert's program of the axiomatization of the sciences, factors which revealed a basic difference in attitude towards

245

M. Heidelberger and F. Stadler (eds.), History of Philosophy and Science, 245–262.
© 2002 *Kluwer Academic Publishers. Printed in the Netherlands.*

mathematical physics. At the bottom of this difference – so I shall argue – stands their conviction that reconciling Ernst Mach's empiricist heritage with modern mathematics required drawing a rigid boundary between mathematics and physics and subscribing to logicism, according to which mathematics consisted in tautologous logical transformations. In this way, and similar to their fellow Circle members studied by Majer, Hahn and Frank missed the substantial difference between the logical structure of a particular axiom system and the axiomatic method as a critical study of arbitrary axiom systems. If this distinction is not properly observed – and admittedly Hilbert himself did deliberately obscure it at places – a core concept of the axiomatic method, "deepening the foundations" (*Tieferlegung*), becomes metaphysical because it might appear as an ontological reduction of basic physical concepts to mathematical ones rather than – as Hilbert intended – an epistemological reduction availing itself of the unity of mathematical knowledge.

To be sure, Logical Empiricists considered the goal of axiomatizing the sciences an important task, but in the way they set it up axiomatization became much more closely tied to a success of the foundationalist program for all mathematics than Hilbert's axiomatic method ever was. Because of this relative independence from the foundationalist program, the process of "deepening the foundations" can be interpreted in a much more pragmatic fashion than in Hilbert. But this was done only in the 1950s by John von Neumann who more than anybody else contributed to putting Hilbert's axiomatization program to work.[6]

I shall, however, focus on the discussions of the 1930s, in particular because Hilbert's 1930 address "Logic and the Knowledge of Nature" received a highly critical answer from his former student Hahn. Frank later even feared that calling modern physics mathematical would give way to a return of spiritual and teleological elements in science. Consequently, Hahn and Frank avoided almost any reference to the philosophical connotations of the Principle of Least Action which figured so prominently in Hilbert's axiomatizations because this principle, and variational calculus as a whole, had frequently been interpreted as a teleological element within the foundations of physics proper. Against this backdrop it becomes clear why Logical Empiricists were not at all pleased by Hilbert's constant assertions that a (non-Leibnizian) pre-established harmony linked physics and mathematics, a harmony that was revealed by successfully "deepening the foundations". In the fourth section I shall analyze the various meanings of this rather wooly concept and investigate which of them could have been palatable to Logical Empiricists, had wording and philosophical heritage not blatantly opposed one of their battle cries: "In science there are no 'depths'; there is surface everywhere."[7]

1. HILBERT'S *KÖNIGSBERG* ADDRESS

In the autumn of 1930, the biennial meeting of the German Society of Natural Scientists and Physicians took place in Hilbert's native city of Königsberg. During the opening session he delivered his last major public lecture on "Logic and the Knowledge of Nature".[8] It is sometimes seen as one of history's ironies that at a small meeting adjoined to the congress, the young Viennese mathematician Kurt Gödel presented a proof that Hilbert's foundationalist program that was intended to justify his optimism about the solvability of every well-formulated mathematical problem, was unfeasible.

But Hilbert's talk was not primarily about foundations. It set out by observing that in the three decades since he had presented his 23 "Mathematical Problems"[9] to the International Congress of Mathematicians in 1900, scientists had witnessed an unbroken chain of discoveries that put them in a position to confidently face "an old philosophical problem, namely, the much disputed question about the share which thought, on the one hand, and experience, on the other, have in our knowledge."[10] This, of course, had been the core theme of the *genius loci* Immanuel Kant.

Not only have the technique of experimentation and the art of erecting theoretical edifices in physics attained new heights, but their counterpart, the science of logic, has also made substantial progress. Today there is a general method for the theoretical treatment of questions in the natural sciences, which in every case facilitates the precise formulation of the problem and helps prepare its solution – namely, the axiomatic method.[11]

After alluding to the model of Euclidean geometry, Hilbert gives a biological example and claims that "the laws of heredity [of the fly *drosophila*] result as an application of the axioms of linear congruence, that is, of the marking-off of intervals."[12] Although it is not clear to me what Hilbert precisely had in mind, this example teaches us that his program was not at all limited to well-entrenched fields of physics. Axiomatization could prove fertile also in preliminary or merely phenomenological theories, and Hilbert himself published an axiomatic treatment of Kirchhoff's law of radiation that was – even in those days – not considered as the last word on atomic processes. This lead to a polemic between Hilbert and the experimentalist Ernst Pringsheim. Max Born – who had been an assistant to Hilbert – attempts reconciliation:

[B]eing conscious of the infinite complexity he faces in every experiment [the physicist] refuses to consider any theory as final. Therefore ... he abhors the word 'axiom' to which common use clings the sense of final truth. Yet the mathematician does not deal with the factual happenings, but with logical relations; and in Hilbert's terms the axiomatic treatment of a discipline does not signify the final assertion of certain axioms as eternal truths, but the methodological requirement: state your assumptions at the beginning of your considerations, stick to them, and investigate whether these assumptions are not partially superfluous or even mutually inconsistent.[13]

Hence, one can conclude with Majer that

[f]rom an axiomatic point of view ... the macroscopic-phenomenological approach is logically just as suited for an axiomatic investigation as the microscopic-molecular one. This shows that both points of views, the axiomatic and the foundational one, are not identical and can be pursued quite independently. Hilbert, however, for the most time in his life had both aims in mind.[14]

In a certain sense, the axiomatic method was even more powerful in this case than for a final theory because here one could analyze more alternative models. Let me quote a longer passage from the introduction to his second lecture on the mechanics of continua in the winter term of 1906/7, the first part of which Frank almost certainly attended.

As goal of *mathematical physics* we can perhaps describe, to treat also all not purely mechanical phenomena according to the model of point mechanics; hence ... on the basis of Hamilton's principle, perhaps after appropriately generalizing it. Physics has ... already gained brilliant successes in this direction ...

Even if the keen hypotheses, which have been made in the realm of molecular physics, sometimes certainly come close to the truth because the predictions are often confirmed in a surprising manner, one has to characterize the achievements still as small and often as rather insecure, because the hypotheses are in many cases still in need of supplementation and they sometimes fail completely. ... Such considerations recommend it as advisable to take *meanwhile* a completely different, yet a *directly opposite* path in the treatment of physics – as it indeed has happened. Namely, one tries from the start to produce as little detailed ideas as possible of the physical process, but fixes instead only its general parameters, which determine its external development; then one can by *axiomatic physical assumptions* determine the form of the Lagrangian function L as function of the parameters and their differential quotients. If the development is given by the minimal principle $\int_{t_1}^{t_2} L \, dt = \text{Min.}$, then we can infer general properties of the state of motion solely from the assumptions with respect to the form of L, without any closer knowledge of the processes ...

In order to give ... an intuition of the new approach I refer to the theory of elasticity, which treats the deformations of solid bodies caused by the mutual influence and displacements of the molecules; we will have to renounce a detailed description of the molecular processes and instead only look for the parameters from which the measurable state of deformation of the body depends at each place. Then one has to determine the form of dependence of the Lagrangian function from these parameters which is, properly speaking, composed of the kinetic and potential energy of the single molecules. ... The presentation of physics just indicated, ... which permits the deduction of essential statements from formal assumptions about L, shall be the core of my lecture.[15]

But in Hilbert's Königsberg talk this subtle distinction between final and (transitory) phenomenological theories was not mentioned. Rather did Hilbert believe that physical science had already reached its mathematical foundations to such an extent that one could draw philosophical conclusions.[16]

"[I]nfinite" has no intuitive meaning and ... without more detailed investigation it has absolutely no sense. For everywhere there are only finite things. There is no infinite speed, and no force or effect (*Wirkung*) that propagates itself infinitely fast. Moreover, the action (*Wirkung*) is itself of a discrete nature and exists only in quanta. There is nothing that can be divided infinitely often. Even light has atomic structure, just like the quanta of action. I firmly believe that even space is only of finite extent, and one day astronomers will be able to tell us how many kilometers long, high and broad it is.[17]

Here we obtain the first parallelism between nature and thought: "the infinite is nowhere realized; it neither occurs in nature nor is it admissible as a foundation in our thought without special precautions."[18] As to the second parallelism, "our thought intends unity and seeks to form unity; we observe the unity of material in matter and we everywhere detect the unity of the laws of nature. This nature in reality greatly accommodates us in our research."[19] Here Hilbert is not far from the Kantian regulative principle of subjective formal teleology (*Zweckmäßigkeit*) that directs our judgment to investigate particular empirical laws according to a "unity such as they would have if an understanding (though it be not ours) had supplied them for the benefit of our cognitive faculties, so as to render possible a system of experience according to particular natural laws."[20] Against this backdrop, it is not altogether surprising that pre-established harmony turns out to be the third parallelism, where Hilbert takes the expression in another sense than Leibniz. This allusion was not really new. Already in 1900 Hilbert had found that "the apparently pre-established harmony which the mathematician so often perceives in the questions, methods and ideas of the various branches of his science, originates in this ever-recurring interplay between thought and reason."[21] In 1930 he looked back to what he had attempted in the field of general relativity.

Here solely through the general demand for invariance together with the principle of greatest simplicity the differential equations for the gravitational potentials are constructed with mathematical uniqueness. This construction would not have been possible without the profound and difficult mathematical investigations of Riemann, which existed long before.[22]

What Hilbert does not say here is that he had to take back a substantial part of his aspirations at a unified field theory, a goal which stood behind his paper's ambitious title "The Foundations of Physics". Not only was his model for matter much too special, but the idea "that a reduction of all physical constants to mathematical constants should be possible"[23] had to be dropped because a major theorem of the first version had to be weakened to what today is known as Noether's second theorem.

The next example in Hilbert's Königsberg address, the foundations of quantum mechanics, would enjoy a better fate, mainly due to von Neumann's book.[24]

Hilbert's general conclusion mentions a key element of his axiomatic method, "elimination" (*Elimination*), which appears to me a new wording for what in his 1918 paper "Axiomatic thought" had figured under "deepening the foundations" (*Tieferlegung*)[25].

We can understand this agreement between nature and thought, between experiment and theory, only if we take into account both the formal element and the mechanism connected with it on both sides of nature and of our understanding. The mathematical process of elimination furnishes, it seems, the resting-points and stations where both the bodies in the real world and the thoughts in the mental world linger and, in this way, present themselves for control and comparison.

But this pre-established harmony does not yet exhaust the relations between nature and thought, and does not yet uncover the deepest secrets of our problem ... [M]odern science shows that the present actual state of physical matter on earth and in space is not accidental or arbitrary, but follows from the laws of physics.[26]

Hilbert sees the specter of Hegelian *Naturphilosophie* rising and avowedly reverts from it: "For what is the origin of the laws of the world? How do we acquire them? And who teaches us that they fit reality? The answer is, that experience alone makes this possible."[27] So far this is quite in line with a Kantian position.

Now I admit that already for the construction of the theoretical frameworks (*Fachwerke*) certain *a priori* insights are necessary ... I also believe that mathematical knowledge in the end rests on a kind of intuitive insight of this sort, and even that we need a certain *a priori* outlook for the construction of number theory. Thus the most general and most fundamental idea of the Kantian epistemology retains its significance: namely the philosophical problem of determining *a priori* that intuitive outlook and thereby of investigating the condition of the possibility of all conceptual knowledge and of every experience ... But we must draw the boundary between what we possess *a priori* and what requires experience differently than Kant: Kant greatly overestimated the role and the extent of the *a priori*.[28]

The classical case in point for the demise of the synthetic *a priori* was, of course, relativity theory. Hilbert concludes that "the Kantian theory of the *a priori* still contains anthropological dross from which it must be liberated; afterwards only the *a priori* attitude is left over which also underlies pure mathematical knowledge: essentially it is the finite attitude which I have characterized in several works."[29]

Looking back to Hilbert's earlier remarks about the impossibility of infinities in physics, one gets the impression that he himself had overestimated the role of the post-Kantian *a priori* in physics. At any rate, Hilbert's aim never was to justify Kantianism but to argue for the central role of mathematics within the sciences. "The instrument that mediates between theory and practice, between thought and observation, is mathematics."[30] Or: "We have not mastered a theory in the natural sciences until we have extracted and fully revealed its mathematical core. Without mathematics, modern astronomy and physics would not be possible; these sciences, in their theoretical parts, almost dissolve into mathematics."[31] Certainly, this was not what Logical Empiricists liked to hear from one of their idols[32], even more as it was so tightly linked to the Kantian *a priori*.

2. HANS HAHN'S RESPONSE TO A FORMER TEACHER

In 1933, a small booklet entitled *Logic, Mathematics, and Knowledge of Nature* authored by Hahn appeared as the second volume of the series *Einheitswissenschaft* (Unified Science). Although Hilbert's name does not appear in the booklet, its title openly alludes to his widely-read lecture. Hahn also responds on the stylistic level. While Hilbert concludes with the separately printed lines "We must know, We shall know"[33], Hahn reverberated the motto of Boltzmann's *Mechanics*: "Put forward what is true; / Write so as to make it clear, / And defend it to the end!"[34]

There is, however, little need for circumstantial evidence because Hahn directly addresses the main topic of Hilbert's talk, the relationship between thought and reality. After a brief historical tour through the philosophical views of empiricism and rationalism, Hahn comes to the point.

The usual view [concerning the relationship in question] can then be described like this: from experience we gather certain facts and formulate them as "laws of nature"; but since by thought we apprehend the most general lawlike connections in reality (of a logical and mathematical nature), our mastery over nature on the basis of facts we have gathered by observation extends much further than our actual observations; for we also know that everything that can be inferred from what we have observed by applying logic and mathematics must be real. ... This view seems to find a powerful support in the numerous discoveries made in a theoretical manner ...

But we are nevertheless of the opinion that this view is completely untenable. For upon closer reflection it appears that the role of thought is incomparably more modest than the role ascribed to it on this view. The view that thought provides us with a means of knowing more about the world than we observed, of knowing what must be unconditionally valid always and everywhere in the world, a means of apprehending general laws of being, this view seems to us thoroughly mysterious. ... Why should what is compelling to our thought also be compelling to the course of the world? Our only recourse would be to believe in a miraculous pre-established harmony between the course of our thought and the course of the world, an idea which is deeply mystical and ultimately theological.[35]

Here we find the main charge against Hilbert: professing the faith of a pre-established harmony amounts to mysticism. To be sure, Hilbert had emphasized that our knowledge of natural laws is of empirical origin, but Hahn required a firmer stand and continues:

There seems to be no other way out of this situation than a return to a pure *empiricist* position, a return to the view that observation is the only source of our knowledge of facts: there is no factual knowledge a priori, *no "material" a priori*. Only we must avoid the mistake of earlier empiricists who would see nothing but empirical facts in the propositions of logic and mathematics; we must look around for a different view of logic and mathematics.[36]

To Hahn's mind, the only way to reconcile a consistent empiricism with modern mathematics is to deny any reality to the concepts of logic and mathematics, and regard them merely as conventions about the use of the symbols of a formal language. They "say nothing about objects and are for this reason certain, universally valid, and irrefutable by observation."[37]

[T]hough logical propositions are purely tautological, and though logical inferences are nothing but tautological transformations, they are nevertheless significant for us because we are not omniscient. Our language is made in such a way that in asserting certain propositions we implicitly assert other propositions – but we do not see right away all the things we have implicitly asserted, and only logical inference makes us conscious of them.[38]

In view of Hilbert's argument that our limited cognitive faculties require an *a priori* finitist attitude, Hahn's reference to our lack of omniscience is puzzling because assuming that the class of all possible systems of scientific tautologies is a meaningful concept one might argue that this cognitive limitation amounts to an *a priori* argument for our need to choose special systems by convention.

Surprisingly, Hahn is still convinced that the logicist foundations of mathematics can be rigorously proven – even three years after his student Gödel's incompleteness results. "Of course, the proof of the tautological character of mathematics is not yet complete on all points; we are here faced with a troublesome and difficult problem; yet we have no doubt that the view that mathematics is tautological is essentially correct."[39] In a paper published three years earlier, Hahn had simply denied that there was any space for Hilbert's foundationalist program.

As regards Hilbert's *formalism*, what must be pointed out from our point of view is above all the unexplained role of metamathematical considerations. What is the origin of claims to metamathematical knowledge? It is certainly not experience! The difficulties we mentioned in speaking of attempts to base logic and mathematics on experience militate against this. Are they logical transformations? Certainly not! For they are supposed to serve as justifications for these transformations. What, then, are they? However, in adopting this sceptical position towards Hilbert's point of departure we do not mean to say anything against the significance of his investigations. On the contrary, I am convinced that many of Hilbert's concrete results will enter into the continuation and improvement of Russell's system.[40]

Two pages before Hahn had called it a major virtue of the logicist view that "the problem of ... the seemingly pre-established harmony between thought and world"[41] dissolves. Interestingly, Hahn seems to believe that only after this rigorous separation can the axiomatization of the sciences fully thrive because all theorems become tautological implications of freely chosen assumptions.

Some chapters of physics have already been axiomatized in the same sense as geometry and turned thereby into special chapters of the theory of relations. Yet they remain chapters of physics and hence of an empirical and factual science because the basic concepts that occur in them are constituted out of the given.

In doing this we may have the following goal in mind: to set up an axiomatic system by which the whole of physics is logicized and incorporated into the theory of relations. If we do this, it may well turn out that, as the axiomatic systems become more comprehensive, ... their basic concepts become increasingly remote from reality and are connected with the given by increasingly longer, increasingly more complicated constitutive chains. All we can do is state this fact, as a peculiarity of the given; but there is no bridge that leads from here to the assertion that behind the sensible world there lies a second, 'real' world enjoying an independent being.[42]

Here another anathema of Logical Empiricism appears, the meaningless metaphysical question about the external world. That Hahn fears this danger is only possible because he assumes the extension of a single (but arbitrarily chosen) axiomatic system of all physics to go hand in hand with conceptual reduction and unification.

Hahn's concept of axiomatization differs from Hilbert's in three respects. First, all basic concepts appear to be on a par within a network of logical relations that covers all of unified science. What Hilbert considered as "deepening the foundations", on Hahn's account, was just choosing a more economical convention. Second, the constitution of the basic mathematical concepts excludes any intuitive justification of mathematical concepts. In his paper "The Crisis of Intuition"[43], Hahn had listed many important concepts of modern mathematics that contradicted our intuition, for instance, non-Archimedian number fields. Hilbert could well have agreed to this criticism. But, intuition still played a significant role in his account precisely because our limited cognitive faculties required a finitist attitude. To be sure, Hahn did not charge Hilbert of relapsing into the infamous Kantian pure intuition as a basis of mathematics, but of placing metamathematics into the empty space between logical tautologies and empirical knowledge once occupied by it. Third, by considering any axiom system exclusively as a system of logical relations plus constitutive definitions of the basic concepts, Hahn made the axiomatic method much more dependent on the success of a foundational program for mathematics than Hilbert.

The reason for Hahn's separating so rigidly logic and mathematics from knowledge of nature had its roots in the Machian heritage, which had been much more important for the 'Austrians' in the Vienna Circle than for Schlick or Carnap. In his *Mechanics*, Mach had criticized Archimedes' derivation of the law of the lever by means of Euclidean geometry because it implicitly presupposed factual knowledge that could only be attained by previous experiences.[44]

Hahn and his colleagues were aware that they had to take Boltzmann's side in the atomism controversy. "Indeed the whole of science is full of propositions which cannot in principle be confirmed by observation because they contain unconstitutable terms; propositions about molecules, atoms, electrons, etc. are not the only propositions of this kind."[45] Attributions of exact numerical values to physical quantities and the universal quantifier occurring in every law of nature belong to this class.

Every time we introduce unconstitutable terms into science we must make sure that they are accompanied by directions for their use ... And these rules must be such that we come in the end to propositions which can be immediately confirmed or refuted by observation. ... Legitimate propositions of science with unconstitutable terms are comparable to adequately covered paper money which can be exchanged for gold at any time at the national bank – whereas metaphysical propositions are like uncovered paper money which is not accepted by anyone in exchange for either gold or goods.[46]

Present-day economists would consider such a restrictive monetary policy to severely hamper economic growth and they would take it rather as a symptom of crisis when people start cashing their banknotes for goods. Mutual support of various theories seems to be rather favorable for scientific growth; and this was a task which, to Hilbert's mind, mathematics could accomplish in virtue of its remarkable internal cohesion.

Hahn, however, closely sticks to Mach by arguing "that there is not a single law of nature of which we know whether it holds or not; laws of nature are *hypotheses* which we state tentatively."[47] By contrast, Boltzmann's pictures – as he called theories in order to avoid realist metaphysics – were attempts at an epistemological reduction and unification. It appears to me that by granting Mach too much at this point Hahn had to fortify the boundary between logical tautologies and hypothetical empirical facts.

3. PHILIPP FRANK AND THE PRINCIPLE OF LEAST ACTION

In 1936 and within the same series as Hahn's criticism, Frank published a booklet entitled *The Fall of Mechanistic Physics*. Its second chapter opposes the thesis that "the new physics is not mechanistic but mathematical."[48] Frank's targets are General Smuts and James Jeans who held that the fall of the mechanical world view and the rise of abstract mathematical entities lead to a return of spiritual elements within modern science, so that "the universe is now more like a great [organic] idea than a great machine."[49] But, so Frank contends, whoever desires to find spiritual analogies will succeed both in classical and in modern physics. Hence, the "assertion that the new physics is not 'mechanical' but 'mathematical' only means that the formulae of relativity and quantum mechanics *contradict* those of the old mechanics or to put it more precisely, agree with them only for small velocities and large masses."[50]

The laws of physics consist of mathematical relations between quantities, as well as of directions on how these quantities can be related to feasible observations, and in this respect nothing has changed even in the twentieth century. The equations have changed, the quantities are different, and the directions, too, are therefore no longer the same; but the general scheme according to which a physical theory is constructed still has the same fundamental character today as it had in Newton's time.[51]

What Frank treated as "mathematical physics" hardly touched the core of Hilbert's axiomatization program.

Against the Principle of Least Action and variational calculus – which were the key tool in Hilbert's axiomatizations of mechanics, continuum mechanics, and relativity theory – there existed even a long historical record of counts of metaphysical teleology, that was broadly discussed in Mach's *Mechanics*. But in this case, Hahn and Frank remained virtually silent. This was rather surprising for a philosophically interested core researcher in the field. According to the historian Wilhelm Frank, "Hahn's publications in this field ... were very positively reviewed in the contemporary journals and often constituted important steps in the development and the simplification of the methods of the calculus of variations."[52]

In Philipp Frank's writings, one has slightly more success. His Ph.D. dissertation of 1906 dealt with the problem of sufficient conditions for a minimum of the Principle of Least Action, a topic that was – as the author repeatedly stressed – typically absent from the treatises of mechanics including the one by his late teacher Boltzmann. That Frank wrote a physics thesis largely from a mathematician's perspective suggests that its topic was proposed by Hahn or Gustav von Escherich. Frank constantly attended their courses[53], in particular Hahn's maiden lectures on variational calculus in the summer term of 1905. In the introduction to his thesis, Frank argued that new mathematical results allowed one to radicalize Carl Gustav Jacobi's ametaphysical stance.

"It is difficult to find a metaphysical cause for the Principle of Least Action, if it is expressed in this true form, as is necessary. There exist minima of an entirely different type, from which one can also derive the differential equations of the motion, which in this respect are much more appealing."[54]

One can give the theorem an even more ametaphysical form than Jakobi's [sic!] by saying: A material point moves according to the Lagrange equations appertaining to the variational problem $J = \int_a^b \sqrt{h-V}\,ds$. This casts off the last remnant of minimum-romance.[55]

In Frank's philosophical works, the Principle of Least Action is almost absent. His seminal book *The Law of Causality and Its Limits* list it as an instance of "another widely spread manner of treating natural phenomena by analogy to human emotional life."[56] Subsequently Frank repeats an example from his Ph.D. dissertation. For the ballistic throw one can prove that J (defined as above) attains a smaller value for a piecewise continuous curve lifting the ball vertically up to maximum height, then transporting it horizontally to a point above the endpoint and finally dropping it, than for that parabola which is the actual orbital curve.

It is not at all characteristic for the orbit a point-mass follows that along that orbit any magnitude assumes its smallest value ... Only a certain mathematical simplification is hidden in the minimal principles of mechanics. With its help the laws of the orbital curves can be expressed in fewer variables ... This has, however, nothing to do with economical measures of nature, since such an expression exists for any group of curves, if only they obey differential equations.[57]

Once again Frank fears opening the door for anti-scientific ideas, in this case for a genuine material teleology within the biological sciences, an idea which he is at pains to criticize throughout the book.

4. On Deepening the Foundations

Both Hahn and Frank thus separated Hilbert's axiomatic method into a general logical part – which they openly approved – and the concrete claims that a particular axiom system revealed a pre-established harmony – which they rejected as mysticism. The present section investigates the element in Hilbert's axiomatic method which defies such separation and accordingly was most suspicious to Logical Empiricists.

In all of Hilbert's axiomatizations and also in the Sixth Problem not all axioms stand on a par, as Hahn's conception of axiomatics suggested; some are rather specializations of a very general framework. This becomes clearer in "Axiomatic Thought" where Hilbert explicitly chooses a bottom-up approach. We start with a certain domain of facts, e.g., mechanics, statics, the theory of money, Galois theory, which are capable of being ordered into a certain framework [*Fachwerk*] of concepts and their logical relations which is not external to this field of knowledge, but "is nothing other than the *theory* of the field of knowledge."[58] Now Hilbert touches upon what Mach had deemed impossible, the need to ground or even "prove" the fundamental propositions (the basic empirical facts) themselves, the parallelogram of forces, the Lagrangian equations of motions, the laws of arithmetical calculations. To be sure, Hilbert does not relapse into seeking an absolute foundation in intuitively true basic axioms, for these reductions

are not in themselves proofs, but basically only make it possible to trace things back to certain deeper propositions, which in turn now to be regarded as new axioms ... The actual *axioms* of geometry, arithmetic, statics, mechanics, radiation theory, or thermodynamics arose in this way ... The procedure of the axiomatic method, as is expressed here, amounts to *deepening the foundations* of the individual domains of knowledge – a deepening that is necessary for every edifice that one wishes to expand and to build higher while preserving its stability.[59]

At the end of the paper, Hilbert is very optimistic about the prospects of his program.

Once it has become sufficiently mature for the formation of a theory, anything which can at all be the object of scientific thought succumbs to the axiomatic method and consequentially to mathematics. By penetrating into deeper levels of axioms ... we also gain deeper insight into the essence of scientific thought and become more and more conscious of the unity of our knowledge. Under the banner of the axiomatic method, mathematics appears to be destined to a leading role in all science.[60]

To be sure, Hahn and Frank could happily applaud to this unification as long as the "leading role" of mathematics was of an epistemological kind. The strict separation between mathematics and the sciences was observed when investigating the independence of the axioms, when proving their internal consistency relative to the consistency of arithmetic by defining appropriate number fields, and when checking that the propositions of a physical axiom system do "not contradict the propositions of a neighbouring field of knowledge"[61] – both empirically and theoretically. But what about "deepening the foundations"?

In elucidating this process which succeeds the analysis of the axioms' mutual dependence, Hilbert deliberately combines examples from pure mathematics and physics among which I can make out at least eight (not necessarily disjoint) types. The first four of them are unproblematic for the Logical Empiricists. But the second four spoil the strict distinction between tautologous mathematics and empirical facts; they can be seen as a more precise version of Hilbert's (non-Leibnizian) pre-established harmony. (i) There is a deepening by *elimination of dependent concepts*. If one axiom is a logical consequence of others, it can simply be dropped; for instance, the axiom of the existence of roots of Galois equations which follows from deeper axioms of arithmetic. (ii) Deepening is achieved by *conceptual simplification*. Hilbert lauds both Boltzmann and Hertz for having deepened the foundations of Lagrange's mechanics containing arbitrary forces and constraints (*Verbindungen*) to either arbitrary forces without constraints or arbitrary constraints without forces. But simplicity is ambiguous. Boltzmann himself opted against Hertz's approach because it was very difficult to find those supplementary conditions which made it at all applicable.[62] Conceptual simplification might also involve the introduction of new concepts, as is the case in the following categories. (iii) If axioms are independent one might also consider a whole class of alternative axiom systems and try to achieve – on this more general level – a *connection with a neighboring discipline*, even before this generalization could itself be grounded in a deeper level. One possible example is Cauchy's introduction of complex paths of integration, which Hilbert had discussed in the "Problems".[63] Deepening here follows from the internal unity of mathematics in a non-reductive way. (iv) Deepening the purely mathematical foundations might yield concepts that are *physically more fundamental*, such as non-Euclidean geometries or if the quantity of action would acquire a core physical meaning. Hilbert's most recent deepening in 1930 was this: "I had developed the theory of infinitely many variables from pure mathematical interest, and had even used the term spectral analysis, without any inkling that one day it would be realized in the actual spectrum of physics."[64] Both linear operators (to become the quantum mechanical observables) and atoms are characterized by their spectra. Although Hilbert here refers to the notorious preestablished harmony, the mathematical deepening has only a heuristic value for judging what is the physically deeper level.

In the remaining four classes this separation between mathematics and physics becomes blurred. (v) According to Majer, in his lectures on classical mechanics Hilbert argued that

[t]he whole of physics shall be set up on the basis of *one* fundamental principle – for example the principles of Hamilton-Jacobi or of Gauss or of Hertz – by combining two modes of division or classification of material systems. (1) The number of mass points respectively particles to be considered is exactly one, many or infinite. (2) Different types of relative motions of mass points are to be distinguished: rest, constant relative velocities (all the same or different), accelerations of different types, linear, circular, etc., and their combinations.[65]

Hilbert's set-up amounts to a deepening of the scientific foundations by a *mathematically guided reordering* of the classes of possible physical systems. (vi) Deepening of a physical theory might deliberately involve a *mathematically deeper* level. By "invoking the theorem that the continuum can be well-ordered, Hamel has shown ... that, in the foundation of statics, the axiom of continuity is necessary for the proof of the theorem of the *parallelogram of forces* – at least given the most obvious choice of the other axioms."[66] Hilbert advocates this type of deepening even at the price of obtaining a physically non-standard formulation, which confirms physicists' prejudice that mathematicians' deepening is rather academic at places.

The axioms of classical mechanics can be deepened if, using the axiom of continuity, one imagines continuous motions to be decomposed into small uniform rectilinear motions caused by discrete impulses and following one another in rapid succession. One then applies Bertrand's maximum principle as the essential axiom of mechanics, according to which the motion that actually occurs after each impulse is that which always maximizes the kinetic energy of the system with respect to all motions that are compatible with the law of the conservation of energy.[67]

(vii) Hilbert treats deepening as a kind of *scientific interpretation* of mathematics which gives rise to an empirical verification of basic geometrical concepts. In the *Foundations of Geometry* he had proven that the Archimedean axiom is independent of all the other axioms.

[This is] of capital interest for physics as well, for it leads to the following result: the fact that by adjoining terrestrial distances to one another we can achieve the dimensions and distances of bodies in outer space (that is, that we can measure heavenly distances with an earthly yardstick) ... The validity of the Archimedean axiom in nature stands in just as much need of confirmation by experiment as does, for instance, the familiar proposition about the sum of the angles of a triangle.[68]

One might consider this as a geometrical version of the cosmological principle connecting local earthbound physics with the laws of the entire Universe. Hilbert moves on to a stronger type of continuity.

In general, I should like to formulate the axiom of continuity in physics as follows: "If for the validity of a proposition [*Aussage*] of physics we prescribe an arbitrary degree of accuracy, then it is possible to indicate small regions within which the presuppositions that have been made for the proposition may vary freely, without the deviation of the proposition exceeding the prescribed degree of accuracy." This axiom basically does nothing more than express something that already lies in the essence of experiment; it is constantly presupposed by the physicists, although it has not previously been formulated.[69]

Unfortunately, Hilbert's axiom of measurement is at the same time too deep to be suitable for any specific physical theory and too restrictive because it *a priori* excludes chaotic systems. If we interpret the term proposition as the value ascription in a measurement "F has value f" the axiom is just fine and expresses the continuity presupposed when reading the pointer of a measurement device, but if we are interested in propositions like "F remains within the range [f,g] *for all times*" the axiom excludes chaotic systems for which there exists a time at which they leave any region even for the slightest variation of the "presuppositions". To be sure, experiments with chaotic systems have their peculiarities, but it appears to me unwise to exclude them *a priori*. As in the case of Hilbert's plea against infinities, we notice the risks of positing mathematically motivated axioms in physics. (viii) When applying the concept of invariance to relativity theory in his "Foundations of Physics", Hilbert had intended to *reduce physical concepts to mathematical ones.*[70] Although he had to take back parts of his claim, the action principle formulation of general relativity represents an important insight.

Summing up, while those types of "deepening the foundations" which respected the borderline between mathematics and empirical science were acceptable for Logical Empiricists, those which did not, sometimes contained important breakthroughs and sometimes *a priori* prejudices. Since Hahn and Frank wanted to make sure that such errors could never occur in a methodologically sober science, their containment strategy against metaphysics had to pay the price of simply ignoring many positive aspects of Hilbert's axiomatic method. In view of the subsequent rise of mathematical physics this was not an entirely fortunate strategy and so the problems centering around Hilbert's "deepening the foundations" are still with today's philosophy of science.

Notes

*   Acknowledgements: This paper emerged from a seminar at the Department for Logic and Philosophy of Science at the University of California at Irvine. I am most grateful to Gary Bell, David Malament, and Paul Weingartner for their questions and comments. I thank Ulrich Majer for providing me with information about Hilbert's lectures. Further thanks go to Volker Halbach and Tilman Sauer for a critical reading.

1.   To be sure, a main reason for this increased interest is that the large body of Hilbert's lectures on physical topics has ultimately become the object of intensive studies. For a general outline and philosophical interpretation, see Ulrich Majer: "Hilbert's Axiomatic Method and the Foundations of Science: Historical Roots of Mathematical Physics in Göttingen (1900-1930)", in: Miklós Rédei, Michael Stöltzner (Eds.), *John von Neumann and the Foundations of Quantum Physics*. Dordrecht: Kluwer 2001, pp. 11-33. More historical details – and until the first volumes of Hilbert's lectures will be published the broadest available overview – are found in Leo Corry: "David Hilbert and the Axiomatization of Physics (1894-1905)", *Archive for History of Exact Sciences* 51 (1997), pp. 83-198; "From Mie's Electromagnetic Theory of Matter to Hilbert's Unified Foundations of Physics', *Studies in History and Philosophy of Modern Physics* 30B (1999), pp. 159-183.

2.   Ulrich Majer: "Hilbert's program to axiomatize physics (in analogy to geometry) and its impact on Schlick, Carnap and other members of the Vienna Circle", in this volume.

3.   In his Curriculum Vitae written for the habilitation (*Personalakt* at the Archive of the University of Vienna), Hahn lists lectures of Hilbert and Minkowski and seminars of Hilbert, Klein, and Minkowski.

4.   In the Curriculum Vitae in the *Rigorosenakt* at the Archive of the University of Vienna, Frank lists Bousset, Hilbert, Husserl, Klein, Riecke, Simon, Wiechert and Zermelo as the Göttingen professors he studied with in 1906. And he presumably did not attend Hilbert's introductory lecture on calculus. For the early activities of Hahn and Frank, see also Thomas Uebel: *Vernunftkritik und Wissenschaft: Otto Neurath und der erste Wiener Kreis*. Wien-New York: Springer 2000, ch. 5.1 & 5.2. Uebel rightly diagnoses a Vienna-Göttingen axis concerning the rejection of pure intuition in mathematics.

5.   Hans Hahn, Ernst Zermelo: "Weiterentwicklung der Variationsrechnung in den letzten Jahren", in: *Enzyklopädie der mathematischen Wissenschaften*. Leipzig: Teubner 1904, vol. IIA, 8a, pp. 626-641.

6.   Cf. Michael Stöltzner: "Opportunistic Axiomatics. von Neumann on the Methodology of Mathematical Physics", in: Rédei and Stöltzner (Eds.), *op.cit.*, pp. 35-62.

7.   Hans Hahn, Otto Neurath, Rudolf Carnap: *Wissenschaftliche Weltauffassung: Der Wiener Kreis*. In: Otto Neurath: *Gesammelte philosophische und methodologische Schriften*. Wien: Hölder-Pichler-Tempsky 1981, p. 305. English translation: Dordrecht: Reidel 1973, p. 8.

8.   David Hilbert: "Naturerkennen und Logik", *Die Naturwissenschaften* 18 (1930), pp. 959-963. An English translation appeared in William Ewald: *From Kant to Hilbert: A Source Book in the Foundations of Mathematics*, vol. II. Oxford: Clarendon Press 1996, pp. 1157-1165. I shall always cite both the German original and the English translation (separated by a "/") which in cases might have been slightly modified.

9.   David Hilbert: "Mathematische Probleme", *Nachrichten von der Königlichen Gesellschaft der Wissenschaften zu Göttingen. Mathematisch-Physikalische Klasse aus dem Jahre 1900*, pp. 253-297. English translation in the *Bulletin of the American Mathematical Society* 8 (1902), pp. 437-479 (reprinted in the new series of the *Bulletin* 37 (2000), pp. 407-436). On the historical influence of Hilbert's Problems, see Jeremy J. Gray: *The Hilbert Challenge*. Oxford: Oxford University Press 2000.

10.   Hilbert, "Naturerkennen und Logik", *op.cit.*, p. 959/1157.

11.   *Ibid.*, p. 959/1158.

12.   *Ibid.*, p. 960/1159.

13.   Max Born: "Hilbert und die Physik", *Die Naturwissenschaften* 10 (1922), p. 90f.

14. Majer, "Hilbert's Axiomatic Method", *op.cit.*, p. 17.
15. David Hilbert: "Mechanik der Kontinua", Transcript of a lecture held in the winter term 1906/7, translation quoted from Majer, "Hilbert's Axiomatic Method", *op.cit.*, p. 26-27.
16. This is not to say that Hilbert had actually changed his mind concerning methodology. As a matter of fact, there is still much historical research ahead of us to obtain clarity about the various phases in Hilbert's work in physics. But, to my mind, this will hardly make Hilbert's axiomatization program come significantly closer to the outlook of Logical Empiricists.
17. Hilbert, "Naturerkennen und Logik", *op.cit.*, p. 960/1159. Notice that the German "Wirkung" denotes both the effect (of a force) and the physical quantity of action.
18. *Ibid.*, p. 960/1160.
19. *Ibid.*, p. 960/1160.
20. Immanuel Kant: *The Critique of Judgement*. Translated by James Creed Meredith. Oxford: Clarendon Press 1991, p. 19 (=A XXV; B XXVII).
21. Hilbert, 1900, p. 257/440.
22. Hilbert, p. 960/1160.
23. David Hilbert: "Die Grundlagen der Physik (Erste Mitteilung)", *Nachrichten von der Königlichen Gesellschaft der Wissenschaften zu Göttingen. Mathematisch-Physikalische Klasse aus dem Jahre 1915*, p. 407. For a detailed analysis of this paper and its context, see Tilman Sauer: "Hopes and Disappointments in Hilbert's Axiomatic 'Foundations of Physics'", in this volume, and his earlier "The Relativity of Discovery: Hilbert's First Note on the Foundations of Physics", *Archive for the History of Exact Sciences* 53 (1999), pp. 529-575.
24. Johann von Neumann: *Mathematische Grundlagen der Quantenmechanik*, Berlin: Julius Springer 1932.
25. David Hilbert: "Axiomatisches Denken", *Mathematische Annalen* 78 (1918), pp. 405-415; English translation in Ewald, *op.cit.*, pp. 1105-1115. Once again, further historical investigations are wanted. Although my identification of "deepening the foundation" and "elimination" is mainly of philosophical kind, it derives further historical support from the fact that most examples occurring in both texts are identical.
26. Hilbert, "Naturerkennen und Logik", *op.cit.*, p. 961/1160-1161.
27. *Ibid.*, p. 961/1161.
28. *Ibid.*, p. 961/1161-1162.
29. *Ibid.*, p. 962/1163. Hilbert here cites his papers "Über das Unendliche" and "Grundlagen der Mathematik".
30. *Ibid.*, p. 962/1163.
31. *Ibid.*, p. 962f./1164.
32. In the manifesto *Wissenschaftliche Weltauffassung* he is mentioned under "Axiomatics: Pasch, Peano, Vailati, Pieri, Hilbert" (*op.cit.*, p. 303/6), and the *Foundations of Geometry* are cited in the references (*Ibid.*, p. 316).
33. *Ibid.*, p. 963/1165.
34. Hans Hahn: "Logik, Mathematik und Naturerkennen", in: *Einheitswissenschaft*. Frankfurt am Main: Suhrkamp 1992, p. 89; English translation in *Unified Science*, ed. by Brian McGuinness, Dordrecht: Kluwer 1987, p. 45.
35. *Ibid.*, p. 64-66/27-28.
36. *Ibid.*, p. 66/28.
37. *Ibid.*, p. 71f./32.
38. *Ibid.*, p. 74/34.
39. *Ibid.*, p. 75/35.
40. Hans Hahn: "Die Bedeutung der wissenschaftlichen Weltauffassung, insbesondere für Mathematik und Physik", in: *Empirismus, Logik, Mathematik, op.cit.* p. 43 (originally in *Erkenntnis* 1 (1930/1), pp. 96-105); English translation in *Empiricism, Logic, and Mathematics, op.cit.*, p. 26. See also Karl Sigmund: "Hans Hahn and the Foundational Debate", in Werner DePauli-Schimanovich, Eckehart Köhler, Friedrich Stadler (Eds.): *The Foundational Debate*. Dordrecht: Kluwer 1995, pp. 235-245.
41. Hahn: "Die Bedeutung", *op.cit.*, p. 41/24.
42. *Ibid.*, p. 44-45/27-28.

43. Hans Hahn: "Die Krise der Anschauung", in: *Empirismus, Logik, Mathematik, op.cit.*, pp. 86-114; English translation in: *Empiricism, op.cit.*, pp. 73-102.
44. Ernst Mach: *Die Mechanik in ihrer Entwickelung. Historisch-kritisch dargestellt.* Leipzig 1883; Section I.1.
45. Hahn, "Logik, Mathematik und Naturerkennen", *op.cit.*, p. 81/39.
46. *Ibid.*, p. 83-84/40-41.
47. *Ibid.*, p. 79/38.
48. Philipp Frank: *Das Ende der mechanistischen Physik*, reprinted in *Einheitswissenschaft.* Frankfurt am Main: Suhrkamp, pp. 166-199; English translation: "The Fall of Mechanistic Physics", in: *Unified Science, op.cit.*, pp. 110-129.
49. *Ibid.*, p. 171/112.
50. *Ibid.*, p. 172/113.
51. *Ibid.*, p. 198/128f.
52. Wilhelm Frank: "Comments to Hans Hahn's contributions to the calculus of variations", in: Hans Hahn: *Collected Works*, ed. by Leopold Schmetterer and Karl Sigmund. Wien-New York: Springer 1996, p. 1.
53. See his *Nationale* (a list of lectures a student enrolled and paid for) at the Archive of the University of Vienna.
54. The quote is from Carl Gustav Jakob Jacobi: *Vorlesungen über Dynamik*, ed. by A. Clebsch. Berlin: Georg Reimer 1866, p. 45.
55. Philipp Frank: *Über die Kriterien für die Stabilität der Bewegung eines materiellen Punktes in der Ebene und ihren Zusammenhang mit dem Prinzip der kleinsten Wirkung*, handwritten Ph.D.-dissertation, University of Vienna 1906.
56. Philipp Frank: *Das Kausalgesetz und seine Grenzen.* Wien: Springer 1932; English translation by M. Neurath and R.S. Cohen *The Law of Causality and Its Limits.* Dordrecht: Kluwer 1998, p. 82/90.
57. *Ibid.*, p. 84f./91f.
58. Hilbert, "Axiomatic Thought", *op.cit.*, p. 405/1108. The German term "Fachwerk" elucidates that the concepts are contained in the factual body of the field of knowledge and are no external scaffolding.
59. *Ibid.*, p. 407/1109.
60. *Ibid.*, p. 415/1115.
61. *Ibid.*, p. 410/1111.
62. Cf. Ludwig Boltzmann: "Über die Grundprinzipien und Grundgleichungen der Mechanik", in: *Populäre Schriften.* Leipzig: J.A. Barth 1905, pp. 253-307, in particular p. 269 (= p. 113 in the English translation in *Theoretical Physics and Philosophical Problems*, ed. by B. McGuinness, Dordrecht: D.Reidel 1974).
63. See Hilbert, "Mathematische Probleme", *op.cit.*, p. 260/444.
64. Hilbert, "Naturerkennen und Logik", *op.cit.*, p. 960f./1160.
65. Majer, "Hilbert's Axiomatic Method", *op.cit.*, p. 24.
66. Hilbert, "Axiomatic Thought", *op.cit.*, p. 409/1110.
67. *Ibid.*, p. 409/1111.
68. *Ibid.*, p. 408f./1110.
69. *Ibid.*, p. 409/1110.
70. See the discussion in Section 1.

Institut Wiener Kreis
Museumstraße 5/2/19
A-1070 Vienna
Austria
michael.stoeltzner@sbg.ac.at

MICHAEL DICKSON

THE EPR EXPERIMENT:
A PRELUDE TO BOHR'S REPLY TO EPR[*]

1. EINSTEIN, PODOLSKY, AND ROSEN'S ARGUMENT

Bohr's (1935) reply to Einstein, Podolsky, and Rosen's (EPR's) (1935) argument for the incompleteness of quantum theory is notoriously difficult to unravel. It is so diffcult, in fact, that over 60 years later, there remains important work to be done understanding it. Work by Fine (1986), Beller and Fine (1994), and Beller (1999) goes a long way towards correcting earlier misunderstandings of Bohr's reply. This essay is intended as a contribution to the program of getting to the truth of the matter, both historically and philosophically. In a paper of this length, a full account of Bohr's reply is impossible, and so I shall focus on one issue where it seems further clarification is required, namely, Bohr's attempt to illustrate EPR's argument by means of a thought experiment. In addition, I shall attempt to clarify a few other points which, however minor, have apparently contributed to misunderstandings of Bohr's position. As the title of this paper suggests, an account of these few points does not constitute an account of Bohr's reply, but it is an important step in that direction.

I shall begin by raising several points about EPR's argument, and especially their example of particles correlated in position and momentum. Some of these points have not been sufficiently noticed in the literature.

Let us begin with a standard, but incorrect, story about EPR's argument. Two particles are emitted from a common source, with momenta p and $-p$, respectively. For simplicity, we assume that their masses are the same. Some time later, particle 1 encounters a measuring device, which can measure either its position, or its momentum. If we measure its momentum to be p, then we can immediately infer that the momentum of particle 2 is $-p$. If we measure its position to be x, then (letting the source be at the origin) we can immediately infer the position of particle 2 to be $-x$. Now, if we assume that the measurement on particle 1 in no way influences the state of particle 2, then particle 2 must have had those properties all along, because it could not obtain those properties merely as a result of the measurement on particle 1. But quantum theory cannot represent particle 2 as having a definite position and momentum, and therefore quantum theory is incomplete.

M. Heidelberger and F. Stadler (eds.), History of Philosophy and Science, 263–275.
© 2002 *Kluwer Academic Publishers. Printed in the Netherlands.*

This argument is not the argument that EPR make. If it were, Bohr's reply could have been fairly short. (He might not have *made* it short, but it could have been.) The short reply is to note that in order to make the requisite predictions, one must know the precise position and momentum of the source. Consider, for example, that you have just measured the momentum of particle 1 to be p. If you do not know the momentum of the source, then in particular you do not know in which frame of reference to apply conservation of momentum. (Above we assumed that the source is at rest relative to us, and so we inferred that particle 2 has momentum $-p$.) Similarly, consider that you have just measured the position of particle 2 to be x. If you do not know the location of the source, then you cannot say where particle 2 is. It is 'the same distance from the source' as particle 1, in the other direction, but how far is particle 1 from the source? Unless you know where the source is, you cannot answer this question.

But if you must know the precise position and momentum of the source in order to make the inferences, then the uncertainty principle will always get in the way of EPR's argument. Suppose, for example, that you know the precise momentum of the source. Then you measure the position of particle 1. The EPR criterion for physical reality says: "If, without in any way disturbing a system, we can predict with certainty ... the value of a physical quantity, then there exists an element of physical reality corresponding to this physical quantity" (Einstein et al., 1935, p. 777). But we *cannot* predict particle 2's position with certainty, because we do not (and under the circumstances, cannot) know where the source is. Of course the same argument applies exchanging position and momentum, *mutatis mutandis*.

Good thing, then, that the 'standard story' about EPR's argument is wrong. We can see immediately that something is wrong with it, because nowhere did that story mention quantum theory, and yet EPR are very concerned to present their argument in quantum-theoretic terms (as they should be). Indeed, the first part of their paper rehearses a number of facts about the formalism of quantum theory, presumably so that they can present their argument in a quantum-theoretic context (which is what they do).

EPR continue by considering a generic system of two particles and a pair of generic (but non-commuting) observables on particle 1, A and B. EPR do not then write down a generic version of the so-called 'EPR state'. Instead, they merely point out that as a result of measuring A on particle 1, particle 2 may be left in one state – call it $\psi_k(x_2)$, as they do – while as a result of measuring B on particle 1, particle 2 may be left in quite another state – call it $\varphi_r(x_2)$, as they do.

At this stage of the argument, EPR might have pointed out that ψ_k and φ_r are eigenfunctions of *some* observables. Hence we would be able to predict, with certainty, the values of two different observables as a result of two different measurements (of A or B) on the first system. One would then have to go on to show that those observables need not commute.

Instead of continuing with this generic case, however, EPR turn to a specific example, using the position and momentum observables. Here they do add the idea that ψ_k and φ_r can be eigenfunctions of position and momentum, respectively. To establish this claim, they suppose that the total system prior to any measurements is in the state

$$\Psi_{\text{EPR}}(x_1, x_2) = \int_{-\infty}^{\infty} e^{(2\pi i/h)(x_1 - x_2 + x_0)p}\,dp, \qquad (1)$$

where x_0 is some constant. EPR then show that (1) is indeed a state of perfect (anti-)correlation between the positions and momenta of the two particles: measuring the momentum of particle 1 (hence collapsing the wavefunction for the compound system!) leaves particle 2 in the relevant eigenstate of momentum, and likewise for position.

So why is the 'standard story' inconsistent with this account? We have already noted that the 'standard story' is not quantum-mechanical, but the more important point for us here is that EPR nowhere describe how the compound system is prepared, nor how it evolves in time. Indeed, the notion of time never enters their discussion. The state Ψ_{EPR} – let us call it the 'EPR state' – is a 'snapshot' of the compound system at a time. Moreover, EPR *could not* give us a dynamical description of the situation, because the EPR state cannot be preserved under Hamiltonian evolution (unless we introduce an infinite potential, a point that I will no longer bother to mention, and presumably did not concern EPR).

The reason is familiar, though not usually mentioned in this context. The (support of the) EPR state has measure 0 in configuration space: $\Psi_{\text{EPR}}(x_1, x_2)$ is zero except when $x_2 - x_1 = x_0$, and so it is a line in the (two-dimensional) configuration space for the compound system. Such a state necessarily spreads under the evolution induced by any Hamiltonian. (We are, of course, ignoring the fact that the EPR state is not in $L^2(R^2)$ in the first place. Neither EPR nor Bohr seem to have been concerned about this point.)

Finally, note that there is no Hamiltonian evolution that can take a generic state $\Phi(x_1, x_2)$ to the EPR state (no matter what Φ is). Only a 'collapse' of the wavefunction can produce the EPR state. Hence we must imagine the EPR state to exist at *and only at* the moment of preparation.

EPR's argument, then, is based on such a state. They point out that upon measuring the position of particle 1, we can predict with certainty the position of particle 2, and likewise for momentum. Of course, only one of the two measurements can be performed, which raises the question whether some modal fallacy has been committed. After all, their argument apparently takes the form:

1. Actually: position is measured for particle 1, and therefore (actually) particle 2 has a definite position.
2. Possibly: momentum is measured for particle 2, and therefore (possibly) particle 2 has a definite momentum.

3. Therefore: particle 2 (possibly? actually?) has a definite position and a definite momentum.

In this form, the argument is clearly fallacious (no matter which modal version of the conclusion you choose). Of course, the notion of 'non-disturbance' is supposed to help patch up the argument: although the circumstances under which we can predict the value of particle 2's position are incompatible with the circumstances under which we can predict the value of particle 2's momentum, the difference between these circumstances is supposed to make no difference to particle 2, due to the space-like separation of the measurement-events for the particles.

Even with the help of some principle of non-disturbance, it is not clear, however, that EPR's argument works. Let us consider, first, a 'weak principle of non-disturbance':

Weak non-disturbance: If momentum is measured on particle 1 and (therefore, by the criterion for physical reality) momentum is definite for particle 2, then: had we not measured momentum on particle 1, particle 2 would still have had a definite momentum (and likewise, substituting position for momentum).

This principle is, alas, not enough to get EPR's conclusion. They need:

Strong non-disturbance: If momentum is measured on particle 1 and (therefore, by the criterion for physical reality) momentum is definite for particle 2, then: had we not measured momentum on particle 1 but instead measured its position, then particle 2 would still have had a definite momentum (and likewise, switching position and momentum).

The weak principle does not entail the strong principle because it might be impossible (without destroying essential features of the situation, in particular, our ability to infer properties of particle 2 from the results of measurements on particle 1) both to measure position on particle 1 and for momentum to be definite for particle 2. (In terms of the 'possible-worlds' semantics for counterfactuals: while the closest 'momentum is not measured'-worlds to the 'momentum is measured and is definite for particle 2'-worlds might all be 'momentum is definite for particle 2'-worlds, those closest worlds may not contain any 'position is measured'-worlds, so that the closest 'momentum is not measured but position is'-worlds to the 'momentum is measured and is definite for particle 2'-worlds need not be 'momentum is definite for particle 2'-worlds. Now say that sentence three times fast.)

Bohr is sometimes understood to deny the strong principle by asserting that the act of measuring position on particle 1 'disturbs' in some strange 'semantic' (and non-local) way the very possibility of particle 2's having a definite momentum. Such a response is (rightly) taken to be uninteresting philosophically. In a longer account of Bohr's reply, I would argue that while Bohr does deny the strong principle, he does so for more interesting reasons. Here, however, I shall only make a few suggestions in that direction. The next section contains several minor observations about EPR's argument and Bohr's reply. These remarks are intended to clear the air of some minor criticisms of Bohr's reply. In the subsequent section, I shall discuss Bohr's thought experiment and make some brief suggestions about how to understand Bohr's reply.

2. SOME CLARIFICATIONS

1. *EPR speak in terms of a 'contradiction'.* Without calling into question Fine's (1986) logical analysis of EPR's argument, we may note that they do speak of a 'contradiction' between their criterion of reality and the completeness of standard quantum theory. At the end of the first section of their paper, Einstein et al. (1935) state their conclusion thus: "We shall show, however, that this assumption [completeness], together with the criterion of reality given above, leads to a contradiction".

As Beller and Fine (1994) argue, Bohr had no problems with EPR's criterion for physical reality, nor with their account of completeness, together understood in a fairly conservative sense (perhaps, in modern terms, as no more than the eigenstate-eigenvalue link). Hence the idea that there might be a 'contradiction' between the criterion and completeness would surely have worried Bohr, and would understandably be the focus of his reply. No wonder Bohr's rhetoric focused on 'soundness', 'rationality', 'lack of contradiction' and 'consistency' (cf. (Beller and Fine, 1994, pp. 3-4).) While we may endorse much of what Beller and Fine (1994) assert to be at the heart of Bohr's general concerns about consistency, the simple explanation seems to be just that EPR do, at least at one point, state their conclusion in terms of a contradiction, a contradiction that was (for reasons that Beller and Fine explore) threatening to Bohr's own position.

2. *The EPR argument focuses on the example.* I mentioned above that EPR begin their discussion in the abstract and could have finished it there, but they do not, instead resorting to the example involving position and momentum. Bohr, too, focuses on the example. Indeed, he takes the example to constitute the argument, writing that "[b]y means of an interesting example, to which we shall return below, they [EPR] next proceed to show that ... [the] formalism [of quantum mechanics] is incomplete" (Bohr, 1935, p. 696). Nobody involved in the debate seems to have thought that this focus on the example is unwarranted or misleading. The point is important for two reasons.

First, it lends greater importance to a proper understanding of Bohr's attempt to realize the example in a thought experiment. From a contemporary perspective, one might be tempted to suppose that the real substance of the EPR argument, and of Bohr's reply, is (and was taken by them to be) in the more abstract discussions (for example, in the early part of EPR's paper and the mathematical footnote in Bohr's reply). While these more abstract discussions can provide important clues to understanding EPR and Bohr's reply, their mutual focus on the example of position and momentum suggests that we too focus on that example in order to understand what is going on.

Second, the focus on the example is, in the end, unwarranted and misleading. Indeed, from a contemporary standpoint, we can see that EPR chose a particularly unfortunate example to make their point. As I shall emphasize again below, the main problem is that position (unlike momentum) is not a conserved quantity, so that correlations in position will in general not be maintained under free (or for that matter, almost any other) evolution. Bohm's (1951) reworking of EPR's argument in terms of a new example (involving incompatible spin observables) fixes the problem (because spin is conserved), and it is unclear whether Bohr's reply could work in this case. (In any case, his thought experiment is mostly irrelevant to the Bohmian example.)

3. *The observables $X_1 - X_2$ and $P_1 + P_2$ can be determined simultaneously.* EPR presume that the total momentum ($P_1 + P_2$) and the distance between the particles ($X_1 - X_2$) can be known simultaneously. There is no obstacle in principle to obtaining such knowledge, since the observables in question are compatible (mutually commuting). Indeed, the EPR state is a simultaneous eigenstate of both of these observables. (Again, we ignore the fact that plane waves and delta functions are not, strictly speaking, states, i.e., not in $L^2(R^2)$.)

But how might one actually prepare the EPR state, or more generally, how might one actually determine $X_1 - X_2$ and $P_1 + P_2$ simultaneously? That is, from a physical point of view, why do these operators commute? Note first – and this point is crucial to an understanding of Bohr's reply – that Bohr insisted that neither position nor momentum observables have any clear physical meaning outside of the specification of some frame of reference. Bohr is acutely aware of the role that reference frames play in relativity theory, and believes that their role in the quantum theory is even more significant – well-specified frames of reference are crucial to the very meaning of 'spatial location' and 'momentum'. Bohr's view seems to have been that only prior to the discovery of the quantum theory, and specifically the 'essential exchange of momentum' involved in any interaction, could one dispense with the insistence that reference frames are essentially involved in the very notion of 'position' and 'momentum'. While a full analysis of Bohr's position on this point (and most especially of his understanding of the term 'essential exchange of momentum') is out of the question here, it is worth noting that Bohr insisted upon the necessary role that well-defined reference frames play in the very definition of the notion of position. He writes:

Wie von EINSTEIN betont, ist es ja eine für die ganze Relativitätstheorie grundlegende Annahme, daß jede Beobachtung schließlich auf ein Zusammentreffen von Gegenstand und Meßkörper in demselben Raum-Zeitpunkt beruht und insofern von dem Bezugssystem des Beobachters unabhängig definierbar ist. Nach der Entdeckung des Wirkungsquantums wissen wir aber, daß das klassische Ideal bei der Beschreibung atomarer Vorgänge nicht erreicht werden kann. Insbesondere führt jeder Versuch einer raum-zeitlichen

Einordnung der Individuen einen Bruch der Ursachenkette mit sich, indem er mit einem nicht zu vernachlässigenden Austausch von Impuls und Energie mit den zum Vergleich benutzten Maßstäben und Uhren verbunden ist, dem keine Rechnung getragen werden kann, wenn diese Meßmittel ihren Zweck erfüllen sollen. (Bohr, 1929, p. 485) [1]

Continuing this line of thought, in his reply to EPR (1935, p. 699), Bohr writes:

To measure the position of one of the particles can mean nothing else than to establish a correlation between its behavior and some instrument rigidly fixed to the support which defines the space frame of reference.

Bohr is careful to discuss position (and momentum) in these terms, not speaking of 'the position [or momentum]' of a system, but its position relative to some other system. For example, at p. 697 of his reply he speaks not of the uncertainty of the position of a particle, but of 'the uncertainty Δq of the position of the particle relative to the diaphragm'. The fact that not only position, but also uncertainty in position, must be discussed relative to a physically defined reference frame indicates the extent to which, for Bohr, such reference frames are involved in the very meaning of 'position'.

These points are important, because failing to appreciate them fully one can be too easily persuaded that passages such as the one above indicate Bohr's adherence to a rather strong form of operationalism. He might, in other words, be suggesting that physical properties are defined by the operations used to 'measure' them. But given the history of Bohr's insistence on the role of (physically specified) reference frames in quantum theory, we can just as well (and indeed, I would argue, more fruitfully) read the passage above and others like it as insisting that a well-defined frame of reference is crucially a part of the notion of position.

4. *The observables $X_1 - X_2$, $P_1 + P_2$, X_1, and P_1 are not mutually commuting.* It is easy to suppose that without losing our knowledge of $X_1 - X_2$ and $P_1 + P_2$, we may go on to determine either X_1 or P_1. (This mistake is all the easier if one conceives of the EPR experiment in terms of the 'standard story' that I outlined above.) The following passage, for example, seems to make this suggestion:

EPR consider a composite system in a state where, at least for a moment, both the relative position $X_1 - X_2$ and the total momentum $P_1 + P_2$ are co-measurable. Moreover, in EPR both of these quantities are simultaneously determinable with either the position or the momentum (not both) of particle 1. (Beller and Fine, 1994, p. 15)

However, X_1 fails to commute with $P_1 + P_2$ and P_1 fails to commute with $X_1 - X_2$. If the EPR situation allowed us to co-determine both $X_1 - X_2$ and $P_1 + P_2$ with either X_1 or P_1, then a great deal more than Bohr's reply would be in jeopardy. If we are to determine X_1, then we must give up our knowledge of $P_1 + P_2$, and if we are to determine P_1, we must give up our knowledge of $X_1 - X_2$.

As Beller (1999, ch. 6) explains, the early Bohr was very concerned to explain *why* it is not possible to observe simultaneous values for incompatible observables. I will suggest, below, that Bohr's reply to EPR continues this discussion, i.e., that he is, in part, attempting to explain why one cannot measure $X_1 - X_2$, $P_1 + P_2$, X_1, and either of X_1 or P_1 simultaneously, within the context of EPR's example. (Here, then, is one sense in which Bohr's reply involves themes and argumentative strategies that he had already used in other cases.)

3. BOHR'S THOUGHT EXPERIMENT

We are now in a position to assess the relevance of Bohr's proposed thought experiment to EPR's argument. Bohr's discussion begins with a rehearsal of two different sorts of experiment. In the first, there is a screen with a single slit, "rigidly fixed to a support which defines the space frame of reference" (1935, p. 697), and a particle is fired at the screen. We assume that the particle's initial momentum is well-defined. Bohr asks whether, after preparing the particle in a state of well-defined position by passing it through the slit (and thereby, according to de Broglie's relation, rendering it's momentum uncertain), we cannot take into account the exchange of momentum between the particle and the apparatus, thereby 'repairing' the loss of initial certainty about the momentum. His answer is 'no', because the exchange of momentum "pass[es] into this common support" which, because it *defines* the space frame of reference, *must* be taken to be at rest, and so "we have thus voluntarily [by fixing the initial screen to the support and taking that support to define the spatial reference frame] cut ourselves off from any possibility of taking these reactions separately into account" (ibid.).

If, on the other hand, we allow the initial screen to move freely relative to the support, then we can indeed measure the exchange of momentum between the particle and the screen, but in so doing, we necessarily lose whatever information we might previously have had about the location of the initial screen relative to the support, and therefore passing the particle through the slit is no longer a preparation of its position relative to the support:

In fact, even if we knew the position of the diaphragm relative to the space frame [i.e., the 'support'] before the first measurement of its momentum, and even though its position after the last measurement [required to determine the exchange of momentum] can be accurately fixed, we lose, on account of the uncontrollable displacement of the diaphragm during each collision process with the test bodies, the knowledge of its position when the particle passed through the slit. (1935, p. 698)

Note that two measurements of the momentum of the screen are *required* (in addition to a prior measurement of the momentum of the incident particle) in order to apply conservation of momentum to the total system, by which we can determine the momentum of the incident particle after it has passed through the slit. Bohr claims that the second measurement of the momentum of the screen disturbs its position relative to the support in an 'uncontrollable' way, thereby preventing us from determining its position (relative to the support) at the moment that the particle passed through the slit.

My aim in making these observations is not to analyze Bohr's claims in detail. Such an analysis would include a deeper discussion of Bohr's notion of a 'reference frame', and his notion of 'uncontrollable disturbance', both of which are crucial to a complete understanding of Bohr's reply. The aim here, rather, is only to remind the reader of the broad outlines of Bohr's understanding of the uncertainty principle, and roughly how he defends that understanding by means of simple thought experiments. The main point here is that Bohr believes that the 'uncontrollable exchange' of momentum and energy between a measured system and a measuring apparatus entails that experimental situations that allow the determination of a particle's position relative to a given reference frame forbid the determination of its (simultaneous) momentum relative to that frame, and similarly, those experimental situations that allow the determination of a particle's momentum relative to a given frame – by means of an application of conservation laws – forbid the determination of its (simultaneous) position relative to that frame.

Let us turn, then, to Bohr's realization of EPR's particular case. He proposes a thought experiment to prepare the EPR state, and to perform the relevant measurements, as follows:

> The particular quantum-mechanical state of two free particles, for which they [EPR] give an explicit mathematical expression, may be reproduced, at least in principle, by a simple experimental arrangement, comprising a rigid diaphragm with two parallel slits, which are very narrow compared with their separation, and through each of which one particle with given initial momentum passes independently of the other. (Bohr, 1935, p. 699)

The arrangement as described thus far allows one to prepare the pair of particles in an eigenstate of $X_1 - X_2$, the eigenvalue being, of course, the distance between the slits (x_0 in EPR's notation). In order to determine $P_1 + P_2$, Bohr proposes the following (a continuation of the quotation above):

> If the momentum of this diaphragm is measured accurately before as well as after the passing of the particles, we shall in fact know the sum of the components perpendicular to the slits of the momenta of the two escaping particles, as well as the difference of their initial positional coordinates in the same direction. (ibid.)

Thus, at this point in the description of the thought experiment, we have determined (or prepared) the values of $X_1 - X_2$ and $P_1 + P_2$ simultaneously.

The crucial question, now, is how one may go on to measure either X_1 or P_1, in order to determine either X_2 or P_2. Concerning the measurement of X_1, Bohr begins

> [T]o measure the position of one of the particles can mean nothing else than to establish a correlation between its behavior and some instrument rigidly fixed to the support which defines the space frame of reference. Under the experimental conditions described such a measurement will therefore also provide us with the knowledge of the location, otherwise

completely unknown, of the diaphragm with respect to this space frame when the particles passed through the slits. Indeed, only in this way we obtain a basis for conclusions about the initial position of the other particle relative to the rest of the apparatus. (Bohr, 1935, p. 700)

Bohr has not yet arrived at his main point, but is here pointing out that, because the initial screen must be allowed to move freely with respect to the support (so that conservation of momentum can be applied to it plus the pair of particles), we do not know where it is relative to the support until we measure the position of one of the particles (relative to the support). After such a measurement, we can learn the position of the screen, because the particles are located where the slits in the screen are located. And once we know where the screen itself is in relation to the support, we can use our knowledge of X_1 to infer the location of the other particle, as Bohr says. Note, in particular, that Bohr nowhere supposes that the measurement of the position of the particle disturbs the screen.

Bohr continues:

By allowing an essentially uncontrollable momentum to pass from the first particle into the mentioned support, however, we have by this procedure cut ourselves off from any future possibility of applying the law of conservation of momentum to the system consisting of the diaphragm and the two particles. (ibid.)

The consequence, as Bohr notes, is that in fact we *lose* the ability to predict the momentum of the second particle, *even if* we were (counterfactually, of course) to measure the momentum of the first particle. In the terms of the first section of this essay, Bohr has rejected 'strong non-disturbance', more or less for the reason suggested there: a measurement of X_1 necessarily destroys an essential feature of the compound system prior to measurement, that feature being the truth of the conditional: if we were to measure P_1, then we could predict (with certainty) P_2. (A more complete analysis of Bohr's position would require a longer discussion of the logic of counterfactuals, which we cannot pursue here.)

From this point of view, Beller and Fine's (1994) complaints against Bohr's thought experiment are not quite right. They make two complaints. First, they are unhappy with the fact that, in Bohr's arrangement, "we have no choice but to measure X_1 at the very moment of passage of the two particles through the first diaphragm" (Beller and Fine, 1994, p. 14). As I have already pointed out, however, there is really no choice. No quantum-mechanical state can evolve into the EPR state, and the EPR state cannot be preserved by any time evolution. Hence it can be the state of a system at, and only at, the moment of preparation. We can hardly fault Bohr for this situation.

Their second complaint arises from the first. They quite rightly point out that Bohr does not describe in any detail how the measurement of X_1 is to occur. Indeed, straightforward physical considerations of the situation seem to imply that any such measurement would involve a disturbance of the diaphragm with the two slits – either indirectly (for how could one interact with the particle without 'touching' the diaphragm?) or directly, by simply fixing the diaphragm to the support. Beller and Fine appear to opt for the latter. After apparently claiming (as I noted above) that EPR's case allows for the simultaneous determination of $X_1 - X_2$, $P_1 + P_2$ *and* either X_1 or P_1, they write:

Bohr's double slit arrangement does not satisfy this requirement. In Bohr's example only one of $X_1 - X_2$ or $P_1 + P_2$ could be co-determined together with the variable [X_1 or P_1] one chooses to measure on particle 1. Indeed, we actually have to change the set-up of the two-slit diaphragm depending on whether we intend to measure position or momentum on particle 1. In the first case the two-slit diaphragm must be immovable; in the second case it must be moveable. (1994, p. 15)

Mainly because of this situation, Beller and Fine refer to Bohr's realization of EPR's argument a "flawed assimilation of EPR to a double slit experiment" (ibid., p. 16).

I suggest an alternative account. According to this account, Bohr completely ignores the fact – even if it follows from simple physical considerations – that a measurement of X_1 implies either a disturbance of the diaphragm or that it is fixed to the support. Instead, he is concerned to point out that a measurement of X_1 involves – in *precisely* the same way that it does in the simpler cases discussed prior to EPR – an uncontrollable exchange of momentum between particle 1 and the support that defines the space frame of reference. Hence the momentum of particle 1 becomes undefined, and hence the total momentum (of the pair of particles) becomes undefined. Or to put the point in more Bohrian terms: conservation of momentum cannot be applied to the compound system, and therefore $P_1 + P_2$ is undefined, because in order for it to be defined, we must be able to apply conservation of momentum to the diaphragm plus the two particles.

At the very least, this account has the merit of following quite closely Bohr's account of the disturbance. He does not say that, in the measurement of X_1, momentum is exchanged between particle 1 and the diaphragm; nor does he ever suggest, in the EPR arrangement, that the diaphragm is fixed to the support. Rather, he says that "momentum [passes] from the first particle into the mentioned support" (Bohr, 1935, p. 700).

Similarly, in his account of what goes wrong when we measure P_1, he claims that such a measurement removes the possibility of determining the location of the diaphragm relative to the support. He could have two arguments in mind. First, along lines suggested by Beller and Fine, one might argue that any measurement of P_1 must involve a disturbance of (exchange of momentum with) the diaphragm, thereby disturbing its position relative to the support, because the measurement of P_1 must occur at the moment of preparation. Second, along the lines that are suggested here, one might argue that since the arrangement *requires* the diaphragm to move freely with respect to the support (lest we be unable to determine $P_1 + P_2$), the only way to determine the location of the diaphragm relative to the support would be to measure the position of one of the particles, relative to the support. But for reasons that were discussed prior to the case of EPR, measuring P_1 'cuts one off' from the possibility of determining particle 1's (and therefore the diaphragm's) position relative to the support.

4. BOHM'S VERSION OF THE ARGUMENT

I finish with a brief comment regarding Bohm's (1951) alternate realization of the EPR state. The main point is that Bohm's realization does not involve position and momentum, but incompatible spin observables. There are two essential differences between this case and Bohr's (and EPR's). First, spin observables, while in a sense dependent on the specification of a spatial frame of reference (because we need to know which direction is, for example, the 'z' direction), are not bound up as closely with the very notion of a frame of reference. In particular, the sort of exchange that must occur between particle and apparatus in a measurement of spin does not seem to involve a disturbance of the very reference frame used to define the notion of 'direction of spin'. Second, spin is a conserved quantity (unlike position), so that the measurement of spin on one particle can be made long after the preparation of the particles.

It remains to be seen whether a Bohrian response of the EPR argument can be worked out in the case of spin. My suspicion is that the Bohrian response would at the least require significant revision. As far as I am aware, Bohr never reacted, publicly or privately, to Bohm's proposed thought experiment. (And, of course, it is more or less Bohm's version that was eventually performed.) However, the investigation of these questions must be preceded by a more complete account of Bohr's reply to EPR, to which the remarks here are at best a partial preface.

NOTES

*   Thanks to audiences at Indiana University and HOPOS 2000 for comments on related talks. Thanks to Arthur Fine for alerting me to some secondary literature. Thanks to Michael Friedman and Scott Tanona for helpful discussions.

1.   In (Bohr, 1934, pp. 97-98), the passage reads:

   As Einstein has emphasized, the assumption that any observation ultimately depends upon the coincidence in space and time of the object and the means of observation and that, therefore, any observation is definable independently of the reference system of the observer is, indeed, fundamental for the whole theory of relativity. However, since the discovery of the quantum of action, we know that the classical ideal cannot be attained in the description of atomic phenomena. In particular, any attempt at an ordering in space-time leads to a break in the causal chain, since such an attempt is bound up with an essential exchange of momentum and energy between the individuals and measuring rods and clocks used for observation; and just this exchange cannot be taken unto account if the measuring instruments are to fulfil their purpose.

   As Michael Friedman pointed out to me, the translation does not perfectly match the original. For example, rather than "an essential exchange of momentum" one should probably say "a non-negligible [nicht zu vernachlässigenden] exchange". These subtle differences are important for a full understanding of Bohr's view and especially (perhaps) its development, but for our purposes here they are not crucial.

REFERENCES

Beller, M. (1999). *Quantum Dialogue*. University of Chicago Press.

Beller, M. and Fine, A. (1994). "Bohr's Response to EPR". In Faye, J. and Folse, H., editors, *Niels Bohr and Contemporary Philosophy*, pages 1-31. Kluwer Academic Publishers.

Bohm, D. (1951). *Quantum Theory*. Prentice-Hall.

Bohr, N. (1929). "Wirkungsquantum und Naturbeschreibung". *Naturwissenschaften*, 17:483-486.

Bohr, N. (1934). "The Quantum of Action and the Description of Nature". In *Atomic Theory and the Description of Nature*, pages 92-101. Cambridge University Press.

Bohr, N. (1935). "Can Quantum-Mechanical Description of Physical Reality Be Considered Complete?" *Physical Review*, 48:696-702.

Einstein, A., Podolsky, B., and Rosen, N. (1935). "Can Quantum-Mechanical Description of Physical Reality Be Considered Complete?" *Physical Review*, 47:777-780.

Fine, A. (1986). *The Shaky Game: Einstein, Realism, and the Quantum Theory*. University of Chicago Press.

Dept. of History and Philosophy of Science
Indiana University
1011 E Third St.
130 Goodbody Hall
Bloomington, IN 47405
U.S.A.
michael@mdickson.com

ANASTASIOS BRENNER

THE FRENCH CONNECTION:
CONVENTIONALISM AND THE VIENNA CIRCLE

INTRODUCTION

In 1929 Moritz Schlick and those scholars he had brought together came to realize that they had given rise to something entirely new, so the text of the *Vienna Circle Manifesto* has it. What was novel was the conception of the world, henceforth scientific. Or as we may put it otherwise: a discipline had been established, the philosophy of science, that is a reflection on science no longer subordinate to traditional theory of knowledge and metaphysics. The text goes on to explain why such a conception arose geographically where it did: "That Vienna was specially suitable ground for [the development of the spirit of a scientific conception of the world] is historically understandable"[1]. The *Vienna Circle Manifesto* proceeds to enumerate the multifarious intellectual movements that were brought together at the beginning of the 20th century in the city of Vienna. Is it irrelevant or untimely to emphasize this cosmopolitan spirit? I believe, on the contrary, that cosmopolitanism provides both a lesson about philosophical creativity and a key for understanding the vitality of Viennese philosophy: the achievements of the Vienna Circle were the result of an exceptional open-mindedness on the part of its members.

Of the various attempts to analyze scientific knowledge at the time, the Vienna Circle certainly provided the largest synthesis. Viennese positivism drew on the philosophical traditions of England and France as well as Germany, each in a significant manner. Rather than promote some national tradition, the aim was to develop a new conception. Singling out one source of influence, I wish to study the relationship between the Vienna Circle and the French conventionalist movement. This relationship has not received, in my opinion, sufficient attention. The impact of Russell's logicism has been commented on time and again. The Austrian antecedents of logical positivism have been carefully excavated. Even the unacknowledged Kantianism of several members of the Vienna Circle has been more recently brought to light. Yet the contribution of French philosophy of science is hardly touched on, despite the fact that the logical positivists themselves referred frequently to Henri Poincaré and Pierre Duhem, ascribing to them several important claims. This lack of interest may not be difficult to explain. While logical positivism was flourishing, an anti-positivist reaction had

277

M. Heidelberger and F. Stadler (eds.), History of Philosophy and Science, 277–286.
© 2002 *Kluwer Academic Publishers. Printed in the Netherlands.*

begun in France, with Bachelard and Koyré. This reaction accounts for the small audience the Vienna Circle attracted there, giving rise to the misconception of a French tradition altogether unwilling to accept positivism. Yet, this does not square with the fact that positivism made an early appearance in France with Auguste Comte. Positivism was influential there, in one form or another, until the First World War, that is well up to the formative years of several members of the Vienna Circle.

Furthermore, a historical examination of the philosophical context of the turn of the 19th and 20th centuries reveals that the endeavor to reformulate positivism preceded the Vienna Circle considerably. Before taking root in Austria, neo-positivism was a French current of thought. Indeed, as early as 1901, Édouard Le Roy published an article entitled "Un positivisme nouveau"[2]. He claimed in this article to perceive the beginning of an intellectual movement and drew up the program of reorienting positivism. This fact raises several questions: to what extent did this movement anticipate the Vienna Circle? To what degree did this movement make it possible to further the project of renovating positivism? By bringing up these questions, I do not intend to make a priority claim. One may acknowledge, from the outset, that the logical brand of positivism is characteristic of the Vienna Circle. The originality of logical positivism, how-ever, cannot be established, if its forerunners are not taken seriously into account[3].

The comparison with conventionalism will make it possible to complete the picture of the major influences acting on the Vienna Circle. How are we to un-derstand the role the concept of convention comes to play in philosophy of science, if we pass over in silence Poincaré's motives for introducing it? Without recalling Duhem's arguments, it is difficult to follow the discussions pertaining to crucial experiment and the testing of theories. How are we to comprehend the complex relationship between the Vienna Circle and positivism if we omit to mention the new positivism of Le Roy and Abel Rey? Conventionalism thus provides several indispensable links in the development of philosophy of science.

1. POINCARÉ AND THE CONCEPT OF CONVENTION

One may begin by mentioning the very concept of convention, which gave its name to the doctrine of Poincaré and related thinkers. The concept was intro-duced by Poincaré in association with the thesis that the hypotheses of geometry are conventional. Poincaré's thesis received a good deal of attention from the logical positivists, and the concept of convention was definitively incorporated into the vocabulary of philosophy of science.

Poincaré seems to have used the concept of convention for the first time in an article dating from 1891, "Les geométries non euclidiennes", which he later took up in *La science et l'hypothèse*. I shall focus on the early version, in which

the author's motives are easier to grasp. Let us work our way back from the conclusion: "The axioms of geometry are neither synthetic *a priori* judgments nor
experimental facts. They are conventions"[4]. If it is obvious that Poincaré is referring here to Kant, one should be no less clear that "experimental facts" is an
allusion to Mill. Both philosophers are mentioned earlier in the text, and, I
believe, together they furnish the key to Poincaré's position. He is examining
here the two major traditional solutions to the problem of the nature of the
axioms of geometry.

How to grasp the reference to Kant? At first glance, we have but meager
evidence at our disposal. Poincaré seldom gives references. But in a later book,
Science et méthode, we learn that Poincaré refused the reduction of the whole of
mathematics to the analytic *a priori*, proposed by the logicists: "For Couturat,
new research, particularly that of Russell and Peano, settled the debate, which
had gone on for so long between Leibniz and Kant. They showed that there are
no synthetic *a priori* judgments"[5]. As to this conclusion Poincaré, however, was
not convinced. He believed that the concept of the synthetic *a priori* is entirely
appropriate in the fields of arithmetic and analysis, and Poincaré used it to characterize the principle of complete induction. The question remains of Poincaré's
relationship with Kant. Is he salvaging Kantianism or subverting it? One understands how philosophers of different views could find here a source of inspiration: neo-Kantians as well as disciples of Mach. Among the readers of Poincaré number Cassirer, Schlick and Carnap before they joined the Vienna Circle;
but also Frank, Hahn and Neurath. Of course, we must add that if Poincaré continues to use the concept of the synthetic *a priori*, the meaning he gives to it is
somewhat removed from that of Kant: what is so designated henceforth is a very
general principle, reflecting a capacity of the human mind to construct formal
systems, not a framework applying to our apprehension of objects.

What does "convention" mean exactly? This becomes clearer when we
examine what Poincaré has to say about Mill in his 1891 article:

Stuart Mill claimed that every definition contains an axiom, since, in defining, one asserts
implicitly the existence of the object defined. This claim goes too far; one seldom gives a
definition in mathematics without following it up with a proof of the existence of the object defined (...). It should not be forgotten that the word existence does not have the
same meaning with respect to a mathematical entity (...). A mathematical entity exists
providing that its definition does not imply contradiction (...). But if Stuart Mill's remark
cannot apply to all definitions, it remains nevertheless pertinent for some[6].

Poincaré is probably alluding to the following passage of *The System of Logic*:
"There is a real distinction, then, between definitions of names, and what are
erroneously called definitions of things; but it is, that the latter, along with the
meaning of a name, covertly, assert a matter of fact. This covert assertion is not a
definition, but a postulate"[7]. Mill applies this theory of definition to geometry.
The definitions of geometry contain in fact two propositions: a definition of
name and an assumption of existence. The proof rests on the latter, for example:

This proposition 'A circle is a figure bounded by a line which has all its points equally distant from a point within it', is called the definition of a circle; but the proposition from which so many consequences follow, and which is really a first principle in geometry, is, that figures answering to this description exist[8].

We have here an empiricist interpretation of mathematics.

Of course, Poincaré does not follow Mill completely. He adds that existence in mathematics means non-contradiction; it does not mean actual existence. But he retains the idea of mutual dependence between the axioms and the definitions. Definitions may hide axioms, and likewise axioms may conceal definitions. What we have here amounts, nonetheless, to a complete overthrow of Mill's conception of mathematics. Poincaré goes on to declare: "However one turns things, it is impossible to discover any reasonable meaning in geometrical empiricism"[9]. By calling on the notion of convention, Poincaré sets himself apart from Mill's empiricism. For if definitions contain axioms, the mathematician's task is to make them explicit. Once all required axioms have been formulated, it remains to understand that by stating them, one chooses a particular geometry. Thus Poincaré's position is similar to the nominalism that Mill explicitly criticizes, when he evokes the term of convention in a premonitory way:

It had been handed down from Aristotle (...) that the science of geometry is deduced from definitions (...). But Hobbes followed, and rejected utterly the notion that a definition declares the nature of the thing (...); producing the singular paradox, that systems of scientific truth (...) are deduced from the arbitrary *conventions* of mankind concerning the signification of words[10].

How are we now to construe Poincaré's position? I believe that he is drawing a parallel between Kant and Mill; they serve as thesis and antithesis in his argumentation. It would be false to think that Poincaré rejects any less Kant's view than Mill's. It is perhaps worth remarking, as the trend has changed today, that Mill was probably more seriously read than Kant in 19th century France, especially among scientists. We should pay attention to what Poincaré is doing: he is borrowing elements from both the rationalist and the empiricist conceptions. Thus it appears that he is aiming at a new synthesis, more in accordance with the state of knowledge at his time.

The logical positivists probably sensed this tension in Poincaré, although they were not primarily interested in giving a coherent interpretation of his philosophy. However that may be, they put his concepts to use. Let us consider the following passage of Carnap's *Aufbau*:

Each constructional step can be envisaged as the application of a general formal rule to the empirical situation of the level in question (...). These general rules could be called a priori rules, since the construction and cognition of the object is logically dependent upon them (...). However, the rules are not to be designated as 'a priori' knowledge, for they do not represent knowledge, but postulations[11].

Several themes are reminiscent of Poincaré: one must redefine the nature and scope of *a priori* knowledge; science contains many implicit conventions, that is unconscious presuppositions, disguised definitions. To be sure, conventions are extended beyond the sphere defined by the French mathematician. Poincaré was criticizing Kant's conception of geometry; Carnap rejected the latter's philosophy of mathematics altogether. But logical positivists were still concerned with a similar set of problems. It is often argued that logical positivists developed an antifoundationalist philosophy. This is all the more plausible as conventionalism had already represented a break with traditional foundationalist theory of knowledge.

There are other aspects of Poincaré's thought which the logical positivists seized upon, in particular his so-called structural realism. But it is important to realize that the latter is not unrelated to the concept of convention. Because Poincaré sought to avoid a radical conventionalism, of which he accused – rather unfairly – Le Roy, he was led to introduce a second-order realism. This doctrine certainly had an influence on the Vienna Circle. But Poincaré's proposal needed to be considerably improved on, and the logical positivists, I believe, could find in the conventionalist movement more precisely fashioned analytical tools.

2. DUHEM'S HOLISM

If the concept of convention makes it possible to avoid some of the difficulties facing traditional philosophy of science, it raises new problems. One must mark out the limits of conventionality, lest one lapse into skepticism. Perhaps the author who noticed this most clearly was Duhem. What has come to be known as the Duhem-Quine thesis is precisely a response to radical conventionalism: speaking of such general principles as the law of inertia, Duhem wrote:

Hypotheses which by themselves have no physical meaning undergo experimental testing in exactly the same manner as other hypotheses (...). There thus disappears what might have seemed paradoxical in the following assertion: Certain physical theories rest on hypotheses that do not by themselves have any physical meaning[12].

Now, this thesis gave rise to several interpretations. Neurath was alone in giving it full force and drawing all its consequences. Carnap accepted it only in a restricted sense, otherwise favoring an inductivism somewhat similar to that of Poincaré. Popper, in turn, exploited Duhem's deductivism, while rejecting his critique of crucial experiments. Yet, Duhem's thesis directed attention to the problem of experimental testing, and this problem was to remain central until postpositivists replaced it with the competition between paradigms.

Furthermore, this thesis is connected with Duhem's conception of physical theory, which logical positivists took up. Let us return to the *Vienna Circle Manifesto*: concerning their sources of influence, the authors of the text assert: "Above all there were epistemological and methodological problems of physics, for instance Poincaré's conventionalism, Duhem's conception of the aim and structure of physical theories"[13]. When Duhem defined physical theory not as explanation, but as an abstract representation, his definition recalled that of Mach. Thus it is not surprising that he adopted the concept of economy of thought. Duhem came into contact with Mach, and the two philosopher-scientists commented on each others works. Their relation of course attracted the attention of those disciples of Mach who were to establish the Vienna Circle. But Duhem went on to add that this representation is also a classification of laws that tends more and more towards a natural classification. In one instance, Duhem is led to make a reservation concerning the central notion of Mach's philosophy:

Logic does not (...) furnish any unanswerable argument to anyone who claims we must impose on physical theory an order free from all contradiction. Are there sufficient grounds for imposing such an order if we take as a *principle the tendency of science toward the greatest intellectual economy*? We do not think so (...). We showed how diverse sorts of mind would judge differently the economy of thought resulting from an intellectual operation[14].

Elsewhere, Duhem observed that intellectual economy concerns theory as well as experimental law. And he clearly distinguished these two levels. Duhem is seeking to understand the structure of theory: it is an organized body, "a system of mathematical propositions". It is also a hierarchical body: propositions are "deduced from a small number of principles." [15] Duhem thus goes beyond Mach. Theory is not only economy of thought. What is important is not so much the phenomenalist definition, but rather the type of analysis made possible by Duhem's idea of structure.

Concerning the problem of the foundations of physics the *Manifesto* gives this characterization:

Originally the Vienna Circle's strongest interest was in the method of empirical science. Inspired by ideas of Mach, Poincaré and Duhem, the problems of mastering reality through scientific systems, especially through *systems of hypotheses and axioms*, were discussed. A system of axioms attains a meaning for reality only by the addition of further definitions, namely the 'coordinating definitions' (...). The changes imposed by new experiences can be made either in the axioms or in the coordinating definitions. Here we touch the problem of conventions, particularly treated by Poincaré[16].

What is being described here is the received view: a scientific theory is an axiomatic system whose application to reality is obtained by means of coordinating definitions or correspondence rules. The problem that the Vienna Circle faced was to explain the mathematical character of the advanced parts of science without sacrificing their empirical base.

Duhem comes close to the received view when he writes:

At both its starting and terminal points, the mathematical development of a physical theory cannot be welded to observable facts except by a translation. In order to introduce the circumstances of an experiment into the calculation, we must make a version which replaces *the language of concrete observation* by *the language of numbers;* in order to verify the result that a theory predicts for that experiment, a translation exercise must transform a numerical value into a reading formulated in experimental language. (...) The method of measurement is the dictionary [vocabulaire] which makes possible the rendering of these two translations in either direction[17].

"Vocabulaire" here has the sense of concise bilingual dictionary; indeed the fundamental concepts of physics are few. Carnap and the Vienna Circle rediscovered this perspective on scientific theory. Undoubtedly, modern logic makes it possible to reach a more precise formulation. Henceforth, three kinds of terms are categorically separated: logico-mathematical, theoretical and observational. The main requirement applying to theory is clearly spelled out: the theoretical concepts are explicitly defined in terms of observables with the help of correspondence rules. It is no less true that conventionalism provided a whole series of themes that came up again: the interpretation of formal systems, the translation between the different languages of science and operational definitions.

3. THE "NEW POSITIVISM" OF LE ROY

Logical positivists thus adopted and discussed several claims made by French conventionalists. It remains to be seen how they perceived this current of thought. This is all the more important as, in some respects, conventionalism anticipated logical positivism. In an effort to develop what he termed already "neopositivism", Le Roy sought to combine the ideas of Poincaré with those of Duhem. Reading Le Roy's characterization of this current of thought, one cannot help thinking of a later movement:

The critical movement of which I am speaking presents this particularity that, far from having been ushered in from without by metaphysical and moral motivations, it occurred within science, under the impulse of internal needs, in close contact with facts and theories. The authors of this movement were practitioners who would not have intended and never did intend to sacrifice the smallest portion of science for anything else. We must consider their endeavor as one aiming at greater sincerity, as an endeavor to reflect more profoundly on their knowledge[18].

What we have here is an attempt to create a new discipline, a new manner of philosophizing. This creation was not easy to bring about, nor was it accomplished overnight. There were resistances to overcome, and it is essential to seize the various stages of this historical process.

Of particular interest is the manner in which Le Roy and another thinker of the time Gaston Milhaud reformulated positivism. If not explicitly Poincaré, at least Duhem, Le Roy and Milhaud agreed that their ideas could be viewed as a sort of positivism. According to Milhaud, an attentive reader of Comte, the latter's third stage of human thought, that is "the positive state", gives rise, in due course, to a "fourth state", characterized by greater freedom and spontaneity. Le

Roy formulated a methodological positivism, which left room for a Bergsonian metaphysics. To avoid ambiguity, Duhem spoke of a "Christian positivism". These authors, as well as Poincaré, were reacting against a narrow version of empiricism. This is clear in their emphasis on the role of theoretical interpretation and the nature of deductive systems. In speaking of the need to go beyond Comte, Milhaud wrote: "We start to understand that things are not so simple and that the fundamental notions of theoretical sciences are not merely the residues of experience, which might always be retrieved by means of an adequate verification"[19]. The comparison with Frank is illuminating. The latter member of the Vienna Circle speaks of mitigating Mach's empiricism with the help of conventionalism, and goes on to describe in this manner the particular contribution of Poincaré:

For him, the general propositions of science, such as the theorem concerning the sum of the angles of a triangle, the law of inertia in mechanics, the law of conservation of energy, etc. are not assertions about reality, but arbitrary stipulations on the manner words such as 'straight line', 'force', 'energy' should be used in the propositions of geometry, mechanics and physics. In consequence, one can never say if one of them is true or false; they are free creations of the mind and one can only ask if such stipulations are convenient or not[20].

In this endeavor, Frank also called on Duhem, and he was not ignorant of Abel Rey's "new positivism", which was derived from Le Roy and Milhaud.

CONCLUSION

The new positivism that Le Roy ushered in did not represent a homogeneous doctrine. It was rather a reaction against earlier conceptions or a transition towards new views. But one must not underestimate its importance. The idea of a critique of classical methodology is apparent in the efforts of several thinkers of the time; it pervades Poincaré's and Duhem's writings. Nor must one forget that Le Roy and Milhaud played their part in establishing philosophy of science in France: the former by incorporating it in his lectures at the Collège de France; the latter by initiating the study of history of science in its relation with philosophy at the Sorbonne.

The logical positivists read Poincaré and Duhem. They learned of the controversies stirred by their conceptions through Rey. Neurath, Carnap and even Popper took up several major theses of conventionalism. Frank characterized French influence thus: the conception of hypotheses made it possible to mitigate empiricism. By escaping the narrowness of Mach's position, the Vienna Circle was able to bring about the renovation of positivism. Thereby Frank pointed to one of the distinctive aspects of the French tradition: the central role ascribed to hypotheses. To the old question of the origin of knowledge, conventionalism provided an entirely new solution. For the first time a claim was made for the existence of an essential feature of knowledge which neither derives from facts nor springs from intuition nor even from the innate, a feature which makes it possible to constitute one system of representation among others. The introduction of the notion of convention opened the way for an assessment of decisional factors in science: the definition of concepts and the construction of theory.

Curiously, French conventionalism had its main impact abroad. For in France this movement was subjected to severe criticism after the First World War, namely by Bachelard and Koyré. Also, philosophy of science never came to have the dominance there that it achieved elsewhere. To a large extent, the influence of conventionalism was mediated by logical positivists. We saw that their understanding of this doctrine, although it may have been in some respect idiosyncratic, was particularly rich and fruitful. The reformulation of positivism carried out by the Vienna Circle was sufficiently profound and vigorous: whereas positivism disappeared in France, it continued to be active in Austria, and the methods of analysis necessary for philosophy of science were more securely established.

NOTES

1. Hans Hahn/Otto Neurath/Rudolf Carnap, "The Scientific Conception of the World: the Vienna Circle", known as the *Vienna Circle Manifesto*, in Neurath, *Empiricism and Sociology* (Dordrecht: Reidel, 1973, pp. 299-318), p. 301. Only the preface is signed. Actually, the document seems to have been prepared by Neurath, revised by Hahn and Carnap, and augmented with observations by other members of the Circle. See editor's note p. 318.

2. Édouard Le Roy, "Un positivisme nouveau", in *Revue de métaphysique et de morale*, 9, 1901, pp.138-153.

3. On conventionalism with respect to Mach and Austrian positivism before the Vienna Circle, see Anastasios Brenner, "Les voies du positivisme en France et en Autriche: Poincaré, Duhem et Mach", in *Philosophia Scientiae*, 3, 1998, pp. 31-42, and Rudolf Haller, "The First Vienna Circle", in Thomas Uebel (Ed.), *Rediscovering the Forgotten Vienna Circle*, Dordrecht: Kluwer, 1991, pp. 95-108.

4. Henri Poincaré, "Les géométries non euclidiennes" (in *Revue générale des sciences*, 2, 1891, pp. 769-774), p. 773 ; *La science et l'hypothèse* (Paris: Flammarion, 1968), p. 75. Translation mine.

5. Poincaré, *Science et méthode* (Paris: Flammarion, 1916), p. 127. Translation mine.

6. Poincaré, "Les géométries non euclidiennes", pp. 771-772; *La science et l'hypothèse*, p. 70; cf. *Science et méthode*, p. 132. Translation mine.

7. John Stuart Mill, *A System of Logic* (in *Collected works*, vols 7 & 8, London: Routledge, 1996) p. 144.

8. *Ibid*, p. 257.

9. Poincaré, *La science et l'hypothèse*, p. 101. Translation mine.

10. Mill, *op. cit.*, p. 144. My emphasis.

11. Rudolf Carnap, *The Logical Structure of the World* (Berkeley: University of California Press, 1967), § 103; cf. § 107. The logical positivists were also impressed by Einstein's "Geometry and Experience". But one must note that Einstein himself refers back to Poincaré.

12. Pierre Duhem, *La théorie physique, son objet et sa structure* (Paris: Vrin, 1981), p. 328; *The Aim and Structure of Physical Theory* (Princeton University Press, 1982), p. 216. On the origins of Duhem's holism, see Brenner, "Holism a Century Ago : the Elaboration of Duhem's Thesis", in *Synthese*, 83, 1990, pp. 325-335.

13. Hahn/Neurath/Carnap, *op. cit.*, p. 325.
14. Duhem, *op. cit.*, p. 149; English trans., p. 101.
15. Duhem, *op. cit.*, p. 24; English trans., p. 19.
16. Hahn/Neurath/Carnap, *op. cit.*, p. 311.
17. Duhem, *op. cit.*, p. 199; English trans., p. 333.
18. Le Roy, *op. cit.*, p. 139. Translation mine. Concerning the relations between Poincaré, Duhem, Le Roy and Milhaud, see Brenner, *Duhem : science, réalité et apparence* (Paris: Vrin, 1990), chap. 1.
19. Gaston Milhaud, *Le positivisme et le progrès de l'esprit* (Paris: Alcan, 1902), p. 140. Translation mine.
20. Philipp Frank, *Einstein: His Life and Times* (London: Macmillan, 1948), p. 80. Cf. Frank, *Modern Science and its Philosophy* (Cambridge, Mass.: Harvard University Press, 1949), "Introduction: Historical Background".

Université de Toulouse-Le Mirail
Département de Philosophie
5 allées Antonio Machado
31058 Toulouse Cedex
France
brenner@univ-tlse2.fr

WYBO HOUKES

CARNAP ON LOGIC AND EXPERIENCE

In recent years, attention for the work of Rudolf Carnap has shifted from polemical discussion to placing Carnap in his intellectual context. Thus, the central question is no longer whether Carnap contributes to solving our current problems, but whether he solved the problems of *his* day and age. This contextualist approach has resulted in a deeper and more refined understanding of, in particular, Carnap's early works and has focused on *Der logische Aufbau der Welt*.

One result of this approach is that the *Aufbau* is no longer understood as a continuation of radical empiricism by logical means, but as a continuation of mainly neo-Kantian themes with greater logical acumen. However, this exegetical shift, brought about by Werner Sauer, Michael Friedman, Alan Richardson, and others, has not led to a more charitable verdict as to the coherence of the *Aufbau* project. Traditionally, it was perceived as faltering on the incompleteness of the reduction of physical terms to those related to sensory experience only.[1] This is no longer perceived as a major problem, since reduction of science to sensory data is not the overriding aim of the *Aufbau* project. Instead, Carnap wants to show that all statements of science are purely "structural", i.e., that they can be characterised by means of the formal properties of relations.[2] This quest for the "dematerialisation"[3] of science culminates in what I would like to call the "complete formalisation" sections,[4] in which Carnap attempts to show that all of science can be cast in formal terms alone. The ideal of complete formalisation contributes to undermining the traditional phenomenalist reading of the *Aufbau*, but the ultimate failure of the project is now also found in it. Both Friedman and Richardson take fault with the ideal, claiming that it threatens to erase the distinction between empirical science on the one hand and logic and mathematics on the other.[5]

Perhaps this merely shows that contextualisation of the *Aufbau* is incomplete. For what could have led Carnap to the ideal of complete formalisation? Surely, it is a strange goal in our contemporary, model-theoretic conception of the relation between logic and experience. But should that not prompt us to find out that Carnap's conception of this relation was such that the ideal of complete formalisation is not absurd? In this paper, I attempt to make a start in doing just that, by sketching some of the context of the *Aufbau*. First, I will consider two views on the relation between formal systems and empirical knowledge, those of Cassirer and Schlick. I will present them as attempting to solve two problems – of objectivity and univocalness – that arose out of the so-called "crisis of intuition" in German philosophy. Then, I will reconstruct Carnap's early views of the relation

M. Heidelberger and F. Stadler (eds.), History of Philosophy and Science, 287–298.

of formal system and experience against this background. Finally, I attempt to show that the complete formalisation sections of the *Aufbau* are driven by Carnap's rejection of Schlick's and Cassirer's views, on the one hand, and the need to account for objectivity and univocalness, on the other.

1. OBJECTIVITY AND UNIVOCALNESS

Before analysing Carnap's views, I will sketch part of the historical context in which he worked. I will focus on two views that were quite popular in German philosophy in the 1920's: Cassirer's modification of orthodox Kantianism, and Schlick's radical formalism. I shall argue that Cassirer and Schlick attempted to solve similar problems in a different way, and that these problems stem from a common root.[6]

The early work of both Schlick and Cassirer may be regarded as the product of what is often called the "crisis of intuition": at the start of the 20th century, many philosophers thought that developments in logic, mathematics, and the sciences had shown that it was no longer necessary or even reasonable to appeal to a faculty of pure intuition in order to provide the foundations of these disciplines. However, the concept of "intuition" solved several epistemological problems in Kant's theoretical philosophy, which were to return with a vengeance. I shall elaborate on only two problems here.

First of all, the Kantian account of the objectivity of knowledge is based on a firm connection between the faculties of understanding and intuition. When intuition is eliminated, this account is undermined and a problem of the objectivity of scientific knowledge arises. Secondly, intuition was presented in the first Critique as an immediate grasp of particular objects, in contrast to mediate, general concepts. If intuition is deprived of its epistemic role, how are we able to cognise individuals? In slightly different terms: does science include knowledge of individual objects, e.g., a way of individuating all of them? Does it give a unique description of reality? I will call this the problem of the so-called *Eindeutigkeit* or univocalness of scientific knowledge.

Both Schlick and Cassirer addressed the problems of objectivity and univocalness, and they did so by means of a common basic concept, that of co-ordination (*Zuordnung*).[7] This is a general notion in German, more or less synonymous with "relation". But, at the time, it was used more specifically, namely as a forerunner of our contemporary notion of mapping: the elements of one domain (e.g., physical events or judgements) were said to be co-ordinated to those of another (e.g., psychological events or states of affairs).

Cassirer and Schlick present similar solutions to the problem of objectivity, based on the notion of co-ordination. They both understand the basic concepts of science as ordering relations, in which the scientist attempts to place all available observational data. In this way, every observational matter of fact is placed in a complicated network of functional dependencies, representing space, time, color, mass, etc. Objects may be understood as the nodes within this network of physical relations and laws. Cassirer uses "co-ordination" ambiguously in the context of this scheme: it designates both the act of placing data in this network, relative to other data, and the act of relating this entire network to reality.[8] Schlick's use is more consistent, since he borrows Hilbert's concept of "implicit definition" to designate the relations between concepts, and uses "co-ordination" exclusively for the relation between concepts (or judgements) and reality. In any case, both authors explain the truth and the objectivity of scientific knowledge by means of the notion of "co-ordination".

Within this "co-ordinative" theory of objectivity, the problem of the univocalness of knowledge can be rephrased as: are all objects *unique* nodes in the system of functional relations? Let me illustrate this new form of the problem with a literary example. In *Invisible Cities*, Italo Calvino gives the following description of the city of Ersilia. Its inhabitants mark relations of "blood, trade, and authority" by connecting their houses with threads of different colors. When the resulting system of threads gets too complicated, they leave their town and rebuild it somewhere else, trying to simplify the previous connections. The problem of univocalness is: if the Ersilians leave their city with a set of instructions and rebuild Ersilia, will they be able to relocate their homes, or is there a chance that they will end up in a house inhabited by someone else in the earlier Ersilia?

With respect to this problem, Cassirer shows his Kantianism by reserving a central role for intuition. As he most clearly states in "Kant und die moderne Mathematik", intuition can no longer be regarded as a source of knowledge – understanding and its network of functional dependencies suffice for all epistemic purposes. Instead, intuition yields a preliminary determination of individual objects, setting an – according to Cassirer unattainable – goal for conceptual activity.[9]

Schlick, on the other hand, chooses a non-Kantian approach.[10] According to him, Hilbert's formalism concerning mathematical concepts can be extended to scientific concepts.[11] Thus, concepts such as "mass" and "simultaneity" are implicitly defined by their role in a system of physical axioms, just as Hilbert defined "point" and "line" in geometry. Only this system of implicitly defined concepts *as a whole* is co-ordinated with states of affairs. Moreover, implicit definition and co-ordination suffice for all epistemic purposes: time and again in his *Allgemeine Erkenntnislehre*, Schlick stresses that there is no need for other cognitive acts, particularly pure intuition.[12] As a consequence, that there is a system which supports a *univocal* co-ordination can at best be a contingent matter, and Schlick's attitude towards the problem of univocalness is implicitly optimistic: if the system of relations is sufficiently complicated, it becomes increasingly unlikely that objects occupy indistinguishable positions.

2. *DER RAUM*

Against the background sketched above, Carnap's doctoral dissertation seems almost reactionary. Written in 1920, two years after the first edition of Schlick's *Allgemeine Erkenntnislehre* and a decade after Cassirer's *Substanzbegriff und Funktionsbegriff*, it uses a Husserlian, quasi-Kantian faculty of intuition in explaining our knowledge of space.

In *Der Raum*, Carnap distinguishes nine notions of space, as combinations of two triplets. I will only discuss the first triplet here. First of all, Carnap defines formal space (R) as any system of relations specified by axioms. Secondly, he introduces intuitive space (R'), which embodies our essential intuition of something spatial. Thirdly and finally, Carnap uses the notion of "physical space" (R''). Physical space is a system of experienced spatial relations between physical objects, such as a cup of water being on top of a table.

Carnap describes the relations between these three spaces in essentially Husserlian terms. Intuitive space is related to formal space by "substitution" (*Einsetzung*) or "de-formalization" (*Entformalisierung*). This is the specification of the purely formal or general in terms of a certain category of objects, in this case spatial ones. Physical space, in turn, is related to intuitive space by "subsumption" (*Subsumtion* or *Unterordnung*).[13] In this way, the tripartite structure of spaces conjoins the formal or highly abstract, the generically spatial, and spatial particulars, and all of these relations are at least partially understood as general-particular relations. As Carnap himself says: "Both the relation of R to R' and that of R' to R'' are those of species to particular, but in different senses."[14] Surprisingly, for one writing in the early 1920s, Carnap does not use the notion of *Zuordnung*. He only mentions it once, in the bibliography, where he notes that Ostwald uses "co-ordination" to refer to both substitution and subsumption.

Where does *Der Raum* stand with respect to the views of Cassirer and Schlick? In one respect, the role of essential intuition is similar to the role of intuition in Cassirer's work. For one of its tasks is to guarantee the univocalness of the formal system:

> [...] we derive from intuition as little theses as possible, but so many that the spatial structure is determined univocally, i.e., that it can be classified (*eingeordnet*) under a single determinate formal ordering structure.[15]

This univocalness is weaker than a specification of *individual* spatial objects, so that Carnap in fact puts less stringent demands on intuition than Cassirer. On the other hand, Carnap reserves a clearly *epistemic* role for intuition, unlike Cassirer. Essential intuition yields a priori knowledge of what is spatial, and even of the necessary "matters of fact" of outer experience. Experience is needed to provide knowledge of particular spatial objects, but essential intuition is a condition of the possibility of experience. Therefore, Carnap's view of the relation between formal system and experience in *Der Raum* is more orthodox than Cassirer's (or Schlick's): the relation is understood to be mediated by intuition.

3. THE EARLY PAPERS

Carnap appears to have discarded essential intuition almost immediately after publishing *Der Raum*. In the four papers published between 1922 and 1928,[16] we find no appeal to intuitive space as an intermediary between formal system and experience. Thus, their relation can no longer be understood as a combination of substitution and subsumption. Instead, this relation is understood as one of co-ordination, in line with Cassirer and Schlick.

In the early papers, Carnap makes a sharp distinction between experience, which he also calls the "given" or "primary world", and a system of axioms, usually taken from mathematical physics. Experience can be ordered in terms of the axioms, and by this constructive activity the "secondary world" of physical objects arises. This ordering of experience in physical terms comes about by a purely formal relation of co-ordination.[17] By "purely formal", Carnap apparently means that co-ordination must not be understood as a synthetic activity of a transcendental consciousness, but as a freely chosen mapping between two domains. Experience can be co-ordinated to many different axiom systems, and only the principle of simplicity decides between various choices.[18]

Now let me turn to the status of the axioms themselves. If essential intuition no longer specifies the formal system in terms of some domain, how does the system come by its content? At the very end of "Über die Aufgabe der Physik", written in 1923, Carnap denies that the axioms have any material content apart from their co-ordination to experience:

The axioms do not have the contents of observation as their objects at all, but they are only formal specifications, which are co-ordinated with the contents of perception. Therefore, one can achieve so-called 'correspondence with reality' for any systems of axioms whatsoever. To this effect, one need only give the relations of co-ordination the appropriate form (the 'valid relations of co-ordination'). The resulting 'valid relations of co-ordination' for the various systems of axioms may differ radically with respect to simplicity.[19]

Consequently, physical events, such as changes in the field tensor or gravitational attraction, are "purely formal complexes ('ordering structures' of the theory of relations)."[20]

So it seems that Carnap has traded his earlier Husserlian view for an outspoken formalism in just a few years time![21] Yet he was to abandon formalism rather quickly. In 1927, Carnap published a paper called "Eigentliche und uneigentliche Begriffe". The main argument of this paper can be reconstructed by considering the problem of univocalness.

In the paper, Carnap distinguishes proper concepts, defined by explicit definitions, from improper concepts, defined by implicit definitions, and he argues against the possibility of defining *all* concepts implicitly. His argument for this goes as follows.[22] Assume that an axiom system has been constructed and implicitly defined. Will we then be able to assert univocally what this system is about? According to Carnap, we cannot, because the system can fail to be monomorphic: it is possible to apply it to systems that do not have a common formal structure. In contemporary terms, implicit definition of its basic concepts does not guarantee that a logical system is categorical, i.e., that all its models are isomorphic. In terms of the Ersilia metaphor, the new city may not be isomorphic to the old one, let alone that everyone will be able to relocate their former homes. So Carnap has found a different problem concerning the relation between formal system and experience than the problem that divided Cassirer and Schlick. Moreover, we find him struggling with the relation between a formal system and its interpretation, and reformulating the problem of univocalness along the way.

As a result, Carnap had to abandon formalism and find a view that goes beyond those of Cassirer and Schlick. As I shall argue next, this leads to the curious dialectics of the complete formalization sections at the end of the *Aufbau*.

4. THE *AUFBAU*

Before turning to the ideal of complete formalization, let me briefly sketch the relation between formal system and experience as Carnap sees it in the *Aufbau*. The main goal of the *Aufbau* is to construct a constitution system and thus to illustrate constitution theory. The constitution system developed in the *Aufbau* uses a type-theoretical logical language with one non-logical term, namely "recollection of similarity" (*Rs*), which holds between elementary experiences. All scientific statements are reconstructed as so-called "purely structural definite descriptions", i.e., descriptions of something in terms of the formal properties of *Rs*.

Having reconstructed sensations from elementary experiences by the procedure of quasi-analysis, Carnap goes on to reformulate physical knowledge in terms of *Rs*. At this stage of the project, we find a use of "co-ordination" similar to that in the early papers. Carnap maintains that there is a one-to-many co-ordination of colour sensations to space-time points.[23] In this way, physical space and the physical world are supposed to be constituted. As Carnap remarks in §125, this constitution presupposes an abstract, formal space, which is part of the logical language of constitution theory. Moreover, the set of sensations can be revised on the basis of the constituted physical world, after which the latter can be reconstituted, *et cetera*. This co-ordination of physical objects to psychological ones is generally regarded as the point where the *Aufbau* project breaks down for a lack of completely determinate explicit definitions.

Here, I intend to uncover a more fundamental feature of the *Aufbau*. This concerns the relation between the logical language of constitution theory and experience. For if co-ordination is primarily used to denote the transition from psychology to physics, or to other higher domains, how are we to understand the relation of constitution theory to the empirical knowledge it reconstructs? In order to answer this question, I will first consider the stated goal of constitution theory, as it may be gathered from §16 of the *Aufbau*. Having introduced the idea of purely structural definite descriptions, Carnap goes on to claim that scientific knowledge can be objective only insofar as its rational reconstruction is limited to such descriptions.

The reason for this emphasis on purely structural definite descriptions is that knowledge starts from subjective elementary experiences. Therefore, the objectivity of knowledge must be found in its logical form rather than in its content. In this respect, Carnap's solution to the problem of objectivity is similar to that of Cassirer and Schlick, and thus we may say that he still understands the relation between formal system and experience as one of co-ordination. Yet this solution of the problem of objectivity now yields curious results concerning the univocalness problem. For Carnap combines it with an outright rejection of intuition, even in Cassirer's non-epistemic sense of setting an (unattainable) goal for conceptual activity:

[Contrary to the view of the Marburg school] I should like to point out that a finite number of specifications suffice for the constitution of the object, and thus for its univocal characterisation among objects in general. [24]

So Carnap does not accept Cassirer's solution to the problem of univocalness; nor does he, from 'Eigentliche und uneigentliche Begriffe' onwards, underwrite Schlick's solution. The resulting tension between purely structural definite descriptions and univocalness manifests itself at the very end of the presentation of the constitution system.

In §§153-55, modestly marked "may be omitted", the *Aufbau* project reaches a curious end, which I will now attempt to reconstruct. Carnap remarks that his constitution system has shown that scientific knowledge may be represented in a language with only one non-logical term, *Rs*. However, he concludes that this does not yet realise the goal he set himself, since *Rs* remains as a primitive term. In order to show that science is purely structural, it must be represented in an *entirely* logical language. In other words: *Rs* must be eliminated. Moreover: Carnap wants to show that constitution theory solves *both* the objectivity problem and the univocalness problem. So the logical language must have scientific knowledge as its *one and only* model.

In my opinion, this suggests, among another things, that Carnap is no longer working in the Kantian tradition, although he has derived the dual problem of objectivity and univocalness from it. As I discussed earlier, for Cassirer, logic had to guarantee the objectivity of knowledge, but univocalness was the province of intuition. For Carnap, logic has to solve both problems at once. So although it seems fair to say that Carnap casts formal logic in a transcendental role as far as objectivity is concerned, I think that his conception of logic in the *Aufbau* can ultimately not be a neo-Kantian one, because of univocalness.

In §§153-155, we find the following dialectics concerning univocalness. First, Carnap argues that recollection of similarity may be implicitly defined – and thus *de facto* eliminated – as that relation that satisfies the term 'Rs' in his constitution system. In this, he appears to share Schlick's implicit optimism, in which univocalness is guaranteed once we take into account sufficiently detailed relations.[25] Next, however, this optimism is undermined. For no matter how many relations we take into account, the relations isomorphic to *Rs* satisfy the system as well. And since constitution theory may contain only purely structural definite descriptions, it cannot distinguish between isomorphic relations, as Carnap remarks in §154. So objectivity and univocalness appear to be incompatible.

Note that this problem is different from that discussed in "Eigentliche und uneigentliche Begriffe". It is not just that the system of purely structural definite descriptions cannot be implicitly defined because it should be monomorphic (or, in modern terms, categorical). In §15, Carnap claims that, contrary to implicit definitions, his "explicit definitions" should characterise a single object, i.e., that they should be univocal.

But how can they the constitution system be univocal, uniquely characterising each object in a finite number of steps? In §154, Carnap introduces a primitive notion, called "foundedness", which is a property of relations. A relation is called "founded" when it can be experienced, if it is "experientiable" (*erlebbar*) or "natural" (*natürlich*). By defining Rs as the *founded* relation satisfying the constitution system, Carnap claims to have characterised it in a univocal manner. This proposal appears to solve the problem of univocalness at the price of objectivity. For is "foundedness" not every bit as subjective as the original recollection of similarity? Carnap does not think so; he speculates that it might be a basic concept of logic.

From our point of view, this is an incomprehensible move, for we can understand "foundedness" as either a restriction on the formal system or an empirical concept. Concerning the former alternative: Carnap appears to lack the distinction between a formal system and its interpretation. For one thing, the very problem of univocalness is, for us, best cast in model-theoretic terms. A formal system is univocal if it has a single model. So univocalness is a property of the interpretation of the formal system rather than of the system itself. But Carnap requires the constitution system to guarantee its own univocalness, by including its own relation to experience as a primitive concept. Further evidence of Carnap's lack of a distinction between formal system and interpretation may be found in his use of four languages to represent the results of constitution theory.[26] Two of these languages are the *Principia Mathematica* system and a language using object terms. Instead of calling the latter an *interpretation* of the former, Carnap appears to regard them as intertranslatable. So introducing foundedness as a restriction of the interpretation would be anachronistic. Yet then, it seems, "foundedness" must be empirical, and it would reintroduce subjectivity at the root of the constitution system. So it seems Carnap is faced with an insoluble dilemma.

Summing up, we can reconstruct the reasoning leading to this strange conclusion as follows. Within his historical context, Carnap was confronted with two problems, those of objectivity and univocalness. He adopted the common, structuralist solution to the former problem, and sought to demonstrate its viability in the *Aufbau*. On the other hand, he rejected two purported solutions to the latter problem: Cassirer's appeal to intuition, and Schlick's appeal to a favourable structure of experience. Instead, he sought a *formal* guarantee of the univocalness of the formal system, thus solving the problem while avoiding intuition. And this is exactly what he claims to do in introducing foundedness as a basic concept of logic.

While this may suffice as a tentative contextualisation of the complete formalisation sections, it leaves one important question unanswered: was Carnap's solution rational at the time, or should he rather have abandoned the goal of univocalness? I conclude this paper with a brief speculation about this question.

In my opinion, the introduction of foundedness as a basic concept of logic may be intelligible, and Carnap's guarantee of univocalness rational, if we consider the influence of the universalist tradition in the philosophy of logic. According to both Russell and Frege, whose influence is palpable throughout the *Aufbau*, logic may be regarded as a universal language, which transcends all domains of objects, and not as a formal system in need of interpretation.[27] As I argued earlier, the *Aufbau* shows a similar lack of distinction between the logical system and its semantics. Moreover, Carnap's reason for calling 'foundedness' a basic concept of *logic* is precisely its universality:

[...] our considerations about characterising the basic relations of a constitution system as a founded relations of a certain kind are valid for every constitution system of any arbitrary domain. Because of this generality we may perhaps regard foundedness as a concept of logic [...][28]

Thus, the constitution system contains only the most general concepts, applicable to the type of relations, such as "founded" and "equivalent". If this universal language is capable of univocally characterising any object of science, Schlick's conundrum of designating individual entities by the most general names is solved. To a universalist, this logicisation of experience was probably no stranger than any conceptualisation, and for Carnap, Cassirer, and Schlick, it would have seemed the ideal solution to the problem of objectivity.

But not all was well. Despite this apparent adherence to universalism, Carnap's two characterisations of logic in the *Aufbau* stress its tautological nature.[29] We find the same token reference to tautologies in Russell's *Introduction to Mathematical Philosophy*, where we also find the famous expression that "[...] logic is concerned with the real world just as truly as zoology, though with its most abstract and general features".[30] It seems that both Russell in 1919 and Carnap before 1928 had some inkling about the significance of the *Tractatus* for their views, but had not yet fully appreciated it.[31] If the analysis presented here is even approximately correct, its consequences for the goal and methods of the *Aufbau* were disastrous.[32]

NOTES

1. E.g., Goodman's and Quine's well-known worry about the "at"-relation employed in §§126-127.
2. Rudolf Carnap, *Der logische Aufbau der Welt*. Berlin : Weltkreis-Verlag 1928, §16; the notion of "structure" is introduced in §12. As has been pointed out by Friedman and others, Carnap's goal in the *Aufbau* is to show that at least one constitution system is possible. This means, among other things, that Carnap is concerned with showing that there is at least one way of casting all of science in a purely structural form.
3. Carnap, *Aufbau*, §12.
4. Carnap, *Aufbau*, §§153-155.
5. Michael Friedman, "Carnap's *Aufbau* reconsidered", in: *Noûs* 21, 1987, p.533; Alan Richardson, *Carnap's Construction of the World*. Cambridge: Cambridge University Press 1998, p.194.
6. My presentation in this section owes a lot to the following papers: Thomas Ryckman, "*Conditio Sine Qua Non? Zuordnung* in the Early Epistemologies of Cassirer and Schlick", in: *Synthese* 88, 1991, pp.57-95, and Don Howard, "Relativity, *Eindeutigkeit*, and Monomorphism: Rudolf Carnap and the Development of the Categoricity Concept in Formal Semantics", in: Ronald N. Giere / Alan W. Richardson (eds.), *Origins of Logical Empiricism*. Minneapolis: University of Minnesota Press 1996, pp.115-164. These papers elucidate the concepts of "co-ordination" and "univocalness", respectively.
7. Göran Sundholm suggested to me that "correlation" may be a better translation of *Zuordnung*. Although "correlation" is used in Blumberg's translation of Schlick's *Allgemeine Erkenntnislehre*, I will stick with the more usual "co-ordination" here.
8. Compare "The connection between members [of a series] is in any case created by means of some general *law of co-ordination...*", (Ernst Cassirer, *Substanzbegriff und Funktionsbegriff*. Berlin: Bruno Cassirer 1910, p.21) with "The manifold of sensations is co-ordinated with the manifold of real objects in such a way that every connection that can be established in one totality indicates a connection in the other." (*ibid.*, pp.404-5) Since the connections co-ordinated in this way are referred to as "co-ordinations" themselves, the ambiguity is clear.
9. "The elements, which we first obtained from intuition, have to be analysed ever more thoroughly, they must be resolved ever more completely in purely conceptual definitions in order to become truly objects of mathematical consideration. Thus intuition, wherever we appeal to it, does not constitute the proper ground of truth of theorems. It is rather like an ultimate unresolved remainder, which awaits further division and operation by thought; it is the goal that points the way for our purely logical formation of concepts." (Ernst Cassirer, "Kant und die moderne Mathematik", in: *Kantstudien* 17, 1907, pp.29-30). All translations from German texts in this paper are my own, unless a translation is mentioned in the reference.
10. That Schlick is aware of the problem of univocalness is clear from: "(...) the individual entity is to be designated uniquely with the help only of the most general names (...)" (Moritz Schlick, *General Theory of Knowledge*. transl. by A.E. Blumberg, Wien: Springer Verlag, 1974, p.14.)
11. Schlick, *op.cit.*, p.32.
12. "The co-ordinating of two objects with one another, the relating of one to the other, is in fact a fundamental act of consciousness, not reducible to anything else. It is a simple ultimate that can only be stated, a limit and a basis, which every epistemologist must ultimately press toward." (Schlick, *op.cit.*, p.383) Schlick is quite outspoken on the elimination of pure intuition: "(...) we

have looked in vain for a pure intuition that might serve as the basis for empirical intuition by supplying it with its form and lawfulness." (*ibid.*, p.358)

13. These terms are listed in the references of: Rudolf Carnap, *Der Raum. Ein Beitrag zur Wissenschaftslehre*. Kantstudien Ergänzungsheft Nr.56, (Berlin: Reuther & Reichard, 1922), p.85.

14. Carnap, *Der Raum*, p.61.

15. Carnap, *Der Raum*, p.23.

16. Rudolf Carnap, "Über die Aufgabe der Physik", in: *Kantstudien* 28, 1923, pp.90-107; Rudolf Carnap, "Dreidimensionalität des Raumes und Kausalität", in: *Annalen der Philosophie und philosophischen Kritik*, 4, 1924, pp.105-130; Rudolf Carnap, "Über die Abhängigkeit der Eigenschaften des Raumes von denen der Zeit", in: *Kantstudien* 30, 1925, pp.331-345; Rudolf Carnap, "Eigentliche und uneigentliche Begriffe", in: *Symposion* 1, 1927, pp.355-374.

17. "(...) physics expresses itself neutrally by means of a purely formal relation of co-ordination" ("Über die Aufgabe", *loc.cit.*, p. 100); "The relation between this kind of experience of the second stage and that of the first stage is produced by co-ordination" ("Dreidimensionalität", *loc.cit.*, p.107)

18. In fact, Carnap shows that the principle of simplicity can be applied in one of two forms ("Über die Aufgabe", *loc.cit.*, pp.103-105)

19. *Ibid.*, p.106.

20. *Ibid*, p.107.

21. Compare Schlick's statement of formalism: "None of the concepts that occur in the theory designate anything real; rather they designate one another in such a fashion that the meaning of a concept consists in a particular constellation of a number of the remaining concepts." (Schlick, *op.cit.*, p.37)

22. The main argument is reconstructed, in slightly different terms than mine, in Howard, *loc.cit.*, pp.156-160; and Richardson, *op.cit.*, pp.43-47. Richardson focuses on the contrast with Schlick's formalism, but he puts little emphasis on the problem of univocalness as the driving force behind Carnap's discontent with it.

23. Carnap, *Aufbau*, §136.

24. Carnap, *Aufbau*, §179.

25. This optimism is expressed elegantly in the well-known railway metaphor of Carnap, *Aufbau*, §14: if the structure of the relation of "railway-connectedness" does not suffice to individuate cities, we can first use relations of connectedness by other means of transportation, then "historical" relations, and finally "all concepts from factual disciplines (*Realwissenschaften*)". Surprisingly, Carnap resolves the remaining lack of univocalness by stipulating that the difference between cities is a merely subjective one. In my opinion, this might show that Carnap has written this part of the *Aufbau* while still under the influence of Schlick's formalism. In §153, apparently written at a later time, Carnap again discusses univocalness in the context of implicit definition, but he claims that a lack of univocalness is highly improbable, not that it would be merely subjective.

26. Carnap, *Aufbau*, §95.

27. Cf. Jean van Heijenoort, "Logic as Calculus and Logic as Language", in: *Synthese* 17, 1967, pp.324-330 for a discussion of universalism in the work of Frege and the *Principia Mathematica*.

28. Carnap, *Aufbau*, §154.

29. "Logic (including mathematics) consists only of conventional stipulations about the use of signs and of tautologies based on these stipulations." (§107); "Logic does not have a domain of its own at all, but it contains those statements that hold (as tautologies) for all objects of any arbitrary domain." (§154)

30. Bertrand Russell, *Introduction to Mathematical Philosophy*, London: Allen and Unwin, 1919, p.169.

31. Russell, *op.cit.*, pp.203-205. For Carnap, there is well-known autobiographical evidence to support his incomplete apprehension of the *Tractatus*; cf. the passage from the Archives cited by Joelle Proust, "Formal Logic as Transcendental in Carnap and Wittgenstein", in: *Noûs* 21, 1987, p.502.

32. I would like to thank Michael Friedman, Jaakko Hintikka, David Hyder, Herman Philipse, and Göran Sundholm for their comments on earlier drafts of this paper and the version read during the HOPOS 2000 conference in Vienna. Any mistakes remaining are, of course, purely my own.

Department of Philosophy
Delft University of Technology
2628 BX Delft
The Netherlands
houkes@letmail.let.leidenuniv.nl

Artur Koterski

Affinities between Fleck and Neurath

First, there are some striking similarities in the history of reception of Fleck's and Neurath's ideas. Due to the style of their writings they were not or often not well welcome in their national philosophical communities, i.e., the Vienna Circle and the Lvov-Warsaw School. This is especially true in the case of Fleck. On the other hand, some prominent logical positivists, like Hempel, are to some extent guilty of making of Neurath a clumsy thinker whose ideas needed to be clear by a more mature philosopher, i.e. Carnap. So their work was often considered as non-scientific. With a lack of response their theories were forgotten for many years, and even if not forgotten they were habitually ridiculed by critics. After a long period of seven thin years they have their renaissance. They are even quite popular. In the last 20 years a number of excellent books and papers on Neurath and Fleck have apperared. But one can hardly find a line of comparison. One might say that this is not surprising at all, as there is nothing to compare there. Neurath was a radical positivist while Fleck's writing were aimed only against neopositivism. Accordingly, these are poles, and all we can say is that they are completely different. But even if Neurath was a follower of radical neopositivism that Fleck was fighting with, there are some common points in their programmes and I will try to show that even if they are in opposition, their paths intersected at certain points.

I. Science as Sociological and Historical Fact

Both Neurath and Fleck stressed that science is an activity of some societies, they called respectively – the republic of scientists and thought-collective. Those groups are not isolated from the rest of the society. They are being influenced by sociological factors, and this must be mirrored in theory of science. So Neurath warns philosophers:

(1.1) "Our thinking is a tool, it depends on social and historical conditions. One should never forget this. We cannot act as prosecutor and defendant at the same time and in addition sit on the judge's bench" ([Neurath 1931a/1983, 46]).

(1.2) "[a]ll science depends on historical conditions [...]" ([Neurath 1931c/1983, 141]; cf. [Fleck 1935, 34]).

Why? According to them:

M. Heidelberger and F. Stadler (eds.), History of Philosophy and Science, 299–306.
© 2002 *Kluwer Academic Publishers. Printed in the Netherlands.*

(1.3) "Every epistemological theory is trivial that does not take this sociological dependence of all cognition in a fundamental and detailed manner [...]"; "[Almost] whole of science is conditioned and could be explained by history of thought, psychology and sociology of thinking" ([Fleck 1935/1979, 43]; cf. [Fleck 1929/1986, 48-9]).

(1.4) "Unified science is the result of comprehensive *collective work* [...]. Unified science [...] is not the work of individuals, but of generation" ([Neurath 1931b/1983, 58 (see also 90)]).

(1.5) "Only through organized cooperative research, supported by popular knowledge and continuing over several generations, might a unified picture emerge" ([Fleck 1935/1979, 22]).

So they say that science is a historical process (1.1, 1.2, 1.3) and it is built in wide co-operation (1.4, 1.5). It would be impossible for particular individuals:

(1.6) "A single, really isolated human being would be condemned to mental sterility" ([Fleck 1960/1986, 155]).

Trying to show the continuity of human knowledge and its historical aspects, they pointed to metaphysical origin of science, which, according to them, shows that its logical reconstruction, research limited to synchronous aspects of science is not enough. To understand science fully one must start not only sociological, but also historical studies.

To exhibit the historical aspects both tried to show also some relations between science and magic (cf. [Fleck 1939/1979, 50]; [Neurath 1931a/1983, 34-36]; [Zolo 1989, 28-29]). They were interested in the genesis of scientific concepts and they studied development of some terms in the language of modern science from the language of magic (as from "taboo of touching dead body" to "infection"). Neurath, in the manner known better from Feyerabend's writings, shows the similarity of scientific and magical procedures (cf. also [Fleck 1938, 194]). These two types of activity are connected by the fact that they have to be tested by appeal to observable events: predictions of a magician are (if not always, quite often) testable. Magic – like science – touches the finite and the empirical; and – unlike theology – its affairs are from this world (cf. [Neurath 1931, 82]). A scientist and a magician are estimated on the basis of their (successful) predictions; in both cases effectiveness is the only proof of the value (cf. [Fleck 1960/1986, 153]). Of course, science and magic are not the same. The most significant difference is that magic is conservative, non-systematical, it lacks regular empirical control – and it lacks great thinkers.

(1.7) "Many very solidly established scientific facts are undeniably linked, in their development, to prescientific, somewhat hazy, related proto-ideas or pre-ideas, even though such links cannot be substantiated" ([Fleck 1935/1979, 23]).

(1.8) "Can epistemology blandly ignore the fact that many scientific positions steadily developed from proto-ideas which at the time were not based upon the type of proof considered valid today?" ([Fleck 1935/1979, 24]).

Just like Fleck Neurath reminds us that relatively many "magical judgments" are preserved in science up to this day, e.g. in psychology and medicine. Neurath and Fleck hold, in anti- or ametaphysical tradition of Mach, that scientific knowledge is an extension of everyday knowledge (cf. [Neurath 1931b/1983, 62 and 64]; [Neurath 1937/1983, 180]; [Fleck 1935/1979, 109]). It is true that Fleck would never say that one could separate science from pseudo-science. On the other hand, Neurath was famous just because his passionate attacks against "M". But it is also true that even he liberalized his stance (cf. [Zolo 1989, 88]).

So, to sum it up, to understand science we have to appeal to history and sociology – that is, to science itself (that is why Fleck wanted to talk about "scientific epistemology"). Fleck's doctrine forbids *a priori* approach to science, and – as it was once pointed out by Toulmin – it is not possible to found theory of science on any Popperian-like criterion of demarcation (cf. [Toulmin 1986, 267]). That is another similarity, because Neurath was of the same opinion: he put *Abgrenzungskriterium* on his *index verborum prohibitorum*.

II. SCIENTIFIC FACT. THOUGHT-STYLES AND ENCYCLOPEDIAS

At the very first glance the two philosophers have absolutely incompatible views here. Neurath did not want to talk about "facts" as it meant for him doubling metaphysics (cf. [Neurath 1934/1983, 113]; [Hempel 1935, 51]; [Scheffler 1967, 100-101]); he preferred to talk about observational or protocol sentences. On the other hand the concept of "fact" is crucial in Fleck. But if we accept Neurath's reservation as a manner of speech we will find another affinities between conceptions of Neurath and Fleck.

In Fleck's theory fact is a conceptual structure. It is theory-laden, or even it is a creation of a theory. When talking about facts we have some theories involved, as theories precede facts. Theory-ladenness of observations is very deep: theories make them possible and organize them – to some extent the outcomes of experiments are determined by theories. And when the theory changes, the facts change as well.

(2.1) "Both thinking and facts are changeable, if only changes in thinking manifest themselves in changed facts. Conversely, fundamentally new facts can be discovered only through new thinking" ([Fleck 1935/1979, 50]).

But those dependencies are bilateral. A theory organizes some empirical material. The states of affairs resist to arbitrariness of the theory, though they never determine it. The choice of set of facts is a question of decision, and something becomes a "fact" when it finds its application in a framework of thought-style (or in encyclopedia). Every new fact changes to some degree other facts. It can finally change the thought-style. So this is an holistic approach.

(2.2) "Facts are never completely independent of each other. They occur either as more or less connected mixtures of separate signals, or as a system of knowledge obeying its own laws" ([Fleck 1935/1979, 102]).

(2.4) "Consequently it is all but impossible to make any protocol statements based on direct observation and from which the results should follow as logical conclusions" ([Fleck 1935/1979, 89]).

The change of thought-style shows that there are no ultimate truths in science.

(2.3) "Knowledge [...] does not repose upon some substratum. Only through continual movement and interaction can that drive be maintained which yields ideas and truths" ([Fleck 1935/1979, 51]).

(2.3b) "One must not forget that there exists no fully completed science but only one that is becoming" ([Fleck 1929/1986, 55]).

Science is something that is happening and it is no surprise that some words change their meanings and that during the development of science scientists introduce concepts which are incommensurable to those that were used in earlier periods. So sometimes it is not possible to compare the old and the new. Sometimes we are not able to understand old theories and its concepts. This will lead Fleck to his theory of truth I am going to talk about later.

The views in this sketch are very similar to Neurath's ideas. First, in the model of encyclopedias every sentence is theory-laden, so it can be corrected or changed (cf. 2.3). After French conventionalists Neurath repeats that there are no "pure facts", or rather "pure protocol sentences".

(2.5) "Our initial observation statements in the sciences are not 'atomic,' but are already imbedded in a body of statements derived from different sources [...]" ([Neurath 1941/1983, 215]).

(2.6) "There exists no observation that would not be forestalled by a directing and limiting readiness of thought" ([Fleck 1946/1986, 123]).

(2.7) "[I] would argue that clear-cut distinction can be made between concrete and abstract elements. The entire classification is based upon a very primitive way of thinking" ([Fleck 1935/1979, 172]).

As the theories change, protocol sentences change too.

(2.8) "Collections of data (sky photograph, etc.), travel journals (for example, Darwin's journal of his voyage around the world is the most instructive about these problems) must of course set out from certain theoretical attitude to make a selection among possible statements feasible [...]" ([Neurath 1932, 204]).

So theory organizes the experience – but it is not independent on it: there is a possibility of empirical control. The choice of protocol statements depends, however, on decision and it is stimulated by pragmatical reasons, mainly by the effectiveness of prediction, because good predictions lead to further development of science. (Note that for Neurath one knows something when he/she is able to predict something else). Accordingly, Neurath (but also Fleck) call into question the role of *experimentum crucis* in science. They both tell us that a scientist should be aware that the change of encyclopedia or thought-style has its pros and cons – there is something we call today a "Kuhnian gap". Both Neurath and Fleck are anticumulativists.

Like Fleck, Neurath thought that the content of concept was a matter of change, though words themselves are often preserved in the language. The most important example here are *Ballungen*, i.e. not well-precised terms of everyday language. They are present in science and this is the main cause of imprecision of the universal slang. Similarly, we have "proto-ideas" in Fleck. But in distinction to Fleck, Neurath says it is exactly this feature of language which makes communication possible for people from different cultural circles and it makes it possible to know and understand the science of the past.

If science is a matter of constant change, it is a dynamic structure, and its development has no natural end. In Neurath we have no final solutions in science, but we have encyclopedias. Every encyclopedia contains the best available knowledge, it is a kind of synthesis, but synthesis that shows us the gaps. And for Fleck any synthesis, so does any encyclopedia, shows them as well.

III. TRUTH

In these two theories the concept of truth has its peculiar place. In Neurath's theory it is a kind of relic, it is a term used just for convenience. In Fleck's work the truth is a historical process. Both were accused of "relativism" and of "throwing out empiricism" (cf. [Schlick 1934]; [Bilikiewicz 1939/1986]).

In encyclopedia there are no incorrigible sentences. This is also true about so-called analytical sentences, especially of logic and mathematics (cf. [Neurath 1936/1983, 146]), and even of physicalism itself (cf. [Neurath 1931b/1983, 62]; [Neurath 1946/1983, 235]). We may call a sentence "true" if it is in good agreement with other sentences we have accepted earlier. Otherwise they are "false" (or, as Neurath used to put it, "isolated"). It is possible that one and the same sentence has a different status in two different encyclopedias. It is then better to say that not particular sentences but encyclopedias are entitled to be called "true". Of course, if we only insist on using that word. If we do we have to remember that truth is not a relation between sentences and facts: "the true encyclopedia" means "highly consistent one". Neurath thought – and this is easiest to find in his correspondence with Carnap – that semantics is incompatible with pluralism in science, and that Tarski's theory leads – via absolutism – straight to theology ("I think him [Tarski] anti-pluralist" – as once (25.09.1943) Neurath wrote to Carnap ([ASP 102-55-03])). It seems Neurath understood "truth" in the strongest sense of the word. His view was supported by some followers of the newest branch of logic who talked about "absolute truth" (cf. [Kokoszyńska 1936, 149ff.]). In his another letter to Carnap, dated 17.07.1943, he wrote:

(3.1) "I really don't understand how somebody may speak about of 'TRUE PROPOSI-TION' without an absolute standpoint; otherwise he has only ACCEPTED proposition with a person index" ([ASP 102-56-04]).

If – he thought – the notion of absolute truth is against pluralism, against a dynamic conception of science, against the thesis of correctibility of scientific propositions, it is purely metaphysical. And, as such, it surely cannot be accepted in encyclopaedic model of science. Note, that Neurath does not say that "all truths are (equally) good" or "the truth does not exist".

For Fleck, putting this briefly, the truth is a three-place relation that holds between 1) judgement, 2) the state of affairs and 3) present state of knowledge. Is this entirely different from Neurath's conception? No, because we my inter-pret this as saying that (1') judgement S is true, when there are (2') protocol statements that confirms S, and S may be (3') embedded within the encyclopedia. Otherwise, S is "false".

Let's ask, if they did call their statement a relativistic theory of truth. In the case of Neurath the answer is no, because he simply refuse to talk about "truth". Fleck also did not want to be recognized as "relativist". He says:

(3.2) "The 'truth' as the current stage of change in thought-style is always only one [...] Variety of pictures of reality is simply caused by variety of objects of knowledge" ([Fleck 1939/1986, 198-199]).

So, the truths of the alchemists in the 16. century should not be considered as false, they are not – or not fully – understandable. Neurath does not talk about a strong version of incommensurability but he agrees that in some cases translation from language of encyclopedia to language of another one may be difficult.

(3.3) "But even the initial statements of successful science are not fixed, since one could begin at the very beginning with different unified languages that cannot be translated into each other straight away." ([Neurath 1935a/1983, 116]).

If we ask if they described their positions as "throwing empiricism overboard", they would disagree – note that we have already said that theories are controlla-ble by (resistance of) facts or protocol sentences.

IV. SOME NEOPOSITIVISTIC FLAVOUR IN FLECK'S BOOK

We could see that Neurath was not so far away from Fleck as it is usually conceived. They are not strangers. On the other hand it seems Fleck was not independent of the neopositivistic tradition of philosophising. Although one can easily find many 'antipositivistic' accents in Fleck's book, it also contains some surprising phrases. So Fleck could write that "scientists proved" (cf. [Fleck

1935/1979, 66]) or "demonstrated" ([Fleck 1935/1979, 68]) something, or that "haemolysis is easier to detect because it may be seen with the naked eye" ([Fleck 1935/1979, 66]). If needed we have unquestionable facts and unquestionable scientific reports: "it became evident that the report was entirely wrong"; we can read also that "pflogiston still haunts in [Löw's] book" ([Fleck 1935/1979, 128]).

Some other passages seem to be surprisingly familiar: "There is nothing less obscure to me than metaphysics" or: "I simply do not understand either the sentence that 'the thing-in-itself' exists in absolute manner or its opposite that 'the thing in itself' does not exist in absolute manner" ([Fleck 1939/1986, 198]). Even if science has "magical" origin, astrology is a branch for "uneducated freaks" ([Fleck 1929/1986, 50]).

One can easily find some passages in Fleck, which are welcomed by postmodernists. They can call him their predecessor. They can, as they think anything goes. But it is not the case with Fleck. He would be deeply disappointed to see them subscribe to his legacy. If questioned whether *anything goes*, Fleck would clearly and firmly answer NO.

V. REFERENCES

ASP
  Archives of Scientific Philosophy, University of Pittsburgh, Pittsburgh PA. Quoted by permission of the University of Pittsburgh. All right reserved.

BILIKIEWICZ, TADEUSZ
  [1939/1986]: "Uwagi nad artykułem Ludwika Flecka 'Nauka a środowisko'", in: [Fleck 1986, 189-197]; 1st edition: *Przegląd Współczesny*, nr 8-9 (1939), 175-167.

COHEN, ROBERT S. and SCHNELLE, THOMAS (eds.),
  [1986]: *Cognition and Fact. Materials of Ludwik Fleck*, Boston Studies in the Philosophy of Science, Vol. 87, Dordrecht – Boston – Lancaster – Tokyo: D. Reidel Publishing Company.

FLECK, LUDWIK
  [1929/1986]: "On the Crisis of 'Reality'", in: [Cohen, Schnelle 1986, 47-57]; 1st edition: "Zur Krise der 'Wirklichkeit'", *Naturwissenschaften*, 17 (1929), 425-430.
  [1935]: "Zagadnienie teorji poznawania", *Przegląd Filozoficzny*, XXXIX (1935), zesz. I, 3-37.
  [1935/1979]: *Genesis and Development of Scientific Fact*, The University of Chicago Press, Chicago 1979; 1st edition: *Entstehung und Entwicklung einer wissenschaftlichen Tatsache. Einführung in die Lehre vom Denkstil und Denkkollektiv*, Benno Schwabe und Co. Verlagbuchhandlung, Basel 1935.
  [1938]: "Dyskusja. W sprawie artykułu p. Izydory Dąbskiej w Przeglądzie Filozoficznym", *Przegląd Filozoficzny*, XLI (1938), zesz. II, 192-195.
  [1939/1986]: "Odpowiedź na uwagi Tadeusza Bilikiewicza", in: [Fleck 1986, 198-202]; 1st edition: *Przegląd Współczesny*, nr 8-9 (1939), 168-174.
  [1946/1986]: "Problems of Science of Science", in: [Cohen, Schnelle 1986, 113-127]; 1st edition: *Życie Nauki. Miesięcznik Naukoznawczy*, t. I, nr 5 (1946), 322-336.
  [1960/1986]: "Crisis in Science", in: [Cohen, Schnelle 1986, 153-158]
  [1986]: *Powstanie i rozwój faktu naukowego. Wprowadzenie do nauki o stylu myślowym i kolektywie myślowym*, Lublin: Wydawnictwo Lubelskie (translation of 1935/1979; some additional papers included).

HEMPEL, CARL G.
[1935]: "On the Logical Positivists' Theory of Truth", *Analysis*, Vol. 2, No. 4 (1935), 49-59.

KOKOSZYŃSKA, MARIA
[1936]: "Über den absoluten Wahrheitsbegriff und einige andere semantische Begriffe", *Erkenntnis*, VI (1936), 143-165.

NEURATH, OTTO
[1983]: *Philosophical Papers 1913–1946*, Vienna Circle Collection, Vol. 16, Dordrecht – Boston – Lancaster: D. Reidel Publishing Company.
[1931]: "Magie und Technik", *Erkenntnis*, II (1931), 82-84.
[1931a/1983]: "Ways of the Scientific World-Conception", in: [Neurath 1983, 32-47]; 1st edition: "Wege der wissenschaftlichen Weltauffassung", *Erkenntnis*, I (1930/31), 106-125.
[1931b/1983]: "Sociology in the Framework of Physicalism", in: [Neurath 1983, 58-90]; 1st edition: "Soziologie im Physikalismus", *Erkenntnis*, II (1931), 393-431.
[1931c/1983]: "An International Encyclopedia of Unified Science", in: [Neurath 1983, 139-144]; 1st edition: "Une encyclopédie internationale de la science unitaire", *Actes du Congrès International de Philosophie Scientifique*, Vol. 2: *Unité de la science*, Hermann, Paris 1936, 54-59.
[1932]: "Protokollsätze", *Erkenntnis*, III (1932/33), 204-214.
[1934/1983]: "Radical Physicalism and the 'Real World'", in: [Neurath 1983, 100-114]; 1st edition: "Radikaler Physikalismus und 'wirkliche Welt'", *Erkenntnis*, IV (1934), 346-362.
[1935]: "Erster internationaler Kongress für Einheit der Wissenschaft in Paris (Congrès international de philosophie scientifique)", *Erkenntnis*, V (1935), 377-428.
[1935a/1983]: "The Unity of Science as a Task", in: [Neurath 1983, 115-120]; oryg. „Einheit der Wissenschaft als Aufgabe", *Erkenntnis*, V (1935), 16-22.
[1936/1983]: "Encyclopedia as a 'Model'", in: [Neurath 1983, 145-158]; 1st edition: "L'encyclopédie comme 'modèle'", *Revue de Synthèse*, 12 (1936), 187-201.
[1937/1983]: "Unified Science and its Encyclopedia", in: [Neurath 1983, 172-182]; 1st edition: *Philosophy of Science*, 4 (1937), 265-277.
[1941/1983]: "Universal Jargon and Terminology", in: [Neurath 1983, 213-229]; 1st edition: *Proceedings of Aristotelian Society*, N.S. 41 (1940-41), 127-148.
[1946/1983]: "The Orchestration of the Science by the Encyclopedism of Logical Empiricism", in: [Neurath 1983, 230-242]; 1st edition: *Philosophy and Phenomenological Research*, 6 (1946), 496-508.

ISRAEL SCHEFFLER
[1967]: *Science and Subjectivity*, Indianapolis – New York – Kansas City: The Bobbs-Merrill Company, Inc.

SCHLICK, MORITZ
[1934]: "Über das Fundament der Erkenntnis", *Erkenntnis*, IV (1934), 79-99.

TOULMIN, STEPHEN
[1986]: "Ludwik Fleck and the Historical Interpretation of Science", in: [Cohen and Schnelle 1986, 267-285].

ZOLO, DANILO
[1989] *Reflexive Epistemology. The Philosophical Legacy of Otto Neurath*, Kluwer Academic Publishers, Dordrecht – Boston – London 1989.

Dept. Of Logic and Methodology
Faculty of Philosophy and Sociology
Maria Curie-Sklodowska University
Pl. MCS 4, 20-031 Lublin
Poland
fishy@ramzes.umcs.lublin.pl

MALACHI HACOHEN

CRITICAL RATIONALISM, LOGICAL POSITIVISM, AND THE POSTSTRUCTURALIST CONUNDRUM: RECONSIDERING THE NEURATH-POPPER DEBATE [*]

"Science does not rest on a rockbed. Its towering edifice, an amazingly bold structure of theories, rises over a swamp," wrote Karl Popper (1902-1994) in the fall of 1932. "The foundations are piers going down into the swamp from above. They do not reach a natural base, but ... one resolves to be satisfied with their firmness, hoping they will carry the structure. ... *The objectivity of science can be bought only at the cost of relativity.*[1] The tower over the swamp represented the end of foundationist philosophy. Objectivity no longer rested on a rockbed but on the turns of scientific experimentation and criticism, as much a matter of vagary and luck as of talent and method. Surely, historians should have written Popper into the hall of fame of nonfoundationist philosophers. They did not. In fact, recent scholarship on the Vienna Circle, especially on Otto Neurath, represents Popper as the foundationist philosopher par-excellence. Some of his followers seem to miss his nonfoundationism, too. *A House Built on Sand* is the title Popperian philosopher Noretta Koertge chose for a spirited collection of essays that takes aim at the follies of science studies.[2] Alas, Popper describes science itself as built on sand (so whatever is wrong with science studies, it cannot be their choice of bedrock). But, then, why should historians and philosophers care about misreadings of Popper? Because they create a distorted picture of interwar Viennese philosophy that obscures, rather than reveals its contemporary relevance. This essay, focusing on the Neurath-Popper debate, attempts to redraw the picture and set the record straight.

During the past decade, historians and philosophers have excavated the logical positivist citadel and discovered sophisticated debates on language and method that pertain to current critiques of Western science and metaphysics.[3] I myself have participated in this quarry: historicization of the postmodern predicament, I assumed, is an effective antidote to false consciousness of its uniqueness. Caution is, however, well-advised. Scholars have used ingenious, but strained interpretations to turn the Vienna Circle into poststructuralists of sorts. In the case of Otto Neurath (1882-1945), they have imposed coherence on an imaginative but unsystematic mind, making him voice each scholar's preferred alternative to traditional scientific philosophy. This is counter-productive. If the past is to inform, rather than vindicate the present, its alterity must be preserved. Popper and the Vienna Circle did address "poststructuralist" prob-

M. Heidelberger and F. Stadler (eds.), History of Philosophy and Science, 307-324.
© 2002 *Kluwer Academic Publishers. Printed in the Netherlands.*

lems, but their answers were different and, in my view, better than ours. They did not echo poststructuralism, but voiced an alternative to it.

Neurath has become the darling of circle scholarship. He forms an attractive figure, catering to current tastes: an imaginative thinker, political radical, and philosophical iconoclast, who berated metaphysics. In contrast, scholars have accorded Popper rough treatment. Nancy Cartwright and Thomas Uebel represent him as a foundationist philosopher, an enemy. Popper's champions have not been successful in defending him against a poststructuralist Neurath because most of them are reluctant to engage poststructuralism head-on.[4] Those who do, like Koertge, end up defending scientific rationality in a manner that Popper would have found objectionable. To Neurath's celebrated metaphor of scientists laboring in a boat on the seas – "we are like sailors who have to rebuild their ship on the open sea, without ever being able to take it apart on the dock and reconstruct it from the best components" – Popper counterposes constructing a tower over a swamp. There is something to be said, I believe, for endeavoring to build over a swamp rather than on the open seas. At least, this essay will make the case for it. I shall present an alternative to both Cartwright and Koertge. In my view, Popper was an original nonfoundationist philosopher, and he shared with Neurath a wide range of assumptions, a common ground ignored both by Neurath and him, and by their interpreters. Where Neurath and Popper disagree, I think Popper had, with few exceptions, the better of the argument. Focusing on the foundation and protocol sentences debates of the mid-1930s, I hope to show that Popper provides the most viable response to "poststructuralist" dilemmas among the disputants: a modified conventionalist, nonfoundationist philosophy that safe-guarded rationalism, but skirted the dangers of absolutism.[5]

The Neurath-Popper debate broke into the open in 1935, but their adversarial positions had been the subject of circle exchanges since the fall of 1932. In an August 1932 meeting of Rudolf Carnap (1891-1970), Herbert Feigl (1902-1988), and Popper in the Tyrolian Alps, Carnap became familiar with Popper's philosophical breakthrough. He had brought a draft of Neurath's "Protocol Sentences" with him to the meeting.[6] Popper, who had just completed an early version of *Die beiden Grundprobleme der Erkenntnistheorie,* learned about recent theoretical developments in the circle, specifically about the early phase of the protocol sentences debate. Later in the fall, after much behind-the-scenes debate, Carnap published "On Protocol Sentences" in *Erkenntnis,* reporting Popper's views. He described Popper's and Neurath's positions as fairly close.[7] Both objected, and Popper claimed that nonfoundationism was his, not Neurath's, discovery. He was gratified by Carnap's promotion of his work, but he resented Carnap's use of his ideas to improve on Neurath. Carnap told him that Neurath, too, claimed to have been first to propound nonfoundationism, and Popper was chagrined.[8] His originality and independence, he felt, were at stake.[9]

Popper had known Neurath for over a decade, albeit not well. He left us a brief memoir of their encounters in 1920 in Vienna. Neurath impressed him as a most unusual personality ... a man who believed passionately in his social, political and philosophical theories, but who believed even more in himself ... a man who ... would not look behind him or, when rushing ahead care very much about whom his big stride might knock down.[10]

For a few months Popper saw him at *Akazienhof,* a pleasant not-for-profit eatery *(Gemeinschaftsküche)* that Schwarzwald opened near the university for students and professors, a meeting place for the radical intelligentsia. They met again eight years later, at Neurath's lecture to the *Verein Ernst Mach,* and, apparently, once or twice when Popper had his class visit the Economic Museum. Nothing about Neurath seemed right to Popper, neither the person nor his philosophy nor his politics.[11] He stood for everything that was wrong with progressive and Marxist intellectuals. Popper shaped both his philosophy of science and political philosophy in confrontation with Neurath.

The Neurath challenge was, I believe, a major inspiration for the last phase of Popper's epistemological revolution – his move to nonfoundationism. During the fall of 1932, Popper wrote a new Kant-Fries critique, containing the first statement of nonfoundationism in his work.[12] He cast himself as a Kantian philosopher. Neurath seems to have clarified to him things that had been only dim in his mind before. Statements can be compared only with statements, language with language. Experience cannot provide an indisputable foundation for science. Popper now understood how nonfoundationism resolved the nagging problem of Kant's synthetic *a priori.* He did not appreciate, however, Neurath's contribution, and ascribed Neurath's position to Jakob Fries (1775-1843), Leonard Nelson (1882-1927), and the contemporary Austrian Kantian philosopher Robert Reininger (1869-1955).[13] They recognized first, he said, the problematic character of comparison between statement and experience. The positivists were rehearsing earlier Kantians, not articulating a new epistemology. His philosophy, he implied, owed much to Kant and Fries, not to Neurath and the circle. He was not wrong, but his reconstruction concealed the positivist contribution to his revolution. He was performing Kantianism to convince a positivist audience that he was not staging their own show.

In November 1932, Popper sent a short note to *Erkenntnis,* summarizing some of his philosophical results, and clarifying his disagreements with logical positivism.[14] This closed the first act in the Neurath-Popper drama. During the first half of 1933, Popper wrote little and corresponded rarely. He was depressed about the prospect of his work. In late June 1933, however, he secured a book contract from Springer. He then went into hiding, as was his practice, to rework *Grundprobleme* into publication, and resurfaced only in late August 1934, in the circle's conference in Prague, to collide with Hans Reichenbach (1891-1953). The book that emerged in the fall of 1934, *Logik der Forschung,* became a classic in the philosophy of science.

As Popper was rushing to complete his book in 1933-34, prolonged tensions in the Vienna Circle broke into the open. Throughout 1933, the circle's head, Moritz Schlick (1882-1936) had grown increasingly agitated about the circle's "left-wing," especially Neurath. Both Neurath's philosophy and politics, he felt, made defending the circle before the hostile academy and government difficult. Finally, Schlick broke the taboo on making circle disputes public. Recovering from a bad flu in April 1934 in Salerno, Italy, he wrote "On the Foundation of Knowledge." "Radical physicalism" was his target, and Neurath the major antagonist. His attack did not remain unanswered. Carnap responded in private, Carl Hempel (1905-1997) and Neurath in public. The summer and fall issues of *Erkenntnis*, the circle's organ, carried the sharp exchange. It became known as the "protocol sentences debate." [15]

Popper was initially oblivious to the debate. *Logik der Forschung* included a sharp critique of Carnap's and Neurath's protocols, but took no account of Schlick's and Neurath's reformulated positions. Popper designed his rhetorical strategies against the positivism he had known in 1932, not 1934. [16] Neurath eventually directed him to the exchange in *Erkenntnis*. Popper did not change his mind. On the contrary: the debate provided him with new ammunition for his attack on positivist psychologism and subjectivism. Neurath's review of *Logik* in 1935 brought out in sharp relief the contrast between protocols and falsification, the positivist-physicalist and critical Kantian programs for philosophical reform. Popper thought little of Neurath, but he was grateful for his critique, all the same. [17] He appreciated the engagement, and delighted in being called a Kantian by a positivist. He was apprehensive lest Carnap's "official" favorable review in *Erkenntnis* would once again assimilate his philosophy into positivism. To Carnap and Schlick, his methodology did not seem revolutionary. "He is completely of our persuasion," wrote Schlick to Carnap. [18] Neurath disagreed. He took seriously their metaphysical differences. His critique illuminated *Logik*'s significance for the protocol sentences debate.

Schlick's 1934 essay moved the foundation problem to the center of circle debates. Most circle members recognized that scientific access to reality was problematic, but held, nonetheless, to verification, without being clear what it entailed methodologically. Popper caught them at their most careless moments when they sounded like old-fashioned empiricists, but they were not all, *pace* Popper, pre-Kantian foundationists. Recognizing the difficulty of confronting theory and experience, Neurath and Carnap tried to confront theory and physicalist protocols. To Popper, they merely translated experience into protocols and remained subjective foundationists. To Schlick, on the contrary, they severed the relationship between theory and experience, science and reality.

Like Neurath and Popper, Schlick recognized the gap between psychological experience and scientific language. Observations confirming predictions generated feelings of certitude – "aha!" experiences – among scientists, but, "as soon as the soul *speaks,* alas, it is no longer the *soul.*" [19] Neither observation nor certitude were fully translatable into scientific language. An "aha!" put into a statement was no longer an "aha!" Schlick resisted, however, the conclusion that statements could only be compared with statements. Assertions, or affirmations *(Konstatierungen)* of "what is immediately observed" were "unshakable points of contact between language and reality."[20] They verified predictions and theories. They constituted both science's end-point and a new beginning, an occasion for forming new hypotheses.

Affirmations did not quite amount to statements. Translated into scientific language, the assertion "here now blue" became a protocol, but lost its affirmative character: "at time x, place y, Schlick perceived blue." Like all scientific statements, protocols were hypothetical. To Schlick, Neurath confused affirmation and protocol, and made science's foundation hypothetical and relative. His suggestion that protocols be accepted or deleted based on their conformity with a system of statements overthrew empiricism. Russell had already criticized formal, or "coherence," theories of truth, said Schlick. They defined truth by consistency of statements, not correspondence to reality, leaving scientists unable to choose among internally coherent, but conflicting scientific theories. If reality could not serve as the final judge of truth, modern science would end up with a very "peculiar relativism."[21] Immediate, certain, and final affirmations provided science with an unequivocal criterion of truth, a foundation of sort.

Neurath responded that "reality" could not sit as a judge in any scientific dispute. In science, reality, too, appeared as a linguistic construct. Competing theories portrayed not "one true world," but several possible worlds. Statements contradicted one another, not "reality." Any talk about "correspondence" to reality was metaphysical. Science did not lay down conditions for "truth," but for the acceptance of statements. Schlick's complaint that "coherence" did not provide an unambiguous truth criterion was thus irrelevant. Indeed, no such criterion was possible. Science never escaped ambiguity. When protocols translated personal experience, or observations, into statements, they did not simply establish facts. They included imprecise concepts *(Ballungen)* and perception terms. Extra-logical criteria, such as economy of time, prompted decisions to adopt, or reject protocols. But this implied no surrender of empiricism. Scientists strove for agreement between hypotheses and as many protocols as possible, amending hypotheses contradicted by protocols. Physicalism purged non-empirical statements. "Certainty" was, however, metaphysical. Affirmations were available to science only as protocols. (The notion that they were untranslatable was mystical. It echoed Wittgenstein's "unsayable.") Translated into protocols, affirmations were first reconciled with each other, then with a plurality of theories. Modern science represented diverse worlds.[22]

Popper and Neurath were epistemologically closer to each other than either was to any other circle member. They represented modified conventionalism, anti-absolutism, and nonfoundationism. Both insisted that language cannot be compared with reality and considered Schlick's "immediate observations" scientifically irrelevant. (Carnap wavered.) Both believed that observation reports were theoretically loaded and regarded them as provisional.[23] Extra-logical factors played a role in decision to accept them. Both sought ways of bridging the gap between experience and language, so that experience informed scientific theory. Popper demonstrated his genius as a methodologist by keeping science and reality separate, and yet allowing experience to arbitrate theoretical questions through falsification, i.e., through a conventional decision, logically and methodologically informed, to accept a falsifying statement. In contrast, Neurath's linguistic control, protocols that testified to their own empirical formation,[24] was cumbersome and ineffective. Popper was so successful because he recognized that the linguistic turn made it impossible for language to guard empiricism. Instead, he sought a measure of empirical control in methodological conventions.

To Popper, positivism grounded science in perceptions and experiences. Schlick's "Foundation" was a perfect example.[25] Placing a premium on the scientist's feeling of certainty, Schlick recapitulated Fries' "immediate knowledge." Such "knowledge" was irrelevant to science. "We must distinguish between our *subjective experience of conviction* ... and *objective-logical relations* among systems of scientific statements."[26] Carnap and Neurath made an unsuccessful attempt to overcome the gap between psychology and logic by translating psychological behavior into physicalist language. Whether phenomenalist or physicalist, their protocols were logical construction of experience, "perception statements," records of sense data, translation of observations into formal speech. They gained nothing by changing the mode of expression. They remained attached to the psychological basis.[27]

Popper's critics claim that he misunderstood Neurath, assimilating his position on protocols to that of Carnap.[28] Neurath, they say, rejected Carnap's phenomenalism. Protocols did not record "immediate experience," but "stimulation states," "changes in certain areas of perception in the brain" in response to the environment. For example: Malachi's protocol at 7 P.M., 24 January 1998: (Malachi's speech-thinking at 6:59 P.M. was: [at 6:58 P.M. there was a UFO in space-region *k* perceived by Malachi].) Neurath thought that behaviorist, or physicalist, language escaped the phenomenalist trap. Malachi's protocol reported physical events and states of affairs, not personal experience. Popper denied precisely that, and Carnap accepted his verdict.[29] Behaviorist trappings notwithstanding, Neurath's protocols related psychological experience. Neurath occasionally spoke of protocols as statements of personal experience *(Erlebnis-aussagen).*[30] He insisted that they include both the observer's name and observational terms (see, perceive). To Popper, this suggested psychologism. His reading of Neurath was neither generous nor thorough, but it was correct all the same. Behaviorism could claim no dispensation from psychologism.

Neurath claimed such a dispensation, but he did not articulate a coherent defense.[31] His protagonists have developed one for him. Uebel regards physicalism as liberal "naturalism." Neurath stipulated that science may invoke events and states of affairs only if they could be naturalistically described and explained.[32] Physicalist, or behaviorist, protocols enforced the provision. By definition, they were anything but experiential. But were they? The aforementioned protocol indicated that Malachi's perception was that "there is a UFO in space-region *k*," and his "speech-thinking" converted it into a protocol. Did this transform the statement into a behavioral description? Or did it, on the contrary, psychologize an intersubjective observation? Popper thought the latter. Neurath showed nothing to justify his high hope for linguistic reform. Why should Popper have made the behaviorist leap?

In contrast to Uebel, Cartwright suggests that protocols were both about individual experience and public matter-of-fact. Information about the conditions of observer and observation served as "entitlement" for making the statement. Neurath required both entitlement and intersubjectivity. (This explains protocols' awkward structure). Popper demanded only intersubjectivity, dropping all reference to observer and observation: "there is a UFO in space-region *k*." "No one would be led into foolish mistakes by Popper's 'shorthand' version," says Cartwright: every scientific statement requires entitlement.[33] Perhaps. For Popper, intersubjectivity (observability) was necessary and sufficient "entitlement." Without reproducible UFO observation, no entitlement would make Malachi's protocol scientific. Cartwright simply refuses to take seriously Popper's radical anti-psychologism (and anti-foundationism). She disregards, at the same time, Neurath's difficulties in moving from individual experience to public matter-of-fact. Neurath used "speech-thinking" *(Sprechdenken)* to explain how experience becomes protocol. The notion embodied psychologism's basic problem: the gap between language and perception. Having separated the two, Neurath could not quite break with the psychological basis, and tried behaviorist reunification. He reintroduced the psychologistic dilemma in speech-thinking protocols.

Popper could therefore rightfully claim to have been the only one to truly break with psychologism. Knowledge was grounded in particular dispositions, biological, psychological, or what not. This was of interest to evolutionary biology and psychology. It had no bearing on logic, or epistemology. Testing, that is, a combination of logical procedures, methodological rules (conventions), and rhetorical persuasion, arbitrated scientific questions. Testing did not stop at protocols, or perceptions, but at easily testable statements which scientists decided to provisionally accept. Science was free of subjective psychological foundation.

Testing held together Popper's science. Without it, language and reality, theory and experience were hopelessly severed. Neither Carnap nor Neurath provided a testing method. How, asked Popper, were decisions on protocols to be made? The Neurath Principle permitted arbitrary deletion of protocols, or amendment of theories. Any theory could be so salvaged, or even confirmed. Protocols were also difficult to test. Malachi's observation of a UFO was testable, but not his brain stimuli.[34] Test statements concerned observable facts, such as movement and position of physical bodies, not psychological experience. Observability *(Beobachtbarkeit)* was intersubjectivity's sole guarantee.[35] Science's empirical dimension consisted in testing theories through observable basic statements. A theory of lower generality (basic statement), closer to experience, but never quite reality –"the thing" – itself, could knock out a theory of greater generality. Protocols provided no comparable method, neither objectivity (intersubjectivity) nor demarcation of scientific (testable) from nonscientific statements. Neurath overthrew empiricism.

Neurath responded to Popper in a 1935 review of *Logik* that has become a classic, "Pseudorationalism of Falsification." [36] Against the backdrop of *Logik,* Neurath made an "encyclopedic turn." [37] Science, he argued, was not an hierarchically structured body of knowledge, a closed, coherent, homogenous system. Rather, there existed multiple sciences, loosely connected, if at all, each with its own practices. To Popper's system of well-defined theories, constructed of clean statements and striving toward simplicity, Neurath counterposed his own model, the encyclopedia. It consisted of porous bodies of statements, linked "sometimes more closely, sometimes more loosely. Systematic deductions are attempted at certain places, but the nexus in its entirety is not transparent." [38] Clarity, simplicity, and coherence were metaphysical ideals because multiplicity, diversity, and indeterminacy characterized science. "'The' system is the great scientific lie." [39]

Physicalist language alone unified science. Falsification could not set methodological rules, because no common rules existed. Testing could not carry the burden Popper put onto it: experiments played little role in many sciences. Popper's *experimentum crucis* flew in the face of practice. Testing provided no certainty of falsification because Duhem's conventionalist stratagems – amending a theory to immunize it against refutation – applied. When a theory served scientists well, an apparent refutation would not convince them to abandon it, only shake their confidence *(Erschütterung).* Popper's view of science as a permanent revolution neither reflected scientific practice nor served it well.

Why, then, did Popper desperately attempt to found scientific logic on falsification? Neurath thought he knew the answer: Popper wanted to maintain at all cost the logical choice between theories. Falsification was a salvage operation for the "old philosophical absolutism," or "pseudorationalism," that appealed to the "real world" to resolve scientific disputes. "The pseudorationalism in Popper's basic view would quickly make us understand why he could feel drawn to traditional philosophy *(Schulphilosophie)* and its absolutism, while his book contains so much of that analytical technique advocated precisely by the Vienna Circle," and why he is so much kinder to Kant than to the positivists. [40]

Alone among the circle, Neurath recognized the limits of Popper's partaking in positivist discourse. Neurath's rhetoric concealed his broad agreement with Popper on conventionalism, but illuminated the chasm between linguistic and critical philosophy. This did not make Neurath a sound judge, let alone an unbiased arbitrator. To him, Popper was positivism's nemesis. He clashed with Popper in the Paris and Copenhagen congresses in 1935 and 1936. Popper then went to New Zealand and was out of touch and without influence. Neurath did not relent. In his correspondence with Carnap, he harped on pseudorationalism becoming a metaphysical fifth column amidst the circle. [41] He fought the legend of the "positivist Popper" tooth and nail from the start. His premature death in 1945 left the legend uncontested in postwar years.

Neurath's protagonists second his judgment of Popper, but they recognize that he lacked method. "Popper is perfectly justified," says Cat, "in noting that 'Neurath gives no method [to] delete a protocol sentence which contradicts a system'."[42] They offer Neurath two divergent ways out. Uebel constructs a Neurath method. He searches for methodological clues in Neurath's works, skillfully establishing rules that Neurath would have enunciated, if only he had put his mind to method.[43] His suggestions are so reasonable that, when he is done, one wonders why Neurath needed obfuscating protocols as empirical control. Uebel moves the debate to the methodological terrain. Popper is there at his best: agile, precise, systematic. Neurath is imprecise, diffuse, unclear.

In contrast to Uebel, Cartwright and Cat argue that Neurath rejected all method, and it is good he did. "Scientific method is logically open-ended," they write.[44] Extra-logical considerations determine scientific decisions. "We always start from historical, natural language. Its sentences are *Ballungen,* and that means mixtures of forms of expression (precise and imprecise concepts)."[45] There are "no fixed connections" between theory and basic statements. "Theoretical hypotheses and protocol statements can[not] endure an unambiguous logical confrontation with testing value," insists Cat.[46] Deduction and comparison of basic statements are impossible. But, then, any comparison of theory and protocol is in jeopardy. Confirmation and shaking are impossible as well! There is no more basis for tentative shaking than for definite falsification. Indeed, how can Neurath even decide whether he has confirmation or shaking? Imprecision must be measured before any conclusion is possible. Logical imprecision can be no reason for tentativeness. Overthrowing method, Cartwright and Cat overthrow Neurath, too.

To Popper, to the extent that method was logical, it was *not* open-ended. Language was "imprecise" in that both universals and undefined terms occurred in basic statements, but this was no obstacle to logical procedures. Method *was* open-ended, because it was conventional, dependent upon decisions. His position was nuanced. He distinguished between logical falsifiability and methodological falsification. Falsification depended on both logical and extra-logical criteria. There was little disagreement between him and Neurath concerning the extra-logical character of scientific decisions. He would gladly second Cartwright's and Cat's view that "every application requires judgment, local knowledge and free decision. ... Every concrete prediction is a matter of construction."[47] Scientists negotiated acceptance and rejection of basic statements, but this did not imply giving up on method. Cartwright insists that science can provide no rule to resolve logically undecidable questions. Popper was more sensible: methodology provided guidelines for negotiation. It was stupid, he said on occasion, to attempt and make such rules precise.[48] It was equally unwise to think that they were dispensable.

Popper's insistence that choice among theories could be informed by rules irks Cartwright and Cat. Any implication that we are not at sea, they say, is absolutist and metaphysical. Indeterminacy is a positive good, determinacy "pseudorationalism." Neurath's great virtue is to have permitted co-existence of multiple encyclopedias as opposed to Popper's one theoretical system.[49] But Popper, too, thought that we are at sea, or, better, that we never know for sure whether we are docked safely at the shore. The boundary between sea and shore is murky, and we better treat any *terra ferma* as moving sands, or swamp. He acknowledged that choice among theories was rarely clear, and not always possible, but he insisted that where tests were possible, a decision may also be. To pursue knowledge meant to seek falsification, not to avoid it. Neurath challenged falsification, but, unlike his protagonists, he did not cherish indeterminacy. Logical indeterminacy was scientific reality. He countered it with linguistic rules. "Open-ended method" is Cartwright's and Cat's, not Neurath's, idea. Seek no rule, they say, negotiate. They are determined to reconfigure Neurath as a poststructuralist.

Neurath claimed to be a consistent empiricist. Poststructuralism has made a laughing matter of traditional scientific and historical empiricism. Yet, Neurath's admirers are all scientifically inclined philosophers, seeking to salvage some empirical practice from the poststructuralist ruins. Having charged Popper with excessive logical determinacy, Cat turns the tables and accuses him of empirical indeterminacy. Falsification, he concurs with Neurath, was a pseudorationalist substitute for protocols. It failed to secure empiricism. "At the level of the empirical basis," he acutely observes, "Popper is no less conventionalist than Neurath."[50] But Neurath's protocols, he insists, secured science's empirical foundation by establishing strict data admission criteria: "statements that wear their own observational genealogy (perception terms) and public doxatic coordinates (institutional encoding) on their sleeve." Popper's test statements "fail to make explicit the ultimate agreeable element on which science can possibly stand, namely our experience."[51] All Popper required was "observability" – vague and undefined. "Decisions to determine what is accepted on an empirical basis are largely immaterial."[52] Excluding empirical "entitlement," Popper overthrew empiricism.

A careful reader, Cat recognizes that Popper was no foundationist, but he misunderstands observability. He confounds preconditions for tests with their results, collapsing observability and falsification. Observability made tests possible; it said nothing about results. It provided no grounds for a basic statement's acceptance. It guaranteed intersubjectivity, not empiricism. Should observability be contested, there was no recourse to experience: another statement would have to be tested. As Neurath had no testing method, he made conformity with linguistic rules precondition for admitting a statement. Cat searches for similar rules in Popper, but he had none. Scientists used whatever (intersubjective) language they preferred. Testing determined acceptance. Only if Cat denies that tests *may* persuade us to accept falsification can he claim that Popper overthrew empiricism. Nothing guaranteed, of course, statements' "empirical character." Why should nonfoundationists seek guarantees? Why should they entrust empirical guards to behaviorist language, translate experience into awkward protocols, try to circumvent *decision* on evidence by lending statements empirical character? Cat's Neurath demonstrates Popper's charge: positivism was in search of an empirical basis.

Neurath condemned falsifiability as "old philosophical absolutism," but offered no demarcation criterion of his own.[53] His protagonists concur with his critique and insist that confronting hypotheses with protocols adequately demarcated science against metaphysics, but they feel ambivalent about demarcation. They share Neurath's antipathy to metaphysics, but not his scientism, his belief in science's exclusive legitimacy. They are apprehensive lest demarcation exclude too much, and prefer blurred boundaries. Uebel argues that Neurath's naturalism was broader than Quine's, recognizing "all human cognitive endeavors." Cat concedes that Neurath's protocols "may be too strict," but, fortunately, they are "not decisive for theory choice."[54] This is all too generous to Neurath, too harsh on Popper. Popper represented their convictions better than Neurath did. There was nothing absolute about falsifiability. "My demarcation criterion," Popper said, is "a *proposal of a convention.* People can differ on its usefulness."[55] Reasonable people *may* accept it. Science had conventional boundaries, and they were anything but clear. Neurath's linguistic demarcation, in contrast, was draconian. He recognized that linguistic choice was pragmatic, but this did not make his physicalist strictures any less dogmatic. He had no doubt what science was and was not. Neurath was the demarcator *par-excellence.*

Metaphysical disagreement underlay the Neurath-Popper exchange. To Neurath, Popper's commitment to the "one and only world" spelled out metaphysics. Behind Popper's methodological realism, he correctly discerned a metaphysical one, but he overlooked his own metaphysical commitment.[56] He deluded himself that protocols avoided the "real world" problem. They did not. Linguistic reform depended on a monistic metaphysics – be it materialist, naturalist, or physicalist. Neurath correctly suspected that linguistic choice often entailed a metaphysical one. In response, he launched a linguistic purge to secure an official language. Popper demurred: argue, do not silence; criticize, for you may persuade. (In postwar years, he recognized that this applied to metaphysics, too, not only to science.) Whether his scientific method was metaphysically neutral, or not, his science permitted a dialogue between statements informed by conflicting metaphysics. Neurath and Popper were both wrong to assume that their philosophy was metaphysics-free, but, then, there is good and bad metaphysics. Popper's realism was liberal, Neurath's monism absolutist.

Popper and Neurath subscribed to similar progressive historical narratives. Both considered science the peak of human achievement. Both recognized, at the same time, that science was conventional. They experimented with liberalizing science, relaxing its boundaries and method. Popper went further than Neurath, because his historical narrative was more contingent, and his politics more liberal. A vicious grand narrative underlay Neurath's linguistic reform. He consigned physicalism's opponents to the dust bin of history, delegitimizing Popper's mode of engaging the past. Popper made some scientific progress, he said, but then fell into serious error, regressing to traditional philosophy. Falsification was a pseudorationalist "residue." The scientific worldview and revolutionary politics both depended on physicalism. He, Neurath, was pronouncing history's judgment. Future science and politics spoke through him. History was moving his way.

Neurath's protagonists ingeniously liberalize him, softening his scientism by loosening his science. They are not "wrong"; rather, they develop tendencies in his philosophy. Still, their interpretations are strained. They override resistance – linguistic dogmatism, monistic metaphysics, historical determinism. Their hostility to Popper makes their partiality to Neurath all the more objectionable. To present Neurath as a tolerant empiricist, pluralist, and pragmatist, and Popper as a regimented metaphysician is unfair and untrue. Of the two, Popper liberalized science more thoroughly. His nonfoundationism was more radical, his method more ecumenical. In postwar years, he declared scientific method simply a systematic application of trial-and-error, the problem-solving approach to life. Unlike Neurath, he rejected scientism. As a person, Neurath may have been more open and tolerant.[57] As a philosopher, Popper was, by far.

Neurath's protagonists cast Popper off because he resisted the linguistic turn, opposed the crusade against metaphysics, and sustained method. His stance on all three issues is currently unpopular. I find him persuasive, all the same. Recognizing the gap between language and reality, he nonetheless refused the myth of the prison house of language. He validated dialogue across languages, in science and everyday life alike. Criticism of empirical evidence and methodologically informed decision were possible in any language. On metaphysics, his philosophy resists the positivist-poststructuralist alliance, demanding that both positivists and poststructuralists acknowledge their own metaphysics (or, to use non-Popperian parlance, their own historicity). At the same time, he admits metaphysics to the critical dialogue. We can argue about metaphysics, and persuade. Critical rationalism shuts no one and nothing out of the conversation, whether on account of language or non-empirical evidence.

Critical dialogue has, to be sure, a few ground rules. They are more explicit in science than elsewhere. Current animus toward rules, even liberal ones, is so great that, for Cat and Cartwright, Popper's method classifies him immediately as the "other." Yet, in indeterminacy's name, Cat is willing to put pluralism itself at risk. "Neurath leaves the door open to the possibility of extra-scientific sources of dogmatism upon scientific practice," he rejoices. "That is one reason why for Neurath it is important that scientists share the same worldview." So much for diversity! Neurath's alternative to liberal rules, it appears, is not Paul Feyerabend's anarchism, but dogma that derives its legitimacy from the community. (Whether Feyerabend's anarchism, or any view rejecting critical dialogue, does not end up in dogmatism is a different question.) Given a choice between Cat's Neurath and absolutist rationalism, I would opt for absolutism. Fortunately, my choice is not so limited. There is Popper. Conflicting perspectives are in vigorous debate, no decision guaranteed, none ruled out.

My critique of Neurath's protagonists does not diminish, but reinforces their achievement. They have made obscure debates in interwar scientific philosophy so movingly speak to the present that we take sides in them. My own work on Popper endeavors to do the same. Their achievement comes, however, at the cost of "presentism." Had they trusted Popper rather than Neurath to lead them out of the poststructuralist conundrum, they would not have violated the past's alterity.

No member of the Vienna Circle subscribed to Cartwright's and Cat's world-view. All remained committed to progressive historical narratives. All believed that science was humanity's crowning achievement. Most made nonfoundationist gestures, but remained empiricists, refusing to severe the relationship between language and experience. None followed the late Wittgenstein's move from scientific language to language-game. None viewed language as a culturally bound artifact, a historical product, whose changing rules were difficult for an outsider to grasp and impossible to translate.[58] None thought local languages deserved protection. Neurath designed educational programs to eliminate any but the universal scientific language. Carnap persisted in efforts to develop a scientific meta-language even after realizing, in 1935, that rules for concepts' use were not syntactic, i.e., logical and formal, but semantic, i.e., dependent on meaning. For good or bad, the Vienna Circle represented late enlightenment *(Spätaufklärung),* not poststructuralism.

Popper weeded illiberalism out of late enlightenment, but preserved its emancipatory potential. His political philosophy targeted progressive grand narratives for criticism, but retained the hopes for progress and freedom. His cultural politics was pluralist. He readily accepted multilingualism – even in science – and multiculturalism, but a diversity of closed cultures was not the best humankind could do. Culture clash in an open society was. Languages had different rules and presented divergent pictures of reality, but they were not self-contained and self-referential. Dialogue across languages was essential. Popper eschewed any certainty, any finality of empirical evidence. He, too, traveled on Neurath's boat, but he thought the boat was navigating murky coastal plains rather than the open seas, at least sometimes. Where they can, scientists should endeavor to construct as solid a building as the moving sands under them would allow. Of course, the structure cannot hold indefinitely. Eventually, it will collapse, nay, scientists will demolish it when they find they can build a better one.[59] But the uncertainty of the structure did not inhibit the growth of knowledge. If decisions on whether, where, and how to build were not always possible, and the solidity of the different structures controversial – indeed even the condition of the moving sands may not be a consensus – criticism of existing and proposed structures was always possible. Openness to criticism helped eliminate error. Conventions set the rules of debate, and traditions set its terms, but they, too, were subject to change.

Popper may have been over-confident about methodology's capacity to guide critical debate and produce a consensus. Neurath's objection that scientific practice never followed Popper's prescriptions remained largely unanswered. Both Neurath and Popper were more sensitive to science's extra-logical dimensions than were other circle members, but Neurath, a social reformer even when speaking of science, emphasized, more than Popper, science's collective character and, already in the 1930s, envisioned an ideal community of scientists. As sociology of science did not yet exist, both Neurath's and Popper's observations were preliminary. *Logik der Forschung* was a book on scientific method, not on scientists, or the scientific community. Popper acknowledged extra-logical criteria for scientific decisions, but they were of no concern to him. Method applied precisely the same way to the individual scientist and to the community. The psychological and sociological conditions making intersubjectivity possible were irrelevant. During his World-War-II exile in New Zealand, he added an important proviso. Political freedom made criticism possible, hence, it was essential to science, but he resisted any temptation to carry the investigation further. Science's institutions and public processes were essential, but he took them for granted – they existed because science did. He assumed the public character of science, but never investigated it.

This was a problem. As Popper discarded foundationism, intersubjective criticism became objectivity's new grounds. Criticism and testing operated by consensus and convention. How did intersubjective criticism really work? It was not clear that access was available to the whole public, certainly not equal access. Once ideas entered the public sphere, who won? Did the logic (or methodology) of science really set the rules of discussion? Without a sociology of science, public criticism remained a regulative ideal at best. Moreover, Paul Feyerabend and Thomas Kuhn charged that Popper could not account for most "correct" historical decisions in favor of better theories. If the key to scientific progress was the psychology, sociology, and routine of scientific communities, not criticism and testing, then Popper's effort to erect a rational edifice of science and explain the growth of knowledge as a rational process was problematic.

Recently, some of Popper's students began inquiring after the conditions facilitating criticism and consensus.[60] This is, I believe, the most promising terrain of investigation for the next generation of Popperians. We may discover that a particular polity is optimal for Popperian science, and it may require democratization of all spheres of life – state and economy, academy and laboratory. Such inquiry need not undermine Popper's belief in learning from error; indeed, such belief is its prerequisite. Our ability to learn from error is the issue dividing Popper from both positivists and poststructuralists. Popper was as instructive in refusing "poststructuralism" as he was in reshaping the legacy of the Viennese late enlightenment.

NOTES

*   This essay draws on my intellectual biography: *Karl Popper – The Formative Years, 1902-1945: Politics and Philosophy in Interwar Vienna* (New York: Cambridge University Press, 2000).

1.   Popper, *Die beiden Grundprobleme der Erkenntnistheorie* [1930-33] (Tübingen: Mohr, 1979), p. 136. (Henceforth: *Grundprobleme.*) Revised, the first half of the quotation (to "they will carry the structure") concluded section 30 of *Logik der Forschung: Zur Erkenntnistheorie moderner Naturwissenschaft* (Vienna: Julius Springer, 1935), pp. 66-67. (Henceforth: *Logik.* The English edition, *The Logic of Scientific Discovery,* trans. Karl Popper [London: Hutchinson, 1959] is quoted as *LSD.*)

2.   Noretta Koertge, ed., *A House Built on Sand: Exposing Postmodernist Myths About Science* (New York: Oxford University Press, 1998).

3.   Nancy Cartwright, Jordi Cat, Lola Fleck, and Thomas Uebel, *Otto Neurath: Philosophy between Science and Politics* (Cambridge: Cambridge University Press, 1996); Jordi Cat, "The Popper-Neurath Debate and Neurath's Attack on Scientific Method," *Studies in History and Philosophy of Science,* 26 (1995): 219-250; Malachi Hacohen, "The Making of the Open Society" (Ph.D. diss., Columbia University, 1993), chap. 6; Thomas Uebel, *Overcoming Logical Positivism From Within* (Amsterdam: Rodopi, 1992); Danilo Zolo, *Reflexive Epistemology: The Philosophical Legacy of Otto Neurath* (Boston: Reidel, 1989).

4.   Yet, they swim against the current, and I find this admirable.

5.   The term "modified conventionalism" belongs to Joseph Agassi: *Science in Flux* (Boston: Reidel, 1975), pp. 365-403. The notions of "foundation" and "basis" were used in the 1930s, but the terms "foundationism" and "nonfoundationism" are of recent vintage.

6.   Neurath, "Protokollsätze," *Erkenntnis,* III (1932): 205.

7.   Carnap, "Über Protokollsätze," *Erkenntnis,* III (1932): 223-28.

8.   At issue was who recognized first the relativity of basic statements or protocols, i.e., that they were not final, but subject to revision and overthrow. Carnap to Popper, 18 October and 28 October, 1932; Popper to Carnap, 22 October and 1 November 1932; Neurath to Carnap, 26 October 1932, all in Carnap Collection. Neurath objected to the circle "going hand in hand with Popper," but he did not question Popper's independence. It is surprising that Carnap reported such a challenge to Popper.

9.   "I almost appear as a good student of Neurath," he complained. Popper to Julius Kraft, c. 2 November 1932 (316, 24).

10. Popper, "Memories of Otto Neurath," p. 52.

11. "Neurath and I had disagreed deeply on many and important matters, historical, political, and philosophical; in fact on almost all matters." Ibid., p. 56.

12. John Wettersten, in his *The Roots of Critical Rationalism* (Amsterdam: Rodopi, 1992), chap. 8, noted the peculiarities of section 11 of *Grundprobleme,* containing the new Kant critique. It is fifty-six page long, by far the longest section, about a fifth of the manuscript. It seems to stand on its own, the transition to the succeeding section abrupt. It includes terminology (empirical basis; basic statement; observation statement) absent elsewhere in *Grundprobleme* I (and standard in *Logik).* Its nonfoundationism conflicts with statements in other sections that accept "singular reality statements" as final. Wettersten concluded that this was the last section to be written, and I agree. The correspondence between Popper and Julius Kraft corroborates his view.

Troels Eggers Hansen disagrees. He examined all of Popper's cross-references to section 11 in the other sections of the manuscript. He counted eight typed references, and twelve inserted in red ink. He concludes that there must have been an earlier version of section 11. (Hansen to author, 20 December 1998 and 31 January 1999.) I concur. However, with the exception of the reference in section 2 (which was rewritten sometime between the late spring and the fall of 1932), none of the typed references touch on testing, or falsification, or the empirical basis, i.e., the epistemological revolution of the fall of 1932. Could Popper have coined the term "empirical basis" and written on it prior to his August meeting with Carnap in the Alps? I doubt that,

322   MALACHI HACOHEN

but his fury when Neurath denied him priority over nonfoundationism would then become comprehensible.

13. Popper, *Logik der Forschung,* sections 25-26, 30, note 4.
14. "Ein Kriterium des empirischen Charakters theoretischer Systeme," *Erkenntnis,* III (1933): 426-7. (English: "A Criterion of the Empirical Character of Theoretical Systems," *The Logic of Scientific Discovery,* pp. 312-14.) The note was first written in July 1932 and slightly revised for publication in *Erkenntnis.* It therefore did not reflect the epistemological revolution of the fall of 1932.
15. Moritz Schlick, "Über das Fundament der Erkenntnis," *Erkenntnis,* IV (1934): 79-99. (English: "On the Foundation of Knowledge," *Philosophical Papers* [Boston: Reidel, 1979], II:370-87.) Neurath, "Radikaler Physikalismus und 'wirkliche Welt'," *Erkenntnis,* IV (1934): 346-362; Hempel, "On the Logical Positivists' Theory of Truth," *Analysis,* 2 (1935): 49-59. The Carnap-Schlick correspondence March-June 1934, Carnap Collection, Archives of Scientific Philosophy, University of Pittsburgh. For an account of the early stages of the debate, see Thomas Uebel's informative *Overcoming Logical Positivism,* chaps. 3-6 and my own rendering in *Karl Popper,* pp. 224-27.
16. See my *Karl Popper,* chaps. 5-6.
17. Popper to Neurath, 28 February 1935, *Neurath Nachlaß, Philosophisches Archiv,* University of Constance. (Original at the *Wiener-Kreis Archiv,* Haarlem, NL.) "Popper und der Wiener Kreis – Gespräch," in Friedrich Stadler, *Studien zum Wiener Kreis* (Frankfurt: Suhrkamp, 1997), p. 536.
18. Schlick to Carnap, 1 November 1934, *Schlick Nachlaß, Philosophisches Archiv,* University of Constance. (Original at the *Wiener-Kreis Archiv,* Haarlem, NL.)
19. This is actually Popper, *Logik,* p. 233, section 30, note. 4 *(LSD,* p. 111), quoting Robert Reininger, *Das psycho-physische Problem* (Vienna: Braumüller, 1916), p. 291: *"Spricht* die Seele, so spricht, ach! schon die *Seele* nicht mehr." Schlick made the same point: "Über das Fundament der Erkenntnis," 97-99.
20. Schlick, "Fundament," 93, 99, respectively.
21. "Fundament," 83.
22. Neurath, "Radikaler Physikalismus," 346-362.
23. Protocols were not, as Carnap thought, "primitive." Popper showed that universal, law-like concepts, such as "glass," could not be reduced to singular "experience." Neurath concurred. But, for him, the problem was linguistic imprecision. Everyday language contained conceptual clusters *(Ballungen),* lacking the precision of axiomatized theories. To Popper, growing precision would never overcome the gap between concepts and experience. Neurath seemed to entertain the hope of eventually closing the gap through linguistic reform.
24. Jordi Cat, "The Popper-Neurath Debate," 234.
25. Uebel argues *(Overcoming,* pp. 214-17) that, since affirmations stood outside the system of scientific statements, their importance was not logical and epistemic, but psychological and motivational. "To call them 'foundations' is to court serious misunderstanding." I share this misunderstanding. Schlick's argument is not only foundationist, but psychologistic and subjectivist. Uebel's efforts to exonerate the circle from foundationism stand in marked contrast to his easy dismissal of Popper as a foundationist. Critics' impatience is understandable: Joseph Agassi, "To Salvage Neurath," *Philosophy of the Social Sciences,* 28 (1998): 83-101.
26. Popper, *Logik,* p. 16; *LSD,* p. 44.
27. *Grundprobleme,* pp. 429-32, 438-39; *Logik,* section 26.
28. Thomas Uebel, *Overcoming,* pp. 265-67. Jordi Cat, "The Popper-Neurath Debate," suggests the same, in a more nuanced fashion.
29. Carnap to Neurath, 13 February 1935, Carnap Collection, Archives of Scientific Philosophy, University of Pittsburgh.
30. Neurath, "Radikaler Physikalismus," 347-48.
31. When Carnap and Popper resisted, Neurath could only say that psychologism was not a bad word. Neurath to Popper, 22 January 1935, *Neurath Nachlaß;* Neurath to Carnap, 18 January and 28 January 1935, Carnap Collection.
32. Thomas Uebel, *Overcoming,* esp. pp. 2-3, 226-28.
33. Nancy Cartwright et al., *Otto Neurath,* p. 201. (Jordi Cat co-authored the chapter quoted.)

34. Jordi Cat, "The Popper-Neurath Debate," proposes that not Malachi's brain stimuli, or speech-thinking, were subject to test, but his observation. Entitlement was not part of the statement. This would make protocols testable, but entitlement irrelevant.
35. Popper insisted that "observation" was a psychological event, but "observability" a logical concept. Not observation's psychology was significant, but its intersubjectivity. I find the distinction persuasive, as do most Popperians. But to Victor Kraft ("Popper and the Vienna Circle," *The Philosophy of Karl Popper*, ed. Paul Arthur Schilpp, 2 vols. [La Salle, IL: Open Court, 1974], I:195-96), the concept remained problematic. Psychological experience convinced people that statements were true. Was the refusal of psychology not question-begging? "When do we see experience as conforming to our expectations (our theory), and when do we see it as conflicting?" ask William Berkson and John Wettersten. "In other words, what is the psychological correlate of the contradiction (between a basic statement and a theory) which Popper describes?" (Berkson and Wettersten, *Learning From Error* [La Salle, IL: Open Court, 1984, p. 20].) These questions had to remain unanswered, if Popper was to solve epistemology's problems. An attempt to reformulate the relationship between logic and psychology would have resurrected the problems that ran his dissertation aground ("Zur Methodenfrage der Denk-psychologie" [Ph.D. diss., University of Vienna, 1928]).
36. Otto Neurath, "Pseudorationalismus der Falsifikation," *Erkenntnis*, V (1935): 353-365. (English: "Pseudorationalism of 'Falsification'," *Philosophical Papers*, pp. 121-31.)
37. Neurath's "anti-system" rhetoric in Prague began the encyclopedic turn ("Einheit der Wissenschaft als Aufgabe," *Erkenntnis*, V [1935]: 16-22), but it became clear only in his review of *Logik*.
38. "Pseudorationalismus," 354. Popper, too, recognized that scientific theories rarely approached axiomatization, but he insisted that scientists aim for the greatest possible clarity and cogency, so as to facilitate the operation of methodological rules guiding negotiation over basic statements. In contrast, Neurath maintained that science remained ambiguous on every level: concepts had rough edges; clean statements did not exist; systems of statements were loose; fields were separate.
39. "Einheit der Wissenschaft als Aufgabe," 17. Popper's and Neurath's scientific metaphors mirrored their intellectual and political life. Neurath took lightly personal risks, cared little about appearance and conditions of life, was talkative, warm, and open. Spreading over disciplines with little patience for detail, he wrote with ease, but not clarity. Popper had little trust in people or the world, led a structured and secluded life, and incessantly rewrote his manuscripts to diminish the prospect of being misunderstood. He wrote with difficulty but achieved incomparable clarity.
40. "Pseudorationalismus," p. 365.
41. Carnap-Neurath correspondence, 1935-37, 1942-45, Carnap Collection.
42. Jordi Cat, "The Popper – Neurath Debate," 227.
43. Thomas Uebel, *Overcoming*, chap. 11.
44. Nancy Cartwright, *Otto Neurath*, p. 205; Jordi Cat, "The Popper – Neurath Debate," 245.
45. Cartwright, *Otto Neurath*, p. 195, quoting Neurath's "Besprechung über Physikalismus," 4 March 1931, Carnap Collection (RC 029-17-03).
46. Cat, "The Popper – Neurath Debate," 243.
47. Cartwright, *Otto Neurath*, p. 222. Cartwright thinks Popper disagreed. She is wrong.
48. Popper to Herbert Feigl, 4 November 1969, Feigl Collection, Archives of Scientific Philosophy.
49. Admittedly, the system metaphor may not best describe science. But Popper focused on competing theories within particular fields rather than on relationships among disciplines. His system of theories was ever changing, not a stable body of knowledge.
50. Ibid., 234.
51. Ibid., 246.
52. Loc. cit.
53. More correctly: Neurath collapsed together falsification and falsifiability, spoke of falsification as a demarcation criterion, and rejected it.
54. Cat suggests that Neurath's linguistic "warrant for theoretical beliefs" was more liberal than Popper's testability. But Popper required no warrant for "theoretical belief." None existed. Theory construction was not subject to scientific control. Who was more liberal?

55. *Logik,* p. 10; *LSD,* p. 37.
56. In *Logik,* Popper claimed metaphysical neutrality. In postwar years, he claimed to have always been a realist. Having demonstrated, by the mid-1950s, that metaphysical theories were criticizable, he validated inquiries into the nature of the universe, previously excluded as a tactical move against positivism.
57. Neurath could also be dogmatic. Against critics of the circle's left wing, he asserted, loud and clear, a party line.
58. Ludwig Wittgenstein, *Philosophical Investigations,* ed. G.E.M. Anscombe and Rush Rhees (London: Macmillan, 1958).
59. "If the structure proves too heavy, and begins tottering, it sometimes does not help to drive the piers further down. It may be necessary to have a new building, which must be constructed on the ruins of the collapsed structure's piers." *(Grundprobleme,* p. 136) Science studies may consider the structure metaphor insufficiently radical, but the piers going into the swamp constitute no Cartesian First Philosophy. In my *Karl Popper* pp. 233-35, I review some of the debates among Popperians on foundationist traces in critical rationalism.
60. Ian Jarvie, *The Republic of Science* (Amsterdam: Rodopi, 2001); Jeremy Shearmur, *The Political Thought of Karl Popper* (London: Routledge, 1996). Shearmur's view that a libertarian polity would be the most conducive to a Popperian public sphere conflicts, however, with my own view.

Department of History
Duke University
609 Colgate St.
Durham, NC 27704
U.S.A.
mhacohen@duke.edu

VERONIKA HOFER

PHILOSOPHY OF BIOLOGY AROUND THE VIENNA CIRCLE: LUDWIG VON BERTALANFFY, JOSEPH HENRY WOODGER AND PHILIPP FRANK [*]

This paper addresses the historical context of Bertalanffy's concept of Theoretical Biology, his early combattants, friends, teachers and the philosophical position of his critical reviewer Philipp Frank. I will describe the characteristics of his theory and how his ideas are embedded into the background discourse of the day. In the following five sections I will show that there are three main historical factors that shaped Bertalanffy's intellectual views: First, his philosophical training with Schlick and Carnap, second, his close connection with the biologists in the *Prater-Vivarium*, and third, his close contact to J.H. Woodger and, through him, with the organicists of the Theoretical Club in Cambridge. This narration also sheds some light on one of the founders of the Vienna Circle, Philipp Frank. I shall provide some historical data about his early interest in biology and discuss his position towards some basic problems in biology.

I. BERTALANFFY'S INTELLECTUAL BACKGROUND

In 1901, Ludwig von Bertalanffy was born into *fin-de-siècle* Vienna of the which harbored so many groups that would substantially contribute to the science of the century. A somewhat scandalous member of one of them lived just next door of the Bertalanffys. The neo-Lamarckian Paul Kammerer not only kindled the young Bertalanffy's fascination in biological science; he also provided the *entré* for him into the experimental-biological laboratory of Hans Przibram which became seminal for 'organismic biology', both experimentally and in the theoretical foundations of Systems Theory in Biology. Together with two other wealthy Jewish biologists, Przibram had founded in 1903 a private research institute that quickly became an internationally renowned center of experimental biology. In 1914 they donated the *Prater-Vivarium* to the Austrian Academy of Science. Przibram was in close contact with outstanding scientists such as Hans Spemann, Jacques Loeb, D'Arcy Thompson, and Sir Frederick Gowland Hopkins, the head of the Dunn Institute for Biochemistry in Cambridge, England. Hopkins was the tolerant head of the biochemists J.D. Bernal and J.S.B. Haldane, both members in the Theoretical Club in Cambridge which was founded by J.H. Woodger in the early 1930's.

M. Heidelberger and F. Stadler (eds.), History of Philosophy and Science, 325–333.

In its research activities, the *Prater-Vivarium* focused on highly interdisciplinary topics in the rapidly growing branches of experimental morphology, defined
as the study of biological structure and forms. They studied developmental
biology under a morphogenetic perspective, in particular the phenomena of
growth. In the Vivarium Wolfgang Pauli senior undertook biochemical investigations with colloids and the endocrinologist Eugen Steinach made his famous
experiments with hormones as a factor for growth and for the sexuality. A
particular characteristic of the institute was its systematic use of mathematics and
its search for inventive and useful applications of theoretical concepts stemming
from physics, chemistry and crystallography, which were highly uncommon in
morphology at that time.

In 1926, four events took place which demonstrate the close connections
between Bertalanffy, the Vienna Circle, and the Cambridge Theoretical Club.
Two years before Paul Kammerer's suicide in 1926, the pioneer in developmental biology Paul Weiss succeeded him as Przibram's assistant in the *Prater-
Vivarium*. In 1926 Bertalanffy defended his Ph.D. dissertation on Fechner having
Schlick as a highly critical second referee. Having already completed substantial
pieces of the *Aufbau*, Rudolf Carnap came to Vienna as Schlick's assistant. Also
in 1926, both the biochemist Sir F.G. Hopkins and Joseph Henry Woodger were
visiting scientists of the *Prater-Vivarium*. Although Woodger was extremely
unlucky in his planned experiments, he received many inspirations in Vienna
that would give his further work a highly theoretical turn.

Bertalanffy's first contact with the Vienna Circle happened in 1924, when he
moved from Innsbruck to Vienna to study philosophy with Moritz Schlick. After
finishing his thesis, Bertalanffy worked from 1929 until 1934 with Carnap's so
called "Studiengruppe für wissenschaftliche Zusammenarbeit", together with
H. Feigl, E. Zilsel, K. Polanyi, E. Brunswik, W. Reich, and P. Lazarsfeld. This
group of young scientists and philosophers of science was organized around the
"Verein Ernst Mach". The latter's aim was the popularization of the ideas of the
Vienna Circle through courses and excursions, public talks etc. The "Studiengruppe" was intended by Carnap to foster a better mutual understanding of the
disciplines and to clarify their position in the multidisciplinary context of the
sciences. We get an even better idea of Bertalanffy's position in the Vienna
Circle, if we notice that he did not just know them and study with them, but furthermore Philipp Frank recommended that he contribute to the Second International Congress for the Unity of Science in Copenhagen in June 21-26, 1936.
The focus of that Congress was "The Problem of Causality – With Special
Consideration of Physics and Biology." From the correspondence among
Neurath, Frank and Carnap we learn that it has been Ph. Frank who proposed
that Bertalanffy talk on the concept of causality in biology. We know that Frank
had a special interest in this problem, first documented in his paper from 1908
"Mechanismus oder Vitalismus. Versuch einer präzisen Formulierung der Frage-
stellung".[1] Bertalanffy did not go to Copenhagen, and he did not manage to
follow the renewed invitation the next year in the Fourth Congress for the Unity

of Science, held at Girton College in Cambridge (England) in July 14 – 19, 1938, because he was constantly short of money.[2]

Although there is no record of Bertalanffy's presence at meetings of the Vienna Circle, there is no doubt about his own words; he relates that his personal contact with Victor Kraft and Friedrich Waismann and the other young philosophers made some strong impressions. His Systems Theory is strongly based on philosophical motivations some of which can be understood as constructed in the light of the methodology of the Logical Empiricists. At the beginning of his career as a philosopher of biology he attempted in his book *Kritische Theorie der Formbildung* (1928) to "elucidate the crises of biology more sharply and at the same time to prepare the proof of the necessity of theoretical thinking in biology"[3]. To this end, he undertook a critical evaluation of the "basic propositions" of current biological theories.

II. METHODS

a. The rejection of the old correspondence theory of truth

Bertalanffy's new concept of theory was inspired by Carnap's *Aufbau*. Bertalanffy sought to construct some biological theories out of mutually isomorphic concepts and isomorphic laws. This operation was intimately connected with the rejection of the old correspondence theory of truth which had claimed to connect single statements to brute facts; it also emphasized linguistic and inner-theoretical criteria, such as consistency and conceptual coherence. Logical Analysis was, above all to Woodger, – the helpful mentor of Bertalanffy all his life long – a key ingredient of biological theorizing. This was supported by adopting Ernst Mach's reformulation of causality in terms of functional dependences because, on that account, both teleological and causal explanation merged into equations between the determinant conditions. Thus, the intimate connection of biological teleology to the existence of specific vital substances or quantities disappeared because now laws simply constituted equations.

b. Bertalanffy's positivist teleology

Bertalanffy attempted to provide teleology with a new systematic place in scientific theory: "Teleology of life is characterized as the result of an empirical law: If we focus on the single parts, the organs, we have to consider them as purposive for a certain function; if we focus on the organism as a whole, this true consideration of purposes which is indispensable in descriptive biology does no longer apply and is replaced by the 'conservation of wholeness', to wit, the preservation of the organism against the assault of the external energies which aspects are the teleologies of the single organs.[4] Bertalanffy hints at concepts, such as adaptation, regulation, regeneration, organization, which are well-entrenched and successful in the mechanistic approach as describing the conservation of a system state called life. Thus Bertalanffy argued that teleology in living systems be understood strictly as a general tendency of conservation of the stationary state.

Accordingly, Bertalanffy simultaneously rejected Driesch's vitalism and insisted on the indispensability of teleological considerations in biology. Already in the *Kritische Theorie der Formbildung*, he emphasizes the central position of the concept of organism because the particular order of processes encountered in living beings constitutes a purposive property that cannot be explained in terms of mechanistic concepts. "The concept of organism allows the description of phenomena in the realm of an animate nature; it does not involve a metaphysical idea just as the physicist does not combine the energy concept with the idea of anthropomorphic action."[5] To Bertalanffy, Woodger, and other organicists, the most important property of teleology is of a formal nature and does not consist in the idea or the knowledge of the intended goal. "The organism is characterized, firstly, by being more than the sum of its parts and, secondly, by the fact that the single processes are arranged in such an order as to preserve the whole."[6]

Thus Bertalanffy's biotheory has another firm grounding, which is the premises of 'organismic biology' that was founded by Whitehead, E.S. Russell, J.H. Woodger, and J.S. Haldane. Theoretical Biology emphasizes the phenomenon of organization that is embraced under the concept of system.

III. Woodger Mediates Bertalanffy, the Vienna Circle, and the Theoretical Club in Cambridge

After his visit to the Vivarium in 1926, Woodger concentrated on exploring the potential resource of the new symbolic logic for biology. He wanted to express the hierarchical order in biology in terms of symbolic logic, "to discern the form of the facts and frame the concepts in accordance with it."[7] Thus, he intended to accomplish a similar conceptual clarification for the central problems of theoretical biology, as the Vienna Circle's logical analyses had undertaken in physics – a program "which Woodger came to adopt at that time as a result of his friendship with Rudolf Carnap and Max Black."[8] Above all, Carnap in Vienna and Gustav Hempel in Berlin provided Woodger the logical means to pursue his program of an axiomatics for biology. This highly abstract and theoretical endeavor was based on Woodger's overall viewpoint as an organicist. He believed that transforming the scope of the scientific method would render it possible to include concepts like "events" or "organism" into the conceptual repertoire of biology – a goal that could not be met with the theoretical tools of the former mechanistic world-view.

During the 1930s Woodger took part in the International Congresses for the Unity of Science that had been launched by the Logical Empiricists. Moreover, he contributed a booklet to their International Encyclopedia of Unified Science. In addition, he translated into English works by Carnap, Tarski, as well as Bertalanffy's *Kritische Theorie der Formbildung*.

IV. THE "BIOTHEORETICAL CLUB" IN ENGLAND

The 'Biotheoretical Club' was founded in 1932 and existed until the beginning of World War II. Among its founding members were Woodger, Needham, his wife Dorothy Needham, C.D. Waddington, J.D. Bernal and Dorothy Wrinch. This group of British scientists was deeply concerned with changes in both science and society. The starting point of their theoretical discussions was again that physics which for a century had held the top position within the hierarchy of the sciences, was subject to deep foundational crises. But, instead of advocating a simple revival of 'strict' empiricism, they strove for a renewal on three levels: First, a new scientific transdiciplinary unity; second, an egalitarian mode of collaboration to counteract isolation by setting up research groups; third, a new balance of power based on informal authority instead of hierarchy. This also meant political activity: some members of the Club were members of the communist party. Their interest in a redefinition of the traditional disciplinary borders above all concerned the relation between the physical and the biological sciences. They advocated a "mathematico-physico-chemical morphology" which "revolved around a joint perspective of the exact sciences on the problem of biological organization, initially at the supra-cellular level of morphogenesis and later at the sub-cellular or molecular level of chromosome and protein structure."[9] This program was indeed very similar to what Bertalanffy was after in those years. "They viewed isomorphisms between organized entities in all disciplines, pluralistic lawfulness and epistemological parity as new themes of scientific unity, to supplant the classical, atomistic and reductionistic view which had been discredited at that time by the combined impact of the relativity and quantum theories."[10] In establishing these isomorphisms across the disciplines, the Club relied on the methods of modern logic.

V. PHILIPP FRANK ON THE VITALISM-MECHANISM-CONTROVERSY

This background story must end with some historical remarks concerning the main trends in biological thought at the turn of the century. Ernst Haeckel had for a long time been the dominating figure in the popularization of evolutionary thinking, but in the late 1880s his monist ideology found serious opposition especially in Germany. August Weismann and Wilhelm Roux both formulated an anti-Haeckelian Neo-Darwinist and Neo-Mechanistic foundation for biology in order to establish within biology those experimental methods that had been considered normal in physics and chemistry. They were strongly motivated in this attempt by the far-reaching results and applications of physiology and its flourishing due to the use of physicalistic methods. On the other hand, the furor with which Haeckel and his followers propagated a mixture of naturalistic and monistic *Weltanschauung* led to a deep demand for empiricism as a philosophical foundation for biology. For that reason many scientists pleaded for a renewed approach to mechanism. Physicists such as Helmholtz, Hertz, Mach, and Poincaré helped establish a new physics and a new mechanistic philosophy whose aim was that causality as the central concept for the old mechanistic philosophy should be freed from anthropomorphical assumptions through skeptical philosophical examination. Functional dependencies moderated and alternated the old concept of causality; the new modest understanding of causality stated only local causal relations and functional relations. As a result, a deeper understanding of the uncertainty of philosophical concepts even in hard science led to a skeptical empiricism.

But empirical skepticism led to a variety of answers, which have also been relevant for the enduring mechanism-vitalism controversy. The opposition to Haeckel's monism yielded two main paradigms, neo-vitalism and neo-mechanism along with a third, which posited itself as a sort of intermediate view, the organicism. On one side there was Driesch as the leading figure of neo-vitalism and on the other side Wilhelm Roux as one of his mechanistic counterparts. All of the three positions took empirical arguments as philosophical and epistemological starting point. In order to grasp the focus of Bertalanffy's biophilosophy, one must understand that neo-mechanists and neo-vitalists concured on what the description of living things entailed. The neo-vitalists as well as the organicists tuned the organismic descriptions up, and the mechanists tuned them down again. The vitalists and the organicists emphasized the organismic features like having a relatively high level of organization, or the feature of the adaptability of organisms, or their special interaction with their environment or their genetic history. The organicists followed Driesch in his formal concept of teleology; they all did not see goal-oriented behavior as a consciously intended realisation of specific aims, but instead they constructed a purely formal concept of the purposiveness of living entities. On the other hand, vitalists like Driesch defended, in addition to the descriptive feature of living things as self-stabilizing, highly organized and goal-oriented, a special "life-making factor" superadded to living matter – his Aristotelian adaptation of entelechy. However this entelechy was a point of contention, since organicists did not feel comfortable at all with this metaphysical factor, which they felt contradicted the neo-materialistic principles. The modern conception of matter as matter-energy, as self-moving, i.e. as inherently possessing a productive capacity and hence not needing to be shaped passively by an efficient cause external to it, formed the core of the new doctrine, which has been very influential for Bertalanffy and the other founders of organicism in biology like Woogder, J.S.B. Haldane, E.S. Russel, and in 1932 J.H. Needham. The model for the new materialism, the charged particles with their relatedness to one another subject to the interplay of dynamic forces, was the background model for Bertalanffy and the members of the new organicism. They regarded order as a necessary and natural attribute of matter, requiring no agent that imposes organization upon the primary chaos. The organicists' view of nature as a continuous system is developed in terms of an accentuated principle of organization and a concept of levels of complexity of objects in nature. This doctrine was the basis for Woodger's and Bertalanffy's claim that biology has special laws, irreducible to physical laws. Although they endorsed a new concept of matter, they asked for a special epistemological treatment for biological problems. In pronouncing the phenomenological difference between the living and the non-living they defended a new form of vitalism, and they did so by stressing the structural and organizational differences between living and nonliving things. The difference is no longer substantive or energetic, it is structural. The laws of physics are not antithetical to those in biology, but just not sufficient; they are incomplete and need to be enriched by purely biological laws. According to Bertalanffy and Woodger, biological entities are natural objects; but life involves something "more" than inanimate matter, and this is irreducible to – although compatible with – physical principles.

Jacque Loeb for example, an arch-mechanist, a close partner in correspondence with Ernst Mach, and a much admired scientist for moderate mechanists like Hans Przibram, bluntly opposed the alleged "Whole-making" feature of the organicists; he denied their description of the organism as not being essential to biology and he saw in this – parallel to Mach and the Logical Empiricists – just a typical imprecision in the formulation of the problem. In Loeb's own words in 1912: "With all due personal respect for the authors of such terms, I am of the opinion that we are dealing here, as in all cases of metaphysics, with a play on words. That a part is so constructed that it serves the 'whole' is only an unclear expression for the fact that a species is only able to live – or to use Roux's expression – is only durable, if it is provided with the automatic mechanism for self-preservation and reproduction."[11] It is noteworthy that Roux, a strong opponent of Haeckel, was highly respected by those men like Ernst Mach and that it was his research program, which Przibram followed in his *Prater-Vivarium*. Loeb is clear enough about his mechanistic basis and states that there is no a priori reason to assume that a tendency toward order should be any more characteristic of the nature of matter than a tendency toward disorder. So he denies that the organization of matter is an attribute which supervenes upon the character of matter as such.

It is interesting to look at Philipp Frank's treatment of the vitalism-mechanism-controversy in the first decade of the century and to look then at how he viewed Bertalanffy's attempt to reformulate the problem of teleology in 1932. And it is even more interesting to explore these issues, in the light of the fact that Frank studied biology quite extensively in the University of Vienna.[12] At the beginning of his studies in 1902 he had 5 weekly lessons "General Biology" with Berthold Hatschek. In his second semester he attended 2 lessons on cell-tissues. After finishing his studies in Physics, he enrolled again in the University of

Vienna to do extra courses in Biology. So he had lessons in 1907 with Hans
Przibram about "applications of mathematics to biology", in the Wintersemester
1907/08 he had lessons with the Botanist Richard von Wettstein and again a lesson with Hans Przibram about "regeneration".

In his 1908 paper "Mechanismus oder Vitalismus? Versuch einer präzisen
Formulierung der Fragestellung" Frank makes clear that there is no logical
argument against Vitalism, if one understands the alternative as whether those
contents, which are sufficient for the causal analysis in physics are also sufficient
for the causal analysis in biology or whether one must supply the causal analysis
in biology with extra-hypotheses or biological constants, "Konstanten" in his
German terminology. In his article, Frank restructures the argumentation of the
vitalists and he takes them seriously in their rejection of mechanistic hypotheses
in causal analysis. He comes to the conclusion that the mechanicistic dogmatism
is to be rejected and that serious work needs to be done by experimentalists and
theoreticians. Besides his own conviction that the mechanistic theory of life deserves the most heuristic and philosophical attention, he thought it would be
dangerous to let vitalism alone dig into the difficult problems of life.

I suggest that this attitude, which he shared with his biological mentor Hans
Przibram, represented the foundation on which Frank criticized Driesch and
Bertalanffy in his book of 1932 "Das Kausalgesetz und seine Grenzen" and
when he criticized Woodger in his correspondence with Neurath, Schlick and
Carnap.

As I understand Frank's position, there is on the one hand Frank's Machian
outlook, that made him quite sceptical towards logical system building that was
not clearly linked to observational results. Przibram's mild mechanism was
acceptable, but Woodger's organicism was not.

On the other hand, Frank's notion of causality was wide enough to consider
whatever meaningful statement organicists could formulate as testable causal
laws. Bertalanffy's attempt to develop vitalism in a positivist fashion thus faced
the following dilemma. Either it was translatable into causal laws, or it just represented an empty tautological statement. The only resort for the vitalist to
escape this dilemma was – according to Frank – to posit an omniscient god-like
being.

NOTES

*   I am indebted to Elliott Sober and Michael Stöltzner for a critical reading.

1.   Philipp Frank, "Mechanismus oder Vitalismus. Versuch einer präzisen Formulierung der Frage-stellung", in: *Annalen der Naturphilosophie* 7, 1908, p. 393-409.
2.   The information about Bertalanffy's involvement in this connection can be found in the correspondence of Otto Neurath. See my *Organismus und Ordnung. Zu Genesis und Kritik der Systemtheorie Ludwig von Bertalanffys*, Ph.Dr. dissertation, University of Vienna, 1996.
3.   Ludwig von Bertalanffy, *Kritische Theorie der Formbildung*, Berlin, 1928, p. 4.
4.   Ludwig von Bertalanffy, "Die Teleologie des Lebens", in: *Biologia Generalis*, Vol 5, 1929, p. 388
5.   Ludwig von Bertalanffy, *Kritische Theorie der Formbildung*, Berlin, 1928, p 74.
6.   *Ibid.*, p. 74.
7.   Ludwig von Bertalanffy, *General System Theory. Foundations, Development, Applications*, New York, 1968.
8.   Pnina G. Abir-Am, "The Biotheoretical Gathering", in: *History of Science*, Vol. 25, 1987, p. 16.
9.   *Ibid.*, p. 10.
10.   *Ibid.*, p. 10.
11.   Jaques Loeb, *Das Leben*. Leipzig, 1911, p.35-36.
12.   See the records at the Archive of the University of Vienna (*Nationale der Philosophischen Fakultät*).

Institute Vienna Circle
Museumstraße 5/2/19
A-1070 Vienna
Austria
veronika.hofer@eunet.at

GERALD HOLTON

B.F. SKINNER AND P.W. BRIDGMAN:
THE FRUSTRATION OF A *WAHLVERWANDTSCHAFT*

The psychologist-philosopher B.F. Skinner and the physicist-philosopher P.W. Bridgman, both dedicated empiricists, initially entered into an intellectual relationship that seemed destined to be warm and fruitful. Yet, it ended up unfulfilled. Since I am now perhaps one of the few who knew both men as colleagues for many years, I might be able to throw some unique light on their interaction, and on what I consider to be one of the missed opportunities in the history of ideas.

SKINNER AND THE "LOST YEARS"

For Skinner's side of the story, I need not go deeply into the biographical details.[1] Entering Hamilton College in 1922, Skinner, somewhat aimlessly, discovered his interest in writing and literature, getting his B.A. in English Literature in 1926. But reading Bertrand Russell and John B. Watson, he absorbed the ideas of science as a means for the reconstruction of society. That is of course an ancient and recurring hope, as expressed for example in the famous 1929 pamphlet of the Vienna Circle, entitled *Wissenschaftliche Weltauffassung: der Wiener Kreis,*[2] issued as part of the *Einstein-Mach-Verein* under the authorship of Rudolf Carnap, Hans Hahn, and Otto Neurath – the very message that helped W.V. Quine, one of Skinner's closest intellectual companions at Harvard from the earliest days, decide to go to Vienna to participate in the activities of the Vienna Circle and then on to Prague to study under Carnap.

When Skinner entered Harvard for graduate study in 1928, he initially still seemed to have no clear direction, and in this searching spirit, he sat in on a course on the History of Science given by the charismatic Lawrence J. Henderson, professor of biological chemistry and head of the Harvard Business School Fatigue Laboratory, and by George Sarton, the father of modern History of Science. I believe it was Henderson who suggested in 1929 that Skinner read Ernst Mach's *Science of Mechanics,* and in that book he encountered Mach's missionizing, antimetaphysical position favoring a coherent world picture and economy of thought. The book had a permanent effect on Skinner. In an interview in June 1988, Skinner stated to me categorically: "I was totally influenced by Mach via George Sarton's course, and quickly bought Mach's books, *Science of Mechanics,* and *Knowledge and Error.*" As E.A. Vargas has pointed out,

M. Heidelberger and F. Stadler (eds.), History of Philosophy and Science, 335–346.
© 2002 *Kluwer Academic Publishers. Printed in the Netherlands.*

Skinner did not stop there: Writing in 1931 to the physiologist W.J. Crozier, he exulted, "I have also been reading Mach's *The Principles of Physical Optics,* which is one grand book. The best summer reading I have come across." [3]

But Skinner had been ready to receive Mach's empiricist message. As he stated in his autobiography, *The Shaping of a Behaviorist* (1979), he could recall only two science books he had read as an undergraduate: Loeb's *Comparative Physiology of the Brain and Comparative Psychology,* and *The Organism as a Whole,* with their largely positivist approach to the study of the behavior of animals. When Skinner came to Harvard University to do his graduate work in 1928, his thesis supervisor, in whose laboratory he remained for 5 years, was W.J. Crozier. It is not accidental that Crozier's own teacher had been Loeb, who in turn had been an admirer and correspondent of Mach. Indeed, it has been said that "it was the ultra-positivistic form of Loebian biology that Skinner encountered at Harvard" (Smith, p. 277).

Skinner's doctoral dissertation at Harvard, dated December 1930, was entitled *The Concept of the Reflex in the Description of Behavior.* The introductory pages read as if they really had been written under the influence of Mach, for they warned that, as Mach had done in his analysis of physical concepts, particularly those of Newtonian mechanics, much of the field of the description of the behavior of intact organisms was beholden to "historic definition, that is to say, vested with extrinsic interpretations, and some of these now appear to embarrass the extension of total behavior." In fact, on the second page, Skinner shows his source:

The reader will recognize a method of criticism first formulated in respect of scientific concepts by Ernst Mach and perhaps better stated by Henri Poincaré. To the work of these men, and to Bridgman's excellent application of the method to more modern concepts [in P.W. Bridgman's book, published three years earlier, *The Logic of Modern Physics,* 1927], the reader is referred for an extended discussion of the method *qua* method. Probably the chief advantage first exploited in this respect by Mach lies in the use of an historical approach ... the second part of the thesis [which is "primarily experimental"] is thus offered as an example of the practicability of the method, and as a partial test of the hypothesis, advocated in the first part.

And as if to make the point quite clear, on page 55, the section "Notes and References for the First Part" begins with just five books: Ernst Mach's *Mechanics* (1883), Mach's *Analysis of Sensations* (1886), Henri Poincaré's *Science and Hypothesis* (1902) and *Science and Method* (1908), and, last but not least, Bridgman's *Logic of Modern Physics* – a confirmation of Skinner's earlier mention of Bridgman's "excellent application of the method ..."

At this point, a puzzle suggests itself. Bridgman, well known to be an admirer of Mach, and in other ways congenial to the kind of epistemological approach Skinner was favoring, was the only one of the three authors Skinner had mentioned who was still alive, indeed at the height of his powers – and right there at Harvard. It would have been eminently reasonable for Skinner, who did not lack courage, to have sought out Bridgman, 18 years his senior but kindly and accessible. It could have been the start of a productive interaction. But apparently it did not happen. Here was perhaps the first of a series of lost opportunities.

In writing his doctoral thesis, young Skinner saw a way of applying the Machian point of view to the clarification of such concepts as the "reflex" of intact organisms, something he considered to be as basic in psychology as, say, mass is in physics. As Skinner recollected, he was "following a strictly Machian line, in which behavior was analyzed as a subject matter in its own right as a function of environmental variables without reference to either mind or the nervous system"; that was "the line that Jacques Loeb ... had taken."[4] In this radically empiricist mode, the study of behavior reduced itself for Skinner, to start with, to the observation of the motion of the foot of a food-deprived rat, pressing down a small lever in an experimental box of standard size. Explanation was reduced to description, causation to the notion of function, and the chief goal was the correlation between observed events. Again, Bridgman would have been sympathetic to this train of thought.

Skinner's PhD was awarded in 1931. Thanks to a National Research Council Fellowship and his election to the elite Society of Fellows at Harvard, he had 5 years, ending in 1936, of postdoctoral freedom. He began a manuscript that was to be his first scholarly book, *Sketch for an Epistemology* (of which only a 1935 article, "On the generic nature of stimulus and response," was published). The project showed, however, that, as Skinner often said, epistemology was his first love. In a way now almost totally foreign to contemporary scientists, epistemology and history of science were part and parcel of his work and undergirded it with enormous strength. His reading in philosophy, science, and literature was extremely wide (he kept a log of what he was reading, which, not surprisingly, included not only authors such as Loeb and W.B. Cannon, but also Einstein). So, when he was busy with finding quantitative regularities and orderliness in the behavior of rats it was *in the context of what science could say to the rest of culture*, as we shall see at more length in a moment. I find it not surprising that, in the same year of 1931, when he developed the final form of the lever-press box, which became his key to the development of the concept of the operant, he also became a member of the fledgling History of Science Society.

By 1936, Skinner left Harvard for an instructorship at the University of Minnesota, then briefly was chair of the Psychology Department at Indiana University. Again, if fate had played him a different card, and if Skinner had stayed on in Cambridge just to the beginning of September 1939, he would undoubtedly have been caught up in a unique and congenial event, the International Congress for the Unity of Science – itself an heir of Mach and his followers – held at Harvard in that month, just as war broke out in Europe. Bridgman, with Philipp Frank, was the main organizer, and Quine served as secretary of the meeting. Henderson, Sarton, S.S. Stevens, Carnap, and Kurt Lewin were among the dozens of speakers presenting papers. Skinner would have felt very much at home there. But even more important was that, of this Congress, Quine could later (in his autobiography) write simply, "Basically this was the Vienna Circle with accretions, in international exile." In fact, this occasion was the beginning of a continuing series of monthly meetings of like-minded colleagues across all fields in the Cambridge area, held most vigorously in the 1940s. The spirit of this interscience discussion group was again characterized by Quine as "a sort of revival of the Vienna Circle," headed by Bridgman and Frank.[5]

The meetings of the group ranged very widely over topics of importance to intellectuals of the time, but above all they were a forum for a merciless striving toward clarity and commonalties. Typically, an announcement for January 8, 1945, was, "Professor Richard von Mises will lead a discussion on 'Sense and Nonsense in Modern Statistics.'" The next meeting's invited speaker was Charles Morris from New York, to make more clear what scientists say when they "seek simplicity or economy in their theoretical work." Talcott Parsons spoke on "Psychoanalysis and the Theory of Social Systems"; Norbert Wiener on "The Brain and the Computing Machine," in which he gave the first draft of cybernetics; George Wald on "Biology and Social Behavior"; John Edsall on "The Life and Work of Walter Cannon." Others who led evening discussions were Wassily Leontief, Hudson Hoagland, I.A. Richards (long a friend of the absent Skinner), Jeffrey Wyman, Crozier, Roman Jakobson, Quine, Edwin Boring, Henry A. Murray, John von Neumann, Oscar Morgenstern, Howard Aiken (on the first electromechanical computer), and Bridgman (on the problem of meaning).[6]

The energy and excitement in many of these discussions were enormous. Being much the youngest of the group, I was persuaded without much resistance to be the secretary of this movement – in charge of keeping track of the meetings, circulating the attendance sheets (many of which have survived), and writing up some of the summaries of the discussions. One can only speculate, of course, how Skinner's thought and career would have been influenced by these meetings, as was the case for many of the participants – not least myself. Skinner's absence was thus again one of those "lost opportunities" with which intellectual history abounds.

After visiting Harvard to give the William James Lectures in 1947, Skinner was brought back in 1948 as a professor, at the instigation chiefly of Boring and Stevens. His great book, *The Behavior of Organisms*, of 1938 had already begun to bring him the recognition that he had been waiting for. His research flourished, chiefly in the style he had discovered during the writing of his thesis. But a new aspect with respect to his general vision of himself became more and more prominent. Unlike most other scientists, who are content to stay at their lab bench, Skinner also saw a function for himself in the wider society. John Cerullo, in his study of Skinner in Harvard[7], properly draws attention to the fact that it would be wrong to think of Skinner as a mandarin in the original sense of the honored and specially selected intellectual, who could be counted on to support the existing societal structure, as in the Chinese empire period. On the contrary, Skinner was more nearly of the type Karl Mannheim defined as the

"Free Floating Intellectual": somewhat of a rebel with respect to the particular contemporary social structure of interests, but at the same time, in the classic sense, a *Kulturträger*, who saw as his function not only to uphold but also to improve the contemporary cultural and social order.

Today, this spirit is largely languishing among the intellectual classes, among whom there are few civil intellectuals in this sense. As Cerullo put it (p. 217), the job of our universities is now to turn out, at best, more mandarins. For Skinner, on the other hand, "His advocacy of scientific method, biological determinism, and social control" (p. 217) was a calling that often put his career at risk in the early stages and certainly went far beyond what was expected of the average experimental psychologist. As Skinner himself wrote to Edward M. Freeman in March 1937, "A society ultimately depends on its top crust of intellectuals. If nationalism prevails, human society will have adopted the principle of the anthill and the beehive. Intelligence must protest and can hardly fail to triumph."[8] Eliminating "mentalisms" was for him only a precondition to becoming an intellectual leader, with a calling to do what the work-a-day psychologist – and, as we shall see, even Bridgman – would have regarded as hubris; namely to "make behaviorism a scientific force" in national life, and "himself a leader of it" (Cerullo, p. 222). Cerullo adds, "Behaviorism was the 'higher and wider truth,' a culture-regenerating – indeed world-transforming – mission to which other activities were properly subordinated" (p. 230). Skinner's book *Walden II*, for example, cannot be understood properly without seeing the author as bent on searching for a radically new base upon which to build the culture of the future – even as many of the original Vienna Circle members, in their own way, also did. In no other manner could one explain Skinner's two pungent chapters in the middle of his book *Beyond Freedom and Dignity* (first published in 1971); that is, Chapter 7, "The Evolution of a Culture," and Chapter 8, "The Design of a Culture" – not to mention his swipe at the "pure" scientists "in the sense of being out of reach of immediate reinforcers" (p. 166).

After Skinner's return in 1948, he was at last put in touch with the inter-science discussion group I mentioned before, which, in his absence, had been in operation for about a decade – even though its energy was beginning to dissipate or diffuse. He did take active part in a public conference, "On the Validation of Scientific Theories," in December 1953, chiefly organized by that discussion group. On that occasion he spoke on the topic, "Critique of Psychoanalytic Concepts and Theories" (indeed a topic that has become ever more relevant).[9]

P.W. BRIDGMAN

At that public conference, with dozens of speakers from near and far away – in some way the high point and also the ending of the operations of that movement in Cambridge – Bridgman was of course also present and active. Let me turn now to the other half of the story of Skinner, Bridgman, and the lost opportunity,

first setting the stage by giving some background about Bridgman, a colleague of
mine, as was Skinner, but one whom I knew longer and better and who might not
be known quite so well by behaviorologists.[10]

I believe that Bridgman could not be fully understood unless one watched
him day after day, as he was doing his lab work, coming in early, usually on his
bicycle, practically before everybody else except the shop crew. He would
change immediately into his well-used lab coat and work on new equipment at
the lathe, making much of his apparatus by himself, and race back and forth
between the equipment and measuring instruments during his daily "runs".
Except for his part-time helper in the shop, and an assistant who, also on a part-
time basis, helped him with measurement readings, he preferred to be alone in
his fairly narrow and cramped surroundings, exuding energy and seriousness and
accomplishing, in a few hours, a typical run that might take me, a student working nearby, a day or more. Watching him, I could not help thinking of a passage
from Goethe's tragedy, *Faust*, in which the scholar opened up the first page of
the Bible, encountered the phrase "In the beginning was the Word," and ener-
getically reacted to it by exclaiming, "In the beginning was the Deed." I was
seeing operationalism hyperactive before my eyes. Bridgman was lean and relatively small of stature but, helped by a rigorous regime of exercise, in splendid
physical condition until his last years. There is a story that he was once in the
hospital with phlebitis, and Stevens, from the Psychology Department, happened
to occupy the bed next to him. Bridgman asked Stevens why he was there, and
Stevens replied, "It is because of phlebitis, owing to the fact that I haven't been
moving around much," to which Bridgman answered, "I too have phlebitis, but
because I have been moving around too much."

Bridgman accepted very few graduate students during his long membership
on the faculty at Harvard (from 1908–1954). Avoiding almost all University
committees, his own activity was reflected in the high and steady output of papers, chiefly on physics but also on philosophy of science. Bridgman published
on the average about six substantial papers a year, many with titles like "The
Resistance of 72 Elements, Alloys, and Compounds to 100,000 Kilograms per
Square Centimeter." His lifetime total was over 260 papers, in addition to over a
dozen books. Most of his writing was remarkably personal, often in the first-
person singular. He was essentially the father of high-pressure physics as a field
of research, and it was for this that he was awarded the 1946 Nobel Prize in
physics. Starting with the maximum obtainable pressure of 6,500 atmospheres,
through his own design of sealing the pressure vessels and other ingenious
inventions, with utter concentration and skill, and despite a remarkably low
annual budget, he drove the experimental range up to an estimated 400,000
atmospheres toward the end of his active research, finding spectacular new
behavior of matter on the way. The materials under study were of course
enclosed in massive steel cylinders, with a couple of long and very thin holes
running the length of the cylinder. The only access to the samples within was via
the wires monitoring the sample. On that point, he once wrote, typically, "It is

easy, if all precautions are observed, to drill a hole [in carbon-alloy steel] ... 17 inches long in from seven to eight hours."

We see an important theme emerging: Bridgman's struggle was always with himself, with his own understanding, with his attempts to think situations through to his own satisfaction. The word "operation," he explained,[11] was first explicitly used in a discussion at a AAAS meeting in 1923 on relativity theory, involving G.D. Birkhoff, Harlow Shapley, and himself (p. 75). His book, *Logic,* was, as it were, a summary of what he had gone through in alleviating his "intellectual distress" regarding relativity theory. The book was written during a half-year sabbatical leave in 1926, in great haste, and he always was aware of its incompleteness, even though the whole book gained from the vast outpouring of energy, and if I may use the word, spirit, which infused it. But one result was that he was under constant external and internal pressure to elaborate his notions, and he had to watch others, mostly to his dismay, "erect some sort of a philosophic system" on his work (p. 76).

Bridgman had not been, and never was or could be, a disciple of any "ism," including operationalism, or positivism, or behaviorism. He wanted to clear his mind, to eliminate metaphysics, as Mach has done, to throw the spotlight on performable action, above all action performed by himself. Ultimately, he was a private man, so much so that he was accused of solipsism, to which he scarcely objected. He gave no courses on his views, which others called a philosophy, although he wrote extensively on them. Even when his point of view was adopted in various versions in other fields, ranging from psychology to economics, he did not applaud this but rather watched with bemusement. There is a significant anecdote involving young Stevens during his graduate years. Stevens, who had been much attracted to Bridgman's ideas, told me that he once asked permission to discuss them with him. Bridgman readily agreed as long as Stevens would keep sitting on one of those high lab stools, so Bridgman could continue to move among his apparatus with his typical intensity and speed. During that discussion, the merits of Bridgman's solipsism became more and more clear to Stevens, who finished the story with a smile: "In the end, I became convinced that one could be sure only of the existence of one person – Bridgman himself."

In keeping with his general attitude, Bridgman was not impressed by those who tried to squeeze his operationalism into a formalism favored by logicians, and he especially refused the concept that others, such as the "scientific community" as a whole, could decide on scientific truth. The existence of those "others," so important in the philosophy of science of empiricism and to sociologically informed views of how scientific truths are finally generated by consensus, counted for him very little. Since he truly had conceptual difficulty with the idea of those "others," his ever-skeptical mind was ready to question what to me were sometimes the most obvious points. For example, after we had become colleagues in the department, and had established a pleasant relationship, he gave me the manuscript of his last book, entitled *The Way Things Are*, asking for my comments. However, he added, "I really wanted to call it, *How It Is*, but the publisher didn't like that. The fact is I am not so sure that 'Things' are."

As he put in his essay "Science: Public or Private," in *Reflections of a Physicist* (1950, p. 56):

The process that I want to call scientific is a process that involves the continual apprehension of meaning, the constant appraisal of significance, accompanied by a running act of checking to be sure that I am doing what I want to do, and of judging correctness or incorrectness. This checking and judging and accepting that together constitute understanding are done by me, and can be done for me by no one else. They are as private as my toothache, and without them science is dead.

One reason for the intensely personal understanding of science was that he realized that all formalisms require for their full meaning the personal background and understanding of the individual. This includes unexamined assumptions that might or might not be adequate, not to speak of what Einstein called, to the consternation of many philosophers, the necessity of a quasi-intuitive *Fingerspitzengefühl* for the subject of scientific inquiry. (As it happened, this was a point which, as a student, I absorbed and later developed into the idea of the often unrealized presuppositions, called themata, that can be shown by documentary evidence to have guided some of the best work of scientists, some of whom maintained their belief even against initially contrary "evidence.")

Without Bridgman's internal psychological strength and confidence, one cannot understand how this essentially lonely man could do so much magnificent work.[12] But he said, "I stand alone in the universe with only the intellectual tools I have with me. I often try to do things with these tools of which they are incapable, and I have often been misinformed and have delusions as to what they are capable of, but nevertheless it is my concern and mine only that I get an answer."[13] He could of course not neglect the stream of commentary, both favorable and unfavorable, that his book and articles launched. He famously had to add "paper and pencil operations" to the more physical or "instrumental" ones in order to accommodate, among other things, mathematics. He constantly had to fight against the idea that there is a "normative aspect to 'operationalism' [or 'operationism,' two terms that he said he 'abhorred'])," which is understood as the dogma that definitions *should* be formulated in terms of operations. "An operational analysis is always possible; that is, an analysis into what was done or what happened ... [it can be given] of the most obscure metaphysical definition, such as Newton's definition of absolute time ... It must be remembered that the operational point of view suggested itself from observation of physicists in action" (*The Validation of Scientific Theories*, p. 79).

On some occasions, Bridgman found himself in the position of having to reassess his own work publicly. One was in 1959, when I asked Bridgman to contribute a self-analysis and review to the quarterly *Daedalus* of his book, *The Logic of Modern Physics*. (I had founded the quarterly in 1957, and Bridgman had kindly agreed to be on the board of editors.) As you would expect, it was a thoroughgoing and rather negative critique of all the things that he now thought he should have added or changed and had misunderstood in the writing of the book. In reading his self-analysis, we begin to see even more clearly the growing distance between himself and Skinner.

Bridgman said there that he was particularly embarrassed to have written in *Logic:*

We should now make it our business to understand so thoroughly the character of our permanent relations to nature that another change in our attitude such as that due to Einstein shall be forever impossible. It was perhaps excusable that a revolution in our mental attitude should occur once, because after all physics is a young science, and physicists have been very busy, but it would certainly be a reproach if such a revolution should ever prove necessary again. (p. 520)

To this he added, in retrospect:

To me, now it seems incomprehensible that I should ever have thought it within my powers, or within the powers of the human race for that matter, to analyze so thoroughly the functioning of our thinking apparatus that I could confidently expect to exhaust the subject and eliminate the possibility of a bright new idea against which I would be defenseless. [After all] *our skulls contain a simply appalling number of undiscovered structures which must condition and limit our thinking. ... In fact, it seems to me that for us here and now the problem of adequately understanding the nature of our minds and what we can do with them is a problem more pressing, and perhaps more difficult, than the problem of understanding the physical world* ... The comparatively new methods of the brain physiologist and the behavioral psychologist are without doubt of the greatest value and should be pushed as aggressively as possible. *But the older methods, methods of 'introspection,' if you like, are not to be discarded.* (pp. 520-521, P.W. Bridgman's "*The Logic of Modern Physics* After Thirty Years," *Daedalus*, Summer 1959 [emphasis added]).

Bridgman was now ready to emphasize more the "mental" operations (p. 522).

To Skinner, all this must have verged on unacceptable "mentalism." To a behaviorist such as Skinner, the mind and consciousness were inaccessible to true scientific study. Moreover, psychologists such as Stevens had become ardent operationalists, but he too had moved far from Bridgman's position when he wrote that "An essential characteristic of all facts admitted to the body of scientific knowledge is that they are public. Science demands public rather than private facts. ... Scientific knowledge has what we may call a social aspect" (S.S. Stevens, "The Operational Basis of Psychology," *American Journal of Psychology*, vol. 47, April 1935, p. 323). In return, the following year, Bridgman wrote about Stevens, in a letter to a colleague, that he admired much about him: "but I simply cannot make him see that his 'public science' and 'other one' stuff are just plain twisted. I have also discussed with him his 'basic act of discrimination' without making much impression, and I have rather washed my hands of him" (May 4, 1936, in Maila L. Walter, *Science and Cultural Crisis*, Stanford University Press, 1990, p. 184).

Indeed, even before Skinner had returned to Harvard, the seeds had been planted for a prolonged debate between him and Bridgman, who had been one of the main guides to his thinking during his thesis-writing years. In an article entitled "The Operational Analysis of Psychological Terms" (*Psychological Review*, v. 52, September 1945, pp. 270-277), Skinner – who, we must remember, had, early in his career and to some degree throughout it, been enchanted with literature, writing, and language – wrote that he considered

the language descriptive of private experience as the product of social reinforcement. Accordingly these private events are inferences supported by 'appropriate reinforcement based upon public accompaniments and consequences,' and being conscious, which is nothing more than 'a form of reacting to one's own behavior', is a social product.

But Bridgman, whose own skull housed a brain that, he believed, was rather impermeable to social reinforcements and products, protested, and his rejoinder was quite sharp: "In the private mode I feel my inviolable isolation from my fellows and may say, 'my thoughts are my own and I'll be damned if I let you know what I am thinking about.'"[14] Even earlier, Bridgman had felt upset by the need to deal with Skinner's objections, writing (in a note in the Bridgman Archives, November 7, 1953), "I shall probably continually have Skinner in the back of my head, imagining that I am discussing or arguing with him."

In fact, Bridgman's crusade was in large part to rescue the individual from what he saw as the collectivization of society all around him. The unique individual, and the uniqueness of the individual – very American ideas – were what lay at the very center of his heart. Certainly, as a member of society, he would act in a "public mode" as a good citizen; but what counted most was that secret inner volcano which, inaccessible to all others, was his true preoccupation, day and night. Resigning from the effort to change his mind, Skinner wrote to Bridgman (May 10, 1956, quoted on p. 192, Walter, undated letter in Bridgman's papers): "My efforts to convince you of the possibility of extending the operational method to human behavior have long since suffered extinction."

There is, finally, another reason why these two genial men were ultimately not more attracted to each other intellectually: the difference in their ultimate agenda, beyond doing good science. This comes out eloquently in one of the few letters from Skinner that the Bridgman archives contain. Shortly after his return to Harvard, Skinner decided to give a freshman course, Psychology 7, entitled simply "Human Behavior." The catalogue description was brief but not unambitious: "A critical review of the theories of human behavior underlying current philosophies of government, education, religion, art, and therapy, and a general survey of relevant scientific knowledge, with emphasis upon the practical prediction and control of behavior."

That description is a simple indicator of a profound difference between our two protagonists. Whereas Bridgman was chiefly interested in getting things straight in his own head, Skinner could be said to have had the active agenda of a researcher and a *Kulturträger* and culture improver, even in his undergraduate courses. On June 13, 1949, Skinner wrote to Bridgman[15]:

I thought you might be interested to look over the inclosed [sic] notes which summarize my lectures in Psychology 7 this past term. The course will be given better another year but I feel reasonably satisfied that I have impressed some of my students with the implications of a scientific attack upon human behavior. I am very serious about this course and although it is cutting into my research time I hope to go on giving it. I don't know of anything more important at the moment than to acquaint the lawyers, politicians, diplomats, business men, journalists and educators of the future with the method of science.

And then he added, in a gesture briefly indicating a willingness for cooperation and conciliation: "I hope to make better use of your own writing along this line when the course is given again."

Apparently there was a lunch meeting between Skinner and Bridgman in June 1958 to discuss their differences face to face. But although they remained friendly colleagues in other matters, the essential gulf remained. At the beginning, there had seemed such a promising potential for a fitting-together, perhaps even a collaboration, of these two seminal minds. They had a common intellectual parentage and a shared scientific and philosophical culture. A bridging between them was not to be, however. Perhaps they would have been drawn closer if they had developed a companionship instead of being separated precisely during those exciting "lost years." The main issue between them was, after all, reminiscent of the division in physics between the world of the palpably macroscopic and the untouchable submicroscopic; and in that case, a complementarity point of view developed, which showed each of the two opposing sides was right within its realm, but that they also fitted together in a larger context. I still feel that in the field on which these two protagonists battled, the eventual solution will be found to be of the same sort. And they, working together, could have been the first ones to find that breakthrough.

NOTES

This article is based on the B.F. Skinner Memorial Lecture, and it is destined to be published eventually in the annual *Behaviorism*.

1. What follows draws on well-known sources, including of course Skinner's own books, and L.D. Smith and W.R. Woodward, eds., *B.F. Skinner and Behaviorism in American Culture* (London: Associated University Presses, 1996), and in particular the essays in it by Nils Wiklander, "From Hamilton College to Walden Two: An Inquiry into B.F. Skinner's Early Social Philosophy," and John Cerullo, "Skinner at Harvard: Intellectual or Mandarin?".
2. Scientific World Conception: The Vienna Circle.
3. E.A. Vargas (1994). Prologue, perspectives, and prospects of behaviorology. *Behaviorology*, 3, p. 112. I have written on the effect of Mach and Skinner in Chapter 1 of my book *Science and Anti-science* (Cambridge, MA: Harvard University Press, 1993).

4. B.F. Skinner, review of Smith (1978) Behaviorism and logical positivism, *Journal of the History of the Behavioral Sciences*, 23, pp. 204-209, on p. 209.
5. For detailed descriptions of these meetings and the movement behind them, see Chapter 1, "Ernst Mach and the fortunes of positivism" in my book *Science and Anti-science*, and in my two articles on the subject, "From the Vienna Circle to Harvard Square: The Americanization of a European world conception," in F. Stadler (Ed.), *Scientific philosophy: Origins and developments* (pp. 47-73) (Dordrecht: Kluwer, 1993); and "On the Vienna Circle in exile: An eyewitness report," in E. Köhler, W. Schimanovich, F. Stadler (Eds.), *The foundational debate* (pp. 269-292; Dordrecht: Kluwer, 1995).
6. Among other participants in these meetings were Karl Deutsch, Frank, Edwin C. Kemble, Hans Margenau, Ernest Nagel (as visitors), Harlow Shapley, Laszlo Tisza, and Paul Samuelson.
7. See endnote 1.
8. From the Harvard Archives' Skinner Collection; still incomplete and being supplemented, it now contains 30 feet of letters and manuscripts.
9. The proceedings of the conference were published in the *Scientific Monthly* in 1954 and 1955 and then republished as a book by Beacon Press in 1956 under the title *The validation of scientific theories*, edited with an introduction by Philipp G. Frank. A glance at the book will show how difficult it would be to bring together today such a high-powered set of thoughtful speakers on such a variety of aspects of the culture of our time.
10. I have written on Bridgman as scientist and philosopher on a number of occasions, most recently in Chapter 11, "Percy W. Bridgman, physicist and philosopher," in my book *Einstein, history, and other passions* (New York: American Institute of Physics Press, 1995).
11. P.W. Bridgman, "The present state of operationalism," in *The validation of scientific theories*.
12. Bridgman's sense of isolation comes through in a letter of March 30, 1938, to Philipp Frank, whom Bridgman expected to see with "great pleasure": "My work is done practically alone. I have no students [in philosophy of science] and have practically no contact with members of the department of philosophy and, in fact, most of them are not at all sympathetic with our point of view. The only young philosopher here whom I have particularly interested is Dr. Quine."
13. P.W. Bridgman, "New vistas for intelligence" in *Reflections*, p. 370. In the end he pronounced that famous definition, "The scientific method consists of doing one's damnedest with one's mind, no holds barred" (pp. 57-58).
14. The quotations are from Maila L. Walter, *Science and cultural crisis*, pp. 188-191. They refer to Bridgman's article "Rejoinders and second thoughts," *Psychological Review*, 52, September 1945, pp. 281-283.
15. Harvard Archives, Skinner Papers, HUG(FP) 60.10, Box 1.

Jefferson Physical Laboratory
Harvard University
Cambridge, MA 02138
U.S.A.
holton@physics.harvard.edu

DAVID STERN

SOCIOLOGY OF SCIENCE, RULE FOLLOWING AND FORMS OF LIFE [*]

1. WITTGENSTEIN ON SCIENCE: "THE POINT IS THAT '"OBEYING A RULE" IS A PRACTICE'"

Ludwig Wittgenstein was trained as a scientist and an engineer. He received a diploma in mechanical engineering from the Technische Hochschule in Charlottenburg, Berlin, in 1906, after which he did several years of research on aeronautics before turning to the full-time study of logic and philosophy. Hertz, Boltzmann, Mach, Weininger, and William James, all important influences on Wittgenstein, are authors whose work was both philosophical and scientific. The relationship between everyday life, science, and philosophy, is a central concern throughout the course of his writing. He regarded philosophy, properly conducted, as an autonomous activity, a matter of clarifying our understanding of language, or investigating grammar. Wittgenstein thought philosophy should state the obvious as a way of disabusing us of the desire to formulate philosophical theories of meaning, knowledge, language, or science, and was deeply opposed to the naturalist view that philosophy is a form of science. In his later work, Wittgenstein rejected systematic approaches to understanding language and knowledge. Wittgenstein's answer to the Socratic question about the nature of knowledge is that it has no nature, no essence, and so it is a mistake to think one can give a single systematic answer:

If I was asked what knowledge is, I would list items of knowledge and add "and suchlike." There is no common element to be found in all of them, because there isn't one. (Wittgenstein, MS 302, "Diktat für Schlick" 1931-33.)

In the *Philosophical Investigations*, one of the principal reasons for Wittgenstein's opposition to systematic philosophical theorizing is that our use of language, our grasp of its meaning, depends on a background of common behaviour and shared practices – not on agreement in opinions but in "form of life" (Wittgenstein 1953, §242.)

Despite these far-reaching links between Wittgenstein's writings and the study of science, the relationship between the two is poorly understood. In part, this is because Wittgenstein wrote relatively little about science *per se*. His explicit remarks on science are for the most part quite short, dispersed widely

M. Heidelberger and F. Stadler (eds.), History of Philosophy and Science, 347-367.
© 2002 *Kluwer Academic Publishers. Printed in the Netherlands.*

throughout his writing, and often focus on the work of figures such as Einstein, Goethe, Frazer, Freud, Hertz, James, Newton, Köhler, Mach, or Weininger. In any case, most interpretations of the implications of his philosophy for the natural or social sciences have paid little, if any, attention to what Wittgenstein had to say about particular scientists, or even his more general views about specific natural or social sciences. Instead, the starting point is usually an exposition of Wittgenstein on rule-following in the *Philosophical Investigations*, and the insights into the nature of science, knowledge and society that are supposed to follow from the "rule-following considerations."

In "Extending Wittgenstein: The Pivotal Move from Epistemology to the Sociology of Science," a contribution to the debate over Wittgenstein's significance for the sociology of science in *Science as Practice and Culture* (Pickering 1992), Michael Lynch states that "Wittgenstein is widely regarded as the *pivotal* figure for a "sociological turn" in epistemology" (Lynch 1992, 218). But it is Peter Winch's interpretation of Wittgenstein that first brought him into the sociological limelight, and David Bloor's critique of Winch that made Wittgenstein into a canonical source for the philosophical agenda of recent sociology of scientific knowledge, commonly known as SSK. The existence of this line of influence is hardly controversial. Bloor's first paper on Wittgenstein makes clear his debt to Winch, and his reading of Wittgenstein has become a point of reference for subsequent debates within the sociology of knowledge over methodology and theory. But the role of Winch and Bloor in the reception of Wittgenstein by sociologists of science is often overlooked. This is, perhaps, because the appeal to Wittgenstein by sociologists of science has for the most part become formulaic and routine, and because it is often taken for granted that the sociology of science's Wittgenstein is taken from Kuhn, not Winch and Bloor. Wittgenstein has joined the august company of philosophical figures such as Aristotle, Hume, or Kant: all convenient historical antecedents, when an intellectual lineage is needed, but rarely read in any detail by those who invoke them in a programmatic way. In the earliest stages of Wittgenstein's reception within the sociology of science – Winch's and Bloor's early interpretations of Wittgenstein – interest in Wittgenstein had to do with the way in which he seemed to open up the possibility of appropriating and transforming the claims of classical epistemology. In more recent work, Wittgenstein has come to play two distinct but interrelated roles for those working in SSK. First, his talk of "forms of life" and the primacy of practice is invoked in order to provide philosophical legitimation for the way in which practitioners of SSK study scientific communities. Second, he has become a focal point for debates about the methodology and philosophical agenda of SSK (Bloor 1973, 1983, 1997; Pickering 1992; Lynch 1993; Friedman 1998.)

It will be helpful to begin by considering some textbook examples of the recent use of Wittgenstein within SSK as providing philosophical legitimation for the turn toward social practice. Jan Golinski's *Making Natural Knowledge: Constructivism and the History of Science* (1998), a book on SSK's methodological implications for the history of science, sums up Wittgenstein's role as follows:

the social collectivity, ignored in the classical model of subject and object, has come to be regarded as critical for the production of knowledge. One source of this is the later philosophy of Ludwig Wittgenstein, with its claim that language finds meaning by virtue of its use in specific "forms of life" (Bloor 1983) ... Analysts who have applied Wittgenstein's notion of "forms of life" have portrayed social formations ([Kuhn's] "paradigms" or [Collins'] "core sets," for example) that are defined by particular configurations of scientific practice (Golinski 1998, 7, 47.)

The master argument for this appeal to Wittgenstein can be summarized very briefly: application of any concept is always indeterminate, and it is only the social collectivity, as studied by SSK, that can resolve this indeterminacy. Here is a representative exposition of this argument from *Science in Context: Readings in the Sociology of Science* (Barnes and Edge, 1982), a set text for Britain's Open University. The quotation is taken from an editorial introduction to Part Two of the reader, on "The culture of science", which contains readings by Kuhn, Collins, Bloor and Pickering.

Every particular case differs in detail from every other and can never be conclusively pronounced identical to any other, or identical in any attribute to any other. Hence, nothing can be unproblematically deduced from a rule or law, concerning any particular case, because there is always the undetermined matter of whether the case falls under the rule or the law – that is, whether it is *the same as* or *different to* those instances which have already been labelled as falling under the rule or law. Formally, this matter of similarity or difference arises at every point of use of a concept, and has to be settled at every point by the using community. Concept application is inherently open-ended (Wittgenstein, 1953; Bloor, 1973; Hesse, 1974). (Barnes and Edge 1982, 70.)

A closely related exposition can be found in Pickering's introduction to *Science as Practice and Culture* (1992), although Pickering adopts a more moderate formulation, on which the master argument is not applied to every case, only new ones:

SSK's perspective on knowledge is, however, typically underwritten by a particular vision of scientific practice that goes broadly as follows ... Since the central problematic of SSK is that of knowledge, the first move is to characterize the technical culture of science as a single conceptual network ... an image of scientific practice follows: practice is the creative extension of the conceptual net to fit new circumstances. And here SSK, following Ludwig Wittgenstein (1953) and Thomas Kuhn (1962), insists on two points. First, that extension of the net is accomplished through a process of modeling or analogy: the production of new scientific knowledge entails seeing new situations as being relatively like old ones. And second, that modeling is an open-ended process: the extension of scientific culture, understood as a single conceptual net, can plausibly proceed in an indefinite number of directions; nothing within the net fixes its future development. (Pickering 1992, 4)

Of course, this conception of scientific practice as indeterminate naturally leads to the question: "what then does produce closure in particular cases?" (Barnes and Edge 1982, 73), or as Pickering puts it, "how is closure – the achievement of consensus on particular extensions of culture – to be understood?" (1992, 4) Pickering provides the following answer: SSK provides a distinctly sociological account by appealing to the instrumental aspect of knowledge, and the interests of scientific actors.

> Introduction of the distinctively sociological concept of interest serves to solve the problem of closure in two ways. On the one hand, actors can be seen as tentatively seeking to extend culture in ways that might serve their interests rather than in ways that might not. And on the other hand, interests serve as standards against which the products of such extensions, new conceptual nets, can be assessed. A good extension of the net is one that serves the interest of the relevant scientific community best. Here, then, is the basic SSK account of practice, and with this in hand we can return to the starting point – the problematic of science-as-knowledge – and articulate a position: scientific knowledge has to be seen, not as the transparent representation of nature, but rather as knowledge relative to a particular culture, with this relativity specified through a sociological concept of interest. (Pickering 1992, 4-5)

In both cases of change and continuity, the distinctively sociological feature is the turn to the interests of the people involved as the factor that takes up the slack in the network model. The interest model is not particularly prominent in Bloor's early work. However, the principal intellectual legacy of the approach he pioneered is the writing of history of science in which group interests take center stage, exemplified in Shapin's (1982) review of the early literature in this area, and Shapin and Schaffer (1985).

Interests are not just supposed to play a part in explaining the particular directions in which actors seek to extend, or modify, existing scientific knowledge, and the standards that are used in choosing between the different choices that are available. Interests are also allocated a comparable role in the maintenance of ordinary, uncontested, practices, in going on in the same way. "The suggestion is that goals and interests are associated with scientific research in all actual situations, and operate as contributory causes of the actions or series of actions which constitute the research. The causes help to explain the problem of the next case, of why a term is applied, or an exemplar extended in that particular way that time." (Barnes, Bloor and Henry 1996, 120) The appeal to interests as an explanatory principle has been irresistibly attractive to many within and beyond SSK, despite forceful objection from both exponents (Woolgar 1981, 1983, 1992) and opponents (Roth 1987, 1996, 1998).

Wittgenstein's term "form of life" has been taken over within SSK as a term for the culture of a specific scientific community, which in turn comprises its practices, interests, and ways of going on. *Scientific Knowledge: A Sociological Analysis*, a recent textbook authored by members of the Edinburgh Science Studies Program, highlights this aspect of Wittgenstein's influence in SSK.

Setting out different ways of presenting the individual as an active agent within sociology of science, they say that one leading alternative is to characterize him or her as a participant in a "form of life":

The term is Wittgenstein's, and its use here is testimony to the relevance of Wittgenstein's work, directly or indirectly, to the work of many sociologists. Those who have taken up the work of Thomas Kuhn have thereby linked themselves to Wittgenstein; so have those who have extended ethnomethods into sociology of science. Harry Collins, who makes the most frequent explicit references to forms of life in science, has used the work of the philosopher Peter Winch as a line of access to Wittgenstein's ideas. [Bloor's] finitist account of the use of scientific knowledge in this book is another version of the same position. (Barnes, Bloor, & Henry 1996, 116.)

This use of Wittgenstein is exemplified in Shapin and Schaffer's widely-read *Leviathan and the Air Pump: Hobbes, Boyle and the Experimental Life* (1985), which makes liberal use of the terms "form of life" and "language-game". In the first chapter of the book, they justify this use in the following terms:

We mean to approach scientific method as integrated into *patterns of activity*. Just as for Wittgenstein "the term 'language-*game*' is meant to bring into prominence the fact that the *speaking* of a language is part of an activity or a form of life," [Wittgenstein 1953, §23] so we shall treat controversies over scientific method as disputes over different patterns of doing things and of organizing men to practical ends. [A footnote here refers the reader to Bloor 1983, ch. 3] We shall suggest that solutions to the problem of knowledge are embedded within practical solutions to the problem of social order, and that different practical solutions to the problem of social order encapsulate contrasting practical solutions to the problem of knowledge. (Shapin and Schaffer 1985, 15)

The principal critique of the appeal to Wittgenstein's philosophy in the sociology of science is Michael Friedman's "On the Sociology of Scientific Knowledge and its Philosophical Agenda" (1998). The central theme of the paper is the tension between "the idea that SSK is an empirical scientific discipline, on the one hand, and its claim to solve the traditional problems of philosophy" (1998, 241). Friedman sums up SSK's use of Wittgenstein as follows:

This philosophical agenda of SSK, in both its theoretical and its applied versions, is explicitly traced to the work of one of the giants of twentieth century philosophy, namely Ludwig Wittgenstein. In particular, the concepts of 'language-game' and 'form of life,' which are central to Wittgenstein's *Philosophical Investigations*, are here interpreted as referring to particular socio-linguistic activities associated with particular socio-cultural groups – where the practices in question are regulated by socio-cultural norms conventionally adopted by the relevant groups. Wittgenstein's insistence on the need for renouncing traditional philosophy in favor of the careful description of particular 'language-games' expressing particular 'forms of life' is then read as the call for an empirical sociological investigation of the way in which the traditional categories of knowledge, objectivity, and truth are socially constituted and determined by the norms, needs, and interests of particular socio-cultural groups. (Friedman 1998, 240-1)

In his critique of this sociological turn, Friedman highlights four leading points on which Bloor and his colleagues misread Wittgenstein. First, SSK aims to be an empirical scientific discipline, while Wittgenstein always said that his philosophy was entirely distinct from natural science (1998, 252-3). Second, Wittgenstein often considers imaginary uses of language, not real ones, which indicates the deeply non-empirical character of his work (1998, 253-4. For an excellent discussion of the weaknesses of Bloor's interpretation of Wittgenstein's use of imaginary examples, see Cerbone 1993). Third, Wittgenstein "shows very little interest in the kind of historical and cross-cultural variation in human linguistic and cultural practices that is the basis and starting-point for the empirically oriented enterprise of SSK" (1998, 254). While he is interested in alternatives to 'our' ordinary practices, and in showing that there is no absolute necessity to those practices, the alternatives are usually imaginary and far more radical than those encountered by ethnologists or historians of science. Fourth, Wittgenstein shows no sign of interest in socio-cultural relativism. While the reference of his "we" is rarely specified – one of the ways in which his writing makes it easy for a socio-cultural relativist to find a foothold – the term is often used interchangably with "humanity" or "all human beings" (1998, 254).

However, Friedman has little to say about how or why this misreading came about, or the role of the master argument as a philosophical argument for the primacy of "forms of life" and a sociological epistemology. But the idea of reading Wittgenstein's philosophy as sociology of knowledge is an audacious one, and the story of how it came about is well worth telling. It is a story of a succession of theories, each claiming to get at what was right about the previous theory, but turning it on its head in order to do so. And it is just the kind of demystifying history of disciplinary myth-making that sociologists of science have so assiduously pursued when studying the sciences, but have for the most part been reluctant to undertake on their own behalf.

At the end of *Wittgenstein: A Social Theory of Knowledge*, David Bloor sums up the relationship between the programme of research he advocated in the sociology of science and Wittgenstein's work in the following terms:

Wittgenstein referred to his work as one of "the heirs to the subject which could be called philosophy" (Blue Book, p. 28). My whole thesis could be summed up as the claim to have revealed the true identity of these heirs: they belong to the family of activities called the sociology of knowledge. ... The point is that "'obeying a rule' is a practice" (Wittgenstein 1953, §202)

In other words, Bloor contends that his sociology of knowledge is a replacement for philosophy, as traditionally conceived. Wittgenstein provides the point of departure for this revolution, a revolution that begins with Wittgenstein's treatment of practice and ends with Bloor's sociology of knowledge. It is in this connection that Bloor cites *Investigations* §199 – "the point is that "'obeying a rule" is a practice'" – as a summary of the main point he takes from Wittgenstein. Certainly, Wittgenstein emphatically insists that obeying a rule is a practice; the significance of practice in the *Investigations* can hardly be overemphasized. But what did Wittgenstein mean by his talk of rule-following and practice, and what were his reasons for insisting on their importance? As there are deep and far-reaching disagreements among Wittgenstein interpreters on just this issue, Bloor's interpretation can best be appreciated if we first review the alternatives. This is the business of section two of this paper. In section three, I look at Winch's reading of Wittgenstein on rule-following, the epistemology he attributes to Wittgenstein, and his claim that "sociology is really misbegotten epistemology." Section four concerns Bloor's reading of Wittgenstein on rule-following, the epistemology he attributes to Wittgenstein, and his reply to Winch: "epistemology is misbegotten sociology." Section five considers what is at stake in this debate about how to read Wittgenstein and how to do philosophy and sociology of knowledge.

2. WITTGENSTEIN ON RULE-FOLLOWING: SCEPTICISM, ANTI-SCEPTICISM, AND QUIETISM

What is it to follow a rule correctly? Taken by itself, the verbal formulation of a rule does not determine its next application, for it is always possible that it will be misunderstood, and any attempt to drag in more rules to determine how to apply the original rule only leads to a vicious regress. Of course, a great deal depends on how one frames and approaches the question about rule-following that I have just sketched so quickly. The standard point of reference in discussions of this issue since 1982 has been Saul Kripke's *Wittgenstein on Rules and Private Language*. Kripke reads Wittgenstein as raising, and attempting to answer, a scepticism about rule following, that is, as replying to someone who holds that we cannot satisfactorily answer the question about what it is to follow a rule. In a sense, Kripke takes Wittgenstein's arguments to lead to scepticism, for he argues that the solution he attributes to Wittgenstein doesn't work, and does nothing to show that a better answer is possible. A few interpreters have actually read Wittgenstein as such a sceptic himself. Michael Dummett's (1959) review of Wittgenstein's *Remarks on the Foundations of Mathematics*, describes Wittgenstein as a "full-blooded" conventionalist who held that every single case of rule-following involves an element of decision, and so it is never necessary to follow a rule one way rather than another. Henry Staten's (1984) reading of the rule-following discussion in the *Investigations* also stresses the role of decision, interpreting Wittgenstein as a sceptic and deconstructionist avant la lettre. On this Derridean reading of Wittgenstein, there is an unbridgeable gap between a rule and its application, an abyss that makes any positive theory about what it is to follow a rule, or a theory of meaning, an impossibility.

While it has few supporters, the sceptical position is important because it is the point of departure for the standard approach to rule-following. Most readers agree with Kripke that Wittgenstein is replying to scepticism about rule-following, but disagree over the right answer. The two main camps are known as "individualists" and "communitarians." "Individualists," such as Colin McGinn (1984) and Simon Blackburn (1984, 1984a) maintain that the resources for a solution can be provided by a single individual. In other words, the practices involved in following a rule may be the practices of an isolated individual. "Communitarians" such as Peter Winch (1958) or David Bloor (1973, 1997), hold that answering the sceptical problem is only possible if one is a member of a community – a group of a certain kind – and so the practices in question must be social, if not community-wide. Before the publication of Kripke's book, it was usually taken for granted that Wittgenstein was offering a communitarian solution. While Kripke himself endorses this reading, he highlighted the importance of the distinction between individualists and communitarians, and the differences between them became a leading issue in the resulting controversy. For instance, Bloor's first book on Wittgenstein (1983) takes it for granted that he was a communitarian, and argues for a sociological construal of "community" and "practice"; his second (1997) is an extended defence of communitarianism.

Kripke and Bloor agree that Wittgenstein begins by arguing that meaning is underdetermined by the available evidence and then provides a community-based solution – meaning is determined by the community's social practices, or "form of life". But Kripke argues that this is only a second-best, 'sceptical' solution, one which does not meet the standards set by the sceptical problem about meaning, while Bloor (1973, 1997) maintains that appealing to a community is a 'straight' solution, one that really does solve the sceptical problem. In other words, Bloor accepts Kripke's starting point, an argument that there is a gap between a rule and its application, but holds that social practices, the forms of human activity studied by the sociologist, provide the answers.

A third alternative is to hold that the debates between sceptics and anti-sceptics, individualists and communitarians, miss the point, which is that Wittgenstein aims to show us that there is no philosophical problem about rule-following, or determinate meaning. On this approach, Wittgenstein aims to get rid of arguments about meaning, not provide a theory of it. Problems about rule-following only lead to scepticism if one approaches understanding a sentence or a rule wrongly, such as thinking of understanding as consisting in giving an interpretation. But Wittgenstein considers such views of rule-following to be mistaken, and without them, the paradox does not get started. There is no "gap" of the kind that concerns both sceptic and anti-sceptics. Admittedly, there is considerable *prima facie* textual support for each of those readings – passages which certainly look as if they formulate and defend sceptical and anti-sceptical theories of meaning – both within the *Philosophical Investigations* and Wittgenstein's other writings. One can read Wittgenstein as either a sceptic or an anti-sceptic, depending on which of those passages one plays up and which one plays down. But neither reading can explain why so much of the material on rule-following, and rule-scepticism, is in the form of a dialogue, a dialogue in which the two positions are played out against each other in such detail. Scepticism and anti-scepticism about meaning fascinated Wittgenstein, but he was always concerned with finding a way out of it, not in defending one side or the other. (For further discussion of this approach, see Stern 1995.)

This reading is often known as "quietism," for its denial that Wittgenstein has anything to say on the subject of grand philosophical theories about the relation between language and world. According to the quietist, Wittgenstein's invocation of forms of life is not the beginning of a positive theory of practice, or a pragmatist theory of meaning, but rather is meant to help his readers get over their addiction to theorizing about mind and world, language and reality. Hilary Putnam, Cora Diamond, and John McDowell are among the leading advocates of this approach. McDowell observes that Wittgenstein's readers often take his talk of "customs, practices, institutions," and "forms of life" as the first steps towards a positive philosophy. The point of the positive views would be to give a non-intentional, or non-normative, justification of our talk of meaning and understanding, by placing it in a broader context of human interaction, interaction which can be described in non-intentional terms.

But there is no reason to credit Wittgenstein with any sympathy for this style of philosophy. When he says "What has to be accepted, the given, is – so one could say – forms of life" ([*Philosophical Investigations*] p. 226) his point is not to adumbrate a philosophical response, on such lines, to supposedly good questions about the possibility of meaning and understanding, or intentionality generally, but to remind us of something we can take in the proper way only after we are equipped to see that such questions are based on a mistake. His point is to remind us that the natural phenomenon that is normal human life is itself already shaped by meaning and understanding (McDowell 1993, 50-51).

Within the sociology of science, one can find echoes of this view in Michael Lynch's (1993) and Jeff Coulter's (1993) ethnomethodological readings of Wittgenstein. Lynch, Coulter and their fellow ethnomethodologists still see room for a scientific study of science that would take its cue from Wittgenstein's insistence on the interwovenness of language, activity, and practice, aiming at "thick description" rather than a formal theory. The debate between Bloor (1992) and Lynch (1992, 1992a) in *Science as Practice and Culture* (Pickering 1992) is over whether to read Wittgenstein with Bloor as a communitarian who provides the basis for a sociological theory of practice, or to follow Lynch in reading Wittgenstein as a quietest precursor of ethnomethodology. Bloor insists that a sociological theory of practice is necessary to show how a rule and its application are tied together; Lynch, following Baker and Hacker, replies that there is a firm logical connection between a rule and its application, and so no need for Bloor's theory. But the best place to begin, if we are to see how Wittgenstein's discussion of rule-following came to be taken up within SSK, is with Peter Winch's *The Idea of a Social Science* (1958).

3. WINCH'S WITTGENSTEIN:
"SOCIOLOGY IS REALLY MISBEGOTTEN EPISTEMOLOGY"

Although nothing could have been further from Winch's intentions, it was his interpretation of Wittgenstein in *The Idea of a Social Science and its Relation to Philosophy* (1958) that proved to be the crucial link in the transformation of Wittgenstein's ideas about rule-following into a new sociology of knowledge. One of the main aims of that very short book was to argue against the view that the method of the social sciences should be the method of the natural sciences, as conceived of by logical positivist and empiricist philosophy of science. It also argued for an interpretive approach to social science that begins from the unre-flective understanding of the participants. Winch's presentation of the case for the distinctively interpretive character of social science presupposes a sharp contrast with a positivistic conception of natural science:

I do not wish to maintain that we must stop at the unreflective kind of understanding ... But I do want to say that any more reflective understanding must necessarily presuppose, if it is to count as genuine understanding at all, the participant's unreflective understand-ing. And this in itself makes it misleading to compare it with the natural scientist's under-standing of his scientific data. (Winch 1958/1991, 89)

Winch maintains that language and action – what people say and do – cannot be understood in isolation from their broader practical and cultural context. Draw-ing on an expression of Wittgenstein's, he calls this context the "forms of life" of the people in question. Because of the way in which what we say and do is embedded within this broader context, language and world are inextricably intertwined. One consequence that Winch draws is that the realist's conviction that reality is prior to thought, that the world is independent of our ways of representing it, is incoherent:

Our idea of what belongs to the realm of reality is given for us in the language that we use. The concepts we have settle for us the form of the experience we have of the world. ... The world *is* for us what is presented through those concepts." (Winch 1958, 15)

Because social institutions embody ideas of what is real and how it is to be un-derstood, Winch holds that causal methods will prove utterly inadequate for the task of understanding our social world: "the central concepts which belong to our understanding of social life are incompatible with concepts central to the activity of scientific prediction" (Winch 1958, 94). Winch is not merely making the familiar claim that the methods of the natural sciences will prove unsuccessful when applied to social questions, but that the very attempt to do so is logically flawed, and strictly speaking, nonsense.

Winch's central argument for these far-reaching conclusions is contained in his exposition of Wittgenstein's account of rule-following in the *Investigations*

(Wittgenstein, 1953 §§ 243ff.; Winch 1958, 24-39). Winch begins by pointing out that words do not have meaning in isolation from other words. We may explain what a word means by giving a definition, but then one still has to explain what is involved in following a definition, in using the word in the same way as that laid down in the definition. For in different contexts, "the same" may be understood in different ways: "It is only in terms of a given *rule* that we can attach a specific sense to the words 'the same.'" (Winch 1958, 27) But of course the same question can be raised about a rule, too: how are we to know what is to count as following the rule in the same way? Given sufficient ingenuity, it is always possible to think up new and unexpected ways of applying a rule. However, in practice we all do, for the most part, unreflectively follow a rule in the same way: "given a certain sort of training everybody does, as a matter of course, continue to use these words in the same way as would everybody else. It is this that makes it possible for us to attach a sense to the expression 'the same' in a given context." (Winch 1958, 31)

An essential part of the concept of following a rule, Winch contends, is the notion of making a mistake, for if someone is really following a rule, rather than simply acting on whim, for instance, we must be able to distinguish between getting it right and getting it wrong. Making a mistake is to go against something that "is *established* as correct; as such, it must be *recognizable* as such a contravention. ... Establishing a standard is not an activity which it makes sense to ascribe to any individual in complete isolation from other individuals" (Winch 1958, 32). Rule-following presupposes standards, and standards presuppose a community of rule-followers.

In a section on the relations between philosophy and sociology, where Winch sums up the results of this argument, he describes it as a contribution to "epistemology." Winch makes it clear that epistemology, as he uses the term, has little to do with traditional theories of knowledge, but is instead his preferred name for first philosophy, that part of philosophy which is the basis for all others. For epistemology, as Winch understands it, deals with "the general conditions under which it is possible to speak of understanding" and so aims at elucidating "what is involved in the notion of a form of life as such" (Winch 1958, 40-1). Thus, on Winch's reading, "Wittgenstein's analysis of the concept of following a rule and his account of the peculiar kind of interpersonal agreement which this involves is a contribution to that epistemological elucidation" (Winch 1958, 41).

Like the "private language arguments" that were so much discussed during the 1960s and 1970s, this one turns on the need for a community if one is to follow rules. In retrospect, Winch's argument is extremely compressed, and a full defence would require that it respond to many of the now-familiar difficulties that have been rehearsed so often in the interim. However, for our purposes the most important point is the *use* Winch made of this argument. His contemporaries took the main force of the private language to be the negative consequences for traditional approaches to epistemology such as Cartesian dualism, scepticism and phenomenalism. Winch, on the other hand, used it to argue for a new conception of epistemology, as the result of following through the implications of his Wittgensteinian grammatical analysis. That epistemology could be positively applied to questions about the nature of society, questions that had previously been regarded as empirical questions for the sociologist:

> ... the central problem of sociology, that of giving an account of the nature of social phenomena in general, itself belongs to philosophy. In fact, not to put too fine a point on it, this part of sociology is really misbegotten epistemology. I say 'misbegotten' because its problems have been largely misconstrued, and therefore mishandled, as a species of scientific problem. (Winch 1958/1991, 43)

To sum up: Winch puts forward a quite general philosophical argument that neither formal logic nor empirical hypotheses are appropriate methods for the study of society. Instead, one must aim at an interpretive investigation of that society's ideas and forms of life, a philosophical investigation that will make clear the kind of work that is appropriate within the social sciences. At first, Winch's argument is extremely general, and primarily concerns what he calls "the notion of human society." But later on, he sets out specific consequences for different modes of social life within a society. Not only is there an internal – logical – relation between the notion of human society and the forms of human life, but social relations between people are also internal, for the structure of our language and our relations with others are two sides of the same coin. "If social relations between men exist only in and through their ideas, then, since the relations between ideas are internal relations, social relations must be a species of internal relation too." (Winch 1958, 123)

As a result, aspects of social life far from what is usually considered the domain of philosophy, such as the significance of a pointed glance or gesture in a conversation or the giving of historical explanations (Winch 1958, 129-30, 133) have to be understood by interpreting the logic of what actors do and say. Thus it turns out that not only central questions about the nature of social phenomena, but also the detailed understanding of particular aspects of our lives, cannot be approached by the methods of natural science, but only by those particularistic, interpretive methods recommended by Winch's epistemology of forms of life.

Winch argues that Wittgenstein's treatment of rule-following shows that we must start in social science with "forms of life," the social practices of human groups. This turn to forms of life, understood as culture-specific practices, is one way of supplanting the central role occupied by representation in traditional philosophy of science – "knowledge that" – with skills or abilities – "know how." But a great deal turns on just how one conceives of this embedding of knowledge in social practice, in "forms of life." Bloor, and others working in the sociology of scientific knowledge, hold that practical context, "language games" and "forms of life," play a central role in Wittgenstein's later philosophy because they function as a background against which determinate meanings are possible. In turn, this is supposed to make possible a *scientific* and *naturalistic* understanding of the relationship between science, practice and culture.

4. BLOOR'S WITTGENSTEIN:
"EPISTEMOLOGY IS MISBEGOTTEN SOCIOLOGY"

In a paper on the relationship between logic and sociology, Jeff Coulter has described Winch as "perhaps the most important figure in the history of the (renewed?) relationship between Logic and Sociology," because "Winch sought to 'dissolve' sociology into conceptual/grammatical analysis of a Wittgensteinian kind" (Coulter 1991, 32-3). In the years immediately following the publication of *The Idea of a Social Science* (1958) and "Understanding a Primitive Society" (1964), the vast majority of professional sociologists in Britain and America considered Winch's views unacceptable. One need look no further than the threat those views posed to the project of an autonomous sociology and the rejection of the possibility of systematic sociological laws.

On the other hand, Winch's respect for the particularity of other forms of life, and the need to understand them from within, was enormously attractive to those who wished to approach scientific cultures along comparable lines, by combining Kuhn's notion of a paradigm with Winch's account of understanding another culture. The crucial move here was to conceive of the culture of a particular group of scientists – one of the senses of Kuhn's famously slippery term, "paradigm," – along lines suggested, if not required, by Winch's discussion of forms of life. Bloor simply takes it for granted that the term "form of life" refers to specific cultural or social groups, social entities comparable to the "primitive societies" discussed by Winch, or Kuhn's scientific research cultures. This way of understanding science as a culture involved a conception of culture and knowledge very different from the thin and formal accounts of the nature of science and knowledge prevalent in the philosophy of science at the time. In this way David Bloor found a way of reading Wittgenstein and Winch as providing the point of departure for a reinvigorated sociology of scientific knowledge. Bloor's guiding insight in this was that Winch's positive conception of sociology as epistemology, as a study of the network of relations between actors and the world, could be prized loose from Winch's claim that "sociology is really misbegotten epistemology" (Winch 1958, 43). In effect, the distinctive feature of this post-Winchian sociology of science is that it sought to reverse the direction of Winch's program, and dissolve Winch's epistemology into sociology.

Broadly speaking, Bloor's reading of Wittgenstein starts from the programmatic conviction that the ahistorical, a priori reasoning about science favoured by the "philosophical tradition" had to be replaced by hard empirical research: a genuinely scientific and naturalistic study of science. Close attention to scientific culture and its characteristic practices led him to see that epistemology could not be done by simply reflecting on the necessary and sufficient conditions for the nature of such concepts as rationality, experimental success, or knowledge. Instead, one had to empirically investigate scientific forms of life and look at the role of social relations in scientific practice. This, then, was to be SSK's conception of "epistemology naturalized": epistemology sociologized, the philosophy of knowledge to be replaced by the sociology of knowledge.

Bloor replaced Winch's sharp distinction between social and natural science, and the contrast between internal relations in the realm of social relations and external relations connecting scientific laws and observation, with a single holistic web, or network, of concepts, concepts connected by the rules that make up forms of life. But where Winch had seen social relations as expressions of ideas, he turned this idea around, and saw ideas as expressions of social relations. Bloor's transformation of Winch's conception of science and society is nicely summarized in Collins' *Changing Order*, a book that explicitly draws on Bloor's reading of Winch. Collins sums up the attractions of the "network model" of concepts by quoting the following words from Winch:

A man's social relations with his fellows are permeated with his ideas about reality. Indeed, 'permeated" is hardly a strong enough word: social relations are expressions of ideas about reality (Winch, 1958, p. 23).

and adds

We must add that the converse is equally true, that ideas are expressions of social relations" (Collins 1985/1992, p. 132).

But what is it for ideas to be expressions of social relations? Here one is faced with a series of vertiginously difficult questions, questions that are for the most part swept under the rug with the presumption that the regress of justification hits bedrock once we arrive at interests and forms of life.

Much as Winch's assumption that the natural sciences uniformly employ positivistic methods makes it easy for him to draw a sharp distinction between them and the hermeneutic social sciences, Bloor's assumption that the "philosophical tradition" is aprioristic makes it easy for him to draw a sharp distinction between that tradition and the new methods of SSK. Following Winch, Bloor conceives of epistemology as the "queen of the sciences," the discipline that characterizes the nature of knowledge and thus lays down the law to the other disciplines. Bloor and Winch agree that knowledge has an essence; their disagreement concerns whether philosophy or sociology is best placed to adjudicate its nature. Thus "Wittgenstein and Mannheim on the sociology of mathematics," Bloor's first published statement of the Strong Programme, an important point of departure for much subsequent work in SSK, ends with the following footnote:

Whereas Winch thinks that much sociology is misbegotten philosophy, the argument of this paper has been that much philosophy is misbegotten sociology. There is an irony about Winch's position which seems to have passed unnoticed. He believes that a proper philosophical understanding will illuminate our understanding of society. The example of philosophical clarity that he appeals to, which is Wittgenstein's analysis of rule following, in fact illustrates the opposite. It shows that a proper grasp of social and institutional processes is necessary for philosophical clarity. Rather than philosophy illuminating the social sciences Winch unwittingly shows that the social sciences are required to illuminate philosophical problems. (Bloor 1973, 191 n 46)

The principal contention of the paper is that Wittgenstein solves Mannheim's problem in *Ideology and Utopia* (1936), namely his inability to give a sociological analysis of logic and mathematics. (For illuminating and, in some ways, parallel discussions of Bloor's use of Mannheim, see Pels 1996 and Kaiser 1998.) Bloor argues that Wittgenstein made possible a reply to a view he calls "Realism," a conception of mathematics on which the mathematician discovers truths about a pre-existing realm that is quite independent of our mathematical activity (1973, 42.) To think of mathematics in this way, Bloor contends, is to conceive of it as a structured and bounded realm, apart from yet somehow connected to the world we live in. It leads to an epistemology that corresponds to that ontology: mathematical knowledge consists in first gaining access to that domain and then moving within it.

Bloor's reply to Realism turns to Wittgenstein's discussion of what is involved in continuing a simple number sequence, such as 2, 4, 6, 8 ... The Realist takes it for granted that "the correct continuation of the sequence, the true embodiment of the rule and its intended mode of application, exists already" (Bloor 1973, 181.) But the Realist conception of rule-following "fails to provide answers to the problems that it was designed to solve. These problems are: How can we make the *same* steps again and again; what makes 'the same' the same; what guarantees the identical character of the steps at the different stages of the rule's application?" (Bloor 1973, 181.) Indeed, how could such an archetype guide someone who was trying to follow it?

For how does the human actor, following the supposed archetype, know that it really is the correct embodiment of the rule he wants? To know that the archetype is correct requires exactly the knowledge that was considered problematic in the first place, *viz.* knowledge of how the rule goes. It emerges that this argument is quite general. It works for any archetype, this-worldly or other-worldly. The trouble with Realism does not lie in the puzzling nature of its ontology but in the circular character of its epistemology. It presupposes precisely what it sets out to explain. (Bloor 1973, 182)

However, this train of argument isn't a knock-down argument against Realism. For it depends on the Realist's agreeing that the actor must select, or identify, the archetype that is to be followed. Once this is granted, the way is open for the Wittgensteinian to reply that the act of selection, or identification, calls for the very skills that the archetype was supposed to explain. On Bloor's naturalistic, causal, framework, there appears to be no alternative. But if one accepts a teleological perspective, one on which either actors have a natural tendency toward the truth, or on which the archetypes actively impose themselves on us, there is no longer any need to invoke the actor's abilities to know what is the same, and what is different. Bloor holds that the teleological approach cannot be refuted, but can be made far less plausible by articulating the "consistent sociological account towards mathematics" (Bloor 1973, 189) he finds in Wittgenstein. On this approach, the meaning of a formula, or rule, is simply a matter of how it is applied, or used, and this, in turn, is "the culmination of a process of socialization" (Bloor 1973, 184). This is how the question concerning the correct application of a rule in a particular instance is to be answered. Moving on to the more general question, as to what makes several applications of a rule consistent with each other, Bloor's answer begins with the same point. People have been trained to behave in certain ways, and so when faced with new circumstances, are likely to respond in a similar way. While there is no external standard of correctness, particular practices can be criticized by appeal to other practices. The appeal to what is natural presupposes a certain background of physical and psychological facts, but these are compatible with considerable variation on the cultural level, and so rule-following is institutional rather than instinctual (Bloor 1973, 186.) In what sense, then, are we compelled by the laws of logic?

Logic compels by the sanctions of our fellow men:
Nevertheless the laws of inference can be said to compel us; in the same sense, that is to say, as other laws in human society. (Wittgenstein 1956 I 116)
Wittgenstein does not deny that logic compels. What he offers is an explanation of the content of the compulsion. ... The importance of the institution explains why we learn to count as we do. ... It is *we* who are inexorable. ... Perhaps the most significant conclusion is that mathematics can now be seen as invention rather than discovery (Bloor 1973, 187-8).

As a result, "calculation and inference are amenable to the same processes of investigation, and are illuminated by the same theories, as any other body of norms. ... The great insight of [Wittgenstein's] *Remarks* is that it treats the grip that logic has upon us as a fact to be explained rather than a truth to be justified" (Bloor 1973, 189-190).

What Bloor misses here, in his determination to find a non-normative and purely causal bottom level of explanation in Wittgenstein's writings on rule-following, is that for Wittgenstein, following a rule does have a normative significance. Wittgenstein's point of departure is that standards for assessment of correctness, of correct and incorrect usage, are part and parcel of our grasp of the rules of logic, or of the language we speak.

"How am I able to obey a rule?" – if this is not a question about causes, then it is about the justification for my following the rule in the way that I do. (Wittgenstein 1953, §217)

If we are to give a causal explanation of what we do when we follow rules, the best we can do is talk about how we learned them, and the conditions under which people conform to them. We can talk about the role of training, and the role of institutions in making sure that certain regularities are maintained. On the other hand, we can not only talk about how one *will* act, but about justification, about what one *ought* to do. In that case, what is at issue is "what actions accord with the rule, are obliged or permitted by it, rather than what my grasp of it actually makes me do" (Brandom 1994, 15). This contrast is a central theme in the rule-following passages in both the *Philosophical Investigations* and the *Remarks on the Foundations of Mathematics*: "Many of his most characteristic lines of thought are explorations of the inaptness of thinking of the normative 'force,' which determines how it would be appropriate to act, on the model of a special kind of causal 'force'" (Brandom 1994, 14). Wittgenstein's principal point here, about the connection between the existence of a community's common responses and the existence of practices is that proprieties of practice *presuppose* agreement in judgements, actions, and behaviour, but are not *reducible* to them.

But we do not need to undermine a non-naturalistic conception of reason in order to give a causal story, and for this reason the Winch-and-Wittgenstein-inspired scepticism with which Bloor begins is unnecessary. According to Bloor, we need to "stop the intrusion of a non-naturalistic notion of reason into the causal story" (1991, 177). This is a widespread conviction within SSK, as Roth (1996, 1998) and Friedman (1998) have observed. Shapin (1982), a frequently cited review of the early SSK-influenced literature in the history and sociology of science, begins with a philosophical response to the threat SSK supposedly faced from philosophical 'realism' and 'rationalism.' Likewise, Collins begins *Changing Order* (1985/1992) with a chapter on Wittgenstein-inspired scepticism and how forms of life provide a sociological solution to epistemological and linguistic scepticism. But we do not need a philosophical argument in order to refrain from normative considerations. We can, in the words of Nancy Reagan, "just say no." There is no need to refute philosophical 'realism' or 'rationalism' before one can legitimately go to work describing what goes on in scientific practice or looking at scientific knowledge from a sociological perspective. Similarly, "we can seek to explain why scientific beliefs are in fact accepted without considering whether they are, at the same time, rationally or justifiably accepted" (Friedman 1998, 245).

5. SSK's WITTGENSTEIN: A MISBEGOTTEN EPISTEMOLOGY AND SOCIOLOGY

If we accept the "compatabilist" resolution of the supposed dispute between philosophy and sociology offered at the end of the previous section, then the philosopher who wants to ask normative and justificatory questions about science, and the sociologist of scientific knowledge who wants to ask descriptive and explanatory questions about science need not disagree. In that case, we may be left wondering what the dispute was about. Why did Bloor and his colleagues think it essential to produce a philosophical refutation of philosophical approaches to the study of science? And why did so many philosophers respond in kind? Why has this dispute been so heated and so prolonged, and why have so few participants even considered the program of pluralist tolerance and peaceful coexistence Friedman proposes? A large part of the answer must be that each side has seen the other as not just mistaken but as an illegitimate claimant to its rightful inheritance, and so threatening to supplant it. Peaceful coexistence has proven impossible because each side can see no way of peacefully sharing the intellectual terrain.

To see how charged the dispute has been from the very beginning, we need go no further than Winch and Bloor's pointed use of "misbegotten", a term that has both a descriptive and a normative significance, to describe the other's discipline. The literal meaning of "misbegotten" is "unlawfully begotten; illegitimate; a bastard"; unsurprisingly, the *Shorter Oxford English Dictionary* also states that it can be used figuratively and as a "term of opprobrium". Philosophers of knowledge have traditionally held that the sociology of knowledge is illegitimate, if not impossible, because knowledge is the concern of epistemology; sociologists of knowledge have replied that epistemology is illegitimate, if not impossible, because knowledge is the concern of sociology. In order to mount a frontal attack on philosophy, sociologists have felt the need for philosophical argument, though they have usually vacillated on the question of whether they were putting an end to epistemology ("epistemology is misbegotten sociology") or setting it straight ("epistemology is really social epistemology"). But to think that we must refute or transform the "philosophical tradition" in order to go ahead is to begin not with practice, but by tilting at windmills. SSK would do better to give up its dependence on a fictional alter ego and the misbegotten arguments that went along with it, and concentrate on doing what it does best: studying what scientists do.

NOTES

* 　I would like to thank the Alexander von Humboldt Foundation and the Department of Philosophy at the University of Bielefeld for their generous support while I was writing this paper, and the audience at my IWK presentation for their constructive comments.

REFERENCES

Barnes, Barry, David Bloor and John Henry: (1996) *Scientific Knowledge: A Sociological Analysis.* Chicago University Press.

Barnes, Barry and David Edge (eds.): (1982) *Science in Context: Readings in the Sociology of Science.* Open University Press, Milton Keynes, England.

Blackburn, S.: (1984) *Spreading the word,* Clarendon Press, Oxford.

Blackburn, S.: (1984a) "The individual strikes back", in Wright, C. (ed.): 1984, *Essays on Wittgenstein's later philosophy, Synthese* 58, #3.

Bloor, David: (1973) "Wittgenstein and Mannheim on the sociology of mathematics" *Studies in the History and Philosophy of Science* 4 173-191. Reprinted in Collins 1982 39-58.

Bloor, David: (1976/1991) *Knowledge and Social Imagery,* Routledge. Second revised edition, Chicago University Press, 1991.

Bloor, David: (1983) *Wittgenstein: A Social Theory of Knowledge.* Columbia University Press.

Bloor, David: (1992) "Left and Right Wittgensteinians" in Pickering 1992, 266-282.
Bloor, David: (1996) "The Question of Linguistic Idealism Revisited" in Sluga and Stern 1996, 354-382.
Bloor, David: (1997) *Wittgenstein, Rules and Institutions.* Routledge, London.
Bloor, David: (2001) "Wittgenstein and the Priority of Practice" in Schatzki et. al. 2001.
Brandom, Robert B.: (1994) *Making It Explicit: Reasoning, Representing, and Discursive Commitment.* Harvard University Press, Cambridge, MA.
Cerbone, David R.: (1995) "Don't Look But Think: Imaginary Scenarios in Wittgenstein's Later Philosophy" *Inquiry* 37, 159-183.
Collins, Harry: (1974) "The TEA set: tacit knowledge and scientific networks" *Science Studies* 4 165-186. Reprinted in Barnes and Edge 1982, 44-64.
Collins, Harry: (1975) "The Seven Sexes: A Study in the Sociology of a Phenomenon, or the Replication of Experiments in Physics" *Sociology* 9 205-224. Reprinted in Barnes and Edge 1982, 94-116.
Collins, Harry (ed.): (1982) *Sociology of Scientific Knowledge: A Source Book.* Bath University Press, Bath, England.
Collins, Harry: (1983) "An empirical relativist programme in the sociology of scientific knowledge", in Knorr-Cetina and Mulkay 1983, 85-114.
Collins, Harry: (1985/1991) *Changing Order: Replication and Order in Scientific Practice.* Sage, London and Beverly Hills, CA. Second revised edition, University of Chicago Press, 1991.
Collins, Harry: (2001) "What is Tacit Knowledge?" in Schatzki et. al. 2001.
Collins, Harry and Trevor Pinch: (1993/1999) *The Golem: What Everyone Should Know About Science.* Cambridge University Press, London. Second revised edition, 1999.
Collins, Harry, and Steven Yearly: (1992) "Epistemological Chicken" in Pickering 1992, 301-326.
Collins, Harry, and Steven Yearly: (1992a) "Journey into Space" in Pickering 1992, 369-389.
Coulter, Jeff: (1993) "Logic: ethnomethodology and the logic of language" in *Ethnomethodology and the Human Sciences* ed. Graham Button. Cambridge University Press, 20-50.
Dummett, Michael: (1959) "Wittgenstein's Philosophy of Mathematics" *Philosophical Review* 68 324-348.
Friedman, Michael: (1998) "On the Sociology of Scientific Knowledge and its Philosophical Agenda" *Studies in the History and Philosophy of Science* 29 239-271.
Fuller, Steve: (1996) "Talking Turkey About Epistemological Chicken, and the Poop on Pidgins" in Galison and Stump 1996, 170-188.
Galison, Peter and David J. Stump (eds.): (1996) *The Disunity of Science: Boundaries, Contexts, and Power.* Stanford University Press, Stanford CA.
Golinski, Jan: (1998) *Making Natural Knowledge: Constructivism and the History of Science.* Cambridge University Press.
Hacking, Ian: (1984) "Wittgenstein Rules," review of Bloor (1983), *Social Studies of Science* 14, 469-476.
Hacking, Ian: (1999) *The Social Construction of What?* Harvard University Press, Cambridge, MA.
Kaiser, David: (1998) "A Mannheim for All Seasons: Bloor, Merton, and the Roots of Sociology of Scientific Knowledge" *Science in Context* 11:1, 51-87.
Kripke, S.: (1982) *Wittgenstein on Rules and Private Language,* Harvard University Press, Cambridge, MA.
Kuhn, Thomas: (1962/1970) *The Structure of Scientific Revolutions.* University of Chicago Press. Second revised edition, 1970.
Lynch, Michael: (1992) "Extending Wittgenstein: The Pivotal Move from Epistemology to the Sociology of Science" in Pickering 1992, 215-265.
Lynch, Michael: (1992a) "From the 'Will to Theory' to the Discursive Collage: A Reply to Bloor's 'Left and Right Wittgensteinians'" in Pickering 1992, 283-300.
Lynch, Michael: (1993) *Scientific Practice and Ordinary Action: Ethnomethodology and Social Studies of Science.* Cambridge University Press.
McDowell, John: (1993) "Meaning and Intentionality in Wittgenstein's Later Philosophy" in *Midwest Studies in Philosophy Volume XVII: The Wittgenstein Legacy.* Notre Dame University Press, Notre Dame, IN, 40-52.
McGinn, Colin: (1984) *Wittgenstein on meaning,* Blackwell, Oxford.

Pels, Dick: (1996) "Karl Mannheim and the Sociology of Scientific Knowledge: Toward a New Agenda" *Sociological Theory* 14:1 30-48.

Phillips, Derek L.: (1977) *Wittgenstein and Scientific Knowledge: A Sociological Perspective.* Rowman and Littlefield, Totowa, NJ.

Pickering, Andrew (ed.): (1992) *Science as Practice and Culture.* University of Chicago Press.

Pickering, Andrew: (1992a) "From Science as Knowledge to Science as Practice" in Pickering 1992 1-28.

Staten, H.: 1984, *Wittgenstein and Derrida: philosophy, language and deconstruction,* Blackwell, Oxford.

Turner, Stephen: (1994) *The Social Theory of Practices: Tradition, Tacit Knowledge, and Presuppositions.* University of Chicago Press.

Ritsert, Jürgen: (1991) "The Wittgenstein-Problem in Sociology or: The 'Linguistic Turn' as a Pirouette" *Poznan Studies in the Philosophy of the Sciences and Humanities* 22 7-38.

Roth, Paul: (1987) *Meaning and Method in the Social Sciences: A Case for Methodological Pluralism.* Cornell University Press, Ithaca, NY.

Roth, Paul: (1996) "Will the Real Scientists Please Stand Up? Dead Ends and Live Issues in the Explanation of Scientific Knowledge" *Studies in History and Philosophy of Science* 27 43-68.

Roth, Paul: (1998) "What does the sociology of scientific knowledge explain?: or, when epistemological chickens come home to roost" *History of the Human Sciences* 7 95-108.

Schatzki, Theodore: (1996) *Social Practices: A Wittgensteinian Approach to Human Activity and the Social.* Cambridge University Press.

Schatzki, Theodore, Karin Knorr-Cetina and Eike von Savigny (eds.): (2001) *The Practice Turn in Contemporary Theory.* Routledge, London and New York.

Shapin, Steven: (1982) "History of science and its sociological reconstructions" *History of Science* 20 157-211.

Shapin, Steven and Schaffer, Simon: (1985) *Leviathan and the Air Pump: Hobbes, Boyle and the Experimental Life.* Princeton University Press.

Sluga, Hans and David Stern (eds.): (1996) *The Cambridge Companion to Wittgenstein.* Cambridge University Press.

Stern, David: (1994) "Recent work on Wittgenstein, 1980-1990" *Synthese* 98 415-458.

Stern, David: (1995) *Wittgenstein on mind and language,* Oxford University Press, Oxford.

Stern, David: (1996) "The availability of Wittgenstein's philosophy," in Sluga and Stern 1996, 442-476.

Stern, David: (2000) "Practices, practical holism, and background practices," in *Heidegger, Coping, and Cognitive Science: Essays in Honor of Hubert L. Dreyfus, Volume 2* Cambridge, MA: MIT Press.

Stern, David: (forthcoming) "The Practical Turn" in *The Blackwell Guidebook to the Philosophy of the Social Sciences,* ed. Stephen Turner and Paul Roth.

Winch, Peter: (1958/1990) *The Idea of a Social Science and its Relation to Philosophy.* Routledge and Kegan Paul, London. Second revised edition, 1990.

Winch, Peter: (1964) "Understanding a Primitive Society" *American Philosophical Quarterly* 1 307-24.

Winch, Peter: (1987) "Language, Thought and World in Wittgenstein's *Tractatus*" in *Trying to Make Sense,* Blackwell, Oxford.

Wittgenstein, Ludwig: (1922) *Tractatus Logico-Philosophicus,* translation on facing pages by C. K. Ogden. Routledge and Kegan Paul. Second edition, 1933.

Wittgenstein, Ludwig: (1953/1958) *Philosophical Investigations,* edited by G. E. M. Anscombe and R. Rhees, translation on facing pages by G. E. M. Anscombe. Blackwell, Oxford. Second edition, 1958.

Woolgar, Steve: (1981) "Interests and Explanation in the Social Study of Science" *Social Studies of Science* 11 365-394.

Woolgar, Steve: (1983) "Irony in the Social Study of Science" in Knorr-Cetina and Mulkay 1983, 239-266.

Woolgar, Steve: (1988) *Science: The Very Idea.* Routledge, London.

Woolgar, Steve (ed.): (1988a) *Knowledge and Reflexivity: New Frontiers in the Sociology of Science.* Sage, London.

Woolgar, Steve: (1992) "Some Remarks about Positionism: A Reply to Collins and Yearly" in Pickering 1992, 327-342.

Department of Philosophy
University of Iowa
Iowa City, IA 52242-1408
U.S.A.
david-stern@uiowa.edu

A.W. CARUS

THE PHILOSOPHER WITHOUT QUALITIES

THOMAS MORMANN, *Rudolf Carnap*. München: Beck 2000.

RUDOLF CARNAP, *Untersuchungen zur allgemeinen Axiomatik*, edited by T. Bonk and J. Mosterin. Darmstadt: Wissenschaftliche Buchgesellschaft 2000.

The revival of interest in Carnap's philosophy over the past two decades has shed much light on particular aspects of his intellectual development and its context. We now have a better appreciation of the background and motivation of the *Aufbau*.[1] The radical nature of the *Syntax* program has finally, more than half a century after its first publication, begun to be acknowledged.[2] And the later Carnap has also been re-assessed; the previously widespread impression that Quine was "right" and Carnap "wrong" in the analytic-synthetic debate has yielded to a more balanced view[3], and the broad outlines of Carnap's late philosophy have begun to emerge.[4]

But what holds all this together? Carnap is difficult to see in the round. From the present literature it is far from clear whether there is an underlying unity to the phases he went through – whether there is a coherent story to be told or only a sequence of disconnected episodes. The transition from *Aufbau* to *Syntax* is ill-understood, for instance, partly because so many of the documents of this period (1928-32) are unpublished. Carnap's long periods of immersion in developing various technical languages give most of his published writings a rather dry, specialized, and dated character; the overall perspective has to be gleaned from a few programmatic statements, and from his replies to critics in the Schilpp volume.

These difficulties of interpretation resemble the problems presented by the scattered writings of Leibniz – of all past philosophers the one to whom, as it happens, Carnap felt the closest kinship.[5] Their works suffered similar fates at the hands of followers: narrow and partial readings (in Leibniz's case most notably by Wolff, in Carnap's case by Ayer) were popularized and became easy targets for criticism in the following generation, thus consigning the original, more sophisticated views to prolonged neglect. Important dimensions of Leibniz's work waited centuries to be rediscovered. By that standard, Carnap has been lucky.

M. Heidelberger and F. Stadler (eds.), History of Philosophy and Science, 369–377.
© 2002 *Kluwer Academic Publishers. Printed in the Netherlands.*

I.

Mormann's introductory paperback is the first attempt at a wide-angle view that puts the recent literature together into a single story. There are some notable omissions (Ricketts, for instance, is not mentioned) but on the whole Mormann's treatment is judicious and touches on the most important issues. The danger of superficiality attendant on such overviews is generally evaded; Mormann is himself a contributor to the literature he reviews, and is in a good position to judge its successes and shortcomings. He gives the uninitiated general reader a skilful summary of Carnap's career, and conveys a sense of why his work should still be regarded as important.

Most successful of all, I think, is Mormann's suggestion for encapsulating the broader significance of Carnap's work in a single formulation: "I want to maintain that Carnap conceives of philosophy as *the science of possibilities*, i.e. as the theory of articulating and exploring conceptual possibilities. Rather than arguing for particular theses, Carnap opens up spaces of conceptual possibilities. This invariably characterizes all phases of his thought." (p. 210) Mormann suggests, moreover, that we understand this "science of possibilities" as something like a species of the "sense of possibilities" Robert Musil holds up as a necessary counterpart to the "sense of reality" in his monumental novel *The Man without Qualities*. This is an inspired comparison, and though Mormann does not dwell on it, there is more to it than might appear. Musil actually read Carnap, and applauded the *Logical Syntax of Language* (the book in which the "open sea of free possibilities" first became an explicit theme in Carnap's work); indeed, when invited by a literary magazine to propose the best book of the year 1934, Musil prepared a satirical response in which the *Syntax* was his first and only choice.[6] Beyond this biographical coincidence, there is a deeper sense in which Carnap and Musil share a remarkably similar stance toward the problems facing their (and our) civilization. They were both men of the Enlightenment, contemptuous of the shallow, anti-scientific cult of feeling then so popular in Central Europe, yet both were influenced and motivated by the utopian yearnings expressed in various ways by the German *Jugendbewegung*, by Nietzsche, and by the various social and artistic reform movements of the early twentieth century.

This is not how Carnap is ordinarily viewed, and for just this reason Mormann's encapsulation, including the comparison with Musil, is a useful corrective. However, the chasm between the prevailing view (even within the more recent literature) and Mormann's suggestion is so enormous that a case has to be made – the burden of proof is surely on the dissenter! Mormann does not shoulder this burden. He does explain and emphasize Carnap's pluralism, both its syntactic form (Ch. 6) and the later versions (Ch. 7); the "principle of tolerance" is even reproduced on the back cover.

But a certain rather subtle misconception pervades Mormann's view of the later Carnap, a misconception that is admittedly, despite the recent literature, not easy to avoid. This is the view that, as Mormann puts it, semantics became for Carnap "the successor discipline to philosophy" *überhaupt*. Mormann does mention in his next paragraph that Carnap followed Morris in dividing the theory of language systems (i.e. what a scientific "philosopher" is concerned with) into *three* parts: syntax, semantics, and pragmatics. But Mormann feels justified in setting aside this complication, as Carnap's treatment of pragmatics remained "underdeveloped" and "never got beyond a few rather aphoristic remarks". (p. 154) This is literally true, but Mormann draws the wrong conclusion from it when he claims that a *cost* of the new pluralism in the *Syntax* was a "dilution of the empiricist content of Carnap's thought", in the sense that:

Themes like the unity of science, the connection between life and science, between every-day knowledge and scientific knowledge, that are still emphasized so strongly in the *Manifesto* [i.e. the pamphlet *Wissenschaftliche Weltauffassung* that Carnap co-authored with Neurath and Hahn in 1929], disappear as explicit themes of Carnap's philosophy. (p. 149)

Should we accept this conclusion? The question goes to the heart of our idea of Carnap's later philosophy, and indeed of his entire philosophical development. Mormann claims, for instance, that Carnap's eventual broadening of the scope of inductive logic to include normative decision theory goes beyond anything that could be included in "semantics" and *therefore* cannot be considered as a culmination or completion of Carnap's thought (p. 191). Worse, we are told that after he moved to America, away from Neurath's beneficial influence, Carnap's "sense of possibilities" became narrow, abstract, formal, and lost all relevance to practical life. (p. 213)

Surely something has gone wrong here! Do we not often hear the later Carnap saying "… this is again only a practical question of language engineering, and therefore ought to be solved according to such practical points of view as convenience and simplicity"?[7] And what is he doing here but referring a question to the realm of pragmatics? Certainly his own development of a systematic pragmatics did not proceed far, but this does not mean that the *use* of pragmatics as a category (the availability of a practical meta-perspective on theoretical language) was marginal for Carnap – on the contrary. If we look closely, for instance, at the central project of his later thought, the project of *explication*, we see that it is at bottom a matter of pragmatics. Explication, as is well known, consists for Carnap in the piecemeal *replacement* of concepts in everyday language (the explicanda) by better and more exact concepts (the explicata). The relation between explicanda and explicata is not one-one in either direction; not only can there be multiple (or no) explicata for a given explicandum, it is also possible that a single explicatum replaces several different explicanda, or does not correspond to any explicandum at all.[8] How then can we tell whether a given explication is satisfactory? The answer, of course, is that this is an *external* question, i.e. a pragmatic one. As Stein has pointed out[9], this puts a *dialectic* between the theoretical and the practical at the very heart of Carnap's conception of scientific and cognitive progress.

Could anything be further from the supposed "disappearance" of questions about the relation between science and life, theoretical and practical, that Mormann discerns in Carnap's later philosophy? It is of course true that Carnap's *mannerisms* were not those of a *Lebensphilosoph*; his concentration on formal languages and dry technicalities blinds superficial readers to his very deep interest (to the end of his life, not just during the Vienna period) in the relation between the theoretical and the practical. But such readers remain at the surface; they fall prey to *Eigenschaften*, the socially defined characteristics or qualities by which people categorize and bewitch themselves – to which Musil was so allergic, and of which he tried to free his main character Ulrich. In philosophy, the *Eigenschaften* we most readily fall victim to are the notorious "isms" by which we pigeon-hole thinkers into standard categories. Carnap's philosophy is no more easily characterized by such crude "isms" than Ulrich is by "qualities". As the recent work has shown, the more closely we read Carnap's texts, the more we find a depth and subtlety that the standard philosophical catch-phrases utterly fail to convey.

II.

This is true even of "logicism", one of the few "isms" Carnap actually endorsed throughout his career: this apparent continuity dissolves when we look more closely. Though the Frege-Russell "genetic" construction of arithmetic did remain associated with Carnap's "logicism", the role and status of this core idea changed quite radically at certain points, especially during the transition from *Aufbau* to *Syntax*. As Bonk and Mosterin indicate (p. 7), the *Syntax* conception of logicism is fundamentally different from the one Carnap had defended in Königsberg a few years earlier. The nature and motivation of this transition has remained obscure, as mentioned above, partly because many relevant documents are still unpublished. Now, at last, one of these documents, the *Investigations on General Axiomatics* (1927-1930), has been made available.

The *Axiomatics* is not easy to describe. At first glance it looks like early model theory, but there is no distinction between object language and metalanguage. Its subject is a central question raised in the second and third editions of Fraenkel's *Einleitung in die Mengenlehre*: do various proposed definitions for the "completeness" of an axiom system (especially those approximately equivalent to what we would now call "categoricity" and "decidability") coincide? The manuscript is baldly technical; there is very little motivation. However, it is clearly intended to give substance to the ideas put forward more discursively in Carnap's 1927 paper "Eigentliche und Uneigentliche Begriffe", which it might have been helpful to reprint along with this manuscript.

The extensive introduction by Bonk and Mosterin provides much helpful background and context. But the siren song of "isms" proves irresistible to them as well. They call the *Axiomatics* "a first step on the path to a new orientation of logicist theses, a path that ended with the publication of *The Logical Syntax of Language*." (p. 1) But in the following ten pages on logicism, there is no further mention of the *Syntax* conception of logicism toward which the *Axiomatics* is headed.[10] "Logicism" is treated, despite the acknowledgement of an evolution or development, as a static and well-defined body of doctrines. And yet even by Bonk's and Mosterin's own description, the connection between the 1920's Frege-Russell logicism they describe and the *Axiomatics* must strike the reader as mysterious. They say, for instance, "Carnap's philosophy of axiomatics can be epitomized in two theses: axiom systems are open sentences (propositional functions) in the system of *Principia Mathematica*, and the derivation of theorems from an axiom system always has the form of a consequence [*Implikationsaussage*[11]]." (p. 37); Bonk and Mosterin tell us that the latter idea grounds Carnap's "logicist interpretation of axiomatically given theories".[12] But they never tell us *how* it does this, or purports to. Nor do they address the connection between a "logicist interpretation of axiomatically given theories" and the main subject of the *Axiomatics*: the *Gabelbarkeitssatz*, which claims that a consistent set of axioms is categorical if and only if it is complete.[13]

What is the explanatory value of "logicism" when it is so ill-defined, when its connection to the text under discussion is left so vague, and when it is acknowledged to be in the process of turning into something completely different (the *Syntax* conception)? Some version of "logicism" clearly has a place in the program of the early Vienna Circle[14], but what that place is, precisely, is a difficult and rather tricky question. (Mormann, perhaps wisely, throws in the towel here, and avoids the subject of "logicism" and "foundations of mathematics" almost entirely.) Was logicism intended as a foundational program? To what extent does the *Aufbau* system rest on logicism? Why did Carnap regard logicism (in the Königsberg discussion, for instance) as the standpoint of the *physicist*? Various theories have been put forward to square the answers to these questions with each other[15], but Bonk and Mosterin neither address these theories nor put forward one of their own. They would appear to have fallen into the trap of regarding the reference to an "ism" as sufficient and self-explanatory. But reliance on philosophical *Eigenschaften* can only distort and coarsen our understanding of a very subtle, self-conscious, and many-dimensional development. Such Procrustean simplification is the route of Ayer; for our collective project of restoring the work of the Vienna Circle to its proper credit and respect, we must be more conscientious.

III.

One section of Mormann's concluding chapter (pp. 199-207) confronts recent German philosophy, and criticizes its systematic neglect and distortion of Vienna Circle positions. There has long been a need for such a survey, and we can be grateful for these few pages. They make a refreshing change from the attitude of conciliation that has often characterized European scholarship on the Vienna Circle. The tendency has been to show how the Vienna Circle was really not very different from the "Continental" philosophy it so intemperately attacked, and by selective attention to use the Vienna Circle as a vehicle for a rapprochement between current "analytic" and "Continental" philosophy. This is understandable as a marketing ploy, or as a survival tactic for the beleaguered minority of analytic philosophers on the Continent; they would naturally want to play down the strident rhetoric of the "left wing" of the Vienna Circle in the 1920's and early 1930's. But it is worth asking whether this approach really does improve our understanding of the Circle's philosophical ideas and ambitions. If not, then however useful it may be in the short term, it can only undermine the long-term viability of the whole project of rehabilitating the Vienna Circle and its philosophical values.

Our present historical perspective puts us in a better position to appreciate the uniqueness of the "analytical philosophy" of the first half of the twentieth century. It was an isolated episode in the history of the academic discipline called "philosophy" – though continuous with certain nineteenth-century tendencies *outside* the discipline (Helmholtz, Mach, Mill), and especially with the eighteenth-century *Encyclopédistes*, who also were not professors of philosophy. The first half of the twentieth century was perhaps the only historical period in which these tendencies gained a strong foothold in (English-speaking) academic departments of philosophy. This period is now drawing to a close. "Analytic philosophy" as practised by Davidson, McDowell, Brandom, or Dummett has reverted to something much closer to what philosophy has been since Descartes or before: the attempt to find an alternative conceptual starting point for our overall understanding of the world from that provided by natural science (in the broadest sense) – the attempt, in short, to find a "first philosophy" prior to science.

From the standpoint of the Vienna Circle, then, there is little to choose between current "analytic" philosophy and current "Continental" philosophy. There are still differences of approach and vocabulary, of course. But in general, analytic philosophy has gone the route that Mormann wrongly accuses Carnap of taking after his move across the Atlantic – it has lost (as Carnap never did) the desire to change the world; it has lost its continuity with the eighteenth-century Enlightenment; it has lost contact with the wellspring of philosophy since Socrates, the urge to criticize complacent, unreflective, and fashionable modes of thought. Analytic philosophy has reverted to the task philosophy professors have excelled at through the ages, which is to justify by detailed, abstruse arguments the unreflective common sense that everyone else already takes for granted. In short, analytic philosophy has become precisely the sort of thing that the Vienna Circle attacked so mercilessly during its vitriolic phase.

One can regret the rather obtuse rhetoric of these attacks without losing the Vienna Circle's sense that confrontation, debate, and intellectual competition are healthier for a pluralistic culture than consensus and passivity toward fashion. Mormann's critical survey of his German contemporaries is therefore a breath of fresh air. He spoils it somewhat, though, by deferring in other passages to the authority of "modern philosophy" or "current opinion", e.g. in his preference for Neurath over Carnap on many issues. Contrasting Neurath's "encyclopedic" model of the unity of science with Carnap's "systematic" model (Ch. 5), he concludes, without much argument, that "recent developments in the philosophy of science" show Neurath to be right, Carnap wrong. But of course Mormann could have taken the same stance with respect to postwar German philosophy: "recent philosophy", he could with perfect respectability have said, "indicates that Carnap was wrong, and Adorno (or Heidegger) was right."

And besides, regarding the unity of science in particular, "recent developments in philosophy of science" have shown nothing of the sort. It is true, of course, that the "disunity of science" is now more widely canvassed than it was in Carnap's day. But which of the conceptions, Carnap's or Neurath's, is better equipped to address the present debate? Neurath's view (as Mormann recognizes, pp. 116-117) was that all of science should be carried out in a single language, continuous with our "conceptual scheme" embedded in natural language. Quine has carried this view forward and elaborated it in more detail. For Quine, as for Neurath, there is no stepping outside this conceptual scheme; Neurath's metaphor of the boat that we must mend from the materials we have on board with us, without being able to start from scratch in dry-dock, became the epigraph to *Word and Object*. The widespread acceptance of Quine's view[16] is no doubt part of what Mormann has in mind when he refers to "recent developments in philosophy of science".

But Quine's view cannot recognize radical differences of conception about the unity of science. For Quine (as for Neurath) we only have the one conceptual scheme we are brought up in; we have no choice. And within that conceptual scheme, the unity of science is, quite simply, *analytic*; there is no arguing about it. Those who reject the unity of science cannot be understood within our conceptual scheme; we can only rule them out of court. For Carnap, in contrast, we *do* have a choice in this matter.[17] We can *decide* on the language in which we articulate the world, based on our goals for humankind (on practical grounds, broadly speaking). We can choose a language in which the unity of science is a constraint[18] or we can choose one without such a constraint. The choice cannot depend on which is "true", because the language is chosen on practical grounds; it makes no more sense to regard the unity of science as true or false than it does to regard realism as true or false, or abstract entities as existing or not existing; these are *external*, i.e. practical questions. Of course Carnap himself proposed that we accept the constraint of the unity of science. But this had the status, precisely, of a *proposal*, and not of an *assertion*. It was a proposal to be argued for or against on pragmatic grounds.

Which of these approaches is better adapted to the present debate about the unity of science? This is itself a pragmatic question, a question of practicalities and of values. If one prefers the firmness shown by Steven Weinberg or Lewis Wolpert in these matters ("science is just plain objectively right, and you guys are wrong, so shut up"), then one will prefer Quine's (or Neurath's) approach. But if one stuffily insists on traditional canons of open and rational debate, respect for one's adversary, and the search for common ground, then one will choose Carnap's approach.

NOTES

1. Michael Friedman, *Reconsidering Logical Positivism*. Cambridge: Cambridge University Press 1999, esp. Part Two (Ch. 5 and 6); Alan Richardson, *Carnap's Construction of the World; The Aufbau and the Emergence of Logical Empiricism*. Cambridge: Cambridge University Press 1998.
2. Thomas Ricketts, "Carnap's Principle of Tolerance, Empiricism, and Conventionalism", in: P. Clark and B. Hale (Eds.), *Reading Putnam*. Oxford: Blackwell 1994; Warren Goldfarb, "Introductory Note to *1953/9*", in: K. Gödel, *Collected Works*, Vol. III, ed. by S. Feferman et. al. Oxford: Oxford University Press 1995, pp. 324-334.
3. Daniel Isaacson, "Carnap, Quine, and Logical Truth", in: D. Bell and W. Vossenkuhl (Eds.), *Science and Subjectivity*. Berlin: Akademie-Verlag 1992, pp. 100-130; Richard Creath, "Introduction", in: R. Creath (Ed.), *Dear Carnap, Dear Van; The Quine-Carnap Correspondence and Related Work*. Berkeley: University of California Press 1990, pp. 1-43.
4. Howard Stein, "Was Carnap Entirely Wrong, After All?", *Synthese* 93, 1992, pp. 275-95; Richard Jeffrey, "Carnap's Voluntarism", in; D. Prawitz, B. Skyrms, and D. Westerståhl (Eds.), *Logic, Methodology, and Philosophy of Science IX*. New York: Springer 1994, pp. 847-866.
5. Abraham Kaplan, "Rudolf Carnap" in: E. Shils (Ed.), *Remembering the University of Chicago*. Chicago: University of Chicago Press 1991, p. 40.
6. Robert Musil, *Briefe 1901-1941*. Hamburg: Rowohlt 1981, vol. 1, p. 664 (draft of a letter to Martin Flinker, 29 October 1935); the question was actually which book had made the "deepest impression", and Musil's published response criticized the question, deleting the mention of Carnap.
7. Rudolf Carnap, "Replies and Systematic Expositions", in: P. Schilpp (Ed.), *The Philosophy of Rudolf Carnap*. LaSalle, Illinois: Open Court 1963, p. 912.
8. H. Stein, *loc. cit.*, pp. 280-281; Carnap's own most detailed exposition of his view of explication is in *Logical Foundations of Probability*. Chicago: University of Chicago Press 1950, Section I (Ch. 1-6, pp. 1-18). Mormann's survey of Carnap's idea of explication (p. 160) does not bring out the aspects stressed here; he appears rather to imply that explication is a purely *semantic* task.
9. H. Stein, *loc. cit.*, pp. 291-292.
10. Most explicitly spelled out on p. 255 of *Logische Syntax der Sprache* (Vienna: Springer 1934; p. 327 of the English translation).
11. That is, in the language of the *Axiomatics*, a "content-related" [*inhaltliche*] consequence; space prevents a discussion of the significance of these terms here, but see S. Awodey and A.W.

Carus, "Carnap, Completeness, and Categoricity: The *Gabelbarkeitssatz* of 1928", in: *Erkenntnis*, 54, 2001, pp. 145-172, esp. pp. 155-156.

12. They claim, indeed (p. 14), that Carnap introduced it, along with the "explicit concept" [*Explizitbegriff*] of an axiom system, in "Eigentliche und uneigentliche Begriffe". This requires some elucidation, as neither term appears in that paper.

13. Which sounds obviously false in view of Gödel's theorem about the incompleteness of elementary number theory published soon after Carnap abandoned the *Axiomatics*. As Awodey and I (*loc. cit.*) have shown, it is not straightforwardly false, but answers a different question from that addressed by Gödel – on whom, however, it evidently had an influence. Nearly the whole of the *Axiomatics* leads up to this theorem; it is decidedly odd to epitomize Carnap's "philosophy of axiomatics" without any reference to it.

14. As articulated, for instance, in what Mormann calls the Vienna Circle "manifesto", the 1929 pamphlet *Wissenschaftliche Weltauffassung*.

15. H.G. Bohnert, "Carnap's Logicism", in: J. Hintikka (Ed.), *Rudolf Carnap, Logical Empiricist*. Dordrecht: Reidel 1975, pp. 183-216; Alberto Coffa, *The Semantic Tradition from Kant to Carnap*. Cambridge: Cambridge University Press 1991, Ch. 15; Warren Goldfarb, "The Philosophy of Mathematics in Early Positivism", in: R.N. Giere and A.W. Richardson (Eds.), *Origins of Logical Empiricism*. Minneapolis: University of Minnesota Press 1996, pp. 213-230; Awodey and Carus, *loc. cit*, parts I and II. Bonk and Mosterín saw the latter as a technical report, and generously acknowledge it. But they neither mention nor respond to the theory put forward there about the relation between logicism and Carnap's *Axiomatics*.

16. Not only by the mainstream of analytic philosophy, but even by more radical critics like Rorty.

17. Jeffrey, *loc. cit.*

18. In the same way, as Ricketts (*loc. cit.*) has shown, empiricism, often regarded as a metaphysical *foundation* for the Vienna Circle, is rather a proposed *constraint* on the language of science.

Suite 2000, 332 S. Michigan Avenue
Chicago, IL 60604
U.S.A.
awc23@aol.com

THOMAS UEBEL

THE POVERTY OF "CONSTRUCTIVIST" HISTORY
(AND POLICY ADVICE)

STEVE FULLER, *Thomas Kuhn. A Philosophical History for Our Time*, Chicago:
University of Chicago Press, 2000.
[Page references to this book are marked "K"]

STEVE FULLER, *The Governance of Science*, Open University Press,
Buckingham, UK, 2000.
[Page references to this volume are marked "G"]

"*I urge that we turn Kuhn on his head and demonstrate that a paradigm is
nothing more than an arrested social development.*" (K402) Notwithstanding the
long debate to which *The Structure of Scientific Revolutions* has given rise since
its publication in 1962, this quote from Steve Fuller's assessment of its author's
legacy suggests an original if controversial project: may a better understanding
of science arise from the ashes of idealist historicism! Yet rather than furnish the
Marx to Kuhn's Hegel, Fuller but manages a pastiche of Proudhon, as it were,
which calls for another inversion of titles as rejoinder in turn. There can be no
doubt that some very important issues are raised in this book, but just as this
reader's expectations were dashed in its course, so the issues end up curiously
diminished; given the significance of the subject matter, it is of interest to find
out why.

Before renewing what will strike some readers as a familiar argument and
repeating recent philosophy (hopefully neither as tragedy or farce), the protagonists and the plot behind the morality play presented as "philosophical history"
must be introduced.

The subject of the study under review, Thomas S. Kuhn (1922-1996), was the
historian of science most closely associated with the rejection of positivist
philosophy of science and the turn to a socio-historical approach in its place. In
his widely read *Structure* Kuhn argued that the history of science exhibits certain
ruptures of development, so-called scientific revolutions, such that between
successor theories there obtain conceptual incommensurabilities which render an
algorithmic choice and claims to straightforward continuity between them
impossible. By contrast, much of normal, non-revolutionary science proceeds by

M. Heidelberger and F. Stadler (eds.), History of Philosophy and Science, 379–389.
© 2002 *Kluwer Academic Publishers. Printed in the Netherlands.*

puzzle-solving within a so-called paradigm (a disciplinary matrix of conceptual frame, experimental procedures and exemplary solutions) whose basic assumptions remain unquestioned except in periods of crisis. While Kuhn's picture of scientific practice is in fact much closer to what some logical positivists believed than what is commonly presumed, it clearly conflicts with Karl Popper's according to which scientists should be constantly trying to refute their best theories. Yet many other philosophers too suspect(ed) Kuhn of jeopardising rational choice in science. What they fail(ed) to see is what Kuhn's essays since 1962 and his postscript to the second edition of *Structure* document, namely, that Kuhn's descriptive history points not to the refutation but to the reconceptualisation of rational theory choice and change.

The author of this study, Steve Fuller, is an exponent of science and technology studies (STS) with a voluminous output to his name. Since the late 1980s he has been championing a research programme of "social epistemology", which is directed against the philosophical orthodoxy about scientific knowledge. Viewed from the sidelines, his criticism of conceptions that exalt the individual over the groups, collectives and networks and say nothing about the institutional hierarchies within which scientists work often seemed justified, though the preferred alternative was hard to establish. Instead, as polemic chased polemic and the Science Wars got under way, Fuller himself has become an institution, a philosopher's gadfly now safely ensconced in the neighbouring discipline of sociology.

Nowadays, of course, stressing that science is a historically situated social practice has lost its power to attract controversy, especially since even some logical positivists have been shown to have argued along such lines. Some would see in this an at least partial vindication of both Kuhn's work and STS (which, after all, emerged as a recognized field of inquiry only in the decades after *Structure* was first published), but not Fuller. Instead his book aims to establish that Kuhn and those who took his point got things terribly wrong. Moreover, in situating his attack historically, Fuller styles himself as vindicating Kuhn's supposedly ill-understood opponent Popper. Clearly then, we're set for some rhetorical fireworks, perhaps even prepared for some mischief: "Counter revolutionary Kuhn Exposed by Popperian Guard Fuller".

Can the book live up to such a billing? To be sure, there is plenty of vitriol flying around: "Kuhn saw as far as he did because he stood on the shoulders of giants ... But it only takes a dwarf standing on the shoulders of giants to see beyond them ..." (Kxii). There's plenty of politically correct indignation: "Kuhn famously updated this elitist myth of humanity's collective quest by associating the great paradigmatic thresholds in the history of science with the names of the great revolutionary geniuses who set the pace for lesser worthies ..." (Kxi) And mixed in with sophomoric rebellion – "that peculiar mutual admiration society we call a community of inquirers" (ibid.) – there's the *Weltschmerz* of squandered youth – "... had I known when I began my academic career that things would turn out this way, I probably would have listened to my mother and gone into law" (Kxv-xvi) – tempered by weary concern: "Nevertheless, I am not without hope." (ibid.). And that's just the Preface which, however, not only sets the tone but also lays the central charge: "the impact of *The Structure of Scientific Revolutions* has been largely, though not entirely, for the worse" (ibid.). How so? "... Kuhn ... not only failed to privilege criticism, but actually went so far as to argue that it should be avoided at all costs until a line of inquiry is saddled with so many unsolved empirical problems that it is forced to ask critical questions about the epistemological foundations the entire enterprise" (Kxv). Kuhn, so the charge, stifled critical rationality in the discourse in and about science.

Can a writer be held responsible for misreadings of his works? Fuller himself notes in the Introduction that "the monumental fecundity of misreadings attached to *Structure* begs for explanation" (K4) and even provides a good one in a footnote: "his readers are really not trying to understand him to any great extent but to use his text as the token in some ongoing disputes". Yet rather than try to do better, Fuller misapplies the principle of interpretative charity to dismiss systematic attempts to get at what Kuhn may or may not have been saying. (Hoyningen-Huene's reconstruction is dismissed for disagreeing with "virtually every substantive commentator on Kuhn's work", while Kuhn's own endorsement of it is discounted as issued "for obvious reasons" (K4-5n)!) Instead, Fuller prefers to turn Kuhn's basically descriptive thesis into a strongly normative one – and then to be off on what looks like his neo-Popperian way.

To be sure, much material is assembled here that is of interest despite the sometimes tenuous relation it bears to Kuhn. Lest the criticisms that follow seem overblown, consider by way of an overview the ambition of Fuller's venture. Chapter 1, "The Pilgrimage from Plato to NATO", quite literally seeks to track through most of Western history the dual character of "the Western conception of reason": what Fuller calls the "critical, libertarian and risk-taking" one of Socrates, the Enlightenment, Mach and Popper, and the "foundational, authoritarian and risk averse" one of Plato, Comte, Planck and Kuhn. With the latter represented as "double-truth doctors" who tell one thing to the elite and another to the rabble, this chapter documents the perceived urgency of Fuller's cause. Chapter 2, "The Last Time Scientists Struggled for the Soul of Science", reads, amongst other things, the notorious Mach-Planck debate as one over the "ends of science", even educational reform, in this light. Chapter 3, "The Politics of Science in the Age of Conant", is the shortest and most informative chapter of the book. It describes Kuhn's mentor James B. Conant marketing "the public value of 'basic science'", delineates the containment strategies developed in and around Harvard's Pareto Circle when faced with dissent within the academy – strategies that informed both Conant's educational ideas (as Fuller puts it,"freedom for the elites and structure for the masses") and Crane Brinton's ideas about revolutions (arising from dissatisfied semi-privileged individuals) – and finally sketches a version of Cold War political realism which Fuller detects in Conant (and without too much ado also attributes to Kuhn). Chapter 4, "From

Conant's Education Strategy to Kuhn's Research Strategy", is dedicated to chronicling the work of transposition attributed to Kuhn first in his work on Conant's General Education in Science program and later in his own historical studies, citing amongst his perceived failings that of "repressing the industrial vision of normal science". Chapter 5, "How Kuhn Unwittingly Saved Social Science from a Radical Future", explores how "Kuhn's disclaimers notwithstanding, social scientists were attracted to his book precisely because it seemed to provide a blueprint for how a community of inquirers can constitute themselves as a science, *regardless* of their subject matter" (K234), and argues that since "Kuhn's ultimate interest was in depoliticizing, not repoliticizing, science", the impact of *Structure* "corrupted all of its left-inspired appropriations" (K251). Chapter 6, "The World Not Well Lost", unfavourably compares Kuhn with C. I. Lewis before reexamining the Kuhn-Popper debate of 1965 and related ones from the Popperian perspective, along the way casting aspersions on the "unsavory affinity between Kuhn and the positivists" (K289). Chapter 7, "Kuhnification as Ritualized Political Impotence", sees Fuller firing several salvoes in the ongoing sectarian wars between different schools in STS which we must neglect here.

Chapter 8 finally presents Fuller's "Conclusions". "*Structure* diverted emerging tendencies in the 1960s to question the role of Big Science in the academy and society at large, while reinforcing the ongoing fragmentation and professionalisation of academic disciplines." (K380) In support, Fuller recapitulates the story of Kuhn's apprenticeship to Conant and surveys his "career of lucky accidents and studied avoidance" by an adversarial interpretation of Kuhn's own autobiographical narrative in the long interview given to A. Baltas, K. Gavroglu and V. Kindi in 1995 – first published in 1997 in the Greek journal *Neusis* and now easily available in Kuhn's *The Road Since Structure* (ed. J. Conant and J. Haugeland, Chicago University Press, 2000) – only to diagnose "culturopathy". This is a new form of academic dementia discovered by Fuller: the "lack [of] reflexive engagement with what [one] say[s] and do[es]", the failure of taking "cognitive responsibility" for one's narratives (K397-8). Fuller then turns to "getting over Kuhn" and duly arrives at the contrast between "citizen science" and "professional science" (K418). His advocacy of the former – a somewhat amorphous amalgam of everything anti-Big Science – is based on what he deems "the practical lessons of STS research": "you do not need to be an expert to understand expertise" and "the experts themselves may not live up to their own standards" (K415). Fuller ends by affirming the "subversive message hidden in" *Structure*, namely that "a forthright attempt to make the past contemporaneous with the present may be the best strategy for progressive thinkers in any field of inquiry to keep the future forever open" (K423).

So what's wrong with the book? Someone impressed with the political agenda may wonder. Isn't it right to seek to stop the advance of Big Science? Since this is not the place to discuss whether Big Science is really all bad, this review will accept Fuller's negative verdict for the sake of discusion. But what may strike one as the book's inconsistency of purpose and execution and its feeble argumentation cannot be so neglected. What turns such flaws into a major cause for irritation, moreover, are the disingenious alliances advertized and the populist posturing in furtherance of doctrines that those nominally called upon would wish to reject most vehemently. To demonstrate this it is necessary to consult also Fuller's second new book which deals with issues in science policy, illustrating my main complaint: Fuller highjacks a legitimate political concern to aggrandize a philosophically bankrupt programme that would leave those concerned by Big Science more incapacitated to resist its advance than ever before.

Let's take these points in turn. Before dwelling approvingly on Harvard's decision to refuse tenure to Kuhn, Fuller denies having "the interest [or] the evidence to deliver a verdict on Kuhn's life [or] indict the man of crimes of the intellect" (K381), yet goes on to suggest in a footnote that if "someone groomed to rule" – like Kuhn in Harvard's Society of Fellows presumably – "fails to provide the expected form of leadership, then that is *prima facie* grounds for believing that such a person has morally failed" (K384n). This is a curiously censorious stance to take, but throughout the book Fuller failed to heed his declared standards and conflated criticism of Kuhn the man and Kuhn the phenomenon and of what he did and what he was widely perceived to have done.

Fuller's project is animated by the promise to socio-historicize Kuhn and thereby put him in his place: his "philosophical history" is not straight history of philosophy or of anything else, but "an explicit pronouncement about the course of history" (K33). Yet to get any conspiracy theory off the ground (be that of the conscious rackets of old or of plots of circumstance with an unwitting protagonist, as here) we need some crime, misdemeanor or indiscretion – anything – to start with. When the dust settles, however, it remains wholly unclear whether Kuhn's involvement in Harvard's General Education in Science program constitutes such indictable behaviour and consequently whether Kuhn merits the charge levelled at him.

I noted at the beginning that Fuller's reading of Kuhn is highly tendentious. That reading, moreover, is presented in a context that treats *Structure* as "an exemplary document of the Cold War era", in particular, as the articulation of "the Cold War political paradigm constructed by James Bryant Conant (1893-1978), the president of Harvard University (1933-53), director of the National Defense Research Committee during World War II (which supervised the the construction of the first atomic bomb), and chairman of the anti-Communist Committee on the Present Danger in the 1950s – as well as the person who introduced Kuhn to the historical study of science, and through whom Kuhn acquired his first teaching post" (K5). By willful association then, so the charge would seem to go, Kuhn got involved in the crime at issue. What must be noted is that the willfulness of the relevant *political* allegiance is never documented but only insinuated on the basis of neutral statements (like Kuhn's admiration of Conant's intellect) or what was said in other contexts (Kuhn's comments on Hiroshima) or what was *not* said at all.

But what was the crime in the first place? "... at the close of the Second World War, both business and government leaders wondered aloud about the benefits that might be reaped from harnessing university-based science to economic interests. In order to tap into this proscience sentiment without having to turn the direction of science over to nonscientists, Conant, [Vannevar] Bush [president of MIT], and other wartime science administrators lobbied for a National Science Foundation by arguing that science could, indeed, yield these benefits, but only as by-products of the autonomous pursuit of research." (K153). Besides the apparent unaccountability of Big Science, Fuller's understandable concern is the concentration of federal funding in "a mere 33 (out of 2.500) institutions of higher education in the US" (K162), but can this latter state of affairs be laid so squarely at the door of the peer review system favoured by Conant, as Fuller goes on to suggest? Moreover, what does this have to do with Kuhn's *Structure*? As for the threat that Big Science poses, Kuhn is criticised for having done nothing to halt its advance: "Conant's key presumption that survives in Kuhn is that the bridge between science and society is best constructed by the public growing more accustomed to science – rather than by scientists growing more accustomed to the public." (K182-3) Granted, Kuhn did not try to resist or roll back the development of Big Science, but there is nothing inherently pro-Big Science about the idea that university graduates should have a modicum of general knowledge about science: its achievements, its history and its methods – or about Kuhn's thesis concerning scientific revolutions.

There's more of course. Elsewhere we read that the impact of *Structure* was to "dull the critical sensibility of the academy": what was lost in its wake "was a public academic space where the general ends and means of 'science' (or 'knowledge production' or 'inquiry') could be debated just as vigorously and meaningfully as the specific ends and means of particular disciplines and research programs" (K7). Kuhn's real crime then was to fail to do what Fuller thinks needs to be done and to have unreflectively joined in Conant's project of protecting science from the nonscientists. Again, quite apart from the issue of whether Kuhn can be held responsible for what some readers may have done with his book, did Kuhn himself do what he is accused of? Here his debate with Popper might perhaps have supplied documentation, but in representing this debate only in Popperian terms as one over ideals of the scientific community as a closed or open society, Fuller absolutises one participant's view and neglects the other. Kuhn was not interested in the ideal but the historical reality of science. Fuller's critique is but Popper's misunderstanding of Kuhn writ large. (That in light of the scientific progress achieved one may consider the paradigm conservatism of scientific communities to be not inherently irrational does not change the original descriptive focus of Kuhn's work. Interestingly, Fuller seems not very interested in the question whether Kuhn got the history of these revolutions right.)

Since some readers may wonder by now whether I'm applying a mistaken positivistic standard to the type of cultural criticism Fuller seeks to effect, let me stress that Fuller's historiography is unsatisfactory not because no "smoking gun" is ever produced (sometimes this is too much to ask), but because the argumentation employed is far too thin to carry his charge.

Suppose – again for the sake of discussion: I've heard Conant described as 'Cold War Liberal' – that Conant may count as a "double-truth doctor", seeking to protect the elite's command over science in the form of Big Science, thus stifling critical rationality in the discourse in and about science by working up canonical scientific case histories demanding universal acquiescence into the rationality displayed. Clearly, however, this does little to undermine the indeed commonsensical claim that in democracies where government plays a role in science (if only by funding it), "each citizen to some degree is involved in scientific decisions and needs some appreciation of the methods of science", as Conant put it in the preface to the 2nd edition of his *Science and Common Sense* (Yale University Press, 12th printing 1967). So while in the case of Conant propaganda for Big Science went hand in hand with efforts at science education, it is clear that the need for science education can be discoupled from the issue of whether one likes Big Science or not: in particular, rejecting Big Science does not amount to an argument against science education as an inevitable tool of double-truth doctors.

Similarly, simply being associated with Conant's science education program does not make one a double-truth doctor or Big Science promoter either. Such is the case with Kuhn. Reading through Kuhn's autobiographical interview one is rather struck by his concern to introduce a hermeneutic element into the General Education in Science course that he felt was missing under Conant, giving an unrealistic impression of the history of science. More pertinent still is the question whether the conception of science that both informs and emerges from Kuhn's historical work is in any way compromised by the mere fact of his teaching that course. Yet what other evidence is adduced for the claim that Kuhn belongs to the "double-truth doctors"? Fuller's charge against Kuhn remains unsubstantiated. Conant's promotion of Big Science does not by itself impugn the General Education in Science program, even less the motives of other teachers participating in it, and still less the latters' philosophies of science – yet nothing more than such vague associations are ever established.

Pressing this issue further one finds that there are two senses of "double truth" that Fuller invokes. There is first the straightforward Leo Strauss-type double truth, which Conant is allegedly guilty of: one story of what science is all about for the elite (a tool for political dominance), another one for the population at large (a method of inquiry). Then there's Kuhn's alleged double truth where "the 'rabble' turn out to be the scientific community (and their philosophical well-wishers), and the 'elite' the historians" (K27). It seems that for Fuller, Kuhn's tolerance of "Orwellian history" (K31) – the official history of science ever rewritten by the current paradigm – indicates that the scientists themselves, not only the general public, need to be fooled about the nature of their work. Only historians supposedly can stand what Fuller takes to be the deconstructionist truth about science that Kuhn discovered but which "he and his admirers have wanted to downplay over the years" (K7).

What emerges here, of course, is Fuller's own version of the double-truth story, one that can be foisted on Kuhn only if one discounts his evident concern to reconceptualize, not reject as chimerical, the rationality of science: Fuller's inside story would appear to be that there is no truth in science, but merely rhetoric. And unlike his Kuhn, of course, Fuller has no intention whatsoever of keeping a lid on it: in place of the double-truth story, he wants to disseminate the no-truth story – an agenda if not forged then certainly honed in the heat of the rather unfortunate Science Wars.

So let's turn to what Fuller wants to argue *for*. Fuller calls on two seeming allies in the course his book: Popper and the Popperians and the Public Under-standing of Science movement. Both are well advised to look closely at what services are offered for their cause. To see why, we must consider just what un-derstanding of science Fuller promotes. Fuller gives relatively little away on this point in his book on Kuhn, though, as noted, the Conclusion briefly revealed his "constructivist scruples", which recalls his self-description in the Preface as "a devout constructivist" (Kxvi). What might this amount to? In the Introduction we read: "Notwithstanding the efforts of Jürgen Habermas, Paul Grice, and others to demonstrate by a priori reasoning that there are incontrovertible foundations to communication, I take Kuhn to have shown that the goal of this quest for foun-dations is simply chimerical; hence, I have increasingly turned my attention to rhetoric as providing insights into how language can motivate collective action without requiring prior agreement on the point of that action." (K7) Never mind how Kuhn is supposed to have shown *that* (he hasn't); focus instead on Fuller's turn to rhetoric. What might this be a turn away from?

The answer is given in Fuller's other new book, *The Governance of Science*, which in part overlaps with the Kuhn book. There Fuller does present himself in very much the way in which STS literature often is viewed by many scientists and philosophers of science, namely as plainly "anti-science". In this book mainly addressed to sociologists, Fuller lets rip like he did not when addressing historians and philosophers of science. (Certainly no time is wasted on explain-ing such flourishes as declaring AIDS a scientific failure or that "science is a destabilizing force in today's world" (G104).) Two points of Fuller's in particu-lar are notable here.

The first should be of concern especially to the Popperians who in this book too are subject to Fuller's advances. (In his comment on his own previous work he writes: "implicit throughout has been a commitment to the republican values associated with Karl Popper's ... original popularization of the open society" (G11)!) Note that Fuller states unambiguously: "'Truth', 'rationality' and 'objec-tivity' are metaphysical hypotheses that recent sociologists of science do not require to understand how science works ..." (G99). For Fuller, what the rhetorical turn is a turn away from is truth and objectivity. One need not view truth as correspondence, as Popper did, to find this worrying, just as one need not be a Popperian to applaud Popper's crusade against false presumptions to possession of truth. Yet being properly sceptical in the latter respect is wholly compatible with thinking that science aims for truth or truth-likeness or even only empirical adequacy – whichever of the three the reader may prefer, Fuller will have none of it.

Fuller is explicit about wishing to knock science from its "pedestral" on which he feels it has been placed (ibid.). But would mere objection to Big Science require this? And does it follow that science is just another discourse regulated from within and without any claim to cognitive distinction from religion, say? The muddle these overreactions leave Fuller in is evident from his calling, on the one hand, on government to effect or fund "the testing of knowledge claims for validity, efficacy and safety" (G105), while on the other hand claiming that "a new finding that cannot be made readily available to competent students operating with radically different assumptions should be seen as problematic until proven otherwise" (G113). While Fuller still talks of a "minimum standard of academic respectability" attaching to what is taught in universities (ibid.), it is utterly unclear what the validity of a knowledge claim could amount to in his scheme of things. (Note also his endorsement of the idea introduced more playfully in the Kuhn book, of "presenting evolutionary theory as compatible with divine creation" (G112).) "I suspect that once systematized," he opines instead (G35), "our current knowledge talk will more closely resemble Ptolemaic astronomy than Newtonian mechanics, in both its restricted utility and its ultimate dispensibility."

The second point to focus on should be of particular interest to those taking an interest in the Popular Understanding of Science, alluded to also by Fuller's occasional uses of the term "democratization of science" in the Kuhn book. Whereas there talk of the "secularization of science" remains opaque amidst associations with public discussion of the "ends of science", here Fuller lets the cat out of the bag. "Like the secularization of religion, the point here is to divest the state's funding of scientific research, while at the same time promoting public access to alternative research programmes, each being allowed to find its own funding constituency." (G97) You read right: "secularized science" would make the distribution of scientific knowledge "public" but make its production "private" – leaving science funding to "corporations, unions, charities and other special interest groups, who should be offered tax incentives for foregoing exclusive rights to intellectual property and thereby keep the fruits of research in the realm of 'knowledge' strictly speaking" (G105, note the scarequotes). Those readers who thought of "democratization of science" as increasing the control of the populace over what science gets done had better take note of Fuller's Thatcherian drive to privatization!

Another idea close to the heart of those concerned with the Popular Understanding of Science is "science literacy". For Fuller, this too is a bad idea: "this
proposal mistakenly assumes that the locus of epistemic privilege lies in the sort
of intellectual qualities that, say, scientists have but ordinary people lack, rather
than in the sort of access to intellectually empowering tools that one group has
but the other lacks" (G45). While Fuller shows an odd sense of consistency here
(if sociologists of science do not need to know the science whose practitioners
they are investigating, why should the public have to know even the meaning of
basic concepts like 'evidence'?), he once again rejects the very means by which
the presumably democratically widened constituency of science could reach rational decisions on pertinent issues. The reason: "increasing the public's science
literacy does not, by itself, open up any new opportunities for citizen participation in the conduct of science." (G46)

As with Kuhn, so with the Public Understanding of Science: Fuller rejects it
for not doing what *he* wants to accomplish. (For a recent overview of the tendencies of and tensions in Public Understanding of Science, see J. Gregory and S.
Miller, *Science in Public. Communication, Culture and Credibility*, Plenum
Press, 1998) There need be little wonder then about his perversion of Popper's
project and rejection of notions like truth and objectivity: criticisms or invoca-
tions of Kuhn, the Public Understanding of Science movement and Popper, for
instance, are but window dressing for Fuller's own programme which, as the
quotes adduced suggest, is not only philosophically bankrupt. Perhaps Fuller's
suggestions, like that to import "to science the sorts of things that makes sports
so compelling for so much of the world's population" – "an easily accessible
canonical accounting procedure", "fair and explicit rules of the game", "something worth contesting" (G149-50) – should be appreciated as a somewhat be-
laboured joke?

It is particularly pernicious, of course, that what amounts to Fuller's cognitive nihilism poses as anti-elitism and is ever again invoked as antidote to the
threat Big Science poses. Yet (again) there is nothing inherently pro-Big Science
about the idea that the general public should have a basic understanding of
scientific methods of investigation, especially of the fallible status of scientific
knowledge claims and the ways in which they can be established and challenged.
Nor is it elitist to think that some knowledge claims are justified better than
others and that the systematic development and appropriate use of concepts and
interactive procedures that seek to explore this differential between knowledge
claims requires some training, indeed expertise.

My conclusion is sadly predictable. It has been remarked before (see, e.g.,
Noretta Koertge, ed., *A House Built on Sand. Exposing Postmodernist Myths
About Science*, Oxford University Press, 1998) that some of the social constructivist literature on science tends to owe its revolutionary appeal to subtle elisions
in argument and not so subtle confusions of related concepts. Needless to say, it
would be wrong to tar all practitioners of either rhetorics or the sociology of

science – even all so-called "constructivists" – with the brush that is wielded so forcefully in this anti-critique. However, for all the learning displayed along the way, Fuller's "philosophical history" seems to me to be very much part of the trend diagnosed there.

Prominent among the more common confusions is that of the concepts of knowledge and knowledge claim. Fuller's "social epistemology" itself trades on the ambiguity of whether "theory of knowledge" delimits a sociological or philosophical domain. But where, say, Barnes and Bloor at least could have claimed merely to delimit the domain of sociological investigation (whether they did is another matter), there Fuller instead revels in the accusation, contrarily. It seems to be precisely his "philosophy" that the philosophical conception of knowledge is bunk. Laying a claim to philosophy incurs a cognitive obligation, however. Only a Pyrrhonian sceptic may not fear self-contradiction, and then only for the price that the display of argument has no hold over those who still believe in the possibility of theory – but Fuller is no Pyrrhonian: he aims to make normative pronouncements about the course of history. Yet, as we saw, he also abjures truth- and objectivity-centred reason. In consequence, it is wholly unclear by what factual considerations, if any, his "constructivist" history – and policy advice – is constrained. The threat of self-contradictions apart, what force does Fuller expect a self-conscious exercise, however elaborate, in mere rhetoric to have to compel the abandonment of truth- or justification-centred reason?

None of these criticisms, it should be stressed, are directed against the premise which Fuller shares with an increasing number of historians of science and philosophy, namely that careful attention must be paid to the socio-political conditions under which scientists and philosophers work. Thus we may note in closing that, with its more tentative conclusions but more incisive analyses, John McCumber's *Time in the Ditch. American Philosophy and the McCarthy Era* (Northwestern University Press, 2001) succeeds where Fuller fails (while still leaving room for disagreements). Rather than perpetuate the extremes of the Science Wars, McCumber employs the socio-political contextualization of analytical philosophy's rise to dominance in post-World War II North America to ground a discussion that seeks to transcend an ongoing and unproductive divide in philosophy. As it happens (Kuhn does not figure as such), the questions dealt with by McCumber – What is the scope of inquiries aiming at truth? Are philosophy and history wholly encompassed by them? – engage more closely and creatively Kuhn's project than Fuller's ill-considered dismissal of it.

Centre for Philosophy
Department of Government
University of Manchester
Manchester M13 9PL
U.K.
thomas.uebel@man.ac.uk

REVIEWS

THOMAS UEBEL, *Vernunftkritik und Wissenschaft. Otto Neurath und der erste Wiener Kreis,* Wien–New York: Springer, 2000.

Thomas Uebel's penetrating book represents the latest outcome of a series of investigations on Otto Neurath and the "forgotten" Vienna Circle he has published over the last ten years. Within the recent galaxy of studies devoted to logical empiricism, the re-evaluation of Neurath's all too neglected work as well as of the "first" Vienna Circle are unquestionably very much indebted to Uebel's contributions, which may be considered, in turn, an original development of Rudolf Haller's pioneering studies on Austrian philosophy and on its leading role in the rise of the Vienna Circle. In particular, the attention paid by Haller to Neurath's surprising epistemological actuality and moreover the "new light" he shed on the "first" Vienna Circle (Philipp Frank, Hans Hahn, and Otto Neurath in the years between 1907 and 1912)[1] constitute the core of Uebel's wide program of historical and, at the same time, philosophical research. An ambitious, but successful programm, that nowadays appears to us as essentially accomplished.[2]

The main topics of this book can be summarized as follows. First of all, it is the quite complex relationship between modernism and post-modernism to represent the framework into which Uebel locates his "contextual reconstruction" of the Vienna Circle (p. IX). According to Uebel, the roots of the first Vienna Circle lie in the culture of modernity (to which the *Wiener Moderne* also belongs) as well as in its crisis, that is, in the transition from the science of the late XIXth century as established paradigm for every epistemological inquiry into the new perspectives opened at the beginning of XXth century by relativistic physics, mathematical logic, and the increasing development of social sciences (see p. 212 with regard to the "inadequacy" of Ernst Mach's philosophy of science). Thus, the key figure is Neurath, who in fact endorses post-modern theses such as the sceptical attitude in contrast to absolute foundations of sciences, the historical, social and pragmatic view of scientific enquiry, a kind of pre- and (to some extent) post-Quinean naturalism, the refusal of the classical theories of truth, and, last but not least, the acknowledgement of a pluralistic view of scientific methods in order to reconstruct the unity of science (p. 23, 30-39). Yet all this by no means signifies that the "post-modern" Neurath has completely abandoned the modernity. The "Enlightenment" as self-awareness of modernity (Jürgen Habermas) remains nevertheless, Uebel suggests, the crucial inspiration of Neurath's work as well as of the work of the first and late, "left" Vienna Circle (particularly Frank). Such an awareness is connected to the critical

391

M. Heidelberger and F. Stadler (eds.), History of Philosophy and Science, 391–417.
© 2002 *Kluwer Academic Publishers. Printed in the Netherlands.*

"reflexion" on the Enlightenment itself (p. 18), i.e. both on the limits of its project and on its intrinsically unfinished nature, so that post-modernism is in turn
discussed as the (self)criticism of enlightened modernity. Uebel gives a stimulating account of this state of affairs in the last chapter of his book, where he
argues that the post-modern program of the Vienna Circle remains still an open
question (in the sense of Neurath's famous metaphor of the ship which must be
repaired during its travel at sea). In any case, according to Uebel, the too often
neglected aim of the "scientific world-view" in its late period was the attempt to
establish a connection of theoretical with practical reason, or of science with life
(p. 359-360), that is to realize a defence of rational thinking without excluding
ethical and political values from the field of human community and reasonable
social behaviour. In other words, the "classical" meta-ethical non-cognitivism of
logical empiricism, and its rejection of any metaphysical foundation of values,
does not imply that scientific rationality can ignore the "other" of reason, i.e. the
search for public arguments in favour of values such as pluralism, democracy,
and so on. Thus Uebel devotes considerable attention to Frank's *Relativity – A
Richer Truth* (1951), whose crucial problem was not the foundation of objective
values, but just how to warrant the public consensus about values (or a pluralistic
range of values) (p. 372)[3].

The second main topic of Uebel's book is the attempt to give a closer interpretation of the tradition of Austrian philosophy. In fact, it is one of the greatest
merits of Uebel's reconstruction that it offers an original overview of Austrian
philosophy, suggesting a richer genealogy of the tradition starting with Bernard
Bolzano. Uebel presents figures (today forgotten, but in their time very influential) such as Robert Zimmermann and Alois Höfler (the founder in 1888 of the
Vienna Philosophical Society, in which Frank, Hahn and Neurath delivered
many lectures between 1906 and 1917) as mediators of Bolzano's philosophy of
logic and mathematics as well as of its anti-Kantianism. The wide echo of these
typical exponents of Austrian philosophy was possible thanks to their handbooks
of logic and philosophical propaedeutic, which had a great diffusion in the Austrian schools and, in the case of Höfler, was well known to Hahn and Neurath,
but probably to Frank as well (p. 135-137). Moreover, the leading spokesman of
Vienna's philosophical milieu at the beginnings of the XXth century can be
considered, as Uebel suggests, the great Ludwig Boltzmann (who played an
important role in the Ph.D. dissertations of Frank and Hahn). From Boltzmann's
philosophy of science, which was certainly different from that of Mach, both
fallibilism and instrumentalism emerge as essential features of scientific knowledge: it is, in fact, a kind of pragmatism, which seems to conflict, on the one
hand, with a central aspect of the Bolzano's tradition, and, on the other hand, still
appears to be of great significance for the early development of logical empiricism (pp. 162-166).

The third crucial and most original point of Uebel's inquiry lies in his
extensive reconstruction of the first Vienna Circle. Here, for the first time, a detailed overview of Hahn's and Frank's *Lehrjahre* is available and the origins of

their scientific careers are widely illustrated, in the fields of mathematical logic and relativistic physics, respectively. Hahn's studies in Göttingen and his acquaintance with Zermelo is crucial in order both to disclose to him the "revolution in logic" brought on by Frege and Russell, and to open a new, until then unexplored, but later fateful front in Austrian scientific philosophy as well as in the Bolzano's tradition (pp. 176-177, 209, 211). Also for Frank Göttingen was the starting-point of his intellectual adventure, since his training as physicist in the University where Hilbert's and Minkowski's teachings prepared him to be very well acquainted with Einstein's theory of relativity before 1910. Frank was thus the first member of the future Vienna Circle to become a "radical relativist" (p. 188) and to endorse an holistic conception of space and time, which signified at the same time the revision of Minkowsi's "metaphysics" and the beginning of a critical assessment of Mach, ending only with the formulation of the general theory of relativity (pp. 190-195). However, both Hahn and Frank were deeply interested in the "crisis of intuition" brought by mathematical logic and theory of relativity, so that their main reference point was more French conventionalism than Mach's economical view of science. According to Uebel, this is exactly the crucial point of the first Vienna Circle. Frank's reading of Poincaré's conventionalism and, to some extent, of Duhem's holism, played a leading role in his anti-realistic as well as in his increasingly instrumentalistic account of physical knowledge. Uebel recognises in the early Frank's epistemological work both a "radical conventionalism" (although overlooking at all Poincaré's "strucutural realism"), and a clear refusal of Kant's synthetic a priori (pp. 232, 234, 238). Thus, one of the most typical features of the late "left" Vienna Circle was already sketched by Frank around 1910. Uebel shows very well that the outcome of a similar conception is a kind of pragmatism which was to be shared by Hahn and Neurath too, but which was surely quite far from Schlick's own understanding of knowledge (and in particular) of Einstein's theory of relativity (pp. 248-262).

The last aspect we want to emphasise here is the rich portrait of Neurath offered by Uebel. Readers acquainted with Uebel's studies on Neurath published in the last years will find in this book, once more, an overview of his extraordinary intellectual biography as well as an excellent survey of all the issues underlying the recent Neurath's rediscovery and stimulating the current scholarship devoted to him[4]. The last goal of Neurath's long way from the first to the late Vienna Circle lies in the reflexive epistemology of social sciences as the new key and the new task of postmodern, but still "enlightened" philosophy (see p. 345-347). The understanding of this way, Uebel argues, requires a careful reconstruction of some central topics, such as Neurath's naturalisation of conventionalism and his pragmatic or pragmatistic view of science, which find their origins, in turn, in Neurath's early focusing of "auxiliary motives" in scientific reasoning and in his criticism of Descartes' subjective foundationalism. Thus, the young Neurath of the first Vienna Circle was already engaged, like Hahn and Frank, with the core of the late logical empiricism: the search of a theory of science already becoming aware of the collapse of the synthetic a priori (p. 289). But Neurath's first steps in the history of Vienna Circle are also connected with his engagement in the educational and ethical projects characterising the culture of Vienna at the beginning of the 20th century – a culture whose leading exponents were figures such as Friedrich Jodl (p. 294-296), whom Uebel efficacioulsy describes, giving instructive quotations from his pages devoted to universal ethics as well as to a "democracy of the spirit" to be realized (it is noteworthy that all the members of the first Circle were well acquainted with Jodl and his lectures at the University of Vienna).

This is, in sum, the content of Uebel's exciting and impressively learned book, which doubtlessly represents a great contribution to the scholarship on logical empiricism. Obviously, there would be some interesting questions to discuss in a more detailed manner; nevertheless we attempt here simply to enumerate them, without any claim to possessing the right answer. To begin with, Uebel seems to underestimate the influence of Herbart's philosophy (especially with regard to the conception of logic as normative science) on the work of Zimmermann and Höfler. This is of a certain importance in order to outline the extent of the "Bolzanian" tradition in Austrian philosophy and its limits (see pp. 116, 125). Secondly, Uebel's interpretation of Frank's first ideas and, particularly, his critical discussion of Poincaré's conventionalism overlooks to a certain degree that Frank, originally, aimed at a revision of the Kantian a priori which was, it may be said, not only inspired by destructive critics. This is surely a very interesting point, since some of the topics of Poincaré's conventionalism Uebel emphasizes lie at the heart of the epistemological debates of the neo-Kantians, e.g., the conception of objectivity determined in terms of relations (pp. 239-241) represents a typical argument of the Marburg School, especially of both Natorp and Cassirer. In short, the history of the "relativized a priori" with its conventionalist and semi-Kantian background is perhaps more complex than Uebel's reconstruction exhibits. Thirdly, it seems sometimes that the great attention devoted to the topic "modernism-postmodernism" comes to constitute, so to speak, a second book within Uebel's wide picture of the first Vienna Circle. This can be considered not only as a problem, but perhaps as a more or less serious difficulty in connecting the historical and epistemological analysis with its more general cultural and critical framework. Fourthly, Carnap is considered a leading figure of the "left" Vienna Circle and his "affinities" with Neurath are frequently stressed (p. 54), although Uebel is well aware of the differences subsisting, for example, between Carnap's logic of science and Neurath's pragmatic of science (pp. 281-282). And yet, surprisingly enough, Carnap's *Aufbau* is little discussed, so that it seems that the one-sided image of Carnap emerges which, to some extent, develops his philosophy more under Neurath's shadow than starting from Russell or neo-Kantianism.

The last critical remark we wish to make is of a general kind. It is partially related to the question to which we have just referred to. One can hardly avoid wondering, whether we can adopt the picture of the Vienna Circle emerging from his reconstruction as the picture of the whole. Obviously, in no way. But, if this is so, how can the history and the assessment of the philosophy of the Vienna Circle be modified as soon as other leading figures such as Carnap and Schlick have been considered more extensively? As Uebel suggests, two opposite perspectives faced each other within the Circle, i.e. the foundationalist one (Schlick, as far as he maintained a categorial distinction between the experience and the form representing it) and the anti-foundationalist one (Neurath, Hahn, Frank, and Carnap, that is, the different manner of relativising the analytical a priori and endorsing an holistic point of view) (p. 277). Moreover, Uebel is right when he emphasizes that the Vienna Circle was a *Denkkollektiv*, so that its unmistakable feature has precisely to be identified with the inner competition among quite different philosophical and epistemological programs (although, evidently, within the common framework of scientific and linguistic philosophy). But for just this reason it would be important to examine the roots and the developments of logical empiricism in the 1920s as a whole, being aware of the intricate character of the "other" side of the Vienna Circle (Schlick and Wittgenstein, on the one hand, and Carnap until the *Aufbau*, on the other, not to mention German logical empiricism in its connection with the "Vienna Station"). A similar view is necessary for understanding – as Michael Friedman recently emphasized[5] – the relationship between logical empiricism and the Kantian as well as the neo-Kantian tradition and, moreover, the connections with other trends of "continental philosophy". Such a historical and systematic task is obviously of extraordinary importance for the positioning of logical empiricism within the history of twentieth century analytical philosophy. To be sure, Uebel's point is that the first and, later, the "left" Vienna Circle with its postmodern enlightened view of science and culture represent not only a historical event in XXth century philosophy, but also a highly stimulating contribution to the philosophical agendas of our times. And yet, could it therefore be said that the "other" Circle belongs exclusively to the past?

NOTES

1. See R. Haller, "New Light on the Vienna Circle", *The Monist*, LXV, 1982, p. 25-37; "Der erste Wiener Kreis", *Erkenntnis*, vol. 22, 1985, p. 341-358; *Neopositivismus. Eine historische Einführung in die Philosophie des Wiener Kreises*, Darmstadt, Wissenschaftliche Buchgesellschaft, 1993, p. 45-60.
2. Here we might remember some of Uebel's main contributions: Th. Uebel (ed.), *Rediscovering the Forgotten Vienna Circle*, Dordrecht, Kluwer, 1991; *Overcoming Logical Positivism from Within. The Emergence of Neurath's Naturalism in the Vienna Circle's Protocol Sentence Debate*, Amsterdam, Rodopi, 1992; "On Neurath's Both", in N. Cartwright, J. Cat, L. Fleck, Th. Uebel, *Otto Neurath: Philosophy between Science and Politics*, Cambridge, Cambridge University Press, 1996, p. 89-166.
3. Of course, a similar insight was clearly stimulated by the discussion on the roots and the rise of totalitarism in the 20th century – a discussion which presupposes, in turn, the tragic crisis of modernity as well as the unaspected revival (according to Ernst Cassirer's analysis, from which – it seems to me – Frank was probably inspired) of the "Myth of the State".

4.   For a short discussion of some recent inquiries on Neurath see M. Ferrari, "Recent Works on
     Otto Neurath", in M. Rédei / M. Stöltzner (eds.), *John von Neumann and the Foundations of
     Quantum Physics*, Dordrecht-Boston-London, Kluwer Academic Publishers, 2001, pp. 319-327.
5.   M. Friedman, *Reconsidering Logical Positivism*, Cambridge, Cambridge University Press,
     1999.

Massimo Ferrari

EDGAR ZILSEL, *The Social Origins of Modern Science*. Ed. by Diederick Raven, Wolfgang Krohn and Robert S. Cohen. Foreword by Joseph Needham. Dordrecht–Boston–London: Kluwer Academic Publishers, 2000 (= Boston Studies in the Philosophy of Science, vol. 200).

Edgar Zilsel is perhaps the least well-known member of the Vienna Circle. Although he was a council member of the official Verein Ernst Mach he never played a central role in the debates of the logical positivists, neither in Vienna before the *Anschluss* nor in exile. While he published only one article in the *Erkenntnis* he contributed a lengthy piece to the Neurath-Carnap-Morris Encyclopedia and published three articles in *Philosophy of Science* (and some others in scholarly journals not affiliated with the camp of logical positivism, all written during the relatively short last period of his life. In 1944, at the age of 53, he committed suicide, six years after his arrival in the American exile.

During his time in Vienna, where he was born in 1891 as the son of a lawyer, his scientific activities were mainly in the field of what we would today call public understanding of science. He wrote some forty reviews for *Die Naturwissenschaften*, a journal read both by lay people and scientists, and he taught at the Volkshochschule when, in 1924, his Habilitation was rejected due to the overwhelming influence of two conservative professors at the University of Vienna. They didn't like his sociologically inspired approach to studying the changing role of genius in the history since the classics. Perhaps Zilsel's opponents were also driven by the then widespread anti-Semitic attitudes within Austria's academic community, but there are no sources to support this claim. At the age of 33 his ambitions to become an academic philosopher were definitely refused: someone whose attempt to become a Privatdozent was turned down by a university committee, built up of highest ranking professors, had to give up those plans. Up to this turning point Zilsel's list of publications was not above average with regard to what other candidates for Privatdozent usually had on record. Zilsel's Ph.D. thesis was published – something very unusual at this time – and he had published another book-length essay *Die Geniereligion*. It's just speculation to claim that the conservative philosophers liked neither the polemic tone nor the rejection of the concept of genius. Given the micro-environmental conditions of high-brow academics Zilsel's decision to submit another, less polemical piece on the emergence of the concept and fact of geniuses in Western philosophy for a formal examination was not very sensible. However Zilsel's turn to the Weltanschauung of the worker's movement might have contributed to his attempt to challenge the conservative elite. A more opportunistic attitude on his part would have been a better choice.

He began contributing articles to the official theoretical journal of Austria's Social Democrats *Der Kampf*, where he published his first article in 1929 *Philosophische Bemerkungen*, a fierce attack on what he then called school philosophy, meaning scholastic and detached from the world thinking, he published in 1926 a presumably re-edited version of his Habilitationsschrift with the title *Die Entstehung des Geniebegriffs. Ein Beitrag zur Ideengeschichte der Antike und des Frühkapitalismus* (The Emergence of the Concept of Genius: A Contribution to the History of Ideas from the Classics and the Early Capitalism). A second was announced but never appeared.

In this comprehensive historical investigation Zilsel laid down what he later elaborated in a series of articles, published during his exile in the U.S.

One has to bear in mind that only after his forced migration Zilsel started a career as a publishing scholar. Between 1926 when his *Geniebegriff* appeared and his flight from Vienna in 1938 he published only very few purely scientific papers besides the above mentioned reviews: one overview about Naturphilosophie for a introductory reader to philosophy, a slightly strange comparative analysis of the concepts of heredity and tradition (re-published in a shortened English version as History and Biological Evolution 1940, reprinted in the volume under review), an obituary for Moritz Schlick and again a polemic against vitalism and quantum physics (in *Erkenntnis*, as mentioned before this is the only article in this journal).

In their well-researched introduction, Raven and Krohn don't make much effort to distangle the striking difference between the Viennese Zilsel, embedded in an intellectually and politically encouraging and supportive environment, both in the Social Democratic movement and the movement of Logical Positivism – the freely chosen title for the Vienna Circle as a movement reminds one to the close interrelationship between politics and philosophy at this time – and a loner in the canyons of Manhattan. The latter is a metaphor Zilsel's son once used to describe his father's feelings.

Historians often begin writing history from back to front. Raven and Krohn were no exemption. Zilsel's suicide overshadows every attempt to reconstruct his living in exile, being disturbed and disoriented, but at the same time and out of the same conditions Zilsel did then what scholars are commissioned to do, i.e. to publish their findings, insights, and theses. Tragedy ultimately prompts historians to search for troubling instances generating the final outcome. And Zilsel's short period of life in America nearly invites someone to take this path. As a side-effect one might be inclined to overlook the contradicting evidence. In Zilsel's case the mainstream interpretation of his demise blames his being a refugee for everything. Lack of recognition, loneliness, disturbing living conditions, the political developments both in Nazi-ruled Europe and the homeland of modern capitalism etc., all seem to be sufficient explanation of his decision to end his life voluntarily. However this reading is not convincing, at least for me.

Besides the fact that we can never fully understand an individual's decision to commit suicide, in this particular case we find at least two contradicting pieces of evidence, which are sometimes mentioned but never fully acknowledged. One is Zilsel's setback when on his arrival in New York City he went to Morningside Heights to see Max Horkheimer and his comrades and ask them for help. To be sure, there existed political affinities between the Frankfurt school and the left wing neo-positivists from Vienna. However, according to Dahms' comprehensive study about the relationship between these two camps, the previously friendly mutual philosophical non-recognition had turned into fierce rivalry by the time Zilsel arrived in New York. Zilsel couldn't know this in advance, but his personal surrender to Horkheimer who increasingly came under the influence of the neurotic anti-positivist Adorno, at least equally accounted for Zilsel's failure in the U.S.

The second point Raven and Krohn overlook is Zilsel's own diffidence in dealing with academic affairs in the U.S. After a somewhat disturbing and exploitative non-collaboration with the Frankfurtists Zilsel started to establish his own connections to American academics, and failed. When Zilsel approached the later famous George Sarton from Harvard (an episode Raven and Krohn missed to mention) he did not realize that Sarton's rejection ("I cannot offer you a desk because I am a marginal man myself") was by no means a personal rejection but a result of Sarton's own marginalization. (His position at Harvard was supported by a special grant from the Carnegie Foundation and he was only a lecturer for history of science and not professor at Harvard.) Due to Zilsel's courtesy (which he brought with him from Europe) he did not insist and could not make the four-hour trip from Manhattan to Harvard Square for personal and perhaps financial reasons. Contingencies of this kind isolated Zilsel in the U.S.

There is a third point which is more ambivalent than the two mentioned above – on that contributed to both Zilsel's failure and success. Before living in the U.S. Zilsel never published as much. The entire book under consideration consists of re-prints of his American publications. One should recognize that the short term contracts forced Zilsel to publish pieces of his never completed grand study. This constraint was completely at variance with the old-fashioned European style of doing science and publishing entire books instead of chapters of a forthcoming one. At the same time he might have recognized that his attempt to re-write the whole history of the emergence of science in the Western hemisphere was simply a bit too much for one author. An aficionado of collectivism in the political sphere, Zilsel was not able to bring together collaborators or to affiliate himself with the tiny groups of sociologists and historians of science that were emerging at the time.

Christian Fleck

MALACHI HACOHEN, *Karl Popper: The Formative Years, 1902 – 1945; Politics and Philosophy in Interwar Vienna*, Cambridge University Press, 2000.

When Karl Popper's *The Open Society and its Enemies* [1] was published in 1945, shortly after the second World War had ended, his work soon became a classic for the defence and support of the western alliance against Soviet communism and Stalinism in the cold war. This interpretation primarily provided by authors of conservative liberalism emphasised Popper's criticism of the historicistic theories of Karl Marx as well as his new contributions for the defence of western democracy, to social philosophy and the philosophy of history. A different interpretation of Popper's work was provided in the early seventies of the 20th century by authors of social-democratic orientation and by those who belonged to progressive liberalism. [2] After having discarded Marxist thought as a theoretical foundation of practical politics, some social democratic and progressive liberal thinkers looked for new theoretical foundations that could serve as new guidelines for policy measures. In actual politics Marxist ideas as guidelines for policy measures had long been substituted by reform policies, and a new theoretical and philosophical foundation was required in order to solve the discrepancy between the theoretical basis and policy measures. Popper's ideas with respect to the social welfare state, western democratic institutions, liberalism, policy measures and the methods of social science seemed to provide such a new theoretical foundation. Some authors even thought that Popper's philosophy of Critical Rationalism ought to be regarded as revival of the so-called Bernstein tradition in (German) social democracy. However, although it seems that some ideas of Critical Rationalism indeed became a kind of theoretical guideline for social democratic policy measures, Critical Rationalism rather was regarded as a support of conservative liberalism and particularly in the late 70ies and 80ies as a kind illegitimate offspring of Hayekianism. Given this spectrum of possible interpretations of Popper's ideas it is not surprising that politicians from the conservative right to the progressive left all quote parts of Popper's work in order to support their positions.

With Hacohen's book the first scientific contribution to Popper's biography appeared that provides an account and an analysis of Popper's intellectual development until 1945 and also a clarification of the above-mentioned situation with respect to the different interpretations of Popper's political and social philosophy. As the subtitle of his book indicates, he investigates primarily those years of Popper's life, where he developed his decisive ideas in epistemology, philosophy, the method of science and social philosophy. This, of course, does not mean that Hacohen believes that Popper's intellectual biography basically stops in 1945 and that afterwards the flow of his orginal and fruitful ideas ended. On the contrary: Hacohen makes it quite clear that he regards Popper not only as one of the greatest philosophers in the 20th century, but as a genius as well who contributed to many scientific areas until his death in 1994. However, he emphasises that many ideas which were developed by Popper only later may be found already in his published and unpublished works of his formative years. Yet Hacohen's book is not only the first and scientific intellectual biography of Popper's formative years. His excellent investigation is carried out in a true Popperian spirit. It is carried by a genuine love of learning and a critical and respectful but non-relativist attitude towards Popper and his philosophy. In particular this attitude is brought to light in Hacohen's (implicit) criticism of Popper's autobiography, where he shows that in this work Popper provides a peculiar though nonetheless most beautiful dehistorizised narrative of his intellectual development that resembles closely a linear model of scientific progress and not one of conjectures and refutations. Hacohen rehistorizises this story, shows where Popper's own story does not match with historical facts and situations and develops the picture, not of a stylised dehistorized, wise and gentle Socrates, but that of an antiauthoritarian and rebellious young genius in his cosmopolitan wanderings between different worlds on the road of unended quest.

The method which Hacohen applies in his investigations is Popper's method of situational analysis, which Popper developed in order to solve some methodological problems of historiography and theoretical social science. With great care, empathy and intuition Hacohen provides a brilliant account of the intellectual atmosphere, the cultural milieu and the social problems in Vienna in the declining Habsburg monarchy and the newly founded republic. In that context, his analysis of the development and the problems of assimilated Jews in Vienna, their role and importance in Viennese culture and society is of special importance, not only with respect to his account of that milieu but above all with respect to a clarification of Popper's attitude to those questions. Avoiding the usual clichés and stereotypes one often finds in historical accounts of that time, Hacohen provides an account that shows the fierce fight between the representatives of social democracy, progressive liberalism and liberal democracy, i.e. the representatives of the so-called *Late Enlightenment*[3], on the one hand, and the representatives of clerical fascism and german nationalism, i.e. the representatives of totalitarian ideas and political conceptions, on the other. And he also shows Popper's terrible disappointment with respect to the policies and strategies the representatives of social democracy followed in order to save the new democratic republic from the authoritarian and totalitarian onslaught of clerical fascism and german nationalism and which he considered to be suicidal. In drawing such a masterly picture of the social, political and intellectual situation that prevailed in Vienna between the two wars, Hacohen illuminates how the philosophy of Popper in principle offered solutions to problems that the representatives of progressive liberalism and socialdemocracy did not then develop and that the representatives of poststructuralism cannot offer today. These considerations, however, make it quite clear that Hacohen's book is not only a book on Popper's intellectual biography but a philosophical and political one as well.

It is a long, a critical and above all a very exciting book that surprises the reader at times with unorthodox historical conjectures, such, that the second volume of Popper's *Die beiden Grundprobleme der Erkenntnistheorie* never existed or with respect to the dates of Popper's solutions to the problem of induction and demarcation.

Although Hacohen's scientific biography of Popper's formative years is an outstanding achievement and most important for the history of ideas I would like to add some minor critical remarks, which, however, do not touch the substance of his book. One such remark concerns Hacohen's appraisal of Popper's analysis of Plato's and Hegel's political ideas and programmes; the other one Hacohen's account of the *Methodenstreit* and Menger's position in particular.

It is well known that Popper's account of Plato's and Hegel's political ideas met severe criticism, saying that Popper treated these two authors unfair and with respect to Hegel not serious. Hacohen shares this criticism, which I found surprising, given his excellent application of Popper's method of situational analysis of Popper's intellectual development in the twenties and thirties. Yet it seems to me that any criticism of Popper's account of the political ideas of Plato and Hegel ought to distinguish two different kinds of questions. The first one concerns Popper's intentions and aims he followed in his *The Open Society and its Enemies*. The other one concerns the question as to whether Popper provided an accurate account of those two authors.

In his *The Open Society and its Enemies* Popper intended to provide a historical sketch with respect to the rise of totalitarian and autoritarian ideas. The book emerged as a byproduct of his investigations with respect to the methods of the social sciences, and although it is a wonderful work, some chapters are, due to their nature of being a historical sketch, uneven and unfinished. Popper's intention was to show that those totalitarian and authoritarian ideas that were used to support the "modern" totalitarian political programmes and conceptions in the twenties and thirties such as those of clerical fascism or even national socialism could be traced to antiquity. Spann [4], a rather prominent figure at the University of Vienna in the twenties and thirties, for instance explicitly based his epistemology and political philosophy on the political ideas of Plato and Aristotle. It is not that Popper wanted to show that Plato was the first representative of fascism, but rather that some of his totalitarian conceptions were still used to support modern totalitarian and authortarian thought. One also should keep in mind that the ideas of Plato and Aristotle, as well as those of authors of the Roman Empire were the bread and butter of the classical curriculum in Austrian *Gymnasien*. (The school reform basically concerned *Volksschule* (primary schools) and *Hauptschule* (secondary schools), but not the *Gymnasium* which was a stronghold of conservativism and antidemocratic thought). In contrast, the question whether Popper provided a fair account of Plato seems to me quite a different one; one that has to be discussed in terms of cultural and philological investigations. But whatever their results are, I do not think that this will invalidate Popper's analysis which is to show the importance and the role of ancient totalitarian conceptions for modern authoritarian thinking and also to explain its success. It seems to me that Hacohen, although he provides such a masterly account of the twenties and thirties in Vienna, does not considerate his own account sufficiently, when discussing Popper's account of Plato.

Similar points may be raised with respect to Popper's account of Hegel. Popper explicitly states, that he does not intent to provide a serious analysis of Hegels philosophy. As I understand it, Popper's intention in the first two chapters of *The Open Society and its Enemies* was to show how ancient totalitarian ideas by way of German Idealism infiltrated modern political thought – often in a very distorted way. Both chapters are an attempt not even to explain the rise of German Idealism, but rather how such a story may be written one day and he was well aware of the fact, that these two chapters were not even the beginning of such an analysis. However, it is interesting to note that in outlining these two chapters Popper follows to a certain extent Fries, who in his history of philosophy provides harsh criticisms of authors such as Hegel, Fichte and Schelling and other authors of German Idealism whose philosophy he regarded as a considerable regress after Kant.[5] Yet it is also interesting to note that Popper in his chapters on the rise of German Idealism does not analyse the role of Ranke whose political ideas also derived from ancient epistemological and political totalitarian thought and became extremely important for the method and philosophy of historiography and thereby for the political conception of the German Reich and for the 3rd Reich as well. This is shown beautifully in the so-called Fischer debate of the late sixties and the literature that spun of it as for instance that on neo-Rankism.[6] Of course, Popper wrote *The Open Society and its Enemies* much earlier than Fischer's revolutionary investigations; but Rankism and neo-Rankism was most important for German and it seems also for Austrian historiography and constitutes an important implicit element of the political and intellectual situation in which Popper happened to find himself, especially for his discussion of the methods of historiography and its political implications.

I would like to conclude with some brief remarks on Hacohen's account of the so-called *Methodenstreit* in economics and in particular on the relations of Menger and Schmoller. Due to Kauder's remarks on Menger's use of the term "das Wesen" some authors hold that Menger held an essentialist position and Hacohen, in my view, unfortunately, follows that line of interpretation. But the use of the term "das Wesen" does not necessarily indicate an essentialist position, since that term, only denotes the definition formula of a definition.[7] In that sense the use of this term is perfectly compatible with methodological nominalism and is often applied so in German language. In order to show that Menger held an essentialist position one needs to show that Menger not only uses the term "das Wesen", a usage which was very common in his times, but that he indeed holds an essentialist philosophy. However, I have tried to argue, that Menger's methodological investigation of the methods of the theoretical social sciences does not contain such an essentialist philosophy and that it is a purely methodological treatise.[8] Emphasising the logical objection to induction and thereby critisising the positions of the older historical school, and in particular that of Roscher, Menger distinguishes between two orientations of theoretical research an emprical realistic one, which is a kind of strict positivism, and an exact one, which he justifies by way of introducing an apriori induction principle. It is striking how close the structure of the internal argument in Menger's first two books of his *Investigations*[9] resembles that of Popper's analysis of the so-called *Normalsatzpositionen* and that Menger bases the distinction between the two orientations of theoretical research on the logical argument against induction. Roscher and later Schmoller did not grasp this argument of Menger's analysis, but rather followed a methodological position which, as Popper has shown, may be regarded as methodological transformation of an Aristotelian essentialist epistemology according to which history is the empirical basis of theoretical social science. Moreover, similar to Popper, Menger tried to show the incompatability of the theory of subjective evaluations and an essentialist position and mentions the impossibility of historical laws of development. Thus I was surprised to find that Hacohen thinks that Schmoller, who is a paradigmatic naïve inductivist, was nearer to Popper than Menger, especially, since Menger did not deny the role of experience as a critical standard of economic theories if they are epistemologically justified in the sense of the empirical orientation of theoretical research. If, however, economic theories are justified according to the exact orientation of theoretical research, i.e. by way of an a priori induction principle, then Menger (like Popper) pointed out that the principle of empiricism has to go.

These remarks, however, do not touch the substance of Hacohen's book in any way; rather, they show the wide range of problems Hacohen has to deal with in his book. Concentrating on certain issues of Popper's methodology of the social sciences and Hacohen's analysis of these issues, I have not even indicated his analysis of Popper's psychological and epistemological investigations, their importance and development. Hacohen's books opens up many problems for the history of ideas in general and with respect to Popper and his biography in particular. Perhaps the "old" Popper would not have liked Hacohen's biography; we will never know; but according the picture Hacohen provides, the "young" one may indeed have done so.

NOTES

1. Karl R. Popper, *The Open Society and its Enemies*, Routledge & Kegan Paul, London 1945 (5th edition 1966/1986).
2. Cf. for instance: *Kritischer Rationalismus und Sozialdemokratie*, (Georg Lührs, Thilo Sarrazin, Frithjof Spreer, Manfred Tietzel, Hrsg.), Internationale Bibliothek Bd. 79, Verlag J.H. W. Dietz Nachf. GmbH. Berlin, Bonn-Bad Godesberg 1975; *Kritischer Rationalismus und Sozialdemokratie II* (Georg Lührs, Thilo Sarrazin, Frithjof Spreer, Manfred Tietzel, Hrsg.), Internationale

Bibliothek Bd. 90, Verlag J.H.W. Dietz Nachf. GmbH. Berlin, Bonn-Bad Godesberg 1976; *Theorie und Politik aus kritisch rationaler Sicht*, (Georg Lührs, Thilo Sarrazin, Frithjof Spreer, Manfred Tietzel, Hrsg.), Internationale Bibliothek Bd. 111, Verlag J.H.W. Dietz Nachf. GmbH. Berlin, Bonn-Bad Godesberg 1978.

3.　Fritz Stadler, "Spätaufklärung und Sozialdemokratie in Wien 1918 – 1938. Soziologisches und Ideologisches zur Spätaufklärung in Österreich", in: *Aufbruch und Untergang. Österreichische Kultur zwischen 1918 und 1938*, F. Kadrnoska, (Hg.), Wien 1981.

4.　Cf. H. Räber, *Othmar Spanns Philosophie des Universalismus*, G. Fischer, Jena 1937.

5.　Jakob Friedrich Fries, "Anhang. Polemische Bemerkungen über neuere große Rückschritte", Jakob Friedrich Fries, *Die Geschichte der Philosophie, dargestellt nach den Fortschritten ihrer wissenschaftlichen Entwicklung*, in: Jakob Friedrich Fries, *Sämtliche Schriften*, L. Geldsetzer, G. König (ed.) vol. 19, Scientia Verlag Aalen, 1969.

6.　Cf. for instance, John A. Moses, *The Politics of Illusion*, The Fischer Controversy in German Historiography, George Prior Publishers, London 1975. Popper esteemed Fischer's work as well as Moses' book so highly that he not only drew my attention to Moses book but sent it to me.

7.　Cf. Karl R. Popper, *The Open Society and its Enemies*, Routledge & Kegan Paul, London 1945 (5th edition 1966/1986), Vol ii, Chap. 11.

8.　Karl Milford, "Menger's methodology", in: *Carl Menger and his legacy in economics*, B. Caldwell (ed.), History of Political economy, Annual supplement to volume 22, Durham and London 1990, p. 215ff.

9.　C. Menger, *Untersuchungen über die Methode der Socialwissenschaften und der Politischen Oekonomie inbesondere*, in: Carl Menger, *Gesammelte Werke*, F.A. v. Hayek, (ed.) vol. 2, Tübingen 1969.

Karl Milford

PAUL RUSNOCK, *Bolzano's Philosophy and the Emergence of Modern Mathematics*, Editions Rodopi B.V., Amsterdam–Atlanta, 2000.

The book is a detailed and penetrating study of the relations between Bolzano's philosophical views and his mathematical achievements. It opens with a *Biographical sketch*, written with a good understanding of the political situation in Central Europe at Bolzano's times. The words of the Emperor Francis II: "I do not need scholars but obedient citizens" (p. 8), and of an English visitor of Austria at those times: "These school-books are the most barren and stupid extracts which ever left the printing press." (p. 9) express the problems of the intellectuals in this part of the world.

Chapter 1, *Introduction*, describes the situation in mathematics and its philosophy at the beginning of the 19th century. The mathematics of the 18th century, dominated by Euler and Lagrange, was based on formal manipulations with symbols. On the other hand, the mathematics of the 19th century, characterized by the work of mathematicians as Cauchy, Dedekind, or Cantor, was mainly conceptual, aimed to replace formal manipulations by precise definitions and exact proofs. Bolzano was among the first who saw the need of this deep change and in his 1810 essay "Contributions to a better founded presentation of mathematics", he aimed at a complete reorganization of the whole of mathematics. From these philosophical views resulted Bolzano's early mathematical writings, among which the most outstanding were the 1816 paper *Binomial theorem*, and the 1817 paper "Purely Analytic Proof". The experience with actual mathematical research revealed to Bolzano a need for new logical and methodological tools. In *Theory of Science*, written during the 1820s, but published only in 1837, Bolzano developed whole new branches of logic and methodology. In the 1830s Bolzano embarked on a new mathematical project, the *Theory of Quantity*, in which he sought to present mathematics using his new logical tools. From this project some of Bolzano's most original mathematical contributions emerged – an arithmetical theory of the real numbers, the construction of a continuous, nowhere differentiable function and the development of the basic concepts of set theory.

Already this short overview of Rusnock's shows an interesting interaction between Bolzano's philosophical ideas and mathematical research. His program of reform of mathematics from his 1810 *Contributions* led him to mathematical research resulting in the 1816 and 1817 papers. It revealed a fundamental weakness of contemporary logic and led Bolzano to write his philosophical masterpiece, the *Theory of Science*. The new tools and insights, contained in that book, enabled him to penetrate much deeper into mathematics and to produce the *Theory of Quantity*. In the chapters of his book Rusnock leads the reader along this path of development of Bolzano's thought and offers a clear, non-technical exposition of its four stages.

Chapter 2, *The Methodology of the Contributions*, discusses Bolzano's early philosophical essay of 1810 "Contributions". Most mathematicians of his time held an epistemic notion of proof, according to which proofs were tools for obtaining certainty. They thus thought it was unnecessary to define concepts already clearly understood or to prove propositions that were obvious. In contrast to this view Bolzano considered truth to be an objective connection, independent of any our subjective recognition of it. Therefore rather than to convince, the aim of science is to display the ultimate grounds of truth. Thus it is necessary to prove many entirely obvious theorems. Similarly, often the most familiar concepts need a proper definition the most. Thus it is fair to say that Bolzano was one of the first to initiate the foundationalist approach in mathematics. He gave a new characterization of the notion of axiom. Instead of characterizing it as a self-evident or intuitive truth, Bolzano defined axioms as judgments composed of simple concepts. Perhaps the most interesting "by-product of his work in the *Contributions*, Bolzano provided a thoroughgoing refutation of Kant's account of mathematical knowledge." (p. 45) An English translation of this part of Bolzano's essay is included in the book as an appendix (pp. 198-204), and so the reader can appreciate Bolzano's arguments in detail. At the end of the second chapter Rusnock concludes: "The originality of Bolzano's position is underscored by the circumstance that Russell and Whitehead were able to put forward remarkably similar views, with an unmistakable air of novelty, some one hundred years later." (p. 51)

Chapter 3, *Early work in real analysis*, contains an exposition of two of Bolzano's above-mentioned mathematical papers ("Binomial Series" and "Purely Analytic Proof"). On the one hand, these papers followed the program set out in the "Contributions", but on the other hand they forced Bolzano to acknowledge that technical parts of his logic required thorough revision. Most mathematicians before Bolzano derived the binomial series using formal manipulations with signs, and either put no restrictions on its convergence at all, or at best restricted the validity of the formula by the condition $|x| < 1$. According to Bolzano this practice was unsatisfactory, because if the formula is valid only for $|x| < 1$, then this condition must be used in the proof. That this was not the case indicated, in Bolzano's view, that the proof was not correct. Bolzano based his proof of the binomial series on a thorough investigation of the partial sums of the series. Rusnock presents the basic steps of Bolzano's proof (p. 64-69) together with the technical details and thus makes it possible for the reader to judge the strengths as well as the weaknesses of Bolzano's approach.

The paper "Purely analytic proof" is concerned with the proof of a special case of the Intermediate value theorem: "any polynomial with rational coefficients which takes on a negative value for $x = a$ and a positive value for $x = b$ has a root between a and b." The intermediate value theorem was considered to be obvious on geometric grounds by almost all of Bolzano's contemporaries and it had remained a gap even in Gauss' attempt to provide a proof of the fundamental theorem of algebra. In order to be able to prove the theorem, Bolzano had to offer a new definition of the continuity of functions and of the completeness of real numbers. So the attempt to prove an obvious theorem led to the progress in understanding of the basic concepts of real analysis.

Chapter 4, *Logic and methodology in the Theory of Science*, presents several topics of Bolzano's philosophical masterpiece. Bolzano was perhaps the first among modern logicians who saw the importance of a logic independent from psychology. Instead of presenting logic, as was customary then, as the study of the powers and of the right use of human understanding, he developed his account around logical objects which were independent of human thought and linguistic expression. In order to illustrate the novelty of Bolzano's views, Rusnock presents a brief sketch of the main traditions in logic, which were influential in Bolzano's times – the Aristotelian, the Port Royal, the Wolffian, and the Kantian ones. All these traditions based their approach to logic on grammatical structure, which was considered to faithfully express conceptual structure, which in turn mirrored ontology. Long familiarity had rendered these casual identifications more or less invisible. "Bolzano was among the first to clearly understand them not as obvious truths, nor as pointing to deeper insights, but as confusions – confusions which led to significant philosophical errors (the idealism of post-Kantian German philosophy prominent among them). In the

Theory of Science he quite deliberately gives logical objects an independent status, detaching them from thoughts and things." (p. 106)

Bolzano's basic logical notions are proposition in itself and idea. The proposition in itself is an undefined basic notion of Bolzano's logic, while an idea is defined as a part of a proposition in itself, which is not itself a proposition. The content of an idea is the unordered collection of its parts. For instance the content of the idea "35" is $\{3, 5, 10, \cdot, +\}$, because 35 is defined as $3 \cdot 10 + 5$. Ideas with no parts are called simple ideas. Ideas whose extension is a single object are called singular ideas. Singular simple ideas are called intuitions. Ideas, which are not themselves intuitions, and which contain no intuitions as parts, are called concepts. These definitions make it possible for Bolzano to characterize mathematics as a purely conceptual science, in which no intuitions occur. We see here a precise articulation of the understanding of mathematics on which Bolzano's mathematical work in the 1810s was based. After introducing the basic notions of Bolzano's logic, Rusnock turns to the presentation of Bolzano's basic logical technique – the technique of variation of parts of a proposition. This consists in considering one or more parts of a given proposition as variable and replacing them by other ideas. This technique made it possible for Bolzano to define many important logical notions as analyticity, compatibility, deducibility, equivalence and exact deducibility. At the end of the chapter Rusnock argues that Bolzano should be viewed as one of the founder of modern logic. The traditional view identifies the beginning of modern logic with Boole and Frege, i.e. with the rise of symbolic logic. According to Rusnock such an identification is unjustified, as symbolization presupposed a great deal of prior conceptual work, especially in mathematics. While this point is absolutely correct, there might be also another line of Bolzano's influence. Riemann's concept of manifold in geometry is created with the help of a method, which resembles Bolzano's method of variation. José Ferreirós in his book *Labyrinth of Thought* (Basel, Birkhäuser 2000) gave an account of the history of set theory, in which he traced the roots of modern set theory back to Riemann and more generally to the conceptual approach in mathematics, which was characteristic of the Göttingen tradition, to which Riemann belonged. Unfortunately, Ferreirós neither discussed Bolzano's contribution to the development of set theory nor his possible influence on Riemann. Maybe the possible connection between Bolzano and Riemann could be indirect, via Herbart, whose philosophical writings Riemann studied thoroughly.

Chapter 5, *Later Mathematical Studies*, presents topics from Bolzano's late mathematical works – *Function Theory* and *Paradoxes of the Infinite*; both published posthumously. From the *Function Theory* Rusnock discusses Bolzano's definition of the concept of a function, continuity and uniform continuity. Then he presents the proofs of Bolzano's three major results, so that the reader can appreciate Bolzano's technical progress by comparing these proofs with the proofs of his earlier mathematical results. Most interesting in this connection is the reconstruction of the concept of uniform continuity. Previous scholars (Rychlik, van Rootselaar, Sebestik) have denied that Bolzano succeeded in forming the concept of uniform continuity. Rusnock challenges this traditional view and shows on the basis of detailed textual analysis that Bolzano had a sufficiently clear notion of uniform continuity. An interesting methodological novelty of Bolzano's *Function Theory* was its use of examples and counterexamples. The first example, discussed by Rusnock, is a function having infinitely many maxima and minima. (Here on page 173 there is a minor misprint. The numerator of the last fraction should read $2^{2n} - 1$ instead of 2^{2n-1}.) The next example, presented with more technical detail is the so called Bolzano function. This function is continuous but has no derivative at any point. Thus this function, dating from 1834, was almost certainly the first example of a continuous, nowhere differentiable function. It is perhaps worth noticing that the graph of Bolzano's function has infinite length. From the picture presented by Rusnock the self-similarity of the function is obvious, and thus it is one of the first examples of a fractal.

The discussion of the *Paradoxes of the Infinite* in Rusnock's book is correct but rather brief. I would like to add that in Prague there is a group of mathematicians around professor Vopenka, that developed the so-called *alternative set theory* based on Bolzano's approach to the infinite. This alternative set theory is interesting because it has only two types of the infinite – countable infinite and the continuum. The whole Cantorian hierarchy collapses. Thus Bolzano's views on the infinite, even if rather strange when compared with Cantorian orthodoxy, have a consistent core and would thus perhaps deserve more attention.

To sum up, the book is very readable, written in clear and precise language. It presents the development of Bolzano's thought both in general terms as well as in technical details. Rusnock persuasively reconstructs Bolzano's heuristic use of philosophy in his mathematical research and of mathematical experience in his philosophical work. Thus the book will doubtlessly fulfill the goal formulated in the introduction: "I hope that students of philosophy will be able to gain from it a better acquaintance with the one of the great philosophers of modern times, the neglect of whose work is as astonishing as it was a hundred years ago" (p. 5).

Ladislav Kvasz

EVELYN DÖLLING, *"Wahrheit suchen und Wahrheit bekennen."* Alexius Meinong: *Skizze seines Lebens,* Amsterdam–Atlanta, GA: Rodopi, 1999 (= Studien zur österreichischen Philosophie, Bd. 28).

There are already numerous monographs and anthologies on Alexius Meinong's philosophical work, but there has, until now, been no comprehensive biography.

Evelyn Dölling has succeeded in filling this gap with her *Skizze seines Lebens* (*Sketch of his life*). Her attempt to delineate Meinongs's life (1853-1920) takes the form of a sketchbook. It consists of several sketches. Some of them are meticulously detailed, as for example the description of Meinong's struggle to establish a psychological laboratory in Graz. Others consist of just a few strokes, and do not go beyond indications and bibliographic notes, as in the case of Meinong's relationship to his younger disciples, for instance Ernst Mally and France Veber.

On the whole, the *Sketch* can be read like a biographical novel. In an authentic and vivid way the author describes the life of a bourgeois academic scholar around the turn of the century. Authenticity is ensured by elaborate quotations from documents, notes and letters – these take the place of dialogues, as it were. Dölling graphically depicts Meinong's family life, the circle of his closest friends and disciples (especially Guido Adler, Christian von Ehrenfels, Alois Höfler, Anton von Oelzelt-Newin), his academic career in Vienna and in Graz, as well as his philosophical work and his achievements within the scientific community. Although Meinong rather shunned social, political and cultural exposure (including the giving of public lectures and participation in conferences), Dölling's biography shows how, in spite of unfavorable circumstances, Meinong was still able to exert considerable influence on the academic community of his time. In particular, she shows to what extent Meinong's great achievements were due to his ambition and perseverance, combined with diligence and a disciplined way of working. It also becomes clear to what extent Meinong's accomplishments were facilitated by a well-organized family life for which his wife Doris (née Buchholz, 1865-1940) bears much of the responsibility and credit.

The coherence of Dölling's descriptions is ensured primarily by the use of chronological context, and at least one theme is treated for each stage of Meinong's life. On the other hand, there are certain motifs that are taken up again and again, in particular Meinong's perennial struggle with his inherited semi-blindness. Although this handicap deteriorated to almost complete blindness in the course of his life, he always tried to conceal it. A further leitmotif is Meinong's quest for the truth. "Seeking truth and bearing witness to truth" (*"Wahrheit suchen und Wahrheit bekennen."*) – it was in this way that Ehrenfels characterized his teacher, colleague and friend. Dölling takes this formulation as Meinong's maxim (p. 213), and she uses it, so to speak as a motto, not only in the title of the whole book but also as the title of the fourth chapter. According to the author, this overall attitude is reflected in Meinong's methodological-epistemological "principle of the critical openness of all knowing" (*Prinzip der kritischen Unabgeschlossenheit alles Erkennens*). (See pp. 2, 10, 24, 60, 210-13. Dölling refers to Meinong's work *Über Möglichkeit und Wahrscheinlichkeit* where – in my opinion – he links an anti-skeptical *prima facie* justification principle with a fallibilistic principle.[1] The former is called the "principle of the self-validity of all knowing" (*Prinzip der Selbstgültigkeit alles Erkennens*) whereas the latter is the already mentioned principle of the critical openness of all knowing. This principle says that critical examination is always possible, since there is always an unverified judgment left at the end of a justification process.)

The chronological account is preceded by a preface (pp. 1-4) and a brief philosophical-systematical introduction (pp. 5-14) that is designed to familiarize the reader with Meinong's terminology. Dölling entitles this introductory chapter "Einleitung oder: 'Über die Philosophie von unten'" (*Introduction or "On the philosophy from below"*) alluding to a quotation from Meinong where he wants to express that he has not created a *system* of philosophy but has rather, in his working method, followed the precepts of a kind of bottom-up thinking. (p. 5)[2] Dölling gives a concise account of Meinong's later contributions to philosophy: of his theory of objects, epistemology, psychology, value theory and ethics. On this point, a critical remark has to be made. Dölling's reconstruction of Meinong's classification of objects with respect to their modes of being suggests that existent and subsistent objects lack *Außersein*. But actually, Meinong attributes *Außersein* to all objects, including both the existent and the non-existent.[3] Thus, it ought to be said explicitly that *Außersein* applies to every object. Accordingly, one should add an "auch" to the last sentence on page 7, so that it reads: "'Außersein' kommt den Gegenständen *auch* dann zu, wenn sie weder existieren noch bestehen" (*"Außersein" applies also to objects that neither exist nor subsist*).

In chapter I – "... unsere Familie ist jederzeit deutsch gewesen" ("... *our family has always been German*"; pp. 15-18) – we are told that Meinong was born on July 17, 1853 in Lemberg (at that time the capital of the Austrian crown estate of Galicia, today Lviv, in the Ukraine). His father was an Austrian Major General and his ancestors came from southwestern Germany. Meinong repeatedly stressed that he was German, and he also joined political organizations of German nationalists. On the other hand, he emphasized his distance from extreme groups like those that had taken up the ideas of Schönerer. (See pp. 24f., 75-77)

In the second chapter – entitled "Die Wiener Jahre" (pp. 19-41) – the author depicts Meinongs years in Vienna, i.e., the period from 1862 to 1882. Among other things, she deals with Meinong's study of history, which he concluded with a dissertation on Arnold of Brescia (*Zur Geschichte Arnold's von Brescia*). She describes Meinong's ambivalent relationship to Franz Brentano (presumably his most important philosophical teacher), whose lectures he attended for more than four terms starting in 1875. She also deals with his activity as *Privatdozent* of philosophy at the University of Vienna, 1878 to 1882, and with some of his friends and disciples (see above). The author mentions Meinong's religious affiliation only in an indirect manner: from certain details of the narrative, it becomes clear that he was a Roman Catholic. Yet Dölling deals thoroughly with Meinong's critical and distanced attitude towards the Catholic Church. (pp. 16f., 21-24, 95) Dölling mentions some of Meinong's early writings in which he – like Arnold of Brescia – advocates the separation of religious and secular affairs as well as the idea of the political equality of all denominations. Furthermore,

Meinong criticized the religious instruction in Austria's "Mittelschulen" (high schools) as inadequate, since he considered it as a restriction on the freedom of opinion. Meinong suggested that historians rather than priests should teach the history of the development of the religions and of the various ways of satisfying religious needs. (p. 22)

"Die Grazer Jahre" is the title of the third chapter, which is also the center-piece of the book. (pp. 43-204) Here Meinong's activities and contributions as an academic teacher and scholar in Graz are outlined. Meinong taught philosophy at the University of Graz for more than 38 years – from 1882 as an associate, from 1889 until his death, November 27, 1920, as a full professor (*Ordinarius*). According to Dölling, Meinong's first book *Über philosophische Wissenschaft und ihre Propädeutik* (1885) plays a pivotal role for the understanding of Meinong's research as well as his teaching activities – a role to which insuffi-cient attention has thus far been paid. (p. 62) Originally, this book was written with the intention to criticize instructions issued by the Ministry of Culture and Instruction to cut down the hours of instruction in psychology, logic and philosophy (*philosophisch-propädeutischer Unterricht*) in high schools. But the book can also be read as a programmatic essay on Meinong's own view of philosophy and psychology. He brings out clearly that he considers psychology as a fundamental discipline of philosophy. Uncertainty in the domain of knowl-edge is not to be considered as an obstacle, but rather as a didactically fertile task and opportunity. (p. 60) Because of this practically-oriented view of philosophy and of science Meinong proposed the establishment of both a Philosophical Seminar and a Psychological Laboratory as university institutions. Dölling depicts graphically and in great detail Meinong's attempts to found these insti-tutions and to maintain them once they were established. (In 1897 Meinong established the philosophical seminar, and already in 1894 the first psychological laboratory in the Austrian monarchy.)

A longer passage of the third chapter is devoted to the discussion of Meinong's theory of objects, where Dölling reports particularly on Meinong's debates with Bertrand Russell and Edmund Husserl (pp. 146-165). In connection with the Russell-debate Dölling refers to Twardowski's study *On the Content and Object of Presentations*[4]. She rightly emphasizes Meinong's remark that you can find a great deal of interesting and useful information in Twardowski's study. It is also correct that Twardowski's book sharpened Meinong's view on non-existent and impossible objects. (pp. 152f.) But Dölling – like other Meinong-interpreters – including John Findlay, for example – does not suffi-ciently take into consideration that Meinong's appreciative remark concerns "the whole difficult and important problem of objects", which involves in particular Twardowski's view on objects – existent and especially non-existent ones – and not exactly Twardowski's distinction between content and object.[5] It is true, that Twardowski's work can be seen as the occasion, or perhaps as the catalyst, for Meinong's beginning to think more deeply about the content-object distinction, and that Meinong did indeed take over some of Twardowski's arguments, even though in slightly modified versions. However, Meinong never thought that Twardowski's distinction nor his conception of the content of a presentation could be understood as a precise conceptual clarification that could be adopted unquestioningly. After 1894, Meinong took psychological contents, i.e., the contents of presentations, of judgments and so on, as something purely mental and concrete. They are as purely mental and concrete as are the mental acts whose contents they are. Therefore, such contents are always something real. Twardowski, however, says that a content is a mental picture on the one hand, but one the other hand he identifies the mental content with the meaning of a name and with Bolzano's objective presentation or presentation as such. For that reason Twardowski takes the content – in contrast to the (real) act – as something that always lacks reality. Meinong, in contrast, became ever more aware of the purely mental nature of the psychological content, and he saw that the mental content was not something abstract or conceptual in the sense of something intensional, that it was not a meaning at all.[6]

The fourth and final chapter (pp. 205-214) has the same title as the whole book and concerns the last year of Meinong's life. Dölling briefly describes his work on the last version of his theory of values and on his autobiographical sketch – both of them are posthumous publications.[7] In using recollections of some of his disciples and pupils (Fritz Heider, Christian von Ehrenfels und Mila Radacović) Dölling concludes her sketch with a retrospective appreciation of Meinong.

The appendix (pp. 231-261) contains materials which support a better and authentic understanding of the main text (the statutes of the Philosophical Seminar, for example), and a collection of useful surveys (of Meinong's writings, his lectures and seminars, a chronology of Meinong's life).

Short personal data of philosophers is helpfully included, and gives the feeling that the philosophers in question are introduced like actors in a play. Such characterizations are, though, susceptible to printing and other errors. For example, Franz Brentano is born in 1838 and not in 1839 (p. 8), Alois Höfler in 1853 and not in 1883 (p. 27), and Eduard Martinak was *Gymnasiallehrer* (high school teacher) in Leoben und Graz, but he also was university professor of pedagogics in Graz (1904-1930), a fact which would have been worth mentioning in this context.

An index of names closes the work – a work that does not only contain a great deal of information about Meinong and his time but also lets us fully participate in significant occasions of his life. Evelyn Dölling's *Skizze seines Lebens* allows us to get a (necessarily incomplete but also) clear and distinct picture of Meinong. The philosopher Alexius Meinong, as both a public and a private person, is contrasted from the other people sufficiently and portrayed in his personality in a well differentiated manner.

NOTES

1. *Über Möglichkeit und Wahrscheinlichkeit. Beiträge zur Gegenstandstheorie und Erkenntnistheorie,* (1915), *GA,* vol. VI., pp. 450-463. (In this review the abbreviation "*GA*" is used for R. Haller, R. Kindinger (eds. together with R. M. Chisholm): *Alexius Meinong Gesamtausgabe,* 7 vols, Graz: Akademische Druck- u. Verlagsanstalt, 1968-1978.)
2. See "A. Meinong (Selbstdarstellung)", (1921), in *GA,* vol. VII, pp. (3)-(62), p. (42f.).
3. See A. Meinong: "Über Gegenstandstheorie", (1904), in *GA,* vol. II, pp. 481-535, p. 492; "A. Meinong (Selbstdarstellung)", (1921), in *GA,* vol. VII, pp. (3)-(62), pp. (18)f., (45).
4. Kasimir Twardowski (1894): *Zur Lehre vom Inhalt und Gegenstand der Vorstellungen. Eine psychologische Untersuchung;* Wien: Hölder (Reprint with an introduction by R. Haller; München and Wien: Philosophia, 1982; transl. with an introduction by R. Grossmann: *On the Content and Object of Presentations. A Psychological Investigation;* The Hague: Martinus Nijhoff, 1977).
5. See Alexius Meinong (1899): "Über Gegenstände höherer Ordnung und deren Verhältnis zur inneren Wahrnehmung", in *GA,* vol. II, p. 377-480 (transl. by M.-L. Schubert Kalsi in her book *Alexius Meinong. On Objects of Higher Order and Husserl's Phenomenology,* The Hague/Boston/London: Martinus Nijhoff, 1978, pp. 137-208), p. 381. John N. Findlay: *Meinong's Theory of Objects and Values,* 2nd ed., Oxford: Clarendon Press, 1963 (1st ed. 1933), p. 8.
6. Concerning the content-object distinction in Meinong and Twardowski, cf. Johann C. Marek: "Meinong on Psychological Content", in: L. Albertazzi, D. Jacquette, R. Poli (ed.), *The School of Alexius Meinong,* Ashgate: Aldershot, 2001, pp. 261-287. (By the way, in this anthology Dölling wrote a biographical sketch of Meinong in English: "Alexius Meinong's Life and Work", pp. 49-76.)
7. "A. Meinong (Selbstdarstellung)", (1921), in *GA,* Bd.VII, pp. (3)-(62). *Zur Grundlegung der allgemeinen Werttheorie,* (1923, ed. by Ernst Mally), in *GA,* Bd. vol. III, pp. (471)-(656).

Johann Christian Marek

DON HOWARD / JOHN STACHEL (eds.), *Einstein: The Formative Years, 1879–1909,* Boston–Basel–Berlin: Birkhäuser 2000 (= Einstein Studies, vol. 8).

The publication of the first two volumes of the *Collected Papers of Albert Einstein* in the years 1987 and 1989 marks a watershed in the history of Einstein scholarship. These volumes put together all available documents relevant to Einstein's early years up to his move to Berne, and they present all his published writings up to 1909, when he would take up his first proper academic appointment at Zurich university. The initiator of the editorial enterprise and editor of these first two volumes, John Stachel, was well aware of the significance of this endeavour. Together with Don Howard, active in the editorial project as well, he also founded the Einstein Studies series whose first volume came out around the same time, in 1989. The series provides a forum for Einstein research, with an emphasis, however, on the history and philosophy of general relativity in those volumes that have appeared up to now. The present volume, number 8 in the series, focuses on the young Einstein and his early work. It has been long in the making. Some papers originated at a conference on Einstein's early years held in 1990, in an attempt to harvest and digest the fruits of the publication of the first volumes of Einstein's Collected Papers. Other contributions were written especially for the volume, one paper was published before and is reprinted here. It was the editors' intention to offer a selection of some of the best recent scholarly studies of Einstein's early years, and the outcome certainly justifies this claim.

Clearly, the challenge of the book is this scandalon of the history of science: the sudden outburst of productive thinking and scientific creativity by a Berne patent expert who set out publishing some articles on thermodynamics and statistical mechanics and then, within a couple of months in the year of 1905, composed three papers, on the relativity principle, on Brownian motion, and on the light quantum hypothesis, each of these now regarded as milestones in the history of science, as revolutionizing our most fundamental concepts and founding new fields of theoretical physics. Plus, a later oft-cited dissertation on molecular dimensions is finished in this same period, and, still not enough, a follow-up to the paper on the relativity principle derives the equivalence of energy and mass, later expressed in the ubiquitous equation $E=mc^2$ that has become symbolic of today's theoretical physics as such. The studies collected in this volume combine forces in an attempt to understand the genesis and assess the significance of Einstein's scientific achievements of the early years, and of his *annus mirabilis* 1905 in particular. They do so by looking at Einstein's individual intellectual development and by analysing the conceptual structure of his works. Other aspects of Einstein's life, e.g. his political convictions, are not dealt with in this book on Einstein's formative years, at least not explicitly. In this volume, one may perhaps say, we remain under the spell of the *annus mirabilis*. Three of the contributions deal with specific influences in Einstein's formative period, with a view of providing background material for an explanation of Einstein's creativity. Four others papers are devoted to a detailed analysis of his early work on statistical mechanics, special relativity, Brownian motion and the quantum hypothesis.

From what is known about Einstein's early reading and intellectual stimuli, an important formative influence of Aaron Bernstein's popular multi-volume *Naturwissenschaftliche Volksbücher* on Einstein's early development may be inferred. Frederick Gregory's contribution provides an introduction to and a brief characterization of the work of this political journalist and popular science writer. His multi-volume opus appeared in various editions and provides a common popular background of 19th century German scientific materialism. From the 750 essays of Bernstein's *Volksbücher*, Gregory also dares to pick out some themes and topics that might have been of direct interest to the young Einstein but he justly refrains from drawing definite conclusions about any specific influence that this reading may have had. The only influence that is confirmed is that pointed out by Einstein himself, of helping him overcome an early religious phase at the age of 12 and of acquainting him, in a qualitative manner, with the full wealth of scientific knowledge of his period.

More specific may have been the influence of Einstein's reading of Kant, claims Mara Beller. She points out that Kant not only was an inseparable ingredient of the culture of German language *Bildungsbürgertum* of the period in question, and in August Stadler's lectures at the Zurich Polytéchnique identifies a specific blend of Kantian philosophy to which Einstein was exposed. She also reminds of various evidence suggesting that Einstein not only studied the Critique of Pure Reason at the age of thirteen, under the supervision of his Jewish friend and teacher Max Talmey, but also time and again reread Kant's writings during his lifetime. The influence, Beller suggests, that this reading may have had is that of providing the basis for a coherency of his epistemological convictions throughout his life. Two Kantian themes in particular are identified in Einstein's own thinking, the dual theory of intuition and judgement, and the concept of the unity of nature as a regulative idea in science.

If Einstein was exposed to science and philosophy in his autodidactic studies as a premature youth, he also had proper academic training as a student at the Zurich Polytéchnique. David Cahan's contribution gives a rich account of the institutional, technological, and intellectual background provided by H.F. Weber's physics institute, and a vivid portrait of Weber and his colleague Pernet. The importance of precision instrumentation and measurement in a Helmholtzian tradition, the introduction to cutting-edge experimental thermodynamics, electricity and magnetism, an acquaintance with the problems of black-body radiation, but also Weber's lack of proper appreciation for some of the latest theoretical developments in the field, Maxwell's theory, are cornerstones of Einstein's early physics education.

Needless to say, the scientific and cultural influences that the young Einstein was exposed to are not at all covered exhaustively by these essays and should be complemented by other studies, both existing ones and others that remain to be written. Nevertheless, these three papers give a fair impression of the richness of the intellectual experience, whether revealed to us by documentary evidence or not, that we have to take into account when analyzing Einstein's early work and his papers of the miraculous year. Other than is the case for the genesis of general relativity, the reconstruction of the emergence of Einstein's early ideas remains a task that must be grounded on frustratingly sparse documentary evidence. The two papers by Jürgen Renn on the origin of Einstein's ideas on statistical mechanics and by Robert Rynasiewicz on the construction of special relativity provide impressive examples of the amount of diligence and precision as well as of imagination and speculation that is inevitably called for when arguing for a coherent and convincing reconstruction of Einstein's early development.

Thus, Renn's paper on the origins of statistical mechanics was triggered by the information contained on the verso of a one-page letter by Mileva to Einstein in 1901 that came to the fore only in 1996 when the Einstein family correspon-

dence was auctioned. The new information concerns Mileva's response to an –
unknown – letter by Albert telling her about a letter – unfortunately lost – he had
received from Paul Drude who in turn was reacting to a letter by Einstein to him
– lost as well. Collecting all available primary and secondary, third-party and
indirect evidence, Renn argues for a speculative reconstruction of this lost correspondence, which, as Renn argues, pertained to Einstein's early work in statistical mechanics, his perception of a logical gap in Boltzmann gas theory that led
to his independent introduction of what became to be known in the reception of
Gibbs' work as the concepts of microcanonical and canonical ensembles. The
relevance of the Drude correspondence, according to this reconstruction, is that it
was the latter's electron theory of metals that was perceived by Einstein as one
other field of possible application of the statistical and atomistic concepts contained in Boltzmann's *Gastheorie*, an application that triggered Einstein's conceptual analysis and generalization of the use of statistical concepts in his first
publications.

Equally well-informed but much more cautious is Robert Rynasiewicz in
reconstructing the construction of special relativity. He also surveys the pertinent
known evidence and tries to make sense of the hopelessly meagre positive
historical information about the order of events in the conceptual breakthrough
resulting in the crucial insight of the relativity of simultaneity. But rather than
offering a positive reconstruction himself he explicates and then calls into question a much too tidy account that might suggest itself at first sight. His paper
essentially tries to formulate most precisely a number of positive queries.
Answers to those queries, even though they may never be given on the basis of
factual evidence, would nevertheless be consequential for any historical reconstruction of the genesis of special relativity.

Somewhat different in character are Sahotrar Sarkar's and John Stachel's
contributions on Brownian motion and the light quantum hypothesis. Sarkar's
systematic interest is in a philosophical understanding of the various forms and
functions of approximations in science. The theory of stochastic processes that
he traces back essentially to Einstein's 1905 paper on Brownian motion serves
him as a case study to exemplify the pertinent conceptual, methodological and
metaphysical distinctions. Less meticulous in historical detail, he sketches some
relevant lines of the historical development, pointing out parallel developments
in physics and economics, and he sketches in some broad strokes the later reception of Einstein's work, providing much valuable insight into the essence of
Einstein's analysis.

John Stachel finally adds an important contribution to recent discussions of
the origins of quantum theory. He argues that it was Wien's work on black-body
radiation rather than Planck's that was the decisive background for Einstein's
work on the light quantum. According to Stachel's interpretation, Einstein took
up Wien's suggestion that there should be a discontinuity between the long-wave
length and the short-wave length behaviour of the radiation due to the molecular
constitution of matter. But it was Einstein who then made the decisive shift of

attributing the short wave length behaviour to an atomistic structure of radiation itself. Analyzing Einstein's reception of Wien's work on the background of a long-standing heuristics of conceiving of a deep-going analogy of radiation and matter, Stachel also offers some answers to his own counterfactual considerations as to why Einstein did not propose his 1924 theory of a quantum gas of radiation a decade and a half earlier.

In summary, the essays provide a many-faceted picture of the young Einstein, his formative years and early work. The book is essential reading for anyone interested in understanding, on the basis of our insufficient documentary evidence, Einstein's early contributions to physics and the, perhaps inevitably miraculous, genesis of his ideas.

Tilman Sauer

ACTIVITIES OF THE INSTITUTE VIENNA CIRCLE

ACTIVITIES 2001

Lecture Series

THOMAS E. UEBEL (University of Manchester)
Vernunftkritik und Wissenschaft. Werk und Wirkung des ersten Wiener Kreises
(January 29, 2001)

GEORGE REISCH (Chicago)
Logical Empiricism and American Politics: 1930s versus 1950s
(December 3, 2001)

PHILIP KITCHER (Columbia University, New York)
The Ends of the Sciences
(December 10, 2001)

9th Vienna Circle Lecture and Keynote Lecture of the International Symposion "The Vienna Circle and Logical Empiricism"

HUBERT SCHLEICHERT (Konstanz)
Moritz Schlick and the Idea of Social Contract
(July 12, 2001)

International Symposion

The Vienna Circle and Logical Empiricism. Re-Evaluation and Future Perspectives of the Research of Historiography (VC/LE)
(July 12–14, 2001)

First Vienna International Summer University

Scientific World Conceptions (VISU/SWC)
"Unity and Plurality in Science"
Organized by the University of Vienna and the Institute Vienna Circle
Main lecturers: DON HOWARD (University of Notre Dame, USA),
ELLIOTT SOBER (University of Wisconsin, USA)

Assistant lecturers: CHRISTOPHER HITCHCOCK (Pasadena), DAVID STUMP (San Francisco)
Guest lecturer: BRIGITTE FALKENBURG (Dortmund)
(July 16–28, 2001)

Cooperation with: Logic Colloquium of the Kurt Gödel Society
in Vienna, August 6–12, 2001

Cooperation with: Symposion of the "Österreichische Mathematiker Gesellschaft / Society of Austrian Mathematicians " (ÖMG)
September 16–22, 2001

ESF-Network:
Historical and Contemporary Perspectives of Philosophy of Science in Europe
European Science Foundation (ESF) 2001–2003
Coordination Committee:
MARIA CARLA GALAVOTTI, Università di Bologna, Dipartimento di Filosofia (Chair)
ARISTIDES BALTAS, National Technical University Athens, Department of Physics
DONALD GILLIES, King's College London, Department of Philosophy
THEO KUIPERS, University of Groningen, Faculty of Philosophy
ILKKA NIINILUOTO, University of Helsinki, Department of Philosophy
MICHEL PATY, Equipe REHSEIS (CNRS and Université Paris 7 – D. Diderot)
MIKLÓS RÉDEI, Lóránd Eötvös University, Faculty of Science, Dept. of History and Philosophy of Science
GEREON WOLTERS, Universität Konstanz, Fachbereich Philosophie und Wissenschaftstheorie
First International Workshop:
University of Bologna, Bertinoro, Sept. 29–Oct. 2, 2001
"Observation and Experiment in the Natural and Social Sciences"

International Symposion
Austrian Exile and Remigration: Politics – Science – Art
The contribution to the Culture of the Second Republic /
«Exil et Retours d'Exil– Quelles contributions à la vie culturelle, politique, scientifique de la Deuxième République d'Autriche?»

Together with:
Université de Paris III
Université de Rouen, Centre d'Etudes et de Recherche Autrichiennes (CERA)

Dokumentationsarchiv des österreichischen Widerstands (DÖW), Wien
Ludwig Boltzmann Institut für Geschichte und Gesellschaft, Wien
Chaired by: PAUL PASTEUR, GERALD STIEG, FRIEDRICH STADLER
Location: Université de Rouen (CERA)
Time: November 21–23, 2001

Book Presentations

Thomas Uebel, *Vernunftkritik und Wissenschaft. Otto Neurath und der erste Wiener Kreis.* Vienna–New York: Springer 2000 (= Publications of the Institute Vienna Circle, Vol. 9)
(January 29, 2001)

Logischer Empirismus und Reine Rechtslehre. Beziehungen zwischen dem Wiener Kreis und der Hans-Kelsen-Schule. Ed. by Clemens Jabloner und Friedrich Stadler.Vienna –New York: Springer 2001 (= Publications of the Institute Vienna Circle, Vol. 10)
(May 19, 2001)

Publications

Friedrich Stadler, *The Vienna Circle – Studies in the Origin, Development, and Influence of Logical Empiricism*, Vienna–New York: Springer 2001

Thomas Uebel, *Vernunftkritik und Wissenschaft. Otto Neurath und der erste Wiener Kreis.* Vienna–New York: Springer 2000 (= Publications of the Institute Vienna Circle, Vol. 9)

Logischer Empirismus und Reine Rechtslehre. Beziehungen zwischen dem Wiener Kreis und der Hans-Kelsen-Schule. Ed. by Clemens Jabloner and Friedrich Stadler. Vienna –New York: Springer 2001 (= Publications of the Institute Vienna Circle, Vol. 10)

Library and Documentation

- Expansion of primary sources and secondary literature on the Vienna Circle and its influence.
- Acquisition of estates and archival material in Austria and abroad.

PREVIEW 2002

Karl Popper 2002

Centenary Congress
Location: University of Vienna
Time: Wednesday, 3 July 2002–Sunday, 7 July 2002
Cooperation with the Programme and Organising Committee
Participation with a special symposium

ESF-Network:
Historical and Contemporary Perspectives of Philosophy of Science in Europe

Second International Workshop, Vienna:
„*Induction and Deduction in the Sciences*"
Location: University of Vienna, University Campus
Time: July 7–9, 2002

Vienna International Summer University

SWC Scientific World Conceptions
„Mind and Computation"
Place: Vienna, University Campus
Time: July 15–26, 2002

Organized by the University of Vienna and the Institute Vienna Circle
Main Lecturers: MICHAEL HAGNER (Max Planck Institute for the History of
Science, Berlin, Germany), BRIAN P. MCLAUGHLIN (Rutgers University, USA)
Assistant Lecturers: t.b.a.
Guest Lecturer: ANTON ZEILINGER (University of Vienna, Austria)
A two-week high-level summer course on questions about the relation between
mind, brain and computation from an historical and epistemological point of
view, with a special focus on quantum physics.

Publications

*The Vienna Circle and Logical Empiricism. Re-Evaluation and Future
Perspectives of the Research and Historiography.* Ed. by Friedrich Stadler et al.
Dordrecht–Boston–London: Kluwer 2002 (= Vienna Circle Institute Yearbook
10/2002)

Wissenschaftsphilosophie und Politik / Philosophy of Science and Politics
Hrsg. von Michael Heidelberger und Friedrich Stadler. Wien–New York:
Springer 2002 (= Veröffentlichungen des Instituts Wiener Kreis, Bd. 11)

Appraising Lakatos – Mathematics, Methodology and the Man. Ed. by Ladislav
Kvasz, George Kampis and Michael Stöltzner. Dordrecht–Boston–London:
Kluwer 2002 (= Vienna Circle Institute Library 1)

*Intellectual Migration and Cultural Transformation: Refugees from National
Socialism in the English-Speaking World.* Ed. by Edward Timms and Jon
Hughes. Wien–New York: Springer 2002. (= Veröffentlichungen des Instituts
Wiener Kreis, Bd. 12)

IN MEMORY OF MARIE JAHODA
(1907 – 2001)

The social psychologist Marie Jahoda passed away on April 28, 2001 at her home in Keymer in Sussex, England. Her death at the age of 94 reminds us again of the Viennese intellectuals from the interwar period who were forced to leave their country by the two dictatorships that came to power in Austria during the 1930s and were never invited to come back.

Jahoda was born on January 28, 1907 as the third of four children to a middle class Jewish family which had lived in Vienna for more than four generations. Her father was a businessman, the brother of the publisher of the famous *Die Fackel*, edited and mostly written by Karl Kraus. Obviously Kraus functioned as a family god for the larger Jahoda family. The other family hero was Josef Popper-Lynkeus, author of, and propagandist for, what he called *Allgemeine Nährpflicht*, a social policy scheme which advocated something which we now would call general subsidence income. These two figures who had a decisive influence on the Jahodas indicate nicely the socio-political atmosphere in which liberal middle-class members formed their attitudes in the final years of the Hapsburg Empire: moderate liberal, socially active, rational in their worldview and Jewish by origin but not religiously.

Young Marie, who was called Mitzi by friends throughout her life, joined the Social Democratic Youth organizations near the end of World War I and became heavily politicised as a consequence of the assassination of Prime Minister Graf Stürgkh by Friedrich Adler. Her political involvement influenced her own decision about what she should study. Since she was completely sure that she would become the first socialist minister for education after the imminent revolution she thought studying psychology would provide her with the best training. When she went to her first lecture given by psychologist Karl Bühler she realized that academic psychology was something different, but, as she recalled much later, "I learnt better."

At that time it was uncommon for women to study; this only changed when the republican government altered the rules after 1918. Jahoda's realism persuaded her to embark upon a second education for occupational reasons. At the Pädagogische Institut, where she studied to become a elementary teacher, one of her schoolmates was the then young Social Democrat Karl Popper.

Because it was then obligatory for students in psychology to take courses in philosophy as well, Jahoda also sat in lecture courses given by Rudolf Carnap whose Aristotelian teaching she remembered vividly during one of our conversations some fifty years after she actually sat in front of Carnap. However, she never became interested in Logical Positivism or any other abstract and therefore unpractical philosophy. Nevertheless she adopted by osmosis, as she called it,

425

M. Heidelberger and F. Stadler (eds.), History of Philosophy and Science, 425–428.
© 2002 *Kluwer Academic Publishers. Printed in the Netherlands.*

the principles of the Vienna Circle philosophy, especially when she worked for a while at Otto Neurath's *Gesellschafts- und Wirtschaftsmuseum*. In 1932, she also concluded her study of psychology with an Ph.D. thesis about the life-cycle of old people living in shelters.

In addition to her political interests, she participated in one of the other scientific endeavours which emerged from the same intellectual background. Together with her first husband for some six years, Paul Lazarsfeld, she was a member of the *Wirtschaftspsychologische Forschungsstelle*, created by Lazarsfeld to earn an income from sources outside the university. The members of the *Forschungsstelle* were never really successful in persuading enough businessmen to commission market research at the *Forschungsstelle* but at this time it seems that it was possible to make a living at a very low level of regular income. The forthcoming contract had to compensate for the losses resulting from the present one.

The most outstanding contribution of the *Forschungsstelle* was the study about unemployment in a small village some thirty kilometres outside Vienna, *Die Arbeitslosen von Marienthal. Ein soziographischer Versuch über die Wirkungen langdauernder Arbeitslosigkeit* appeared at the most inappropriate time one could imagine: 1933, published by the Leipzig based publisher Hirzel. To avoid anti-Semitism the book appeared without the Jewish sounding names of its authors. Nevertheless, the small book received a warm welcome, even only outside Germany.

After the ban of the Social Democratic Party and all other left-wing organizations by the Catholic *Ständestaat* regime Jahoda joined the underground movement of the Revolutionary Socialists, an anti-Communist successor of the discredited old party. During the next years Jahoda divided her time between illegal political activities and the *Forschungsstelle*, which could go ahead with its activities in spite of the fact that all its members where in opposition to the new government. Jahoda's political activities came to an end involuntarily when the police arrested her for her participation in the underground movement. After more than a half year in prison she was released because of protest from abroad on the condition to leave the country.

Jahoda went to London where thanks to the help of social scientists there she could start anew. Her first field study brought her to South Wales where she conducted an investigation of a subsistence scheme for unemployed mine workers, initiated and organized by Quakers. Since the results of her investigation did not fit into the high brow expectations of the philanthropists who hoped their activities were appreciated by the workers and because the leading man had helped Jahoda's family members to escape Nazi occupied Austria, Jahoda decided not to submit her manuscript to a publisher. When I approached her some forty years later she still hesitated to break her promise but finally she thought publishing it now wouldn't hurt anyone now. The book's appearance was announced as a follow-up study to Marienthal, which indeed it was.

During World War II Jahoda lived in England, participated in Austrian exile politics and joined the propaganda division of the Ministry of Information where she participated in *Radio Rotes Wien*, an endeavour to strengthen the residence forces in Austria, unsuccessfully as we know today. Besides this she won the prestigious Pinsent Darwin Fellowship for a three-year period of independent research and did some social research in factories, department stores and with school girls.

Near the end of the war Jahoda had to make up her mind about to her own future. Though still a committed Social Democrat who was interested and willing to return to Vienna to help restructure the country, she received discouraging letters from Vienna. Therefore she took the opposite direction and migrated to the United States, a move also prompted by the fact her daughter lived there. From then on Jahoda gave up her active role in politics, which does not mean that she ceased being a *homo politicus*.

In New York City where Jahoda lived for the next twelve years, she experienced a exciting and successful time as a researcher. First as a research associate at the American Jewish Committee where she joined the staff of Max Horkheimer, the head of the Frankfurt Institute for Social Research who had become the research director of the AJC for a series of investigations that became known as *Studies in Prejudice*. Jahoda contributed one co-authored volume to this series and acted as a research assistant for some of the other studies. Some years later when she prepared a critical re-evaluation of the best known studies from this series, *The Authoritarian Personality*, Jahoda got into trouble with the two leading men from the Frankfurt School – Horkheimer and Adorno. Both tried to persuade her to withdraw some of the chapters out of fear they would get into trouble with Joseph McCarthy's investigators. The ridiculousness of this apprehension could not persuade someone who has experienced real oppression. The relationship to the Critical Theory masters cooled down considerably thereafter.

For a while Jahoda entered what was a sort of follow-up institution of the *Forschungsstelle* on the Upper West Side of Manhattan, the Bureau of Applied Social Research at Columbia University co-directed by her former husband Lazarsfeld and the theoretician Robert K. Merton. Her collaborative research with Merton resulted in co-authored manuscripts which, however, never appeared in print because they did not find the final imprimatur by the grand master of sociological theory. They lacked a particular theoretical underpinning but were interesting from a more applied point of view about race relations and voluntary segregation in housing.

At the age of 40 Jahoda was appointed for the first time in her career to a professorship. At New York University she became professor of psychology and social relations, which at that time was synonymous for the interdisciplinary cooperation between sociology and psychology. The ten years at NYU were highly productive years. Together with two colleagues she wrote and edited a book in Methodology of Social Research, then and later on a widely used textbook which was re-issued at least four times until the late sixties.

Her own research focussed on topics like race relations, prejudice, mental health, college education, and in particular on the socio-psychological consequences of the McCarthy hysteria. Jahoda and Stuart Cook were the first who did serious empirical research on the consequences of this mood to those who never would have become victims of the witch-hunters. The depressing result of this study was that people scaled down their language and opinions, stopped reading, borrowing or buying dangerous journals, papers, or books, and abandoned social relations to the usual suspects.

In 1957 Jahoda left the United States – not for political reasons but strictly private ones. She re-married and her husband, a Labour MP, could not leave his country without becoming unemployed. Back in London she entered the staff of Brunel College to head the department of psychology, where she immediately started an investigation of the student body of Brunel, a unit where students got at the same time an academic and a occupational education as technologists. Jahoda did not have the wish to leave Brunel, but the founders of the newly established University of Sussex persuaded her to join their faculty where she was one of the few woman and the first female chairperson in social psychology in the United Kingdom. After reaching retirement age Jahoda continued to collaborate with others from Sussex, especially with economist Chris Freeman from the Science Policy Research Unit.

When, in the eighties, unemployment became a concern again Jahoda became heavily involved in doing research about this topic again. She published a book-length review about social scientific findings about unemployment and its consequences and wrote piece after piece about the psychological need to have regular work.

In addition, she published a study about the interrelationship between academic psychology and Freud and contributed to evaluations of the world future projections which became famous during the 1970s.

After the founding of the Institute Vienna Circle its director Friedrich Stadler invited Jahoda to speak about her own research and her views with regard to philosophy. It was typical of the ignorance of Austria's university professors that only few of them attended her Vienna Circle Lecture in 1996, but the younger people present there learned better, to paraphrase Jahoda's description of her encounter with Bühler.

Anyone who ever had the opportunity to meet Jahoda and chat with her (she was extremely open-minded to people who did not reek of Nazism or opportunism) will retain a vivid memory of her forever.

Christian Fleck

IN MEMORY OF WESLEY C. SALMON
(1925 – 2001)

On April 22, 2001, Wesley Salmon died in a car accident, leaving a great void in the community of philosophers of science all over the world. Born in 1925, he studied under Reichenbach to become one of the most distinguished philosophers of science of the last century. After a long and productive career Salmon was University Professor Emeritus of Philosophy and History and Philosophy of Science at the University of Pittsburgh, and was working on a number of different projects at the time of his death.

Salmon's contribution to philosophy of science is outstanding, and so is the philosophical legacy he left behind. His works, written in a systematic and crystal-clear style, cover a wide range of topics, including logic, the philosophy of space and time, the foundations of probability and scientific inference, and scientific explanation, which had been the leading theme of his production for more than thirty years. His introductory books *Logic* (1963) and *The Foundations of Scientific Inference* (1966) are pearls of clarity and rigour, while his works on explanation, from *Statistical Explanation and Statistical Relevance* (1971) to *Scientific Explanation and the Causal Structure of the World* (1984), *Four Decades of Scientific Explanation* (1989) and the recent collection of essays *Causality and Explanation* (1998) are milestones in the literature on the subject.

Starting with a revision of Hempel's "received view" of explanation, Salmon maintained that genuine explanation is causal, and that in order to explain why something happened one has to produce information on the causal mechanisms responsible for its occurrence. He combined this idea with the belief that our knowledge is, in most cases, probabilistic, and that causality ought to be defined in probabilistic terms. In an attempt to match a purely empiricist approach with the ideal of explanatory objectivity, Salmon developed a probabilistic version of mechanicism, inspired by the conviction that causality cannot be defined only in terms of human expectations, as Hume did, but refers to the physical world. His theory of causality in terms of processes underpins a theory of explanation that aims at being objective, and at the same time realistic, in the sense of accounting for mechanisms at the level of everyday material objects. This view of causal explanation goes hand in hand with a frequentist interpretation of probability of the kind advocated by Reichenbach, and substantiates a notion of rationality that takes causal knowledge of facts to be not only indispensable to our understanding of the world, but also crucial for rational action.

The originality and fruitfulness of Salmon's approach to explanation and causality is shown by the vast debate which has arisen around it. In trying to cope with various objections, he modified his ideas substantially. Salmon's writings of

429

M. Heidelberger and F. Stadler (eds.), History of Philosophy and Science, 429–430.
© 2002 *Kluwer Academic Publishers. Printed in the Netherlands.*

the Seventies demonstrate his trust in the possibility of working out a general model of explanation entirely in terms of empirical concepts, free from the epistemic relativity of statistical explanation typical of the Hempelian approach. His faith in the accomplishment of such a project weakened progressively, leaving room for the admission that explanation is generally context dependent and that all sorts of pragmatic considerations have to be taken into account when formulating a causal explanation. What he never abandoned was the conviction that causal considerations are essential for explanation, and that explanation has to retain an objective character. He lately put forward the idea that there is such a thing as a "complete causal structure", some aspects of which are caught by the explanatory accounts formulated in various contexts. Such a structure is much too complex to be described in detail, but causal explanations can approximate it. This view has a great heuristic impact on the search for causes. On the other hand, knowledge of causes is most useful in all sorts of practical situations both in everyday life and science, because it guides detection of relevant factors. The notion of "complete causal structure", contained in a forthcoming paper entitled "A Realistic Account of Explanation", written during Salmon's stay in Kyoto as a Visiting Professor in the Spring 2000, is new to Salmon's perspective, and will certainly provoke further debate and nurture fresh reflection on explanation and causality.

During his life, Salmon received many honours, visited many institutions and was fellow (and at times chairman) of prestigious associations and academies. Not only a leading academic, an illuminating writer and a profound philosopher, Wesley Salmon was also a dedicated teacher and a man of rare qualities, openminded and curious about other people's ideas, generous and kind in all circumstances. For those who had the good fortune to know him and be his friend, his death is a tremendous loss, but remembrances of him are a great treasure.

Maria Carla Galavotti

IN MEMORY OF PAUL NEURATH
(1911 – 2001)

When Paul Neurath passed away in New York on September 4, 2001, shortly before his ninetieth birthday, it was a surprise for most people close to him. Only a few weeks before his death he was still planning the lectures he was going to give in the upcoming semester at the Institute of Sociology at the University of Vienna. After the tragic loss of his son that had devastated him and his wife Margarethe at the beginning of June in New York, he was planning to move his library from New York to Vienna. He needed his books for the many projects he was still working on – one of them a book on "Otto Neurath and the Beginnings of Pictorial Statistics". Even at his advanced age he was still actively involved in scientific work, and he loved to teach. Everyone who ever attended his lectures knew with how much humor, joy and talent for explaining things he helped his students overcome the small and big hurdles of statistics or conveyed to them population-theoretical problems of poor countries.

Paul Neurath was born on September 12, 1911. Already his grandfather Wilhelm Neurath (1840-1901) was professor of political economy at the University of Agriculture in Vienna. His father was Otto Neurath (1882-1945), a political economist, sociologist and philosopher who tried to dissuade his son from studying sociology after completing secondary school, for the former believed that sociology was not a real science at that time. Instead Paul Neurath studied law but also "given the uncertainty of the time (also) did something which students normally don't do"[1], e.g., underwent training in electrical engineering and machine construction and took courses in gas welding which was to prove useful when he was an emigrant in Sweden. By the time Paul Neurath completed his doctorate in law in 1937, his father Otto Neurath and his second wife Olga Hahn-Neurath had emigrated to Holland. Paul had a very warm ties with Olga and always spoke of her with the greatest of admiration, e.g., in his biographical introduction to the book on his father, which was published in 1994 of which more later.

In 1934, when the Dollfuss regime seized power, Otto Neurath happened not to be in Vienna and thus escaped being arrested as a prominent social democrat. Paul Neurath remained in Vienna to complete his studies. He, too, was an active social democrat and was for many years involved in the youth organisatons of the party. Immediately after Nazi troops marched into Austria on March 13, 1938, Paul Neurath narrowly escaped being arrested by the Gestapo but four days later he was captured a few kilometers from the Czech border and taken to the Dachau concentration camp with the first transport of Austrians. When he was released in March 1939 he emigrated to Sweden where he became a metal worker. In 1941 he succeeded in obtaining a place on a boat to New York. There

431

M. Heidelberger and F. Stadler (eds.), History of Philosophy and Science, 431–433.
© 2002 *Kluwer Academic Publishers. Printed in the Netherlands.*

he came into contact with Paul Lazersfeld whom he knew from his days as a
socialist high school student in Vienna and who got him a small job at what at
the time was the Office of Radio Research (later the Bureau of Applied Social
Research), and helped him to embark on what he had always wanted to do: study
sociology. He completed his studies in sociology and statistics with a dissertation
at Columbia University in which he examined the social structure that had
evolved among the inmates at the concentrations camps in Dachau and Buchen-
wald. From 1943, he taught statistics at the School of Business at City College in
New York but his real academic life did not begin until he received an appointment to Queens College where he was a faculty member of the Department of
Sociology from 1946 until 1977 and taught sociology and statistics along with
special subjects such as the methodology of social research, population problems, etc. Parallel to this, he taught at the Graduate Faculty of the New York
School of Social Research in New York from 1949 until 1967.

Of his many activities I would like to mention in particular his two Fulbright
Professorships. The first was at the Tata Institute of Social Science in Bombay
(1955-1957) where he had been asked to establish social research as a new major
field of study. In addition, he developed special radio programs for improving
agriculture and for introducing hygiene in villages; the ideas presented in these
programs were implemented by groups of listeners with great success. For his
second professorship he was invited by René König to Cologne which also
marked the first step of his "return home". René König asked him to write a volume on *Grundlegende Methoden und Techniken* (Fundamental Methods and
Techniques) for the *Handbuch der empirischen Sozialforschung* (Manual of
Empirical Social Research) (volume 3b, Stuttgart: Enke 1962) on which, in turn,
his great book *Statistik für Sozialwissenschaftler* (Statistics for Social Scientists)
(Stuttgart: Enke 1966) was based. From 1961 onwards his "return home"
brought him again and again to Vienna where he began to teach, on and off, at
the Institute of Sociology at the University of Vienna until a permanent professorship was established in 1979/80. From then on he taught alternately in New
York and Vienna and established the Paul F. Lazarsfeld Archives.

Paul Neurath was awarded the Golden Medal of Honor by the City of Vienna
and, just a few months before he died, the Honorary Cross for Special Achievements in Science and Art in Science and Art. Only a few months ago Paul
Neurath and his wife Margarethe decided to settle permanently in Vienna instead
of moving back and forth between Vienna and New York.

Towards the end of the eighties Paul Neurath asked me to collaborate with
him on a book on his father Otto Neurath. For me this offer was an honor and
marked the beginning of a particularly pleasant experience. Work on the book
was often not very easy, for from Otto Neurath's extremely diverse and extensive oeuvre, we wanted to select texts that were relatively representative of his
work. The mixture of tenacity, humor and plain good sense with which Paul
Neurath pursued this project contributed significantly to making a reality what
on occasion had seemed impossible to us. We presented the book *Otto Neurath*

oder die Einheit von Wissenschaft und Gesellschaft (Otto Neurath or the Unity of Science and Society) (Vienna – Cologne – Weimar: Böhlau 1994) at the Institute Vienna Circle with which Paul Neurath was linked in a number of ways. Whenever he could he attended the events organized by the Institute. The panels of pictorial statistics from Otto Neurath's collections exhibited there, likely to the last original ones from the period before his emigration to England (that is before 1940), were donated to the Institute by Paul Neurath.

NOTES

1. A fascinating autobiographical article can be found in Friedrich Stadler (ed.): *Vertriebene Vernunft. Emigration und Exil österreichischer Wissenschaft*, 1930-1940: Paul Neurath's "Wissenschaftliche Emigration und Remigration", pp. 513-537.

Elisabeth Nemeth

INDEX OF NAMES

Not included are: Figures, Tables, Notes, References

VIENNA CIRCLE INSTITUTE YEARBOOK

The *Vienna Circle Institute* is devoted to the critical advancement of science and philosophy in the broad tradition of the Vienna Circle, as well as to the focussing of cross-disciplinary interest on the history and philosophy of science. The Institute's *Yearbooks* provide a forum for the discussion of exact philosophy, logical and empirical investigations, and analysis of language. Each volume centers around a special topic which is complemented with a permanent section with essays arising from the scientific activities at the Institute and reviews of recent works in the history of philosophy of science or others with a particular relation to the tradition of logical empiricism.

1 [1993] F. STADLER (ed.), *Scientific Philosophy: Origins and Developments.* 1993.
          ISBN 0-7923-2526-5

2 [1994] H. PAUER-STUDER (ed.), *Norms, Values, and Society.* 1994.
          ISBN 0-7923-3071-4

3 [1995] W. DePAULI-SCHIMANOVICH / E. KÖHLER / F. STADLER (eds.), *The Foundational Debate.*
          Complexity and Constructivity in Mathematics and Physics. 1995.
          ISBN 0-7923-3737-9

4 [1996] E. NEMETH / F. STADLER (eds.), *Encylopedia and Utopia. The Life and Work*
          of Otto Neurath (1882-1945). 1996.
          ISBN 0-7923-4161-9

5 [1997] W. LEINFELLNER / E. KÖHLER (eds.), *Game Theory, Experience, Rationality.*
          Foundations of Social Sciences, Economics and Ethics. In honor of John C. Harsanyi. 1998.
          ISBN 0-7923-4943-1

6 [1998] J. WOLEŃSKI / E. KÖHLER (eds.), *Alfred Tarski and the Vienna Circle.*
          Austro-Polish Connections in Logical Empiricism. 1999.
          ISBN 0-7923-5538-5

7 [1999] D. GREENBERGER / W.L. REITER / A. ZEILINGER (eds.), *Epistemological and Experimental*
          Perspectives on Quantum Physics. 1999.
          ISBN 0-7923-6338-X

8 [2000] M. RÉDEI / M. STÖLTZNER (eds.), *John von Neumann and the Foundations of Quantum Physics.*
          2001.
          ISBN 0-7923-6812-6

9 [2001] M. HEIDELBERGER / F. STADLER (eds.), *History of Philosophy of Science.*
          New Trends and Perspectives. 2002.

KLUWER ACADEMIC PUBLISHERS – DORDRECHT / BOSTON / LONDON

www.ingramcontent.com/pod-product-compliance
Ingram Content Group UK Ltd.
Pitfield, Milton Keynes, MK11 3LW, UK
UKHW030640130325
456173UK00006B/34